C0-ATA-187

Telecommunications Transmission Engineering

Telecommunications Transmission Engineering is published in three volumes:

Volume 1 — *Principles*
Volume 2 — *Facilities*
Volume 3 — *Networks and Services*

Telecommunications Transmission Engineering

Third Edition

Technical Personnel
Bellcore and
Bell Operating Companies

Volume 1
Principles

ST–TEC–000051

Operating and regional company employees may obtain copies of this volume, order the Bellcore Technical References cited in this volume, as well as subscribe to Bellcore's DIGEST of Technical Information by contacting their company documentation coordinators.

All others may obtain the same copies as well as subscribe to the DIGEST by calling Bellcore's documentation hotline 1–800–521–CORE or (201) 699–5802.

Order each *Telecommunications Transmission Engineering* volume by the ST number as follows.

Volume 1, *Principles*	ST–TEC–000051
Volume 2, *Facilities*	ST–TEC–000052
Volume 3, *Networks and Services*	ST–TEC–000053

Prepared for publication by
Bellcore Technical Publications.

This is the third edition of the *Telecommunications Transmission Engineering* three–volume series previously copyrighted by AT&T and AT&T Bell Laboratories.

ISBN 1–878108–01–8 (Volume 1)
ISBN 1–878108–04–2 (three–volume set)

Library of Congress Catalog Card Number: 90–62180

Telecommunications Transmission Engineering

Introduction

Telecommunications engineering is concerned with the planning, engineering, design, implementation, operation, and maintenance of the network of facilities, channels, switching equipment, and user devices required to provide voice and data communications between various locations. Transmission engineering is that part of telecommunications engineering that deals with the channels, the transmission facilities or systems that carry the channels or circuits, and the combinations of the many types of channels and facilities that help form a network. It is a discipline that combines skills and knowledge from science and technology with an understanding of economics, human factors, and system operations.

This three-volume reference book is written for the practicing transmission engineer and for the student of transmission engineering in an undergraduate curriculum. However, the material was planned and organized to make it useful to anyone concerned with the many facets of telecommunications engineering. Of necessity, it only represents the current status of communications technology being used and deployed today by the Bell operating companies in their intraLATA (local access and transport area) networks, which provide exchange telecommunications and exchange access services, as well as their official networks. The reader should be aware of the dynamic nature of the subject.

Volume 1, *Principles*, covers the transmission engineering principles that apply to communications systems. It defines the characteristics of various types of signal, describes signal impairments arising in practical channels, provides the basis for understanding the relationships between a communication network and its components, and provides an appreciation of how transmission objectives and achievable performance are interrelated.

Volume 2, *Facilities*, emphasizes the application of the principles of Volume 1 to the design, implementation, and operation of the transmission systems and facilities that are used to form the public networks.

Volume 3, *Networks and Services*, builds on the principles and facilities discussed in Volumes 1 and 2 and shows how the principles are applied to facilities to form networks that are used by the exchange carriers to provide services for various users.

The authors use a generic approach throughout all three volumes. However, they often use specific examples and illustrations that are most familiar to them to help clarify a concept. These examples are not intended to recommend any equipment or to imply that there is only one solution.

The material has been written, edited, prepared, and reviewed by a large number of technical personnel from the Bell operating companies and Bellcore. Thus the book represents the cooperative efforts and views of many people.

As a point of reference for the reader, a brief description of these organizations follows. At divestiture, on January 1, 1984, 22 Bell operating companies were transferred from AT&T to seven newly formed regional companies: Ameritech, Bell Atlantic, BellSouth, NYNEX, Pacific Telesis, Southwestern Bell Corporation, and U S WEST. These regional companies, through their operating telephone companies, were empowered to provide local exchange telecommunications and exchange access services. They were also called upon to provide a common central point to meet the requirements of national security and emergency preparedness. Finally, they were allowed to create and support a centralized organization for the provision of engineering, administrative, and other services. This *central point* and organization is Bellcore.

The regional companies are both the owners and the clients of Bellcore. Bellcore's mission—to provide research, technical support, generic requirements, technical analyses, and other services to the Bell operating companies—allows them to provide modern, high-quality yet low-priced services to their customers. Bellcore provides neither manufacturing nor supplier recommendations, as procurement is a function of the regional companies.

Irwin Dorros
Executive Vice President –
Technical Services
Bellcore

Volume 1 — Principles

Preface

This volume, comprising five sections, covers the basic principles involved in transmitting signals over exchange carrier intraLATA (local access and transport area) communications facilities. Section 1 provides a broad description of the transmission environment and an overview of how transmission parameters affect the performance of the network. The second section consists of a review of most of the mathematical relationships involved in transmission engineering. A wide range of subjects is discussed, from an explanation of and justification for the use of logarithmic units (decibels) to a summary of information theory concepts.

The third section is devoted to the characterization of the principal types of signals transmitted. Voice, nonvoice, digital and analog data, address, and supervisory signals are described. Multiplexed combinations of signal types are also characterized. The fourth section describes a variety of impairments suffered by signals transmitted over practical channels, which have imperfections and distortions. Also discussed are the units in which impairments are expressed and the methods by which they are measured and controlled. The fifth section discusses the derivation of transmission objectives, gives many established values of these objectives, and relates them to requirements applied to system design and operation.

Contents

Figures

Tables

Telecommunications Transmission Engineering

Section 1

Background

Telecommunications networks in the United States provide a wide variety of services that affect the lives of everyone. Telecommunications services that enable people to talk to each other involve computers that communicate and a variety of voice, data, and signalling functions. A communications system must supply the means to accomplish this in an ever–evolving environment. Such a system is made up of a complex aggregate of electronic equipment and transmission media that provide a multiplicity of channels over which speech and digital messages, along with associated signalling information, can be transmitted. Vast and complex telecommunications systems have grown and are still growing rapidly. This expansion has been driven by customer growth, technological discoveries that promote innovations in equipment design and facility usage, and the political and economic climate in the United States. These changes have resulted in the orderly growth of telecommunications networking and the rendering of communications services throughout the United States, North America, and indeed the world.

Prior to World War II there existed three basic information–handling networks in the United States. First, the public switched telephone system, by far the largest of these networks, was used for speech communication. Second, nationwide record message service was furnished over a separate dedicated telegraph network. Third, radio broadcast stations were connected by a network of audio channels into four major national radio systems.* In addition to these networks, which were available to the general public directly or indirectly, there were a comparatively small number of point–to–point facilities leased by government entities, corporations, and individuals for private use. Transmission paths for all of these services were provided by a

* NBC Red, NBC Blue, CBS, and MBS.

complex combination of electronic equipment and a network of cable and open–wire facilities.

After World War II, there was a tremendous demand for telephones and a resultant increase in telephone traffic. Also, the rapid expansion of the television industry, through the proliferation of manufacturers as well as increasing sales of television sets, created a demand for nationwide television service. This growth in both communications media resulted in a cohesive, cross–country web of microwave radio routes. Carrier systems*, operating over microwave radio networks and a smaller coaxial cable network, ultimately replaced the prewar cable and open–wire circuits used for long–distance services. Satellite facilities were introduced in overseas communications services in the 1960s and became a major force in intercontinental and domestic long–haul services. Fiber optic systems, introduced in the late 1970s, afford strong competition to both microwave and satellite operations.

In the early days of telephony, available technology and economics severely limited the size of the area served by a single switchboard or switching system. Facilities in the area served by the switchboard or switching system were known as local exchange plant. Interconnections between these serving areas developed and increased in number partially because of technological advances that improved the quality of service on the transmission facilities. Such advances made possible the extension of connections to an increasing number of more remote central–office areas. As a result, subscribers could reach larger numbers of people, which made telephone service more valuable to them. Greater connectivity, better quality, and improved reliability promoted telephone use, thus feeding equipment and facility growth as well as reducing the per unit cost of providing service.

As the numbers of interconnecting facilities increased and the distances covered became greater, larger and more complex switching and transmission systems were designed and installed.

---

* Carrier systems expand the capacity of transmission media by combining a number of individual channels into a common frequency band or into a common bit stream (multiplexing). Multiplexing equipment increases the initial cost of a telecommunications system, but, ultimately, the increased capacity results in an economic benefit.

Charges for these calls out of the local area were usually on a per unit or toll basis. This calling segment of the network thus became known as the *toll plant*.

The introduction of automatic switching of toll circuits permitted rapid and economical call–switching, beginning in 1947, and eventually provided a customer direct–dialing nationwide telephone network. This network used an integrated, unified direct distance dialing (DDD) plan. Also, the population shift from urban areas to the suburbs that resulted in the growth of towns and supporting industry promoted complex interoffice connections and networking. These networks, characterized by high–density interoffice trunking, were developed from metropolitan service and trunking plans.

Digital carrier* made its appearance in the early 1960s and has almost replaced physical cable wire pairs as the facility for short–haul interoffice connections. While carrier systems had been in limited use in loop plant for many years, their use did not become popular until the late 1970s when digital carrier technology was introduced for loop service.

In its 1968 Carterfone decision, the Federal Communications Commission (FCC) set aside restrictions by the then Bell System for the interconnection of private systems and customer equipment to the public switched network. During the period 1968 to 1975, the FCC and state tariffs recognized various classes of customer–provided equipment that could be interconnected with the switched network, and authorized standardized interfaces between the customer–owned equipment and the network. During the middle 1970s, the FCC issued a series of orders establishing a registration program for all types of equipment that would be connected to the public switched telephone network. Telephone equipment manufacturers, seizing this opportunity, developed products that allowed every telephone subscriber the option to purchase and install station equipment containing more functions at a lower price from a variety of sources.

---

* Digital carrier combines and converts a number of transmission paths into a single serial stream of pulses, transmits the data over a common facility, and decodes and reconstitutes the signal into its original form at a distant terminal.

Concurrent with the changes in station equipment, technological advances in transmission media, and the evolution in switching systems, the telecommunications industry has experienced a period of remarkable growth in the message telecommunications service (MTS) network as well as in point–to–point and switched private lines and networking. This expansion of private–line and switched services has been accelerated by the divestiture of the Bell System companies on January 1, 1984, which segmented the nationwide network and accelerated competition in the interexchange arena. Coincident with the separation of the operating telephone companies (OTCs) from American Telephone and Telegraph Company (AT&T), the federal court sanctioned the establishment of seven regional holding companies. The court also recognized local access and transport area (LATA) territories to be served by the OTCs, or in some cases, non–Bell local carriers. Interexchange carriers (ICs) were permitted to provide interLATA telecommunications services to customers. The divestiture settlement specified certain actions to aid equal access to IC networking for the customers served by local switching systems.

Out of the evolving structure of systems and subsystems has grown a large number and variety of services other than the familiar MTS network. These include digital data, video, high–capacity digital, facsimile, and program services available to an expanding clientele over an ever–changing and growing network whose ownership and operational responsibility is divided.

While no one can predict the structure of the telecommunications network of the future, most observers envision an all–digital voice and data network from customer location to customer location. Therefore, growth of the telecommunications network is predicted to be a migration of services from an analog environment to an all–digital network.

As background, Chapter 1 provides an overview of the public switched MTS network as it exists in the late 1980s with an emphasis on the transmission and switching facilities that provide nationwide telephone service. Also included is an outline of expected changes in the character of the network into the 1990s. Equipment used for other services is briefly discussed.

Chapter 2 provides an introduction to telecommunications signals and transmission concepts. Brief descriptions of telephone, program, video, and data signals are presented, transmission terminology is defined, and basic techniques and modes of transmission are explained. Some specialized equipment, used to improve plant performance, is described to illustrate the interactions of various parts of the network.

The text is intended to cover the basic engineering principles in sufficient detail to provide the reader with an appreciation of the interrelation of transmission objectives and achievable performance.

Chapter 1

The Transmission Environment

Telecommunications systems of the world have their roots in telegraph systems developed during the mid 1800s, but owe their twentieth–century status to the invention and proliferation of the telephone toward the end of the nineteenth century. Although person–to–person voice communication is expected to grow in volume in the years ahead, the major growth in the remainder of the twentieth century is forecast to be in the transfer of digital information.

Whether the information to be transferred is voice, digital data, or video, the supporting networks contain a large number of interconnected and interrelated systems and subsystems, each of which must be designed to operate within specified limitations established for its segment in the overall network. The design, operation, and maintenance of each segment of the system is essential to the day–to–day quality and reliability of the services as well as the orderly expansion of the telecommunications network to accommodate growing traffic volumes and the introduction of new services.

The basic service provided by the exchange carriers is voice communication using the message telecommunications service (MTS) network. However, services such as telegraph, facsimile, and voiceband data also use this network. In addition, other services (e.g., point–to–point private line, video, digital data, high–capacity digital, telephoto) are provided, some of which require special arrangements.

The provision of transmission paths, or channels, and the flexible interconnection of these paths by switching are the two principal functions performed by the MTS network, the largest service category that uses the exchange carrier plant. The facilities (multiplexers, optical fibers, radio systems, etc.) for these transmission paths are shared by many other services. The network transmission paths, highly variable in length, are of three major types: intraLATA (within the local access and transport area), switched access [extending from local users to interexchange carriers (ICs)], and interLATA (provided to ICs, except for internal use by the exchange carrier).

1-1 TRANSMISSION PATHS

Transmission paths are designed to provide economical and reliable transmission of signals between terminals. The designs must accommodate a wide range of applied signal amplitudes and must guarantee that impairments are held to acceptably low levels so that received signals can be recovered to satisfy the needs of the recipient, whether a person or a computer–like device.

Transmission paths may be designed to operate over two–wire or four–wire facilities. Most two–wire facilities are arranged to provide both directions of transmission over the same wire pair while four–wire facilities provide separate paths for the two transmission directions. Carrier–derived facilities, by their nature, operate in the four–wire mode. Since all interLATA services use carrier facilities, they are considered four–wire.

The major elements of the switched network are station terminal equipment (telephone sets, data modems*, etc.), customer loops (cable pairs or carrier systems that connect the station terminal equipment to the serving central office), interoffice trunks (may include intraLATA, switched access, or interLATA facilities between switching systems), and the paths established within the switching systems.

A switched voice connection may consist simply of two station telephone sets, two customer loops only a few hundred feet in length, and a path through a local central–office switching system. Other customer loop connections may be provided by digital loop carrier, fiber optic systems, or rural radio systems extending fifty or more miles from the serving central office. Customers,

---

* The term modem is derived from the data set functional operations of a *mo*dulator and a *dem*odulator.

served from different central offices using the MTS network, will use at least one interoffice trunk and may use paths through several or more trunks and switching systems owned, operated, and maintained by more than one exchange carrier. These paths may encounter a mix of analog and digital technology.

Station Terminal Equipment

Station terminal equipment accepts a signal from a source and converts it to an electrical form suitable for transmission to a receiver that reverses the process at the distant end. The station terminal equipment may be a telephone set that converts human speech energy into an electrical signal (analog or digital). The signal is then transmitted over the loop and network facilities toward the distant telephone receiver where it is converted to an audio signal that is a faithful reproduction of the original speech. Other types of station terminal equipment convert digital signals originating from computers to an analog signal suitable for transmission over voice–grade facilities, while station equipment such as digital private branch exchanges (PBXs) may connect to suitable digital facilities and digital switching systems in a direct bit stream. This type of digital connectivity has grown rapidly as digital facilities and digital switching systems become available.

Customer Loops

The station terminal equipment is connected to the central office by means of a transmission and signalling facility called a customer loop. Historically, this connection to the central office was made by means of a pair of insulated wires bundled together with many other pairs into a cable carried overhead on poles, underground in ducts, or buried directly in the earth. Carrier systems have been used as customer loops under special conditions (primarily in rural areas) since shortly after World War II, but it was not until the 1970s that applied digital technology made it economically feasible for their widespread use. Since 1980, a large percentage of new customer loop facilities have been carrier–derived using digital loop carrier. As the demand for digital services increases, it is expected that digital loops and digital loop carrier will make up the primary loop plant using optical fiber cable.

Loops are busy (i.e., connected to trunks or other loops) only a small percentage of the time—in most cases, less than one percent. This led, where calling patterns permitted, to the installation of line concentrators between station terminal equipment and the central office. The concentrator is a small switching device that allows a number of loops to be connected to the central office over a smaller number of loop pairs which are, in effect, trunks. It is placed either in or on a building, on a pole, or in a ground–mounted cabinet nearest to the majority of customers.

Early concentrators were simple switch units capable of selectively connecting a number of subscriber lines via concentrator trunks to the serving switching system where all the line–to–line switching functions were carried out. More recently, remote switch units (RSUs), under control of the serving switching system, are being installed at or near concentrations of subscribers that perform all the switching functions of the host switching system. Many RSUs can act as stand–alone switching units providing local service to their connected subscribers even when cut off from their host systems by failure of the connecting trunks.

The loop plant can be physically reconfigured to satisfy the requirements of providing circuits to customers in a serving area. It may also be rearranged to accommodate a variety of services, for example, wire pairs inductively loaded for analog PBX–CO trunks, or with loads removed for digital data services.

In some services, the loop may be intermixed with trunk facilities to provide an extended loop for voice or digital transmission, or may be used as part of a channel between customer locations for the transmission of wideband signals.

The loop must satisfy the transmission requirements for the particular service to be carried. In addition, the customer loop must accommodate signalling and supervisory requirements for most voice operating systems and control functions. Therefore, the loop plant may require special treatment (i.e., loop electronics, specialized amplifiers, signal extension units) depending on the electrical characteristics of the loop, distance from the serving central office or switching system, or the type of telecommunications service that will use the facility.

Switching Systems

For the switched MTS, the customer's loop connects the station terminal equipment to a switching system in the serving central office, which enables connections to be established directly to other local stations or, through trunks and other switching locations, to any other station terminal on the network. The various classifications of switching offices that make up the MTS network are illustrated in Figure 1–1.

Nearly half of the serving switching *systems* in operation in 1986 were electromechanical, most of which were step–by–step (ratchet–driven switch directly controlled by the user's dial) or crossbar (matrix–type switch controlled by electrical logic). However, a large majority of customer *loops* are connected to central offices equipped with electronic switching systems using stored program control. Modern switching systems are digital, using time–division switching techniques. All of these systems connect directly to predominantly two–wire customer loops*. However, while the switching is equivalent two–wire, the switching paths within the system may be equivalent two–wire or four–wire.

Digital switching systems were introduced into the operating switched network in the 1970s. The first commercial applications were in PBXs followed shortly by switching systems designed for tandem trunk applications. Digital switching systems were introduced in North America to local plant starting in 1975 and account for almost 100 percent of new installations today. These switching systems provide equivalent four–wire switching, although most line circuits are connected to two–wire analog or digital loops.

Transmission paths are provided through switching systems in a variety of ways and by a number of different mechanisms ranging from step–by–step switches to digital time–slot interchangers (TSIs). They have one purpose; to connect any one of several thousand terminals to any other one in the same system. To minimize customer annoyance, switching systems are designed to complete requests for connection with a minimum of delay

* Some electronic and/or digital PBXs connect via four–wire loops or digital carrier directly.

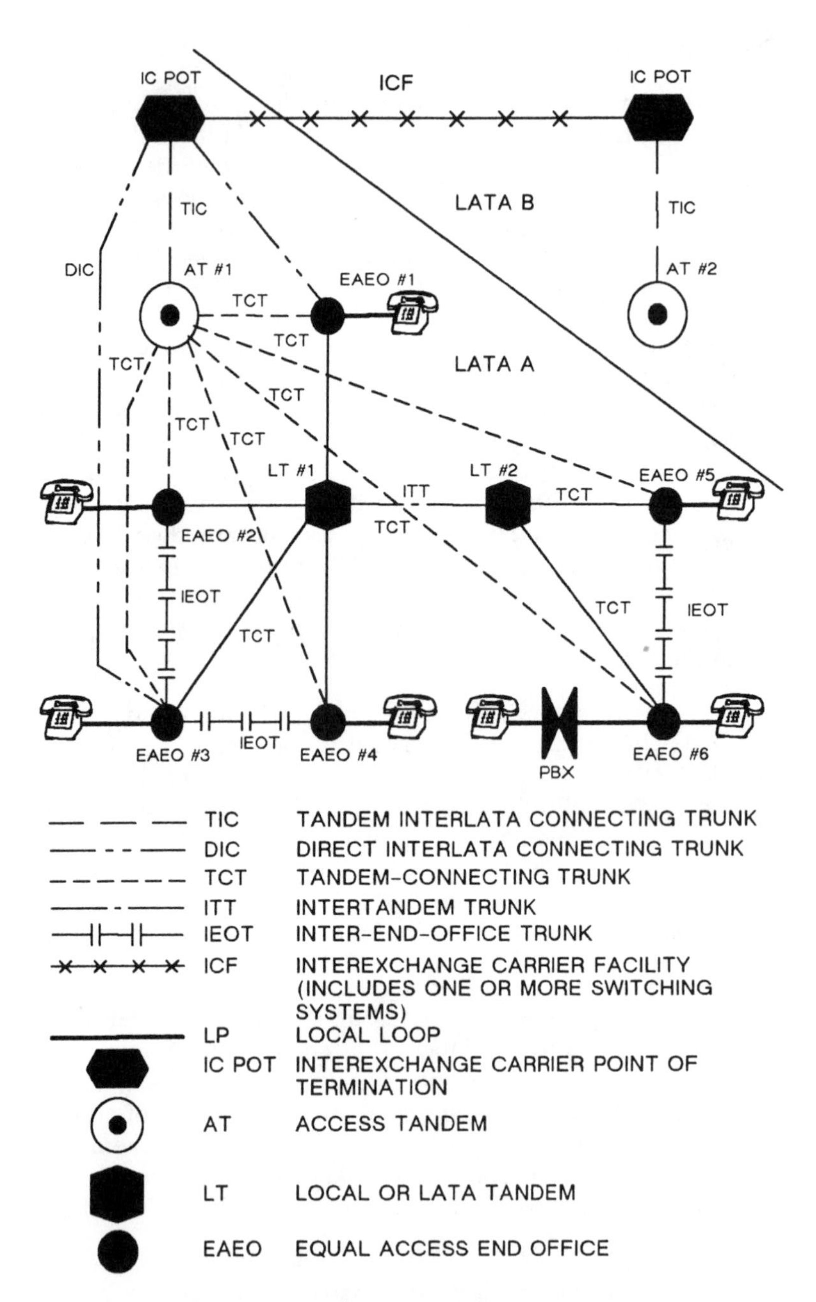

Figure 1-1. Analog switched MTS network.

during the busiest time period. However, occasionally during a busy period, a small percentage of calls may not be completed as a result of all paths being busy between the calling party and the called party. In addition, each of the transmission paths through the switching system is designed to provide good transmission quality for all the services it is to accommodate.

Digital switching systems may be arranged to connect directly to digital loop carrier systems. In this mode of operation, the connection is switched four–wire. The conversion to two–wire is made at the remote terminal where the digital–to–analog conversion of the signal takes place. The circuit then extends on a two–wire basis to the customer's location. As more digital station equipment and digital PBX switching systems are installed, these loops may be installed as four–wire digital* local facility from the central office to the customer. However, integrated services digital network (ISDN) technology permits effective four–wire digital channelization over a two–wire loop.**

While a major portion of the interoffice trunks in metropolitan areas and some intraLATA intercity trunks may continue to be switched on a two–wire basis for some years, the development and use of digital switching promise that most facilities will switch to effectively four–wire. Thus, forecasts for the 1990s predict that intercity connections will be effectively all four–wire from end office to end office. Also, if the switching systems are digital, the connections may be effectively channelized as four–wire from customer premises to customer premises.

Record message networks designed to transmit text have been in existence for many years, but usage has been restricted mainly to business applications. The reduction in the cost of computer memory devices and the development of relatively inexpensive personal computers have spawned a host of information and retrieval systems and networking. Figure 1–2 illustrates a packet switching system (PSS) network that is one solution to the

---

* Large–volume users have found high–capacity local facilities economically desirable in connecting a large digital PBX to a digital switching system or a distant PBX.

** ISDN can use a two–wire loop for simultaneous bidirectional digital services.

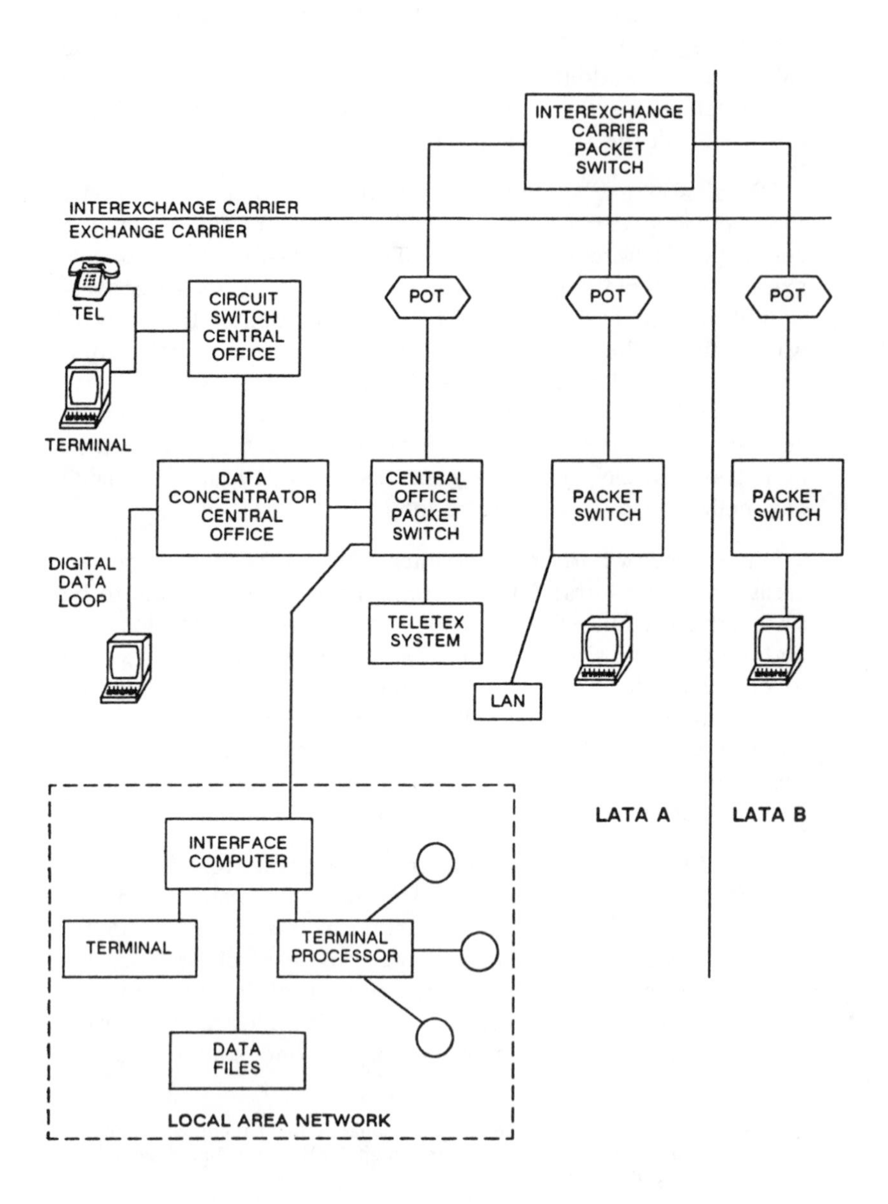

Figure 1–2. Public packet-switched data network.

transmission and switching of large amounts of data economically and reliably. This network and others are addressed in Volume 3 of this series.

Trunks

The transmission paths that interconnect switching systems are called trunks. One essential difference between a loop and a trunk is that a loop is generally associated with station terminal equipment, whereas a trunk is a common connection shared by many users. There are several classes and types of trunk depending on signalling features, operating functions, classes of switching offices interconnected, etc.

Broadly speaking, there are four types of trunk in the LATA switched network. Inter–end–office trunks directly connect end offices for local service, without intermediate switching. Tandem–connecting trunks extend from an end office to a tandem office, that is, one having the function of connecting trunks together. Intertandem trunks join two tandem offices. LATA access (interLATA connecting) trunks extend from an end or tandem office to an IC to give its services a connection to the local area. These trunk types and the switching offices that they connect are illustrated in the simplified network of Figure 1–1.

All trunks must provide transmission and supervision in both directions simultaneously. Supervisory functions may be provided on the same facility as the associated transmission path or on a data link. On digital facilities, supervisory signals are most frequently sent by "borrowing" bits from each message channel at periodic intervals.

Trunks are designated "one–way" or "two–way" according to whether address signalling can be provided in only one direction or in both directions. Two–way signalling is available on interLATA trunks so that calls can originate on the trunks from the switching offices at either end and only one trunk group is potentially needed to service all traffic between the two points. One–way signalling has been most frequently used on intraLATA and switched access trunks. These conventions are rooted in the economics associated with the traffic loads and characteristics between the two points being connected.

Generally, a small two–way trunk group will handle more traffic than an equal number of trunks split between two one–way trunk groups. Offsetting this gain in trunk usage efficiency is the relative complexity, physical size, and cost of the two–way trunk equipment units needed to terminate the trunks on analog electromechanical switching systems. Digital switching systems and the evolving common–channel signalling (CCS) schemes have eliminated the previous disparity in complexity between one–way and two–way trunk equipment and it is likely that two–way trunk operation will increase in popularity.

Any trunk may be assigned to use carrier transmission systems. In fact, all interLATA and a majority of tandem–connecting and switched access trunks use carrier paths. While there will be interend trunks using two–wire cable pairs for some time in the future, economics and the technical demands of the network will combine to drive all trunking to four–wire (or effective four–wire using ISDN) digital operation. The transmission media providing the carrier paths for network trunking include paired cable, coaxial cable, microwave radio, satellite radio, and optical fiber cable. All of the transmission media may be used for interLATA, switched access, and intraLATA services, although satellite radio systems are normally used for interLATA services and paired cable is, in practice, used for inter–end–office trunks.

1–2 SWITCHING ARRANGEMENTS

The service provided by the MTS network consists fundamentally of providing transmission capability "on demand" between two or more points in the network. The implication of "on demand" means that the switching network must be capable of finding the desired distant end (or ends) of a connection and complete the connection between the originating and distant ends promptly and accurately. This is accomplished by a large number of switching systems connected together and organized around considerations of geography, concentrations of population, communities of interest, and dispersion of facilities.

Figure 1–1 illustrates a theoretical switching and trunking plan that might serve a medium size metropolitan area. As shown, equal access end office (EAEO) 1 through 6 plus LATA tandem

(LT) 1 and 2 provide service in LATA A, and access tandem (AT) 1 provides access service to and from another LATA for interLATA service.

While Figure 1–1 shows LATA tandems as separate entities, many are collocated with an end office. Similarly, AT 1 would probably be collocated with LT 1 and, in fact, both tandem switching functions would reside in a single switching system. The trunks provided for local intraLATA service and those provided for LATA access services may then be operated as segregated trunk groups or may be designed as a single combined group. The decision is made on the basis of economics, bearing in mind that should a common analog trunk group be provided, all trunks in the group must meet the transmission performance specifications for the service having the most severe requirements.

The Metropolitan Switching Hierarchy

Figure 1–3 illustrates the various degrees of complexity that may involve switching within a metropolitan area. The simplest connection in the MTS network is from one station telephone set to another through a switching system as shown in Figure 1–3(a).

A connection in a multioffice area might be set between two local end offices in a number of ways as shown in Figure 1–3(b). Within the metropolitan area illustrated in the figure, connections may be established directly between the two end offices by using interend (interlocal) trunk A. A connection may also be established between customer A and customer B via LT 1 using trunks B and E. An alternate arrangement, during peak traffic conditions, could be a connection established through both tandem offices including trunks B, C, and D.

Metropolitan trunking schemes using more than two tandem switching systems and three trunks in tandem may be used in the United States. However, this arrangement is not common due to the high cost of multiple stages of switching, the difficulty of insuring high customer satisfaction through multiple analog trunks, and the greatly increased trunk capacity of modern digital switching systems.

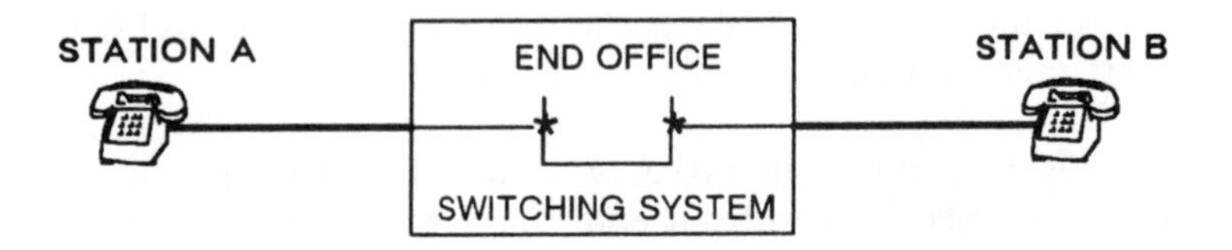

(a) Station-to-station connection through same end office.

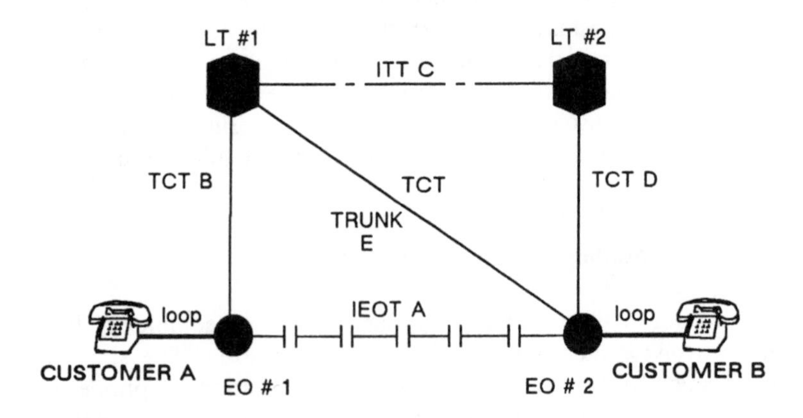

(b) Station-to-station using metropolitan area trunking.

Figure 1-3. Illustrative telephone connections in a metropolitan area.

InterLATA Switching Network

The switching plan used to interconnect the interexchange network strongly influences the performance requirements of both the intraLATA and interLATA transmission systems. This is readily understood when one notes that both systems must be designed to connect to the same local loop plant and serve customer populations having the same characteristics. Thus, their performance parameters are bound by the same limits. Second, geographical conditions and economic considerations frequently dictate that interoffice trunks between the same two points use the same routing, share the same facility, and be designed to meet the same transmission requirements.

The service quality of the predivestiture Bell System toll network was frequently referenced as the benchmark for postdivestiture performance during the legal discussions that culminated in the judicial approval of the Modified Final Judgment with AT&T. Thus, it is appropriate to examine this network and the postdivestiture arrangements for interLATA service. Figures 1–4 and 1–5 show simplified sketches depicting these two network switching hierarchies.

Figure 1–4 illustrates the five–level switching hierarchy of the predivestiture Bell System switched network. The figure shows the probable trunk routing that was available to process and connect a call originated by customer A served by end office 1 (EO1) to customer B served by EO2. Each of the toll switching systems was capable of offering alternate routes on all traffic. Thus, toll center 1 (TC1) could route a sample call directly to TC2 using the high–usage trunk group (shown by the dashed line between the two offices). If all the trunks in this group were busy, the call might have been routed to primary center 1 (PC1) using the final trunk group (denoted by the solid line joining the two offices) connecting TC1 and PC1. PC1 would attempt to connect the call by selecting one of the high–usage trunk groups indicated by (1) and (2). If PC1 failed to find an idle trunk in the high–usage routes, it would advance selection to the final trunk group, noted as (3), connecting PC1 to sectional center 1 (SC1).

If it took all final trunk groups to complete the call for this TC1–to–TC2 connection, a total of seven intertoll office trunks would have been connected in tandem. In addition to the intertoll segment, the connection always included the two toll–connecting trunks that connect EO1 and EO2 to their respective toll centers. Thus the maximum number of trunks in a given connection could have reached as high as nine. This situation very seldom occurred because a number of factors worked to reduce the

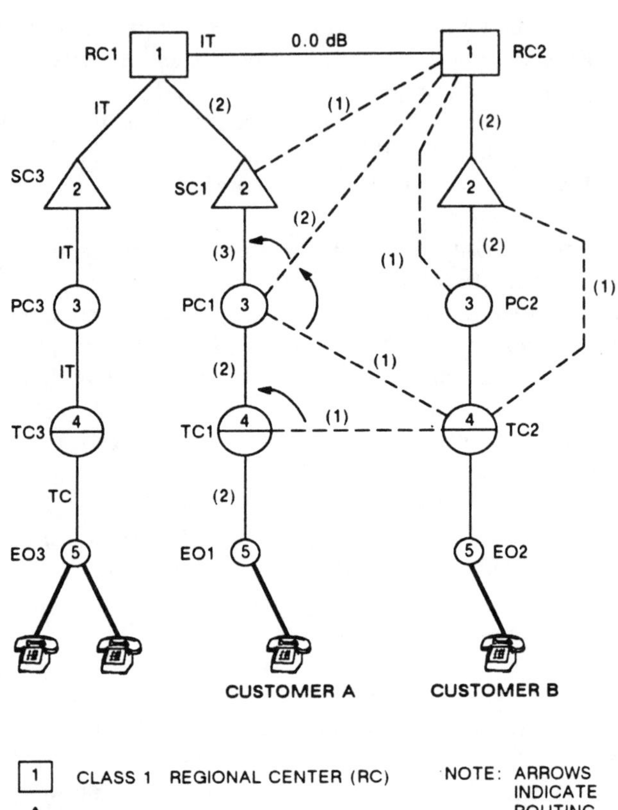

Figure 1–4. Predivestiture toll switching hierarchy.

number of trunks connected in tandem for any given end–to–end connection.

The main factors that served to minimize the average number of trunks connected in tandem follow.

(1) High–usage trunk groups were liberally provided.

(2) The majority of customers made calls to the same or adjacent town or county. Many originated and terminated in end offices that homed on the same toll center and involved no intertoll trunks. Most calls involved only a single high–usage trunk group between adjacent toll centers. Customers served by this arrangement were effectively seeing a two–level switching network.

(3) Most primary centers (class 3 offices) and some sectional centers (class 2 offices) also served as toll centers with end offices homing directly on them.

Network traffic studies made over a period of years confirmed that as a result of design and operational policies, the average toll connection in the predivestiture network included two intertoll and two toll–connecting trunks.

Toll–connecting trunks were designed with 1000–Hz losses ranging between 2 and 4 dB. In practice, the majority operated over digital carrier and were designed with a 3–dB loss. Intertoll trunk design losses range from 0.0 to 2.9 dB (regional–center–to–regional–center trunk loss is 0.0 dB) with the majority of losses falling in the area of 0.5 to 1.0 dB. This design, coupled with an active maintenance plan, resulted in end–office–to–end–office connection losses (based on sample measurements taken each calendar quarter throughout the United States) averaging in a range between 6.5 and 7.0 dB, with standard deviations ranging from 1.8 to 2.0 dB. Data from the same connection appraisal plan placed the end–office–to–end–office connection grade–of–service rating for the network at 82 percent good or better (GoB) and 4 percent poor or worse (PoW) for the years prior to divestiture.

Figure 1–5 illustrates the two–level hierarchical switched network designed to meet the postdivestiture equal access

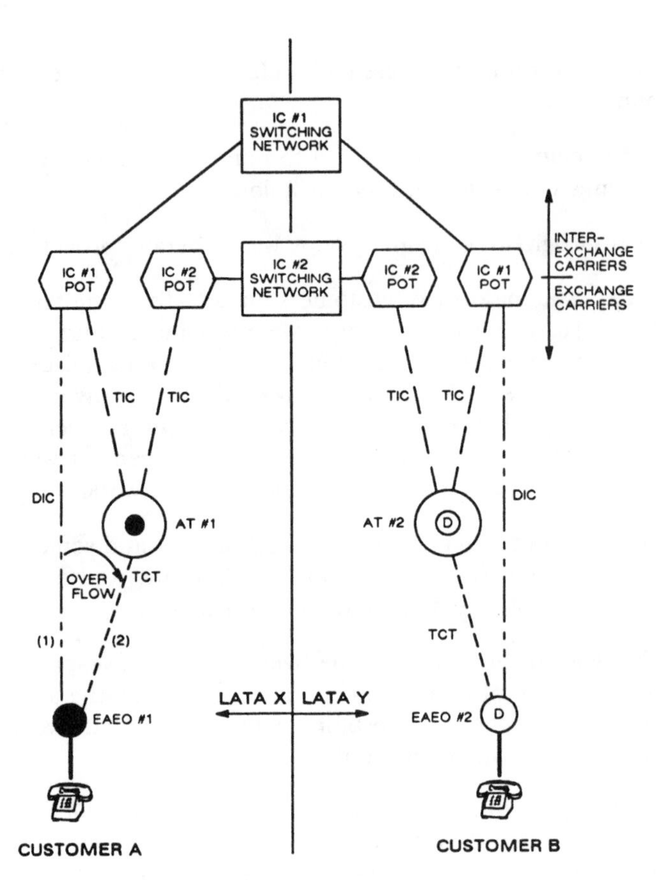

AT	ACCESS TANDEM
DIC	DIRECT INTERLATA CONNECTING TRUNK
EAEO	EQUAL ACCESS END OFFICE
IC	INTEREXCHANGE CARRIER
LATA	LOCAL ACCESS TRANSPORT AREA
TCT	TANDEM-CONNECTING TRUNK
TIC	TANDEM INTERLATA CONNECTING TRUNK
●	ANALOG—END OFFICE—DIGITAL (D)
⊙	ANALOG—ACCESS TANDEM—DIGITAL (D)
POT	IC POINT OF TERMINATION

Figure 1-5. Postdivestiture two-level equal access switching hierarchy.

requirements for ICs wishing to connect to switched access subscribers. With these arrangements, connections to an IC facility at a point of termination (POT) can be provided by:

(1) routing traffic directly from an end office

(2) concentrating traffic by routing via an access tandem

(3) establishing a high–usage trunk group between the end office and the IC point of termination, with overflow traffic routed via the access tandem.

The three access methods are designed to handle the range of traffic volumes each IC may generate from the various sizes of end offices. For example, an IC with a small traffic volume from a given end office would be served through the access tandem. With large traffic volumes, a direct end–office–to–IC trunk group would prove economical. In the latter case, especially if the peak–to–average traffic load ratio is high, it may be possible to minimize costs by tightly engineering the high–usage group. It may also be possible to take advantage of alternate routing to overflow the peak period traffic to the AT where it would be grouped with traffic from other offices that might not have the same peak busy–hour traffic load.

The access tandem, from a transmission standpoint, takes the place of the toll center or class 4 office in the predivestiture switching plan but its switching capability is limited to routing IC traffic to the IC–designated POT within the LATA in which the tandem is located.

The interexchange segment of the connection must contain at least one switching system but may contain as many as the individual IC chooses for efficient service to its clients. Thus, the end–office–to–end–office connection will contain at least four segments if direct end–office–to–IC trunking is used or six segments if the connection is made via the access tandems.

In practice, most POT interfaces are made at the DS1 [1.544 megabits per second (Mb/s)] or higher level digital bit streams. This arrangement, from a transmission point of view, provides a continuous trunking facility between the IC switching system and either the access tandem or end office, depending on whether the direct or tandem access plan is being used. Thus, the insertion of the IC–POT connection in the trunking network need not affect the performance of the interconnecting system but, as will be discussed in later chapters, does complicate operation and maintenance procedures.

Exchange carriers normally design direct interLATA connecting (DIC) trunks between the IC switch and the end office with a 3–dB loss at 1004 Hz. The same overall loss is obtained for traffic routed via the access tandem by designing the EO–AT tandem–connecting trunk for a 3–dB loss and the AT–to–IC link so that it can be operated as part of a zero–loss facility. Each IC is free to design the interLATA facilities to its own specifications, but most design their switch–to–switch path to operate with losses between 0.0 and 0.5 dB. Thus, the overall EO–to–EO connection losses on the postdivestiture network compare favorably with the predivestiture arrangements.

In addition to the predivestiture and equal access plans described, three other connection plans for connecting local or intraLATA facilities to the toll or interLATA facilities are in use today. The oldest of these plans is presently called Feature Group A (FGA). This plan provided for the connection of interexchange facilities to an end–office line appearance in the same manner as local subscribers are connected to their serving switching system. A distant party being switched to this incoming line appears to the local switching system as a local subscriber and can be connected to any other subscriber having local service in the exchange (Figure 1–6, FGA).

In addition, Figure 1–6 shows the Feature Group B* interconnection plan. This plan provides for connecting the interexchange facility to a trunk appearance at either an end office or an intraLATA tandem office. The trunk appearance at the end office allows the trunk to be switched to any subscriber line served by the office. Connection via an intraLATA tandem office provides access to all subscribers served by offices having trunks to the tandem office.

---

* Both Feature Groups A and B provide traffic downstream from the first switch below the POT to be grouped with local or metropolitan traffic and routed over local trunking facilities.

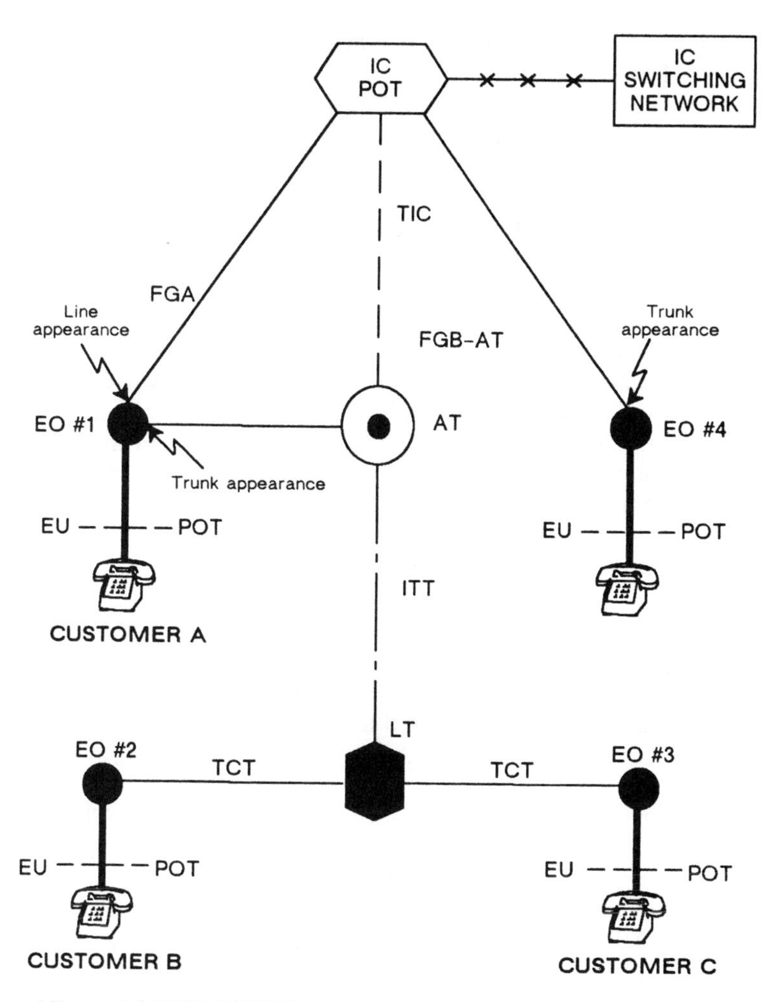

AT - ACCESS TANDEM
EO - END OFFICE
EU - END USER
FGA - FEATURE GROUP A
FGB - FEATURE GROUP B
IC - INTEREXCHANGE CARRIER
ITT - INTERTANDEM TRUNK
LT - LATA TANDEM
POT - POINT OF TERMINATION
TIC - TANDEM INTERLATA CONNECTING TRUNK
TCT - TANDEM-CONNECTING TRUNK

Figure 1-6. Feature Group A and B access arrangements.

Feature Group C provides a trunkside connection that can be provided only through a non–equal–access end office to all lines terminated directly at the EO. FGC is provided on an interim basis until the EO is converted or replaced.

Switched Digital Data Networks

Switching hierarchies for voice services evolved slowly in the past but services are expected to increase rapidly as digital switching systems and digital facilities are installed and orchestrated into an all–digital network. Besides increased facilities in the handling of digital data, radical changes are expected in the handling of voice traffic. The transformation from an analog to an all–digital network is being driven by demands for the transmission of digital data at higher speeds.

Switched–data speeds in the 1980s range from 300 b/s to 1.544 Mb/s. The most common rates of transmission are 1.2, 2.4, 4.8, 9.6, and 56 kb/s. Networks may switch circuits or data pockets and may provide such value–added services as error control, access to "gateways" connecting to information providers, or conversion of operating speed and data protocol between terminals.

Switched–data speeds above 56 kb/s have generally been confined to local area networks (LANs), which have limited application to a single building, building complex, or campus. LANs, which have grown rapidly since 1980, normally operate at speeds below 10 Mb/s, but equipment is available to operate at higher speeds.

Integrated services digital networks (Figure 1–7), covered in Volume 3, are being deployed presently. They allow subscribers to select a digital bit rate for the type of service desired (i.e., data, voice, facsimile, etc.). The subscriber may use the ISDN access line to combine more than one service (e.g., a user with 144 kb/s of access capacity might use it for a single high–quality voice circuit plus a 64–kb/s data channel). Hardware has been developed and made available, standards have been recommended, and various transmission schemes and software are being tested in trial networks dedicated to specific customers.

However, the ultimate parameters of public ISDN–based services will be determined by developing market demands and technological trends.

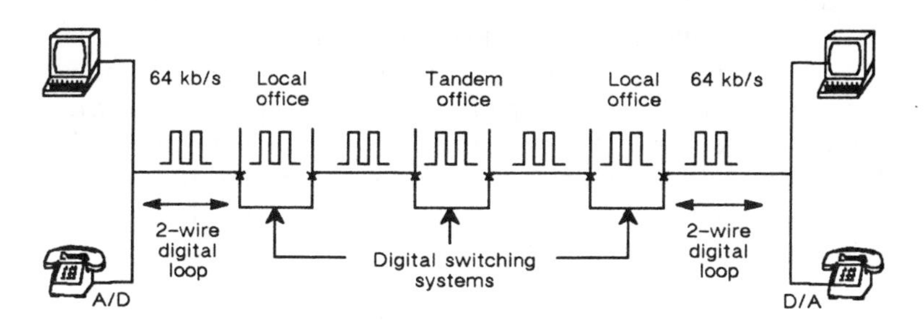

Figure 1–7. ISDN digital end-to-end connectivity.

Private Switched Networks

Private switched networks, most of which interconnect with the MTS network, have been in use for many years. Some operators of these networks, such as railroads, public utilities, pipeline companies, and governmental agencies, own and maintain their own transmission facilities in addition to leasing circuits from exchange carriers and subscribing to public switched (MTS) services. Switching may be accomplished by PBXs owned by the private network operator, or by PBXs or switching systems (including centrex) owned and operated by exchange or interexchange carriers.

Private networks may be simple, involving only two PBXs connected together by trunks. Other networks may consist of thousands of stations scattered over an appreciable area, involve several switching systems equivalent to those used in the MTS network, and include a multilevel switching hierarchy similar to that found in the public network.

Digital Switched Network

The MTS network is evolving into an all–digital switched network. It is a gradual evolvement coordinating the ever–growing number of digital switching systems with the existing analog plant. In time, all will be replaced with an architecture that will transport digital signals through the network without modification. In the interim, many all–digital connections will be possible with robbed bit signalling as the only signal modification through the network connection. Currently, all–digital connections may also use digital pads in the end office to provide traditional network loss performance (no bit integrity). The design of such a network would use a two–tier hierarchical backbone, with digital connectivity to interexchange points of terminations and digital PBXs.

Figure 1–8 illustrates an all–digital network in which the loss is zero between encode and decode points where the only loss in the digital connection is the XDL decode value. In order to achieve the desired grade of service, connection loss is controlled by specifying standard encode and decode levels at the analog–digital and digital–analog interfaces. Thus, a fixed–loss transmission plan would replace the via net loss plan used for many years. The fixed–loss plan will improve the signal–to–noise ratio in the network, afford greater control of echo and return loss, and provide end–to–end bit integrity when robbed bit signalling is eliminated. An all–digital network would support the functions of ISDN and extend digital connectivity to digital PBXs. The goal of an all–digital network plan is to simplify the engineering, installation, and maintenance of systems and circuits.

Transmission design of these systems is discussed in detail in Volume 3 of this series. It suffices to note here that a marginally designed communications circuit used exclusively and regularly by a small number of people may be considered as providing acceptable service, since people have become accustomed to its vagaries. However, if the facilities are to be randomly accessible to the public or a large employee group, certain design criteria must be rigorously followed. These design criteria are generally the same as those applied to the design of facilities for the public switched network.

1–3 IMPACT OF NETWORK PERFORMANCE

The provision of customer–to–customer communication channels can involve a multiplicity of terminals, facilities, and systems

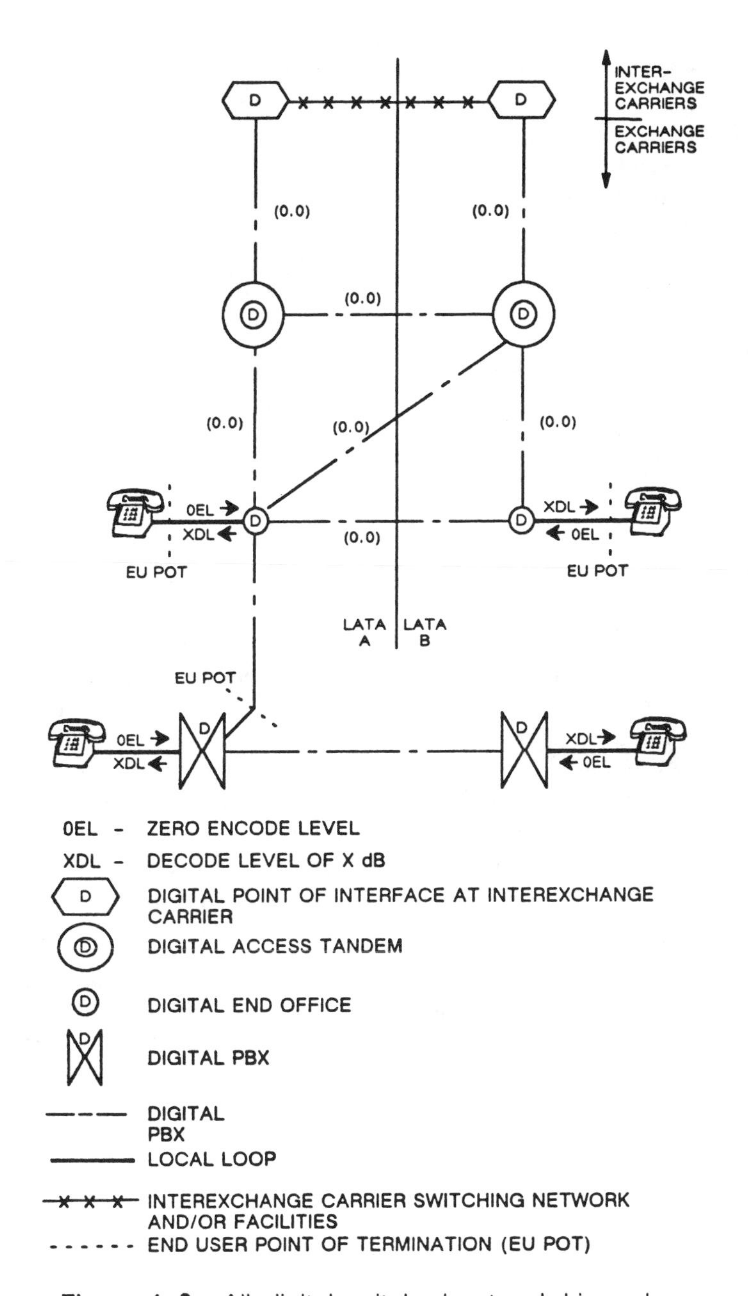

Figure 1-8. All-digital switched network hierarchy.

interconnected in many ways. Station terminal equipment, loops, and end offices are particularly important, especially in the MTS network, since they are used in every connection. Also important when calling outside the local area are the tandem–connecting trunks and the access tandem offices owned and operated by the exchange carriers, and the interexchange facilities and switching systems owned and operated by ICs. Together they make up a complex configuration of plant items whose interactions give rise to several broad problems in total network design and operation.

The first problem is the accumulation of performance imperfections or impairments (loss, noise, jitter, echo, etc.) from a large number of interconnected systems. This results in severe performance requirements for individual equipment and facility units. Also, the ways in which these imperfections accumulate complicate this problem.

The second issue is the variable makeup of the systems forming overall connections. The problem of economically allocating tolerable amounts of impairment among these systems is quite complex. Deriving objectives for a connection of fixed length and composition involves customer reactions and economics. However, when these objectives are to be met for connections of widely varying length and composition, which are made up of segments owned, operated, and maintained by several companies, the problem of deriving objectives for a particular system requires an even more complex statistical study involving knowledge of plant layout, operating procedures, and the performance of other systems.

A third problem involves the satisfactory operation of each part with nearly all the other parts of the network. Compatibility is particularly important when new equipment and new systems are being developed. The new must interact with the old requiring standards to be set and parameters published. Also, plant growth must take place by gradual additions and phased in with transition plans rather than by massive replacement. This problem is particularly acute when the responsibility for providing and maintaining facilities for a service is split between two or more companies.

A fourth concern is that of reliability. Only a small percentage of outage time is acceptable for communications provided to the public. Considerations of outage time must include all causes such as equipment failure, natural or manmade disaster, and operating errors.

Finally, any discussion of the transmission environment must take into account that exchange carrier plant and power distribution systems share the same geography, either aerially or underground. This fact is important for safety as well as for its effect on transmission quality over telecommunications facilities. Power systems may come in contact with exchange carrier plant as the result of storms, traffic accidents involving aerial communication or power plant, or excavation equipment digging into buried telecommunications or power cables. These accidental contacts may endanger customers, employees, and property unless protective measures are developed and applied. Inductive coupling from paralleling power lines may cause low-frequency noise in telecommunications cables, and in some cases, sufficient 60–Hz ac voltages to endanger company personnel and damage equipment.

1-4 MAINTENANCE

The switching patterns that have been described impose strict requirements on all transmission circuits. For example, an interLATA connection may have as few as three trunk links or as many as nine if the five-level hierarchy shown in Figure 1–1 is accessed. IntraLATA connections may involve no interoffice links or may use as many as three or four tandem trunks. Connections may also include trunking links that are a part of a private network. Additional maintenance complexities are added by alternate routing plans, a long-time fixture in the interLATA message networks but only recently incorporated widely in intraLATA switching plans.

As a result of the many facility combinations possible in a connection between two end offices, some performance variance is likely to occur. Design and maintenance plans must be coordinated in order to prevent the normal variations in quality from becoming objectionable to the user. The objective is to maintain transmission performance on all connections between two network points so the user will notice no difference between the quality provided by successive connections between these same points. An all–digital network with a fixed–loss plan promises to meet this objective.

Additional Reading

Abate, J. E., L. H. Brandenberg, J. C. Lawson, and W. L. Ross. "The Switched Digital Network Plan," *Bell System Tech. J.*, Vol. 56 (Sept. 1977).

Browne, T. E. "Network of the Future," *Proceedings of the IEEE* (Sept. 1986).

Members of Technical Staff. *Transmission Systems for Communications*, Fifth Edition (Murray Hill, NJ: AT&T Bell Laboratories, Inc., 1982).

Notes on the BOC IntraLATA Networks—1986, Technical Reference TR–NPL–000275, Bellcore (Iss. 1, Apr. 1986).

Chapter 2

Introduction To Transmission

The movement of intelligence from one point to another is the basic task of a telecommunications system. The intelligence to be moved can be called a *message*, regardless of the form it takes or its purpose. The most common form, of course, is speech, and the telephone system was initially developed around the need for voice communications. Over the years, however, many other types of message (such as facsimile, program, video, and data) have evolved. The automation of modern business offices and the development of affordable personal computers have opened up a new demand for data service that may well rival the voice network traffic volume.

Transmission technology has advanced in parallel with this evolution, providing a means of translating these messages into electrical signals and developing the telecommunications channels that make it possible to transmit the message in recoverable form via existing transmission media. Extension of the capabilities of the existing multilink plant and the development of new plant that is compatible with the old and with new customer needs are among the problems confronting the transmission engineer.

This chapter describes various message signals and channels. It brings together some basic concepts of transmission and transmission systems in preparation for later chapters. The functions and operation of some specific types of ancillary equipment are also described briefly in order to show how they interact with system concepts.

2-1 TRANSMISSION MEDIA

The fundamental factors in the development of transmission systems are the efficient use of the available transmission capacity and the physical capabilities and relative costs of the available transmission media. Early telegraph and telephone lines consisted of a single wire or pair of wires and carried signals from direct current (dc) to perhaps 2500 Hz*. As the number of users grew, the proliferation of poles, crossarms, and wire created an intolerably crowded situation on city streets, particularly in the vicinity of telephone switching offices. Paired cable was developed to relieve these situations.

Until the advent of electronics, the transmission characteristics of wire and cable restricted the length of loop plant that could be served from a central office and also determined the quality of the voice messages being sent over interoffice connections. Electronics not only made it possible to extend the length of the transmission path by making amplification or gain available, but greatly improved the quality of the circuit, thus making the service more pleasing and desirable to the user. Electronics also made it possible to send more information over a given facility, e.g., by carrier techniques that use a greater portion of the available line capacity. As more bandwidth became usable, more efficient transmission media, such as coaxial cable and microwave radio, were developed. The discovery of the transistor and subsequent development of solid-state, miniaturized electronic equipment made more transmission capacity economically usable, opening the way for the transmission of information via satellite and through optical fiber cable.

Figure 2–1 shows the frequency-handling capabilities of cable plus the frequency bands allocated to various North American radio services.

Paired Cable

A cable pair is made by twisting two insulated conductors together. Conductors for communications use are usually made of copper in one of three wire gauges: 26, 24, or 22 gauge. However, small amounts of 19-gauge and coarser wire remain in use. The insulation may be wood pulp, paper, or polyethylene. The cable is formed by twisting the desired number of pairs, each with

* 1915 transcontinental service cut off at 1600 Hz.

These frequency spreads correspond to both analog and digital carrier—the analog systems cut off much lower; the digital systems extend to the full top frequencies shown.

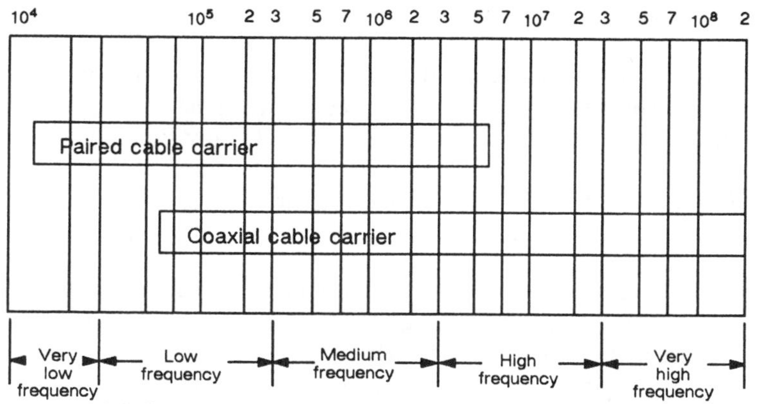

(a) Paired cable and coaxial cable carrier frequencies

Frequency band (GHz)	Bandedges (GHz)	Bandwidth (MHz)
2	2.11 - 2.13	20
	2.16 - 2.18	20
4	3.7 - 4.2*	500
6	5.925 - 6.425#	500
11	10.7 - 11.7+	1000
18	17.7 - 19.7	2000

(b) Common carrier frequency allocations for terrestrial radio relay systems

Frequency band (GHz)	Uplink (GHz)	Downlink (GHz)	Downlink bandwidth (MHz)
4/6	5.925 - 6.425#	3.7 - 4.2*	500
12/14	14.0 - 14.25	11.7 - 12.2	500
	[Domestic]		
	14.0 - 14.5	10.95 - 11.2+	250
	14.0 - 14.5	11.45 - 11.7+	250
	[International]		
19/29	27.5 - 31.0	17.7 - 21.2	3500

(c) Frequency bands for satellite radio point-to-point services

* Used in terrestrial service and as a downlink in satellite systems.
# Used in terrestrial service and as an uplink in satellite systems.
+ Parts of this band may be used as a downlink in satellite systems.

Figure 2-1. Cable carrier and radio frequency assignments.

a specified length of twist, about themselves to form a spiral rope called a unit. A cable consists of one or more units spiraled together and covered with a sheath made of plastic or plastic and metal bonded together.

Paired cable is currently the most pervasive transmission medium in the network, although it has been supplanted by radio and optical fiber cable for all long–haul and many short trunking installations. In addition, optical fiber has begun to supplant copper in loop distribution. Nevertheless, paired cable still maintains continued, albeit declining, usage for the immediate future.

Paired cable facilities in the loop plant provide voiceband channels and paths for analog carrier systems operating at frequencies up to about 150 kHz. These cables also support 144–kilobit–per–second (kb/s) integrated services digital network (ISDN) "basic" access at 160 kb/s (including overhead) and digital carrier systems transmitting bit streams at 1.544 megabits per second (Mb/s). Special shielded pair cables have been used to extend baseband frequencies from microwave radio terminals to television station locations and exchange carrier central offices.

Trunk installations using analog carrier formerly operated extensively at frequencies between 36 and 268 kHz. Digital carrier in the same service transmits binary bit streams over 3 Mb/s. In addition, special low–capacitance cable has been used, which will handle binary bit stream rates at 6.3 Mb/s.

Coaxial Cable

Coaxial cable, which came into use in the 1930s, assumes many different physical forms and sizes and comes with electrical properties designed for specialized telecommunications services. The material is used extensively by cable television distribution systems and to connect radio and television antennas to transmitters and receivers. Short sections are also used to interconnect carrier equipment in exchange carrier central offices and to interconnect other devices using high frequencies that require shielded conductors to minimize radiation. Coaxial cable also provides transmission media for carrier systems (up to 4000 miles) capable of providing up to 13,200 simultaneous voiceband circuits per pair of coaxial tubes using the frequency band between 3.252 and 64.844 MHz. Coaxial cable has also been used to a limited degree for digital transmission at 274.176 Mb/s.

Physically, the coaxial tube consists of an outer conductor in the form of a hollow cylinder and a single inner conductor separated from the tubular outer conductor by an insulator or separator. The outer conductor may be solid or may be of a braided construction. The inner conductor may be a small hollow tube or a solid or stranded copper wire. The insulator or spacing material between the inner and outer conductor may be either a solid extrusion or a uniformly spaced beaded arrangement. The coaxial cable used for intercity facilities in the United States consists of a copper tube 0.375 inches in diameter and a solid copper wire about 0.1 inch in diameter held in place by polystyrene discs 0.1–inch thick and spaced on 1–inch centers along the tube.

Waveguide

A waveguide is a metallic tube or duct used to confine and guide the propagation of an electromagnetic wave. There are three main types of waveguide: rectangular, circular, and elliptical. Each construction form has its strong and weak points, which usually dictate their field of use. These are discussed in some detail in Chapter 5, Part 5.

If a coaxial cable and a waveguide were designed to transmit the same frequency spectrum, the coaxial tube would have about three times the loss per unit length as the waveguide. However, the physical dimensions of a waveguide are dictated by the wavelength of the signal to be carried. This factor normally limits their use to wavelengths less than thirty centimeters (frequencies of 1 GHz or more)*.

The first practical applications of waveguide appeared during World War II when radar systems using centimeter wavelengths (e.g., SP–1 radar at 3000–MHz) were developed. Today, the

* FAA ARSR–1 radar uses waveguide at 1200 MHz. Also, waveguide is used for 400–MHz space radars and upper–UHF TV stations (above 600 MHz).

sole use of waveguides in the telecommunications industry is to connect microwave radio antennas to transmitters and receivers. Theoretically, a single circular waveguide having a diameter of about 2.5 inches could provide an intercity pathway for over 200,000 two-way digital voice circuits. Such a system was developed and field-tried, but has been eclipsed for commercial use by fiber optic systems.

Optical Fiber Cable

The theory behind transmitting light through a strand of transparent material (e.g., glass) has been known for many years, but only recently have the materials and methods been available to construct a practical fiber optic system. Optical fibers are waveguides that have several properties that make them attractive as a communications medium. Primary among these factors is the low loss to light, which allows relatively long cable runs between terminals without a need for regenerators. A second factor is the enormous bandwidths the facilities make available (30,000 GHz or more).

Fiber optic transport systems are in use for such diverse purposes as interconnecting segments of a digital switching system within an office to digital loop carrier (DLC) to long-haul trunking facilities stretching thousands of miles across the country or across oceans.* Most telecommunications applications use the fibers to transmit information in a digital mode at speeds ranging from 45 Mb/s to 810 Mb/s. However, optical fiber and equipment are becoming available to support bit rates above 1100 Mb/s and laboratory systems have been tested successfully at rates above two gigabits per second (Gb/s). Regenerator spacings have grown from only three or four miles to almost 100 miles today. Optical fiber has become, in fewer than ten years, the medium of choice for new plant.

Microwave Radio

Until the recent improvements in fiber optic transport systems and the somewhat earlier advent of satellite radio facilities, the

---

* Optical switching devices, under development, are expected eventually to facilitate switching of high-speed fiber optic transmission systems.

two main long–haul transmission systems used were based on coaxial cable and terrestrial microwave radio relay. The latter is used more because it has generally been less costly per circuit route–mile for all applications except those involving long, very high–density, fast–growing traffic routes.

There are two overriding features of microwave radio that make it unique. First, it makes use of a "free" transmission medium (the troposphere) between repeaters and terminals that have characteristics that are not ideal and do not remain constant with time. Furthermore, variations with time are neither controllable nor 100 percent predictable, since they depend to a large extent on the vagaries of the weather.

Second, the troposphere is not a self–contained medium like waveguide or optical fiber cable. Governments consider it a natural resource to be used for the common good. Users of electromagnetic energy in the atmosphere can interfere with one another. There are more potential users than there is spectrum available. Therefore, governments allocate sections of the available bandwidth for specific purposes by designated users. As a result, the use of microwave radio in urban areas is severely limited by frequency congestion.

Satellite Radio

Satellite radio systems consist of microwave radio relays located on satellites placed in geostationary orbit 22,300 miles above the earth's equator. The satellites provide a platform in space for mounting antennas and a group of transponders (channels) that receive and rebroadcast signals transmitted from earth stations. This arrangement can link widely separated or remote areas where construction of conventional radio relay systems would be uneconomical. In essence, satellites make the line–haul cost of a connection between any two points equal, as long as both terminals have a line–of–sight view of the satellite.

The ability to connect any two points within line of sight of the satellite opens up the possibility of dynamically managing a network by using satellite circuit assignments made temporarily surplus by periodic light traffic conditions on their regularly assigned routes. These idle paths can be reassigned to relieve temporary overload conditions on other routes. Thus, a limited number of satellite circuits can potentially provide relief to several trunk routes if their busy hours do not coincide.

Satellite communications systems were initially available to North American users only for international service. Later, they were widely used for domestic intercity services as well as private–line services. However, because of the time–delay impairment inherent in satellite facilities and because of the attractiveness of digital optical fiber facilities, use of satellites for domestic telephone trunking is diminishing. Communication via satellite is ideally suited to one–way transmission and its use is somewhat limited for two–way domestic speech and data. In addition to voice and record services, domestic satellites are used to distribute network television program material to network affiliates, special news and sports programs to cable television systems, and movies, etc., to hotels and other subscribers. Additional services provide mobile radio and paging systems using satellites to relay their signals.

2-2 MESSAGE SIGNALS

The characterization of transmitted message signals is essential to an understanding of how such signals interact with the channels over which they are transmitted. The message signal is defined as an electrical representation of a message. This signal can be transmitted in an electrical or light form from source to destination. Qualitative descriptions of the more common signals found in telecommunications networks are given here. More detailed quantitative characterizations of all these signal types are given in Chapters 12 through 16. The signals described in this chapter include voice, program, video, facsimile, data, and control signals. The latter, usually classified as signalling and supervision, are used in the transmission of all messages presently offered as network telecommunications services. Any of the signals may be transmitted in either digital or analog form; the choice depends in some cases on the termination facilities available.

Speech Signals

Currently, the most common signal transmitted over message network facilities is speech, an electrical signal generated in customer telephone terminal equipment as an analog of the acoustical speech wave created in the voice box, or larynx, of the speaking telephone user. This signal carries most of its information in a band of frequencies between 200 Hz and 3500 Hz. Most of the energy is peaked near 500 Hz, while most of the articulation is above 800 Hz. There are higher and lower frequency components, but since most of these occur outside the normal passband allotted for a voice circuit, they are not transmitted to the receiver. A speech signal is extremely complex, not only because of the large number of frequency components it contains, but also because of the wide range of amplitudes that any component may have and the rapidity with which the frequencies, amplitudes, and phases of its components change.

Another complexity is the time relationships inherent in the speech signal. Care must be taken to precisely define the signal duration when circumstances demand it, for example, when considering system loading, crosstalk effects, speech synthesis, or adaptive signal processing. By one definition or criterion, the signal duration might be measured from the time the connection is established until it is broken. By another criterion, the signal duration might be defined as the speaking interval during a typical telephone connection—each party speaks about half the time and listens the other half. But the situation is even more complex. There are short intervals, sometimes only milliseconds in length, during which a speaker pauses for breath, thinks, listens to the other person, etc. Signal duration could be defined as covering the time between those pauses. It is thus a matter of definition depending on circumstances.

Program Signals

Program signals are those associated with the distribution of radio program material, the audio portion of television program material, or "wired music" systems. These signals are normally transmitted over one–way channels having a wider bandwidth than the voice–frequency message channel. Usually, material being fed to AM radio stations will be sent over channels capable of handling frequencies up to 5 or 8 kHz. Those providing service to FM transmitters require 15–kHz channels. These channels may be provided over conditioned cable pairs or through the use of special program channel units on either analog or digital carrier systems. Program channel units normally displace two or more regular voice channels. The number of carrier channels displaced is determined by the bandwidth needed and the number of coding bits needed to give suitably low distortion.

Video Signals

Video transmissions over networks in the United States fall into two basic categories: broadcast and cable television, and closed–circuit teleconferencing video services. Each of these services has a specific set of transmission requirements that will be addressed in later chapters of this book.

Broadcast television local service involves connecting the broadcaster's studios, master control center, and transmitter together. Connection to the rest of the world for network service may consist of facilities joining the master control center to a satellite ground station, microwave radio terminal, or to an exchange carrier television operating center for further extension to a point of termination (POT) with an interexchange carrier (IC). Cable television systems instead use satellite circuits and microwave facilities to bring in video material from other markets to supplement off–the–air pickup of local telecasts and studio–originated material. These signals are then distributed within a franchised territory using a coaxial cable network, possibly including optical fiber feeders.

A great deal of technical and marketing research effort has been expended in recent years to produce a switchable public video network that would provide a product desired by the user at a price that would attract enough customers to be profitable. Technical research in connection with the project has spawned several methods (e.g., slow scan and signal compression) of reducing the transmission bandwidth needed to provide an acceptable video picture. These advances, along with the rapid spread of optical fiber cables, with their promise of wide bandwidths, and the increase in satellite capacities, coupled with the introduction of video conferencing services in some local area networks (LANs), may be the catalyst needed to popularize the service and make it economically viable. Finally, broadband ISDN is being contemplated with the premise of making video transmission a universal service.

Data and Facsimile Signals

The basic data signal usually consists of a train of pulses that represent, in coded form, the information to be transmitted. Such signals are processed in many ways to make them suitable for transmission over analog private lines or the message telecommunications service (MTS) network. To represent coded values of the signal, the amplitude may be changed, the frequency may be varied, or the phase of the carrier may be shifted. Combinations of the basic modulating techniques may also be used to increase the amount of data that may be transmitted over a given facility and reduce the number of errors received.

The speed with which changes are made, no matter which parameter is changed, determines to a large extent the bandwidth required to transmit data signals. Transmission speeds vary from a few bits (binary digits) per second for supervisory control channels, to thousands of bits per second for digital data, to 90 megabits per second for digitized television transmission.

With the blossoming of personal computer (PC) use in business, local area networks were developed to connect PCs together at low cost. LANs are located in a single building or within several thousand feet of each other. They use a common medium with a wide bandwidth, are shared by all users, and operate at high speed. This is an economical facility arrangement for communicating data in a physically restricted environment.

Another type of transport for data information is the packet switching system (PSS). PSS formats multiple messages destined for various end users into packets of information that are multiplexed onto a transmission medium and routed through a network on a store–and–forward basis. It is a service provided by exchange and interexchange carriers to part–time users that is economical because trunking is time–shared on an "as–needed" basis.

The Digital Data System (DDS) provides private–line service at 2.4, 4.8, 9.6, 19.2, and 56 kb/s over multiplexed networking. Other networks offer speeds from 1.5 to 45 Mb/s.

Most data services, as well as digitized voice, are expected to migrate to ISDN. This concept provides end–to–end digital connectivity using digital facilities and digital switching systems. Local loops may serve an end user with up to two bidirectional 64–kb/s channels and one signalling channel on one wire pair. ISDN can accommodate telephone, alarm, teleconference, videoconference and data on a switchable, as–needed basis.

Facsimile, as a public network service, predates television by decades. Originally strictly an analog system used primarily by the National Weather Service and military to distribute weather maps, the equipment is now widely used by business establishments to quickly transmit copies of documents across town or around the world. The growth of digital transmission systems and trends in office automation have resulted in improving the speed and accuracy with which a document copy can be transmitted, raising the usefulness of the service. Since modern facsimile machines produce standard digital signals, their transmission requirements are included with the discussion of data signals in Chapter 14, Part 3.

Some two–valued facsimile signals (black and white facsimile) closely resemble asynchronous binary data signals and may be compared with them in many ways. Multivalued facsimile signals are more like video signals; they produce pictures at slow speeds with gradations of grey between black and white. Such facsimile signals are often regarded as special forms of data because channel requirements are quite similar to those for data transmission.

Data signal durations depend on the length of the message to be sent and the rate at which the bits or bytes of data pulses are transmitted. Both of these factors vary widely. Thus, data message durations are highly variable and may extend from a few seconds to an hour or more. Facsimile messages, with modern terminal equipment, characteristically last less than one minute per page of material.

Control Signals

To implement the functions of any switched network it is necessary to transmit three types of control signal: (1) alerting, (2) address, and (3) supervisory. Traditionally, these signals have been transmitted directly on local loops. Current planning contemplates an expansion of digital services involving increasing segments of the loop plant. These services are proposed as integrated voice–data according to the ISDN plan. This scheme provides a signalling and supervisory path either separated from the message transmission path or included as part of the bit stream.

Most intraLATA trunks currently use either digital carrier or metallic facilities, both of which transmit control signals over the connection facility. One major IC currently uses a version of CCITT* Signalling System No. 6 called common–channel interoffice signalling (CCIS) to provide trunk signalling between its computerized switching systems. Other communications companies, including the operating telephone companies, have begun the application of a common–channel signalling (CCS) system for their trunking networks and other services. These installations will conform with the CCITT Signalling System No. 7 recommendations.

Alerting signals include the ringing signal, which is supplied to a loop to alert the customer to an incoming call on the line, and a variety of signals that are used to alert operators to a need for assistance on a call. Addressing signals, transmitted over loops and trunks, provide information concerning the called number and sometimes the identification of the calling number. Supervisory signals are used to indicate a demand for service, the termination of a call, and the busy or idle status of each loop or trunk.

Many forms of addressing and supervisory signals are employed. These include pulsing of the direct current supplied on loops for talking purposes or on voice–frequency trunks for supervision, changes in state of direct current supplied on voice–frequency trunks, and single–frequency or multifrequency alternating current signals, which may be transmitted within the

---

* International Telegraph and Telephone Consultative Committee, Geneva, Switzerland.

voiceband of a carrier or voice–frequency circuit. The more important of these signals are described and characterized in Chapter 13.

Many other types of signalling information are transmitted for subsidiary functions. These include dial tone, audible ringing tone, coin signals (deposit, return, and collect), and busy and reorder tones. None of these relate importantly to transmission work, however, and so are not described in detail.

The duration of information signals varies widely. Addressing signals last for only a short time, one to several seconds. Supervisory signals, on the other hand, may be present for minutes or hours when, for example, a trunk is not called into use. Address signals may be regarded as transient by nature, and supervisory signals are steady state.

2–3 CHANNELS

A channel is defined as a frequency band in the frequency domain, or its digital time–slot equivalent in the time domain, established to provide a communication path between a message source and its destination. In most cases, the exchange carrier provides a channel between a customer's POT and the serving central office or another POT. The characteristics of the signal derived from the message source determine the requirements imposed on the channel regarding bandwidth, signal–to–noise performance, etc.

In the MTS network, a variety of channels are provided on a full–time, dedicated basis in the form of loops, local trunks, tandem–connecting trunks, and interexchange trunks. Each such channel is a well–defined entity between its terminals, existing for long periods of time.

The end–to–end frequency band established between station terminal equipment in a built–up telephone call is also considered a channel. This frequency band is dedicated and maintained only for the duration of the call. In this case, the built–up channel is made up of other tandem–connected channels, interconnected loops, and trunks used to establish the connection.

Historically, changes in an individual channel makeup or configuration could be made only by changing wire cross–connections or by patching within a jack field. In recent years, jack fields have been omitted from trunk groups and short–haul facilities. The installation of equipment providing switched access to a circuit trunk or facility for test purposes led to the obsolescence of test jack fields. Digital cross–connection systems are now in wide use for software–controlled reconfiguration of circuits.

Time–division switching and its integration with digital transmission bid well to change another time–honored conception, namely, the end–to–end concept of a channel in a built–up telephone call. Most present systems maintain the integrity of the channel in the time–domain equivalent of the analog channel, but it is possible to increase the capacity of a trunk cross section to handle message traffic by using time–division methods such as packet switching and time–assignment speech interpolation (TASI) systems. TASI, which is described later in this chapter, has been used for many years in its analog version to increase the voice traffic handling capacity of international telephone facilities. The cost of the analog TASI system precluded its general application for domestic service.

As discussed above, channels in the MTS network may be regarded as fixed, changeable, or switchable. In any case, each type of channel must be designed to have a transmission response that will satisfy the objectives set for the type of service to be provided. That is, they must be of sufficient bandwidth, must have gain/frequency and phase/frequency characteristics that are controlled, and must not be contaminated by excessive noise or other interference. These parameters are discussed quantitatively in later chapters.

Channels may also be regarded as one–way or two–way. Carrier systems are operated on a four–wire basis, involving a separate path for each direction of transmission. On one such path, the dedicated time slot (i.e., the channel) carries signal energy in one direction only, so each path represents a one–way channel. Voice–frequency circuits (loops and trunks), on the other hand, are usually operated so that both directions of transmission are carried on the same wire pairs as a two–way channel. In any case, in the MTS network, loops and trunks must be capable of two-way simultaneous usage.

Definitions involving channels in the MTS network have been stressed. It must be recognized that many other types of channel are provided by the network. These include very wideband channels for high-speed data or video signal transmission; full duplex ISDN digital loop channels transporting voice, data, and signalling; channels of greater bandwidth than speech channels for radio and television program signals; voiceband channels that are specially treated to meet data or facsimile transmission objectives; and very narrowband channels for low-speed data signal transmission.

2-4 VOICE-FREQUENCY TRANSMISSION

Digital carrier and digital switching systems, driven by the decreasing cost of digital technology and the increasing requirement for data transmission capability, are replacing physical cable plant and analog switching in the local plant arena. Nevertheless, because voice transmission continues to represent the major public service requirement, the local loop plant will continue for the foreseeable future to be served primarily by paired cable operated in a two-wire mode. However, with the carrier serving area (CSA) concept, digital loop carrier will replace most of the long-wire cable plant (12 kft and beyond). In the trunking arena, the two-wire mode on physical cable pairs is now confined to a few groups of very short trunks and is doomed to extinction. Cable pairs are being replaced with digital carrier facilities as a result of the advent of digital switching, optical fiber cables, and the growth of DDS requirements.

Because voice transmission over cable pairs is subject to end-to-end loss and other impairments, these impairments must be controlled within established limits in order to provide satisfactory service. When losses in an analog channel exceed established limits, compensation must be provided, usually in the form of voice-frequency repeaters. In addition to the voice circuit handling capability, an analog customer loop must be able to provide low-frequency paths for ringing and dialing signals as well as the direct current needed to operate telephone transmitter and central–office supervisory circuitry. Hence, the cable wire size chosen represents an economic compromise between the desirable and minimum acceptable ranges for signalling, supervision, and transmission.

Modes of Voice–Frequency Transmission

Telephone station equipment is basically a four–wire device, one that requires two wires for the transmitter and two for the receiver. If the four–wire nature of the set were extended into the entire local plant, including both loops and trunks, four wires would have to be provided for every connection, including the transmission paths through the switching machines. Such an arrangement, illustrated in Figure 2–2(a), would be inordinately expensive in a purely paired cable environment, since it would double the amount of copper required for transmission paths. To avoid the expense associated with the four–wire loop, most station telephone circuitry combines the transmitter and receiver conductors so that only one pair of wires is needed for transmission in both directions. This arrangement, called two–wire transmission, is illustrated in Figure 2–2(b).

Three factors are changing the above economic equation. First, DDS requests are growing, especially from customers using devices requiring four–wire service. Second, in–place costs of digital loop carrier compare favorably with cable pair costs for an increasing number of installation conditions. Third, the decrease in the relative cost of digital switching units has made them the normal choice for local switching. Also, they may be connected to digital carrier loop plant more cheaply than to two–wire cable pair loops.

Four–wire station equipment and loops in conjunction with digital switching units would allow better transmission performance in the network and, if the current technological trends continue, may also become economically justifiable on an overall basis. Nevertheless, the economics and political problems associated with replacing the present local office switch units and customer–owned station terminal equipment are such that two–wire analog cable loops, somewhat modified, will likely be around for a long time. Also, new technology using echo cancellers reduces the advantages of four–wire loop plant. While very few four–wire voice–frequency circuits use physical cable pairs in the MTS network, cable pairs may be used to piece out or extend a four–wire carrier channel from a central office to the customer location.

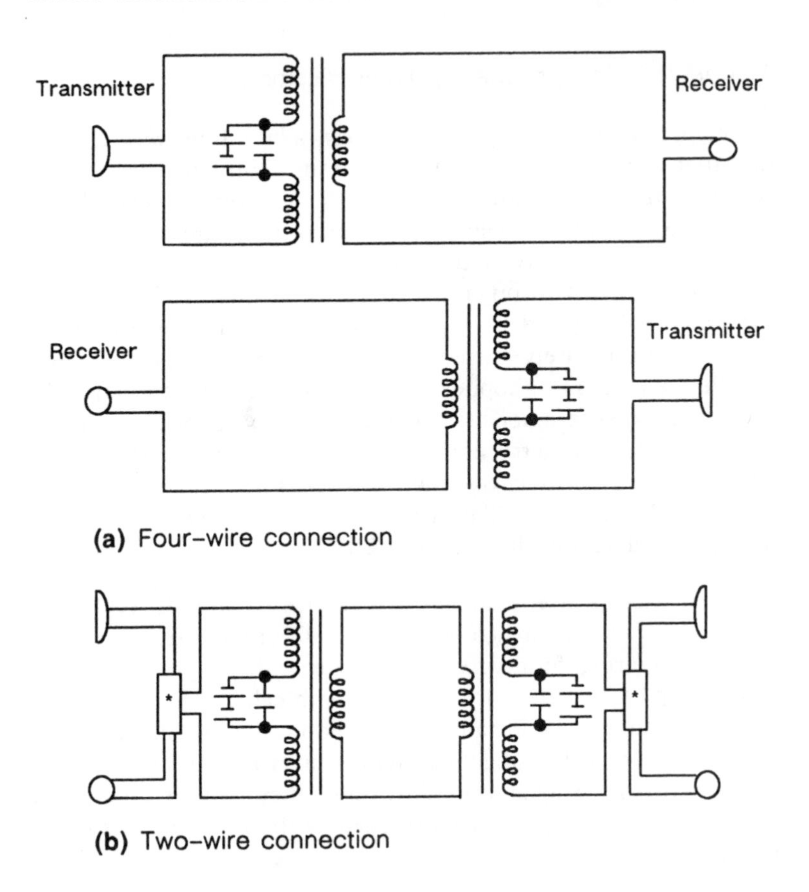

Figure 2–2. Voice–frequency modes of transmission.

Voice–Frequency Repeaters

Voice–frequency application problems are partly matters of selecting an optimum combination of wire gauge, mode of transmission (two–wire or four–wire), and repeaters to meet transmission objectives. Two basic repeater types are used in voice–frequency circuits. One type, known as the negative impedance repeater, was widely used on two–wire trunk facilities from the mid 1950s until digital carrier supplanted physical cable pairs as the economic choice for local trunks. Later designs use a two–wire repeater, for example, the metallic facility terminal (MFT) in Figure 2–3, consisting of two hybrid circuits and a separate amplifier for each direction of transmission. While the number of two–wire repeaters on trunks and special circuits continues to dwindle, they are still used to provide gain on long subscriber loops and to compensate for transmission impairments such as those that occur when the third party is bridged to the connection via the "add–on" feature in a two–wire analog switching office.

Figure 2–4 shows the block schematic for a four–wire repeatered line. The arrangement shows the use of simple amplifiers along the line for each direction of transmission and four–wire–to–two–wire conversion at each circuit end. This system, once widely used for intercity trunks, is now rarely used in the form shown. The four–wire–to–two–wire conversion equipment, normally called a four–wire terminating set, and a pair of amplifiers are often found in one form or another on the customer end of the cable pairs used to extend a four–wire private–line circuit from the carrier terminal in a central office to the customer premises.

2–5 CARRIER SYSTEMS

From its introduction in 1917, carrier system usage has been predicated on the installation of electronic equipment in lieu of building new wire or cable routes whenever the overall cost of operating the carrier system was less than the cost of new wire facilities. Since the message–handling capability of carrier systems has increased each year and the relative cost of electronic equipment has decreased, carrier facilities have steadily displaced physical facilities in additional service areas.

Carrier systems (which for the purpose of this discussion include wire–cable, microwave radio, satellite, and fiber optic systems) are high–capacity, multichannel facilities that operate

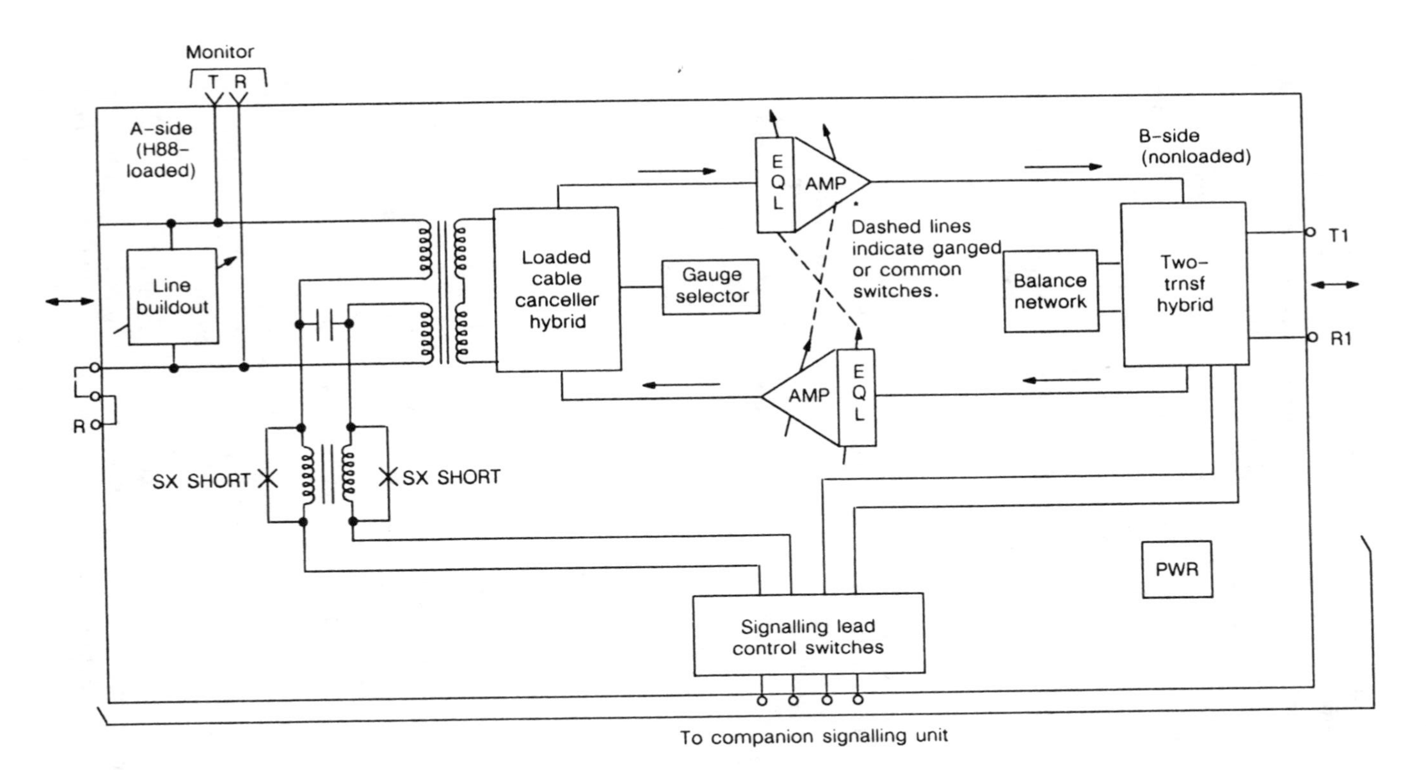

Figure 2-3. Metallic facility terminal using a two-wire intermediate repeater.

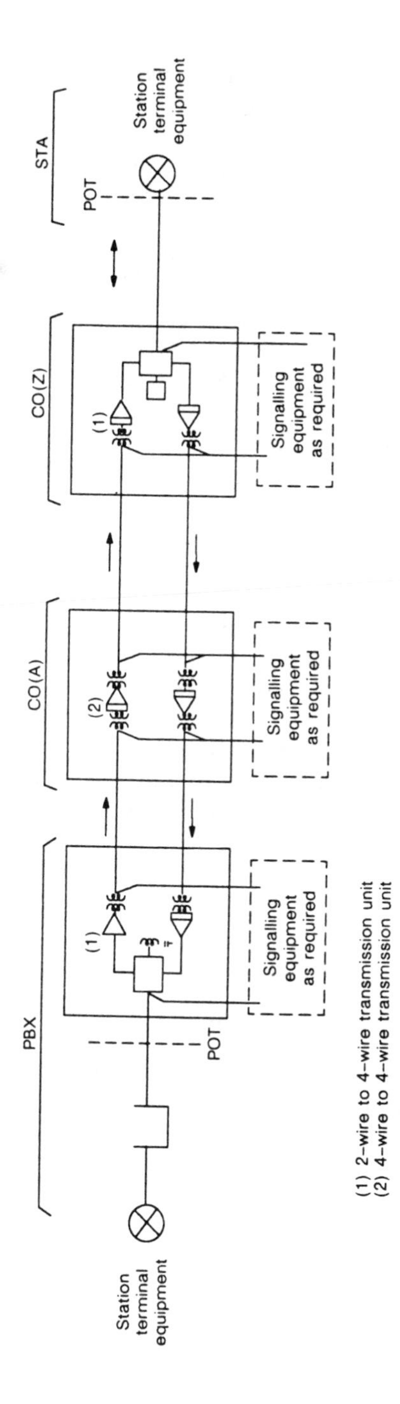

Figure 2-4. Four-wire repeatered line with two-wire terminations.

four–wire. Carrier use is economical because it allows one base facility to be used for a multiplicity of lower–capacity channels (for speech, data, or other signal transmission), thus spreading the cost of the cable or radio route infrastructure over many channels and services.

Carrier systems operating on cable facilities normally use a set of equipment for each direction of transmission, with each set operating at the same frequency or bit rate and connected to separate pairs, coaxial tubes, or fibers in the same cable. Other short–haul analog systems provide the two transmission directions over a single cable pair or coaxial tube using different frequency bands to obtain the appearance of four–wire operation. This is called "equivalent four–wire."

A carrier system may be regarded as having three major parts: (1) high–speed line or radio relay equipment that, with the transmission medium, provides a broadband channel of specified characteristics to permit the simultaneous transmission of multiple communication signals, (2) modulating and coding equipment that processes signals (voice, data, TV, etc.) from whatever form they are received to a form suitable for transmission over a broadband channel, and (3) multiplexing equipment that combines, at the system input, and separates, at the system output, the various signals sent through the system. To achieve efficient equipment packaging, some of the coding and multiplexing equipment are usually combined. Together, these three major parts provide transmission channels having fixed gain, acceptably low noise and distortion, and high velocity of propagation.

High–Frequency/High–Speed Line Equipment

To provide a broadband channel, high–frequency line equipment must perform a number of functions that differ depending on the type of system. The basic function found in all systems is that of amplification to compensate for losses in the medium between repeater points. Such compensation may require the gain to be a nonflat function of frequency. As such, it is often considered as the first step in equalization.

In analog coaxial cable systems, amplification and basic equalization are the primary functions of the high–frequency line equipment. Other functions, such as regulation to maintain the overall system gain constant in the presence of temperature changes, and mop–up equalization to compensate for small deviations in the transmission response are also provided, but for present purposes they may be regarded as secondary. Another secondary function is that of protection line switching; in the event of failure or for line maintenance reasons, service may be switched automatically or manually to a spare line.

The repeaters in the high–frequency line equipment in microwave radio systems also provide gain as their primary function. Secondary functions include: (1) modulation to translate the signals from high radio frequencies to intermediate or baseband frequencies (where amplifier designs are simpler), (2) frequency frogging between radio frequency bands, (3) high–speed switching to provide alternate paths for the signals to overcome the effects of fading in the medium (the troposphere) or to overcome the effects of equipment failure, and (4) in later systems, dynamic or adaptive equalization to offset the impairments introduced by selective fading.

The high–speed line equipment of digital systems functions to provide gain. In addition, a digital regenerator detects the presence of a pulse and transmits a new one to the next regenerator. The regeneration process must also include a timing function so that the pulses are transmitted in correct time relation to one another. This applies to both cable and radio systems. However, fiber optic systems presently require the optical signal to be converted to an electrical signal, then regenerated and converted back to an optical signal.

Modulating Equipment

Input signals to a carrier system must be processed to make them suitable for transmission over the line equipment. The processing is usually referred to as modulation. There are several forms of modulation [e.g., amplitude modulation (AM), frequency modulation (FM), pulse code modulation (PCM), and phase modulation (PM)]. They may be used singly or in combination according to the needs of the system.

The process involves modifying the signal in some reversible manner to prepare it for combining with other signals, for transmission over the carrier line or both. This may be accomplished by varying (modulating) a carrier in light intensity, amplitude, or frequency in accordance with the amplitude and frequency variations of the input signal (sometimes called the baseband signal); or the input signal, regarded as a continuous wave, may be sampled in time and then coded into a stream of pulses, as in digital transmission systems. At the receiving terminal the message signal is recovered from the signal on the carrier line.

Data and Speech Compression

Time–assignment speech interpolation is a capacity expansion system that has been in use for a number of years to increase the efficiency of international trunk groups using analog carrier over submarine cable. TASI operates as a high–speed switching system to allow a number of talkers to share a smaller number of trunks in the submarine cable high–frequency line. The switches are voice–operated and allow a channel in the submarine cable system to be taken from a speaker during breaks in his or her conversion. These breaks occur when a user is listening, rather than talking, and during other pauses in normal conversation.

Digital versions of the original analog TASI equipment have been developed that are more efficient and less expensive than earlier analog versions. In addition, the application of a time assignment unit to a digital bit stream carrying speech signals is not only less expensive than the analog counterpart but can provide significantly better service under overload conditions. Thus, although the application of TASI units has previously been confined to overseas submarine cable circuits, the emergence of digital processing technology may make its use feasible on domestic systems.

Digital devices, called statistical time–division multiplexers (STDMs), use microprocessors to concentrate both synchronous and asynchronous data with a combined aggregate input speed that normally would exceed the basic analog bandwidth or digital throughput of a transmission channel.

Bit compression, a method of encoding digital bit streams using redundancy–reduction and prediction techniques, is used for high–efficiency voice and video services permitting greater throughput on a given facility.

Multiplex Equipment

Multiplexing means the combining of multiple signals for simultaneous transmission over a common medium. The simplest form of multiplexing might be called space–division multiplexing. It occurs when many signals are transmitted over separate pairs of wire in the same cable. The term is usually applied, however, to the two broad categories called frequency–division multiplexing (FDM) and time–division multiplexing (TDM). Additionally, fiber optic systems can be arranged to use optical wavelength–division multiplexing (WDM).

In an FDM system, a number of message signals modulate carriers, each of which uses a different segment of the frequency spectrum. The modulated carriers may be transmitted simultaneously over a common medium provided that (1) the bandwidth of each modulated carrier covers a part of the broadband spectrum of high–frequency line equipment different from all other modulated bands, and (2) the total bandwidth does not exceed that of the high–frequency line equipment. Such signals are combined in electrical networks in the transmitting terminal of the system and are separated by frequency–selective networks at the receiving terminal.

In a TDM system currently used for voice–coded transmission, the pulses, which are formed for each of several signals in the modulating equipment, are interspersed in a regular time relationship at the transmitting terminal. Timing pulses, transmitted with the signal information, permit the operation and control of gate circuits at the receiving terminal. These circuits separate the signals from one another so that they may be processed, or demodulated, individually at the receiving terminal of the system. There are three basic classifications of TDMs: bit– or byte–interleaved, character–interleaved, and statistical.

In a fiber optic system employing TDM, the system's information–carrying capacity is limited typically by its high–speed modulating and demodulating circuitry. However, by coupling light from several modulated laser transmitters operating at different wavelengths into a single fiber, and separating them at the far end to different photoreceivers, additional multiplexing steps can increase the capacity of a fiber optic system. This is called optical wavelength–division multiplexing. WDM accommodates different, even nonstandard, optical line signals.

Communication users can realize considerable savings on facility and equipment costs by using high–capacity facilities and multiplexers. The popularity of high–capacity services represents a realistic step toward all–digital, fiber optic networking.

2–6 ANCILLARY EQUIPMENT AND FUNCTIONS

This section contains descriptions of a number of types of equipment and functional arrangements that are importantly related, yet subsidiary, to transmission.

Compandors

The word compandor is made up of syllables taken from the words *com*pressor and ex*pandor*. The performance of some carrier systems is improved for speech signals by the use of these devices. At the transmitting terminal of the system, the low–level voiceband signal amplitudes are amplified more than the higher level signals, compressed into a narrower than normal level range, and then restored by the expandor at the receiving terminal. The result is a significant improvement in the signal–to–noise ratio during periods of small signal transmission and during quiet intervals. These are the periods when noise is most objectionable to the user.

Frogging

The performance of analog transmission systems is often improved by transposing or frogging. The latter terminology originated in the railroad industry where a frog is a special section of rail used to cross one track over another.

Frogging may consist of changing the physical assignment of transmission systems (e.g., exchanging the east–to–west and west–to–east transmission paths) within a cable at periodic intervals, or changing the relative positions of channels in a common frequency spectrum. The latter is usually termed "frequency frogging" and is applied to both cable carrier and microwave radio systems to reduce the noise, crosstalk, and distortion resulting from the characteristics of the transmission media, repeaters, and terminal equipment.

Echo Suppressors and Cancellers

In addition to compensating for losses incurred in the medium, the design of telephone circuits involves dealing with and overcoming various other impairments suffered in the course of transmission. One such impairment is echo. If a circuit is so designed that the likelihood of a disturbing echo is high, the circuit is equipped with an analog echo suppressor. This device acts as a pair of voice–operated switches; while one subscriber is talking, the echo suppressor inserts high loss in the opposite direction of transmission to attenuate the echo before it is returned to the speaker.

In taking advantage of new electronic technology, echo cancellers have been developed to overcome the deficiencies inherent in echo suppressors, especially when the suppressors are applied to satellite–derived circuits. Basically, echo cancellers sample the signal arriving from both the near and far ends of the circuit and compare the respective levels. If the level difference is greater than a specified amount, the canceller adjusts the level of the incoming signal sample to approximate the sampled near–end signal (echo plus noise). Then the canceller inverts the incoming speech sample, delays it in time equal to the round trip delay from the canceller location to the echo source, and adds it to the near–end signal sample. Ideally, this effectively cancels the echo leaving only the noise originating in the near–end terminal. Figure 2–5 is a block diagram of an echo canceller. Where the signal is already digital, as in a digital switch or digital carrier system, echo cancellation is particularly easy to supply.

Two–way two–wire transmission of ISDN voice, data, and signalling capability is available using digital echo canceller transceivers at both ends of a nonloaded facility (exchange termination and network termination). The echo canceller transceiver uses an adaptive digital echo canceller, filters, adaptive digital equalizers, and decoders for the received signal. The terminating unit converts binary information into a line format suitable for transmission over cable pairs (digital subscriber line).

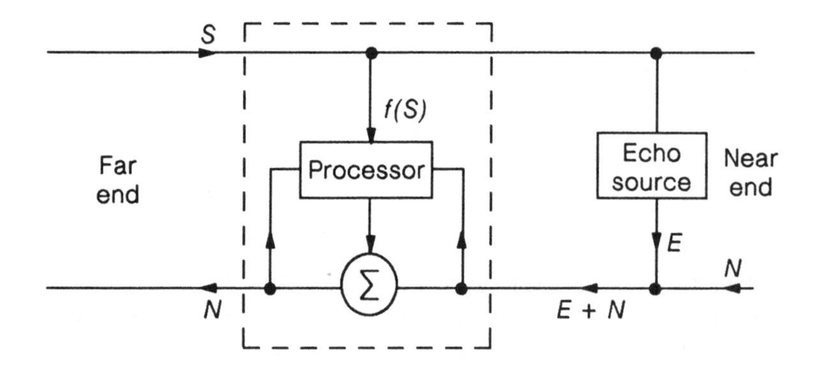

S	Distant talker signal
f(S)	Far-end signal S sample
E	Echo
N	Near-end system noise

Figure 2-5. Basic echo canceller schematic.

Additional Reading

Members of Technical Staff. *Transmission Systems for Communications*, Fifth Edition (Murray Hill, NJ: AT&T Bell Laboratories, Inc., 1982).

Reference Data for Engineers: Radio, Electronics, Computer, and Communications, Seventh Edition (Indianapolis, IN: Howard W. Sams and Company, Inc., 1985).

Telecommunications Transmission Engineering

Section 2

Elements of Transmission Analysis

While this book has been written for persons with an electrical engineering background, it must be appreciated that many of the subjects covered in this section have been worthy of entire textbooks at both the undergraduate and graduate levels of study. For the most part, the mathematics used is presented without apology, without proof, and without the thorough mathematical development to satisfy a mathematician. For additional background information, the reader is referred to the literature listed at the end of each chapter.

Chapter 3 provides a transition from the "Background" section of the book to the more theoretical subjects that follow. Some terminology is defined, and justification for the use of logarithmic units in transmission work is presented. The concept of transmission level points is discussed, and measurements of certain types of signal and interference are described.

Chapters 4 and 5 cover the related subjects of four–terminal linear networks and transmission line theory. The material in these chapters includes discussions of the basic Ohm's and Kirchoff's laws and their application; the analysis of networks and their interactions, impedance relationships, return loss and reflections; and the theory and applications of transformers and hybrids. Transmission lines are treated in terms of primary electrical constants, equivalent circuits, characteristic impedance, velocity of propagation, and loading.

Chapter 6, on wave analysis, is presented in order to increase the reader's understanding of the Fourier series and Fourier transform. This permits a more general understanding of time–domain and frequency–domain relationships between signals and transmission channels.

Chapter 7 covers negative feedback amplifiers from the points of view of how design compromises are made to accomplish objectives and how these objectives are related to the performance of transmission systems in the field. The principal benefits of feedback are discussed and means for providing feedback, as well as the manner in which feedback mechanisms interact with each other, are described.

In Chapter 8, a number of signal processing methods are described in order to show how signals are modified for more efficient transmission over existing media and then restored to their original form for final transmission to the receiver. Various amplitude, angle, and pulse modulation processes are covered.

Probability and statistics is the subject of Chapter 9. The application of this branch of science to transmission system design and operation is among the most important aspects of transmission work. Without the application of probability and statistics, the operating telephone companies could not provide service economically and perhaps not at all. The terminology and symbology of this branch of mathematics are first described. Then examples of statistical and probabilistic analyses illustrate how such techniques may be used to solve transmission problems.

Chapter 10 covers a brief history and description of information theory and its application in transmission engineering. Mathematical expressions show the theoretically maximum channel capacity for both ideal (distortion-free) channels and for typical noisy channels. While the subject is of great concern to development and research workers, an understanding of the principles also should enhance the work of the transmission engineer in the field.

Chapter 11, the last chapter of this section, consists of a presentation of the more important aspects of engineering economy studies. Transmission problems usually have more than one technically sound solution; the selection of one of several alternative lines of action can often be made on the basis of economic comparisons of the alternatives.

Chapter 3

Fundamentals of Transmission Theory

Transmission systems for telecommunications are made up of a large number of tandem–connected two–port (four–terminal) discrete networks and distributed networks such as transmission lines. In the analysis of transmission systems, the properties of these networks must be defined mathematically. Logarithmic units are commonly used because the ratios of currents, voltages, and powers found in these networks are large and awkward to manipulate. If the input–output relations, or *transfer characteristics*, of the individual two–port networks are determined, the transfer characteristics of the tandem connection of several such networks can be found by taking a product of the appropriate network transfer characteristics.

Transmission parameters of telecommunication systems are measured in a manner consistent with mathematical analysis techniques. Thus, many types of telecommunications test equipment are designed to measure signal and interference amplitudes in logarithmic units (decibels). Other test equipment types measure conventional parameters, such as volts or amperes. Some test instruments are arranged to display signals as functions of either time or frequency, while more sophisticated units, containing microprocessors, measure, calculate, evaluate, and print out detailed results of tests.

3–1 POWER AND VOLTAGE RELATIONS IN LINEAR CIRCUITS

Some of the mathematical relations necessary for evaluating of system performance can be explained in terms of the simple circuit diagram of Figure 3–1. The transducer in this circuit is assumed to be linear; i.e., the relation of the output signal to the input signal can be described by a set of linear differential equations with constant coefficients.

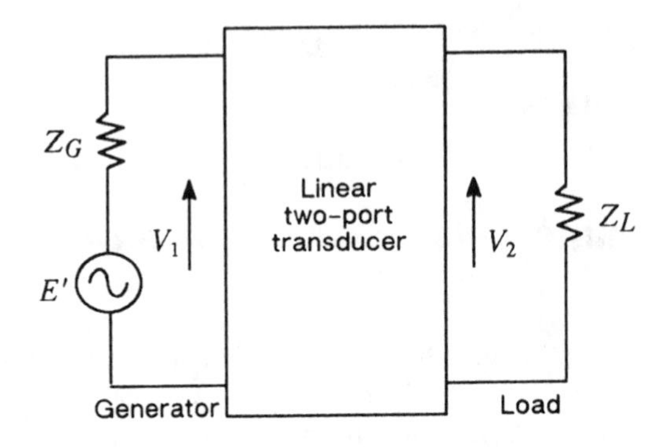

Figure 3–1. Terminated two–port circuit.

Energy is transferred from generator to load via the linear two–port transducer. The transducer may assume a wide variety of forms, ranging from a simple pair of wires to a complex assortment of cables and amplifiers, modulators, filters, and other circuits. The four terminals are associated in pairs; the pair connected to the generator is commonly called the input port and the pair connected to the load is the output port. If the energy at the output is greater than at the input, the transducer is said to have gain. If the energy is less at the output than at the input, the transducer is said to have loss.

As illustrated in Figure 3–1, a generator may be characterized by its open–circuit voltage, E', and its internal impedance, Z_G, and a load by its impedance, Z_L. If the signal produced by the generator is periodic, it may be represented by a Fourier series, $V = V_1 + V_2 \ \ldots \ + V_k + \ \ldots \ V_n$, each term of which has the form

$$V_k = E_k \cos(\omega t + \phi_k), \tag{3–1}$$

or, more conveniently,

$$V_k = Re\,[E_k\, e^{\,j(\omega t + \phi_k)}]. \tag{3–2}$$

Re (real part) is usually omitted for convenience. In Equations 3–1 and 3–2, the subscript k represents the kth term of the Fourier series, V_k represents its instantaneous voltage, and E_k

represents its peak voltage. The input–output relations of the transducer are not dependent on the presence or absence of other similar terms in the series nor of their magnitudes. The Fourier series signal representation is thus convenient for this type of analysis.

For example, if the generator voltage of Figure 3–1 is a single-frequency signal represented by

$$E' = E_S e^{j(\omega t + \phi_S)}, \qquad (3\text{–}3)$$

the ratio of V_1 to V_2 is given by

$$V_1/V_2 = [E_1 e^{j(\omega_1 t + \phi_1)}]/[E_2 e^{j(\omega_2 t + \phi_2)}]$$

$$= (E_1/E_2) e^{j(\omega_1 t - \omega_2 t + \phi_1 - \phi_2)}. \qquad (3\text{–}4a)$$

Where the radian frequency, ω, is the same for E_1 and E_2 (generally true except for modulators), this equation may be written

$$V_1/V_2 = (E_1/E_2) e^{j(\phi_1 - \phi_2)}. \qquad (3\text{–}4b)$$

As mentioned above, the ratios encountered in telecommunications transmission are often very large, and the numerical values involved are awkward. Moreover, it is frequently necessary to form the products of several ratios in order to express the gain or loss of a network or a tandem connection of networks. The expression and manipulation of voltage or power ratios are simplified by the use of logarithmic units. The natural logarithm (ln)* of the ratio of Equation 3–4b is a complex number.

$$\ln (V_1/V_2) = \theta = \alpha + j\beta = \ln (E_1/E_2) + j(\phi_1 - \phi_2). \qquad (3\text{–}5)$$

The real and imaginary parts of Equation 3–5 are uniquely identifiable, which is to say

$$\alpha = \ln (E_1/E_2), \text{ the attenuation constant,}$$

and

$$\beta = \phi_1 - \phi_2, \text{ the phase constant.} \qquad (3\text{–}6)$$

When this measure of voltage (or current) ratio is used, α is said to be expressed in nepers and β in radians. The term *neper* is an

* Natural log may also be noted as $\log_e$.

adaptation of Napier, the Scottish mathematician credited with the invention of natural logarithms.

The Decibel

The logarithmic unit of signal ratio, which is the most widely used transmission unit in the telecommunications industry, is the decibel (dB). The decibel is equal to 0.1 bel, a unit named in honor of Alexander Graham Bell, whose investigations of the human ear revealed its logarithmic response to sound intensity. Strictly speaking, the decibel is defined only for power ratios; however, as a matter of common usage, voltage or current ratios also are expressed in decibels. The precautions required to avoid misunderstanding of such usage are developed in the following.

If two powers, p_1 and p_2, are expressed in the same units (watts, microwatts, etc.), then their ratio is a dimensionless quantity, and as a matter of definition

$$D = 10 \log (p_1/p_2) \quad \text{dB} \qquad (3\text{–}7)$$

where log denotes the common logarithm (to the base 10) and D expresses the relative magnitude of the two powers in decibels. If an arbitrary power is represented by p_0, then

$$D = 10 \log (p_1/p_0) - 10 \log (p_2/p_0) \quad \text{dB}. \qquad (3\text{–}8)$$

Each of the terms on the right of Equation 3–8 represents a power ratio expressed in dB, and their difference is a measure of the relative magnitudes of p_1 and p_2. Thus, the value of this difference is independent of the value assigned to p_0. However, it is often convenient to use a value of one milliwatt for p_0. The terms $10 \log (p_1/p_0)$ and $10 \log (p_2/p_0)$ are then expressions of power (p_1 or p_2) in decibels relative to one milliwatt, abbreviated dBm. Note, however, that their difference is in dB, not dBm. In short, Equation 3–7 is a measure of the difference in dB between p_1 and p_2. Nepers and decibels may be related by the expression nepers/dB = 20 log 2.718 = 8.686. This relationship is derived from the definitions of Naperian and common logarithms.

Voltage and current ratios are often expressed in decibels as a matter of common usage. Such relationships are simple and direct when the impedances are equal at the points where the voltages or currents are measured. If the impedances are not equal, errors may be introduced unless care is taken to use appropriate correction factors. Therefore, if there is a root mean square (rms) drop of e volts across a complex impedance ($Z = R + jX$ ohms) as a result of an rms current of i amperes flowing through the impedance, the power dissipated in the impedance may be written

$$p = i^2 R \quad \text{watts.} \tag{3–9}$$

The rms voltage and rms current are related by Ohm's law, discussed in Chapter 4, Part 1, in such a way that $i = e/|Z|$. By substituting this value in Equation 3–9 and expanding $|Z|$, where

$$Z = \sqrt{R^2 + X^2}\ ,$$

and multiplying and dividing by R^2,

$$Z = \frac{R^2\sqrt{R^2 + X^2}}{R^2}$$

and

$$Z = R\sqrt{1 + \frac{X^2}{R^2}}\ .$$

Then

$$p = \frac{e^2}{R(1 + X^2/R^2)} \quad \text{watts.} \tag{3–10}$$

Using appropriate subscripts to indicate two different measurements, the value of p from Equation 3–9 may be substituted in Equation 3–7 to yield

$$\begin{aligned} D &= 10 \log (p_1/p_2) = 10 \log (i_1/i_2)^2 + 10 \log (R_1/R_2) \\ &= 20 \log (i_1/i_2) + 10 \log (R_1/R_2) \quad \text{dB.} \end{aligned} \tag{3–11a}$$

This may be written

$$\text{dB} = 20 \log \frac{i_1 \sqrt{R_1}}{i_2 \sqrt{R_2}}.$$

If, in both cases, the impedances R_1 and R_2 are equal, they will cancel and the following formula results:

$$D = 20 \log \frac{i_1}{i_2} \quad \text{dB}. \qquad (3\text{–}11\text{b})$$

Similarly, the value of p from Equation 3–10 may be substituted in Equation 3–7. This gives

$$D = 10 \log (p_1/p_2) = 10 \log \frac{e_1^2/R_1(1 + X_1^2/R_1^2)}{e_2^2/R_2(1 + X_2^2/R_2^2)}$$

$$= 20 \log (e_1/e_2) - 10 \log (R_1/R_2)$$

$$- 10 \log \frac{(1 + X_1^2/R_1^2)}{(1 + X_2^2/R_2^2)} \quad \text{dB}, \qquad (3\text{–}12\text{a})$$

and if $R_1 = R_2$,

$$D = 20 \log \frac{e_1}{e_2} \quad \text{dB}. \qquad (3\text{–}12\text{b})$$

The terms beyond 20 log (i_1/i_2) and 20 log (e_1/e_2) in Equations 3–11 and 3–12 give rise to serious error unless they are included when expressing voltage and current ratios in decibels, except when the impedances Z_1 and Z_2 are equal. The extent of these errors may be illustrated by some simple examples.

Example 3–1:

Let

$$p_1 = 2 \text{ mW} = 0.002 \text{ watt},$$

$$p_2 = 1 \text{ mW} = 0.001 \text{ watt}.$$

Then, from Equation 3–7

$$D = 10 \log (p_1/p_2) = 10 \log 2 = 3 \text{ dB}.$$

Let

$R_1 = 10$ ohms, $R_2 = 10$ ohms, $X_1 = 10$ ohms, $X_2 = 10$ ohms.

Then, from Equation 3–9

$$i_1 = (p_1/R_1)^{1/2} = (0.002/10)^{1/2} = 0.014 \text{ ampere}$$

$$i_2 = (p_2/R_2)^{1/2} = (0.001/10)^{1/2} = 0.01 \text{ ampere}$$

and from Ohm's law

$$e_1 = i_1|Z_1| = 0.014\sqrt{10^2 + 10^2} = 0.2 \text{ volt}$$

$$e_2 = i_2|Z_2| = 0.01\sqrt{10^2 + 10^2} = 0.14 \text{ volt.}$$

From Equation 3–11a

$$D = 20 \log (i_1/i_2) + 10 \log (R_1/R_2)$$

$$= 20 \log (0.014/0.01) + 10 \log 1$$

$$= 3 \text{ dB.}$$

From Equation 3–12a

$$D = 20 \log (e_1/e_2) - 10 \log (R_1/R_2) - 10 \log \frac{(1 + X_1^2/R_1^2)}{(1 + X_2^2/R_2^2)}$$

$$= 20 \log (0.2/0.14) - 10 \log 1 - 10 \log 1$$

$$= 3 \text{ dB.}$$

Thus, no error results from computing the current or voltage differences in dB simply by taking 20 times the log of the current or voltage ratio. This is because the impedances Z_1 and Z_2 are equal and all terms after the first in Equations 3–11a and 3–12a reduce to zero.

Example 3–2:

In this example, the assumption is again

$$p_1 = 2 \text{ mW} = 0.002 \text{ watt}$$

$$p_2 = 1 \text{ mW} = 0.001 \text{ watt.}$$

Then

$$D = 10 \log (p_1/p_2) = 10 \log 2 = 3 \text{ dB.}$$

Now, assume R_1 =10 ohms, R_2=20 ohms, and X_1=X_2=0.

From Equation 3–9

$$i_1 = (p_1/R_1)^{1/2} = (0.002/10)^{1/2} = 0.014 \text{ ampere}$$

$$i_2 = (p_2/R_2)^{1/2} = (0.001/20)^{1/2} = 0.007 \text{ ampere.}$$

From Ohm's law

$$e_1 = i_1|Z_1| = 0.014 \times 10 = 0.14 \text{ volt}$$

$$e_2 = i_2|Z_2| = 0.007 \times 20 = 0.14 \text{ volt.}$$

From Equation 3–11

$$D = 20 \log (i_1/i_2) + 10 \log (R_1/R_2)$$

$$= 20 \log 2 + 10 \log (1/2)$$

$$= 6 - 3 = 3 \text{ dB.}$$

From Equation 3–12

$$D = 20 \log (e_1/e_2) - 10 \log (R_1/R_2) - 10 \log \frac{(1 + X_1^2/R_1^2)}{(1 + X_2^2/R_2^2)}$$

$$= 20 \log 1 - 10 \log (1/2) - 10 \log 1$$

$$= 0 + 3 + 0 = 3 \text{ dB.}$$

Once again the three expressions for D give the same answer, 3 dB. Note, however, that in this example significant errors would occur if D were computed for current or voltage ratios without concern for the impedance of the circuits. In the case of the current ratio, the answer would have been 6 dB; in the case of the voltage ratio, the answer would have been 0 dB.

Example 3-3:

Once again, assume

$$p_1 = 2 \text{ mW} = 0.002 \text{ watt}$$

$$p_2 = 1 \text{ mW} = 0.001 \text{ watt.}$$

Then

$$D = 10 \log (p_1/p_2) = 10 \log 2 = 3 \text{ dB.}$$

Now, assume $R_1 = 10$ ohms, $R_2 = 20$ ohms, $X_1 = 20$ ohms, and $X_2 = 10$ ohms.

From Equation 3-9

$$i_1 = (p_1/R_1)^{1/2} = (0.002/10)^{1/2} = 0.014 \text{ ampere}$$

$$i_2 = (p_2/R_2)^{1/2} = (0.001/20)^{1/2} = 0.007 \text{ ampere.}$$

From Ohm's law

$$e_1 = i_1|Z_1| = 0.014\sqrt{10^2 + 20^2} = 0.31 \text{ volt}$$

$$e_2 = i_2|Z_2| = 0.007\sqrt{20^2 + 10^2} = 0.16 \text{ volt.}$$

Then, from Equation 3-11

$$\begin{aligned} D &= 20 \log (i_1/i_2) + 10 \log (R_1/R_2) \\ &= 20 \log 2 + 10 \log (1/2) \\ &= 6 - 3 = 3 \text{ dB.} \end{aligned}$$

From Equation 3-12

$$D = 20 \log (e_1/e_2) - 10 \log (R_1/R_2) - 10 \log \frac{(1 + X_1^2/R_1^2)}{(1 + X_2^2/R_2^2)}$$

$$= 20 \log 2 - 10 \log (1/2) - 10 \log (5/1.25)$$

$$= 6 + 3 - 6 = 3 \text{ dB}.$$

As in Example 3–2, the value of D is 3 dB no matter how computed. However, the importance of including impedance factors in the computation is demonstrated.

Loss, Delay, and Gain

There are several different methods of describing the transfer characteristics of a two–port network. Such characteristics require specification of four complex quantities representing input and output relationships. However, in many cases where the network environment (such as source and load impedances) is controlled, the transfer can often be characterized more readily by one frequency–dependent complex number describing the loss (or gain) and phase shift through the network. Several different means of expressing the transfer characteristic have come into use, each having merit for a particular set of circumstances and each depending in part on the definition of the network parameters involved.

Insertion Loss and Phase Shift. In the circuit of Figure 3–1, assume that it has been determined that power p_2 is delivered to the load Z_L when the open-circuit voltage E' is applied. Next assume that the two–port network is removed, the generator is connected directly to the load, and the power delivered to Z_L is p_0. The difference in dB between p_0 and p_2 is called the *insertion loss* of the two–port network; i.e.,

$$\text{Insertion loss} = 10 \log (p_0/p_2) \quad \text{dB}. \qquad (3\text{–}13)$$

If the impedances are matched throughout, there is no ambiguity in expressing insertion loss as a voltage or current ratio. The instantaneous voltages, V_0 and V_2, corresponding respectively to p_0 and p_2, may be expressed in terms of peak values, E_0 and E_2. By proceeding as in the development of Equation 3–6, the insertion loss and a definition of the *insertion phase shift* may be written:

$$\text{Insertion loss} = 20 \log (E_0/E_2) = 20 \log (I_0/I_2) \text{ dB}; \quad (3\text{–}14)$$

$$\text{Insertion phase shift} = 57.3(\phi_0 - \phi_2) \text{ degrees} \quad (3\text{–}15)$$

where ϕ_0 and ϕ_2 are given in radians.

If the transducer of Figure 3–1 furnishes gain, then $E_2 > E_0$ and the insertion loss values are negative. In order to avoid confusion when talking about negative loss*, it is customary to write

$$\text{Insertion gain} = 20 \log (E_2/E_0) \quad \text{dB.} \quad (3\text{–}16)$$

If complex gain is expressed in the form of Equation 3–5, the phase shift will be the negative of the value found in Equation 3–15. Unfortunately, there is no standard name that clearly distinguishes between the phase shift calculated from a loss ratio and that calculated from a gain ratio. The ambiguity is entirely a matter of algebraic sign and can always be resolved by observing the effect of substituting a shunt capacitor for the transducer. This gives a negative sign to the value of ϕ_2 and a positive change in the phase of Equation 3–15.

Phase and Envelope Delay. The phase delay and envelope delay of a circuit are defined as

$$\text{Phase delay} = \beta/\omega$$

$$\text{Envelope delay} = d\beta/d\omega$$

where β is in radians, ω is in radians per second, and delay is therefore expressed in seconds. In accordance with the sign convention adopted previously, both the phase and the envelope delay of an "all–pass" network are positive at all finite frequencies. The above expressions show that the envelope delay is the rate of change, or slope, of the phase delay curve. If the phase delay is linear over the frequency band of interest, the envelope delay is a constant over that band.

* A negative loss is in reality a gain.

For cables or similar transmission media, the phase shift is usually quoted in radians per mile. In this case, phase delay and envelope delay are expressed in seconds per mile. Their reciprocals are called *phase velocity* and *group velocity*, respectively, and the units are miles per second.

Available Gain. The maximum power available from a source of internal impedance, Z_G, is obtained when the load connected to its terminals is equal to its conjugate, Z_G^*, i.e., if

$$\left.\begin{aligned} Z_G &= R_G + jX_G \\ \text{and} \qquad & \\ Z_G^* &= R_G - jX_G. \end{aligned}\right\} \qquad (3\text{–}17)$$

It should be noted that maximizing available power is not necessarily the optimum relationship because, when conjugate impedances are interconnected, large reflections occur. As a result, other impedance relationships are preferable. For an open–circuit generator voltage having an rms value, e, the maximum available power is

$$p_{aG} = e^2/4R_G. \qquad (3\text{–}18)$$

The power actually delivered to Z_L in Figure 3–1 is also maximized if the output impedance of the transducer is conjugate to Z_L. Designating this power as p_{a2} leads to a definition of *available gain*, g_a, as

$$g_a = 10 \log (p_{a2}/p_{aG}). \qquad (3\text{–}19)$$

Transducer Gain. Ordinarily the impedances do not meet the conjugacy requirements, and it is necessary to define the *transducer gain*, g_t, of the two–port circuit as

$$g_t = 10 \log (p_L/p_{aG}) \qquad (3\text{–}20)$$

where p_L is the power actually delivered to the load. Transducer gain depends on load impedance and can never exceed available gain. Transducer gain is equal to available gain only when the load impedance is equal to the conjugate of the network output impedance.

Power Gain. Finally, *power gain,* g_p, is defined as

$$g_p = 10 \log (p_L/p_1) \tag{3-21}$$

where p_1 is the power actually delivered to the input port of the transducer. The power gain equals the transducer gain of a network when the input impedance of the network equals the conjugate of the source impedance. The power gain equals the insertion gain of the network when the input impedance of the network equals the load impedance. The inverse is true when the network exhibits a loss

$$l_p = 10 \log (p_1/p_L).$$

3–2 TRANSMISSION LEVEL POINT

In designing transmission circuits and laying them out for operation and maintenance, it is necessary to know the signal amplitude at various points in the system. These values can be determined conveniently by use of the transmission level point (TLP) concept.

> *The transmission level at any point in a transmission circuit or system is the ratio, expressed in decibels, of the power of a signal at that point to the power of the same signal at a reference point called the zero transmission level point (0 TLP).*

Thus, any analog point in a transmission circuit or system may be referred to as a transmission level point. Such a point is usually designated as a $-x$ dB TLP, where x is the designed loss from the 0 TLP to that point. Since the losses of transmission facilities and circuits tend to vary with frequency, the TLP is specified for designated frequencies. For analog carrier systems, the frequency in the carrier band must be specified. For voiceband circuits, this frequency is usually nominally 1000 Hz. Actually, in the analog environment, a test signal of 1004 Hz is used to verify continuity and transmission power. However, in the digital environment, continuity and signal integrity of connections through the digital networks are verified by using the digital reference signal (DRS). The DRS is a digital representation of 1004 Hz at 0 dBm power.

When decoded at a digital–to–analog 0 TLP in the network, it will produce a 1004–Hz sine wave at a power level of 0 dBm, also known as 0 dBm0.

The TLP concept is convenient because it enables circuit losses or gains to be quickly and accurately determined by finding the difference between the TLP values at the points of interest. This principle may be extended from relatively simple circuits, such as message trunks, to very complex broadband transmission systems where the TLP values often vary with frequency across the carrier band.

The TLP concept is also a convenience in that signals and various forms of interference can easily be expressed in values referred to the same TLP. This facilitates the addition of interference amplitudes, the expression of signal–to–noise ratios, and the relation of performance to objectives in system evaluation. These important advantages are apparent where various types of signals and interferences are involved.

TLPs are applied within the switched message telecommunication network and special–services networks. Similar concepts are applied to wideband services such as television and wideband data. The channels used for these services are given specially defined transmission reference points.

Confusion often arises because the word *level* is used (properly and improperly) in many ways. Frequent references may be found to such things as power level, voltage level, signal level, or speech level. To add to the confusion, the word *level* is often used interchangeably with the word *power*. Here, *level* is generally used only as a part of the phrase *transmission level point*. Signal power and voltage are referred to in appropriate units such as watts, milliwatts, dBm, volts, or dBV.

A troublesome correlation exists between transmission level point and power. When a test signal of the correct frequency is applied to a properly adjusted circuit at a power in dBm that corresponds numerically with the TLP at which it is applied, the test signal power measured at any other TLP in the circuit corresponds numerically with the designated TLP value. Careless use of terminology often leads to referring to a TLP as the x dBm level point. It should be stressed strongly that *this is improper terminology even if it happens that a test signal of x dBm is measured at the x dB TLP*. This correlation is unfortunate in that it has led to some confusion. On the other hand, when properly used, TLPs simplify loss computations.

While *0 TLP* is used in this book as the abbreviation for the reference transmission level point, it should be pointed out that other terminology is sometimes used elsewhere. These include zero level, zero–level point, 0–dB point, 0–dB TL, 0 SL (system level), 0 dBr, and 0 ELP.

Commonly Used TLPs

Application of the TLP concept must begin with the choice of a common datum or reference point and the arbitrary assignment of 0–dB transmission level to that point. Other TLPs in the trunk or system are then related to the reference point by the number of dB of gain or loss from the reference point to the point of interest. If (in a properly adjusted circuit) a signal of x dBm is applied or measured at the reference point and if that signal is measured as y dBm at the point of interest, the point of interest is designated as the $(y-x)$ dB TLP. For example, if y is 0 dBm and the circuit exhibits 16 dB of loss, then $x = 16$, resulting in a −16 dB TLP at the point of interest:

$$-16-(0)=-16 \text{ dB}.$$

The 0–dB transmission level point is so defined as a matter of convenience and uniformity. It would be convenient also to have the 0 TLP available as an access point for connecting probes and measuring equipment. However, there is no requirement that such an access point be available. In fact, as a result of changes in circuit arrangements resulting from changing objectives, the 0 TLP is seldom available physically in the toll plant.

Originally, the 0 TLP was conveniently defined at the jack of a toll switchboard in the outgoing direction. Analog intertoll trunks were equipped at each end with 4–dB pads that could be switched in or out of the circuit to suit the needs of a particular application. As technology improved and the need for better performance increased, these pads were reduced to 2 dB; later, under the via net loss (VNL) design plan, they were eliminated entirely from the intertoll trunks.

With these changes in intertoll trunk design, it would have been possible to redefine the reference TLP. However, this would have resulted in changing all TLP values. It was instead deemed desirable to maintain the original 0 TLP concept as well as other important TLPs. As a result, the analog switch to which a trunk is connected is designated −2 dB TLP and the end–office switch is defined as 0 TLP. When a −2 dB TLP was used, it was necessary to reduce the conventional 0–dBm test level to −2 dBm either within the test equipment or by providing an external test pad of 2 dB, called TP2, at analog tandem switching offices.

In the layout of four–wire trunks on analog carrier, a switched access maintenance test point is usually provided to facilitate testing and maintenance between the trunks and the switching system terminations. TLPs at these test points have been standardized for all four–wire trunks. On the transmitting side the TLP is −16 dB, and on the receiving side the TLP is +7 dB.

In four–wire circuits, the TLP concept is easily understood and applied because each transmission path has only one direction of transmission. In two–wire circuits, however, confusion or ambiguity may be introduced by the fact that a single point may be properly designated as two different TLPs, each depending on the assumed direction of transmission.

For trunks involving digital connectivity, the analog–to–digital interface at a digital switching system is designated the encode level point (ELP). In the other direction of transmission, the digital–to–analog interface is known as the decode level point (DLP). The difference in levels between an ELP at one switching system and the DLP at another connected switching system is defined as trunk loss.

In a switched digital network, the 0 TLP is known as 0 ELP transmitting at the interface in an end office, while receiving at a −6 DLP for an overall loss of 6 dB. Transmission levels in digital switching can be controlled by software within some digital switching systems. Other digital switching systems may have no control of transmission levels so that various ELPs may be found in a connection (e.g., 0 ELP, -3 ELP*, and -6 ELP).

Figure 3-2 shows that when the combined analog-digital switched network migrates to an all-digital network, the CCITT recommendation of -6 ELP may become the standard in most switching networks. The principal advantage, with no loss between digital switches, is that the capability of achieving bit integrity throughout the digital path can be realized. Also, transmission impairments due to digital pads and digital-to-analog-to-digital conversions are not added to the connection. To ensure satisfactory signal power to the user, the overall loss can be set at the last digital-to-analog point in the connection (including digital PBXs). A -6 DLP is used for many connection arrangements providing a 6-dB overall loss in the digital path.

Illustrative Applications of TLP, ELP, and DLP

The transmission level point concept is applied to an individual trunk as illustrated in Figure 3-3. The circuit elements within each trunk are interconnected to produce predetermined gains and losses so that each point in the trunk may be assigned a transmission level value. Some of the important transmission level points discussed earlier and the assignment of transmission level point values within a tandem trunk are illustrated in the figure.

Starting in the upper left corner of the diagram, the outgoing side of the switch is designated as the 0-dB TLP. As the circuit is followed from left to right, the carrier channel unit transforms the circuit from two-wire to four-wire. The diagram shows an office loss of 0 dB. The channel unit on the D4 channel bank has TLPs of -8.5 transmit (XMT) and +4.0 receive (RCV) that correspond to a DRS in the digital facility; i.e., a -8.5 dBm signal at the XMT jack causes a DRS to be sent over the digital facility to the tandem office, and a DRS sent from the tandem office over the digital facility will result in a +4.0 dBm test signal at the RCV jack. As the connection is traced toward the right, the

* -3 ELP is used at digital-to-analog interfaces in some digital tandem switching systems requiring a 3-dB test pad (TP3).

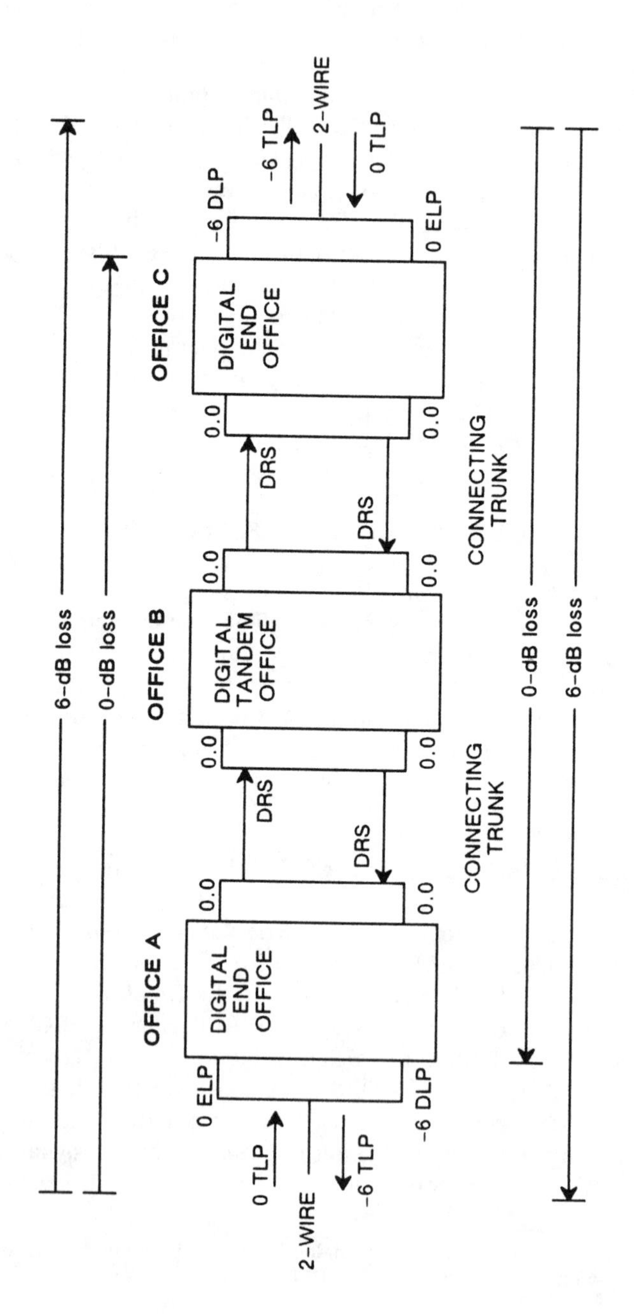

Figure 3-2. All-digital connectivity, end-office-to-end-office.

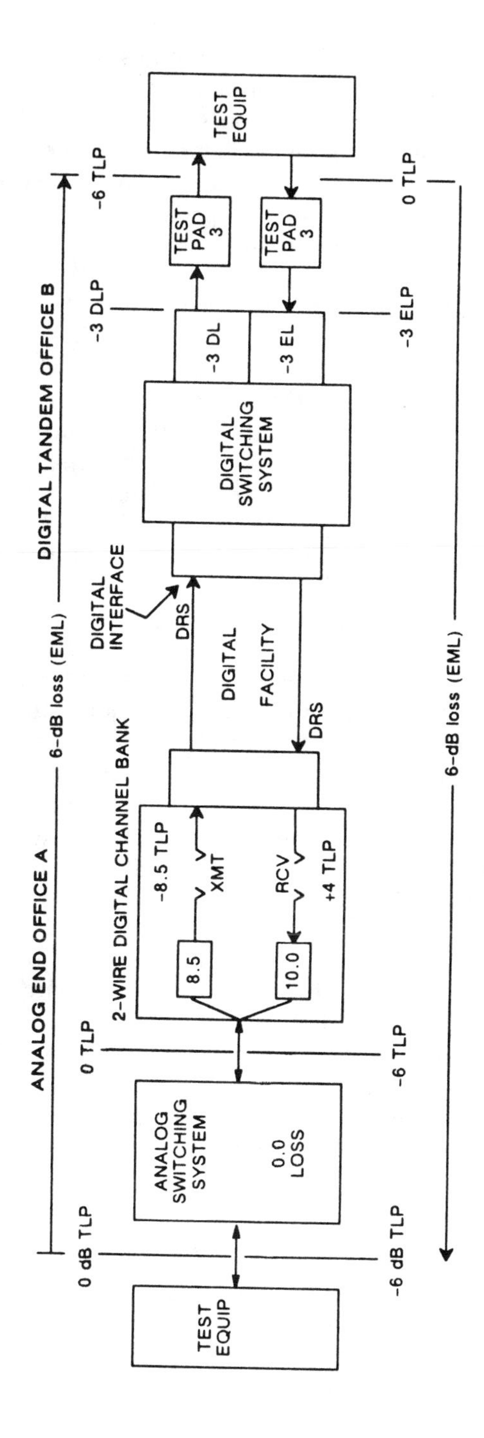

Figure 3–3. Illustrative loss plan of analog–digital connecting trunk.

channel unit has 8.5 dB of loss between its two–wire port and the XMT jack and, in effect, has an 8.5–dB gain from the XMT jack to the DRS digital representation on the line facility. The DRS exists until the digital tandem switching system decodes with 3 dB of loss (–3 DL) which, when added to an analog pad loss of 3 dB, results in an overall 6–dB expected measured loss (EML)* as measured by the test equipment.

If the circuit is followed from right to left, the analog test pad provides 3 dB of loss (a –3 TLP) and the –3 EL provides 3 dB of gain so that the DRS is generated in the switching system and facility. It exists until the D4 channel unit at the end office decodes it with 4 dB of gain at the +4 TLP RCV jack. Loss of 10 dB to the two–wire port provides a net channel bank loss of 6 dB (–6 TLP) for an overall loss at the test equipment of 6 dB. Note that at closely related points (four–wire input and output), the transmission level points are quite different for the two directions of transmission. In the circuit, the same point (electrically), e.g., the end of the trunk, has the same values of TLP.

Figure 3–4 shows a built–up connection between a digital end office and a digital tandem office (test equipment added) illustrating the same TLP (ELP and DLP) concepts with slight variance in the method of accomplishing the transition between DRS in the digital tandem switching equipment and the test equipment. Depending on the type of switching system employed and the types of channel units or digital interfaces used, TLPs must be redefined for each trunk.

By common usage, TLP values are found by determining the gain or loss between TLPs at 1004 Hz in the circuit of interest. The modulators in the terminals of frequency–division multiplex carrier systems shift the frequency from 1004 Hz in the original circuit to some higher frequency in the carrier system. The TLP value can be determined at the higher frequency and related to the 1004–Hz value in the original circuit to obtain the TLP in the

* EML is an engineering prediction of circuit losses as measured between appropriate test equipment. The calculation of EML anticipates select impedances and the resultant transducer loss between and through equipment and facilities that make up a circuit.

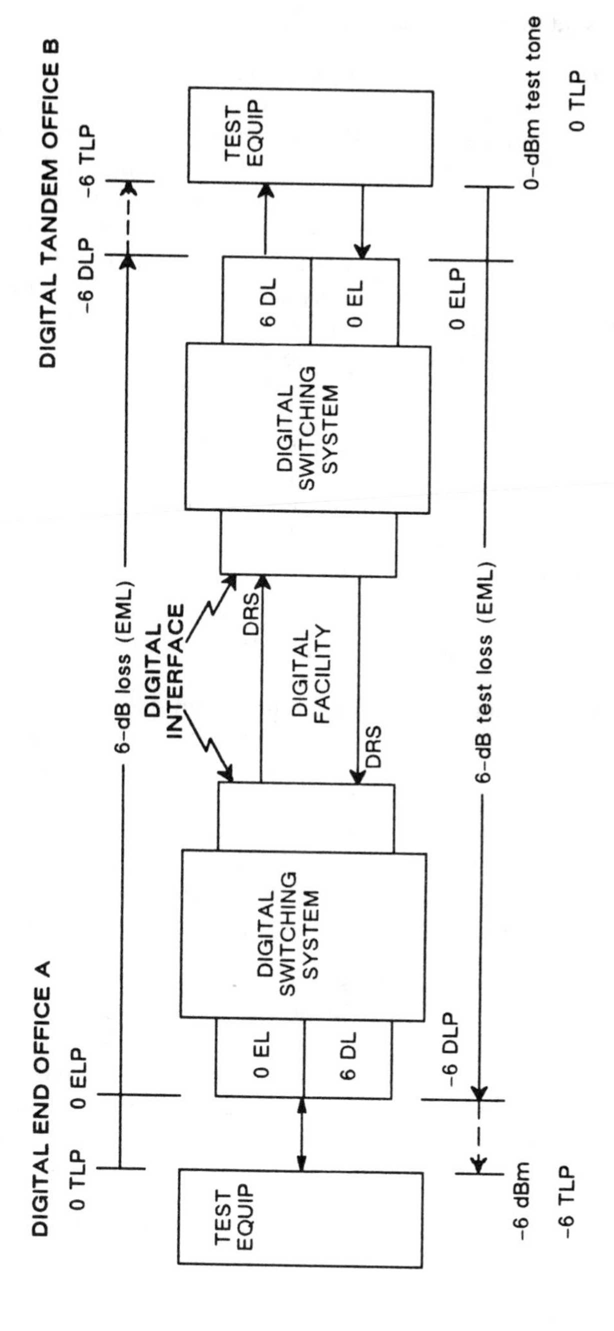

Figure 3–4. Digital trunk between digital end office and tandem office.

carrier system. Similarly, the DRS can be decoded from its digital representation as a 1004–Hz signal.

If the value of a transmission level point is not known, it can be determined by measurement. The process depends on the direction of transmission and on having proper values of pad losses and amplifier gains in the circuit between a known TLP, *A*, and the TLP to be determined, *X*. If the unknown is to be established by transmitting from *A* to *X*, a 1004–Hz signal (or equivalent in a carrier channel) may be applied at *A* and measured at *A* and *X*. The TLP at *X* is determined by subtracting from the TLP value at *A* the loss (in dB) from *A* to *X*. If the TLP at *X* is to be determined by transmitting from *X* to *A*, the value at *X* is the value of the TLP at *A* plus the loss (in dB) from *X* to *A*.

In order to avoid overloading analog transmission systems, the applied test signal power (in dBm) should be at least 10 dB below the TLP value at any point. Since signal power is often expressed in terms of its value at 0 TLP, the unit dBm0 is used as an abbreviation for "dBm at 0 TLP."

3–3 SIGNAL AND NOISE MEASUREMENT

The TLP concept is valuable in system design, operation, and maintenance in that it provides a means of calculating signal and interference amplitudes at given points in a system, as well as the gains or losses between TLPs. Nevertheless, operating systems must be checked at times by actual measurement to see that signal or interference amplitudes are being maintained at the expected, or calculated values. When there is excessive gain or loss in a system, measurement is also a means of locating trouble.

In telecommunications systems, there are complex signals and noises to be measured. Simple instruments are inadequate, particularly since they do not take into account any of the subjective factors that determine the final evaluation of a circuit. Both the instruments and units of measure used in telecommunications for signal and noise measurements must be adapted to the special needs involved.

Since these circuits operate with signal and interference powers that rarely are as large as 0.1 watt and that may be lower than

10^{-12} watt, the use of the watt as a unit of measurement is unwieldy. A more convenient unit is the milliwatt, or 10^{-3} watt. An exception is in radio transmitter work, where output power is frequently measured in watts.

Many other types of equipment are used for evaluating transmission quality and facilitating maintenance. These include oscillators, voltmeters, and transmission measuring sets. The parameters measured, the units of measurement, and the techniques involved are all important aspects of transmission engineering. The oscilloscope and spectrum analyzer are among the most powerful of these specialized instruments in that the parameters of interest can be displayed for study and analysis.

Volume

The amplitude of a *periodic* signal can be characterized by any of four related values: the rms, the peak, the peak–to–peak, or the average. The choice depends on the particular purpose for which the information is required. It is more difficult to deal with *nonperiodic* signals such as the speech signals transmitted over circuits where the rms, peak, peak–to–peak, and average values and the ratio of one to another are all irregular functions of time; one number cannot easily specify any of them.

Regardless of the difficulty of the problem, the amplitude of the signal must be measured and characterized in a fashion that will be useful in designing and operating systems involving electronic equipment and transmission media of various kinds. Signal amplitudes must be adjusted to avoid overload and distortion, and gain and loss must be measured. If none of the simple characterizations is adequate, a new one must be invented. The unit used for expressing speech signal amplitude is called the volume unit (vu). It is an empirical kind of measure evolved initially to meet the needs of radio broadcasting and is not definable by any precise mathematical formula. The volume is determined by reading a volume indicator, called the vu meter, in a carefully specified fashion.

The development of the vu meter was a joint project of the predivestiture Bell System and two large broadcasting networks.

Its principal functions are measuring signal amplitude to enable the user to avoid overload and distortion, checking transmission gain and loss for the complex signal, and indicating the relative loudness of the signal when converted to sound.

The vu meter can be used equally well for all speech, male or female. There is some difference between music and speech in this respect, so a different reading technique is used for each.

The meter scale is logarithmic, and the readings bear the same relationship to each other as do decibels; however, the scale units are in vu, not in dB. The transient response (damping characteristic) of the meter movement prevents the meter needle from registering very short high–amplitude impulses such as those created by percussive sounds in speech. A correlation between talker volume and long–term average power and peak power can be established. Also, a vu meter reading for a sinusodial signal delivered to a 600–ohm resistive termination is numerically equal to the power in dBm delivered to the termination (either internal or external). Such correlations are valuable, but the fact that they exist should not be allowed to confuse the real definition of volume and vu. Simply, a −10 vu talker is one whose signal is read on a calibrated volume indicator as −10 vu. The vu meter has a flat frequency response over the audible range, and is not frequency–weighted in any fashion. Some meters calibrated in dB can be used to read vu; however, the transient response of the meter movement must meet certain carefully defined specifications as in the vu meter.

The vu meter, while still a standard instrument, suffers from its need for subjective interpretation. Various peak and quasi–peak reading meters have been proposed in recent years as replacements; it is likely that the traditional vu meter will be supplanted in future years by improved designs.

Noise

The noise measurement of a telecommunications channel, like the measurement of volume, is an effort to characterize a complex signal. The measurement is further complicated by an interest in how much it annoys the user rather than in the absolute power. Consider the requirements of a meter that can measure the subjective effects of noise.

(1) The readings should consider that the interfering effect of noise is a function of frequency as well as of amplitude.

(2) When dissimilar noise components are present simultaneously, the meter should combine them in the same manner as do the ear and brain to measure the overall interfering effect.

(3) When different types of noise cause equal interference as determined in subjective tests, use of the meter should give equal readings.

A noise–measuring set is essentially an electronic voltmeter that meets these requirements, respectively, by incorporating (1) frequency weighting, (2) a detector approximating an rms detector, and in some cases, (3) a transient response similar to that of the human ear.

The first of the requirements for noise measurement on a voice channel involves annoyance and the effect of noise on intelligibility. Since both are functions of frequency, frequency weighting is included in the set. To determine the weighting characteristic, annoyance was measured in the absence of speech by adjusting the amplitude of a tone until it was as annoying as a reference 1000–Hz tone. This was done for many tones and many observers, and the results were averaged and plotted. A similar experiment was performed in the presence of speech at average received volume to determine the effect of noise on articulation. The results of the two experiments were combined and smoothed, resulting in the C–message weighting curve shown in Figure 3–5. The experiments were made with a 500–type telephone; therefore, the weighting curve includes the frequency characteristic of this telephone as well as the hearing of the average subscriber. The remainder of the telephone plant is assumed to provide transmission that is essentially flat across the band of a voice channel. Therefore, the C–message weighting is applicable to measurements made almost anywhere except across the telephone receiver.

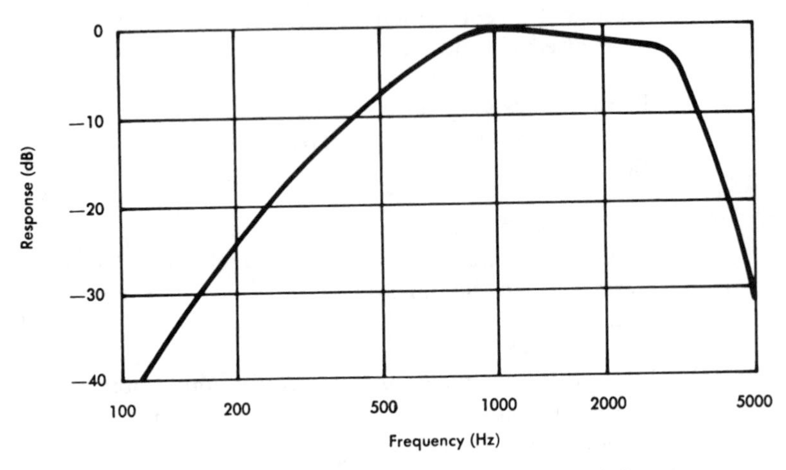

Figure 3-5. C-message frequency weighting.

To illustrate the significance of the weighting curve of Figure 3-5, a 200-Hz tone of given power is found to be 25 dB less disturbing to a listener than a 1000-Hz tone of the same power. Hence, the weighting network incorporated in the noise meter has 25 dB more loss at 200 Hz than at 1000-Hz.

Other weighting networks may be substituted in noise-measuring sets. For example, a 3-kHz flat network may be used to measure the power density of Gaussian noise. This network has a nominal low-pass response down 3 dB at 3 kHz and rolls off at 12 dB per octave. The response to Gaussian noise is almost identical to that of an ideal (sharp cutoff) 3-kHz low-pass filter. For measuring noise on digital transmission channels where the presence of a signal causes quantizing noise, a special "C-notched" network is commonly used. This network provides the C-message weighting curve plus a response notch near 1004 Hz. It is thus usable for measuring noise in the presence of a 1000-Hz "holding" tone.

The second factor affecting the measurement of the interfering effect of noise involves the evaluation of simultaneous occurrences of noise components at different frequencies and with different characteristics. Experimentally, narrow bands of noise were used in various combinations. It was found that the closest agreement between the judgment of the listener and the reading of the noise–measuring set was obtained when the noises were added on an rms, or power, basis. Thus, for example, if two tones having equal interfering effect when applied individually are then applied simultaneously, the effect when both are present is 3 dB worse than for each separately.

The third factor that affects the manner in which noise must be measured is the transient response of the human ear. It has been found that, for sounds shorter than 200 ms, the human ear does not fully appreciate the true power in the sound. For this reason, the meter on the noise–measuring set (as well as the vu meter) is designed to give a full indication only on bursts of noise longer than 200 ms. For shorter bursts, the meter indication decreases.

These three characteristics of most noise–measuring sets—frequency weighting, power addition, and transient response—essentially prescribe the way message circuit noise is measured for speech signal transmission. This is not yet enough; a noise reference datum and a scale of measurement must also be provided.

The chosen reference is 10^{-12} watt, or –90 dBm. The scale marking is in decibels, and measurements are expressed in decibels above reference noise (dBrn). Table 3–1 illustrates how a 1000–Hz tone at a power of –90 dBm gives a 0–dBrn reading regardless of which weighting network is used. For all other measurements, the weighting must be specified. The unit dBrnc is commonly used when readings are made using the C–message weighting network.

As with dBm power readings, vu and dBrn readings may be taken at any TLP and referred to 0–dB TLP by subtracting the TLP value from the meter reading. Thus, a typical noise reading might be 25 dBrn at 0–dB TLP, abbreviated 25 dBrn0. Similarly, values of dBrnc referred to 0 TLP are identified as dBrnc0.

Other noise–measuring instruments have been designed to evaluate the effects of noise on other types of signals or in other types of channel (e.g., impulse noise on high–speed digital data circuits).

Table 3–1. Relationship Between Power Reference Expressions

dBm	dBrn	Watt
0	90	10^{-3}
–10	80	10^{-4}
–20	70	10^{-5}
–30	60	10^{-6}
–40	50	10^{-7}
–50	40	10^{-8}
–60	30	10^{-9}
–70	20	10^{-10}
–80	10	10^{-11}
–90	0	10^{-12}

Display Techniques

Among the many specialized measurements that are needed in the evaluation of transmission circuits and signals are those of signal or interference amplitudes. Two types of display are commonly used. One type shows amplitude as a function of time (oscilloscope). The other shows amplitude as a function of frequency (spectrum analyzer). These are referred to as time–domain and frequency–domain displays. They may be presented on a cathode–ray tube or, using a personal computer with specialized software, plotted automatically.

3–4 ADDITION OF POWER IN dB

The merits of expressing power and power ratios in dBm and dB, respectively, have been demonstrated by the preceding discussions. A difficulty arises, principally in noise and interference studies, when it is necessary to find the sum of two or more powers that are given in dBm. Although the necessary steps are straightforward (the values must be converted to milliwatts, added, and then reconverted to dBm), they are time consuming. Specifically, suppose powers P_1 dBm and P_2 dBm are being dissipated in a circuit and it is desired to determine the sum, also in dBm.

The expression for the sum of the two powers is

$$p = p_1 + p_2 \text{ milliwatts.}$$

This may also be written

$$p = p_1(1 + p_2/p_1). \qquad (3\text{–}22)$$

The sum and each of the individual powers may be expressed as P, P_1, and P_2 dBm and the summing expression may be represented by the shorthand notation

$$P = P_1 \text{ “+” } P_2$$

where $P = 10 \log p$, $P_1 = \log p_1$, and $P_2 = 10 \log p_2$.

Equation 3–22 may be written in logarithmic form as

$$10 \log p = 10 \log p_1 + 10 \log (1 + p_2/p_1)$$

or

$$10 \log p = 10 \log p_1 + 10 \log s_p$$

where $s_p = 1 + p_2/p_1$. Then,

$$P = P_1 + S_p \quad \text{dBm} \qquad (3\text{–}23)$$

where $S_p = 10 \log s_p$ dB.

It is convenient to assign the symbol p_1 to the larger of the two powers to be added so that s_p lies in the range of $1 \leq s_p \leq 2$ and S_p is in the range of $0 \leq S_p \leq 3$ dB. The value of S_p is shown in Figure 3–6 as a function of the difference between P_1 and P_2 in dB. Thus, the sum of two powers can be determined by first finding the value of $P_1 - P_2$, next estimating the value of

S_p from the figure, and then adding S_p to P_1 as in Equation 3–23. It should be noted that this method may be applied to any two powers, such as dBW or dBrnc, expressed in dB relative to an absolute value.

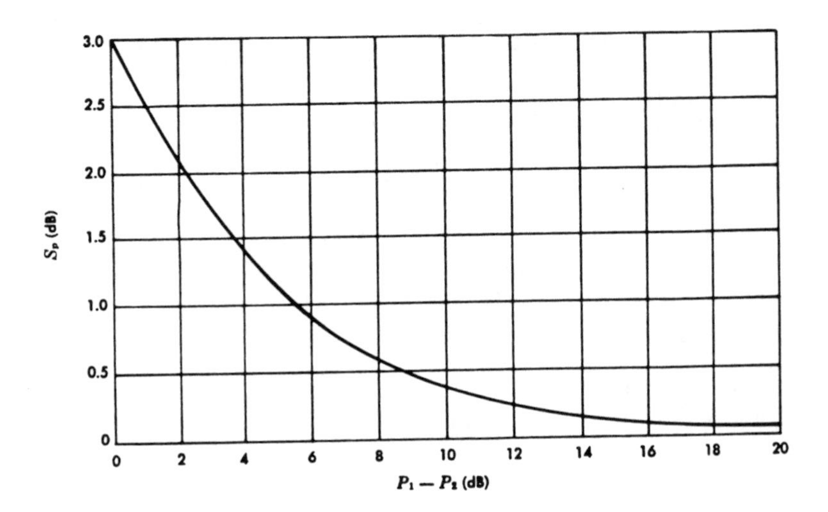

Figure 3–6. Power sum of two signals, both expressed in dBm.

In the foregoing, it is assumed that the two powers to be added act independently and that the resultant is the linear sum of the two components. Such an assumption is valid, for example, when two sine waves of different frequencies or a sine wave and a band of random noise are to be added. It is not true when two sine waves of the same frequency are to be added. In this case, the two are said to be coherent, the resultant power depends on the phase relationship between them, and the summing process must be treated somewhat differently.

For two sine waves of the same frequency, the power sum may be written as

$$p = (\sqrt{p_1} + \sqrt{p_2} \cos \theta)^2 + (\sqrt{p_2} \sin \theta)^2 \text{ milliwatts}$$

where θ is the phase angle between the two sine waves. The above equation may be written

$$p = p_1 + p_2 + 2\sqrt{p_1 p_2} \quad \cos\theta$$

or

$$p = p_1(1 + p_2/p_1 + 2\sqrt{p_2/p_1} \quad \cos\theta). \qquad (3\text{–}24)$$

This expression may be converted to a logarithmic form similar to Equation 3–23 and is then written

$$P = P_1 + S_v \quad \text{dBm} \qquad (3\text{–}25)$$

where $S_v = 10 \log s_v$

$$= 10 \log (1 + p_2/p_1 + 2\sqrt{p_2/p_1}\cos\theta) \quad \text{dB}.$$

Of primary interest in noise and interference studies is the case in which $\theta = 0$, i.e., the case representing in–phase addition of interferences. As in the earlier analysis, it is convenient to assign the symbol p_1 to the larger of the two interference signals to be added so that s_v lies in the range $1 \leq s_v \leq 4$ and S_v is in the range $0 \leq S_v \leq 6$ dB. With this choice ($\theta = 0$), the value of S_v is shown in Figure 3–7 as a function of the difference between P_1 and P_2 in dB. The sum for such in–phase addition may thus be found by determining P_1 and P_2, estimating the value of S_v from the figure, and then adding S_v to P_1 as in Equation 3–25.

The subscripts p and v applied to S_p and S_v are used to denote "power" and "voltage" addition as these processes are commonly called. The shorthand notation used to represent in–phase addition is usually written

$$P = P_1 \text{ "“+”" } P_2,$$

a form analogous to that used earlier to represent power addition.

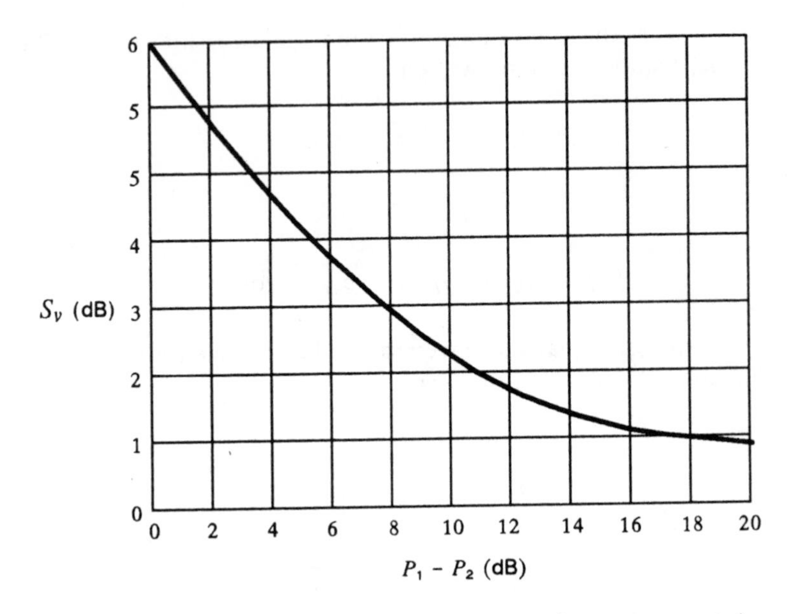

Figure 3-7. Sum of two in-phase signals expressed in dBm.

Additional Reading

Brady, P. T. "Equivalent Peak Level: A Threshold–Independent Speech–Level Measure," *Journal of the Acoustical Society of America*, Vol. 44 (Sept. 1968).

Brady, P. T. "Statistical Basis for Objective Measurement of Speech Levels," *Bell System Tech. J.*, Vol. 44 (Sept. 1965).

Cochran, W. T. and D. A. Lewinski. "A New Measuring Set for Message Circuit Noise," *Bell System Tech. J.*, Vol. 39 (July 1960).

Members of Technical Staff. *Transmission Systems for Communications*, Fifth Edition (Murray Hill, NJ: AT&T Bell Laboratories, Inc., 1982), pp. 14–18, 23–28, and 56–61.

Notes on the BOC IntraLATA Networks—1986, Technical Reference TR-NPL-000275, Bellcore (Iss. 1, Apr. 1986), pp. 7–26 to 7–32.

Chapter 4

Four-Terminal Linear Networks

The transmission of an electrical signal from sending to receiving terminal is accomplished by transferring energy from one electrical network to the next until the receiving terminal is reached. An understanding of the complete transmission process requires an understanding of the general principles of linear alternating-current networks and of how they interact when they are tandem-connected to form a complete signal path. Linear networks are those whose output voltages or currents are directly proportional to the input voltages or currents. The networks may or may not be bilateral.

This chapter presents a number of basic and important theorems involved in analyzing the performance of linear four-terminal networks. Impedance relationships and interactions are examined thoroughly. Some mathematical tools used in network analysis and synthesis are then discussed briefly. Finally, some specific network components and their principal uses are described. These components include transformers, series and parallel resonant circuits, and electric wave filters.

Network computations are approached differently depending on whether the problem is one of analysis or synthesis. In analysis, the stimulus and the network are given, and the problem is to determine the response of the network to the stimulus; i.e., the problem is to determine the output given the input and the network configuration. In synthesis, the stimulus and response (input and output) are given, and the problem is to determine the network configuration and component values that satisfy the given input-output relationships. Since the synthesis process is of interest only to the network designer and developer, this chapter is concerned only with analysis.

The material in this chapter is important because, in considering the layout and application of transmission circuits, it is essential that the transmission engineer recognize basic limitations in performance including circumstances in which these limitations are or are not involved and what corrective measures may be applied to overcome the basic limitations. The situations in which such judgments must be made may be as simple as interconnecting station terminal equipment with its loop or as complex as judging the effect of adding a new trunk in a built–up connection covering thousands of miles. This chapter provides some engineering tools to enable making such judgments. While the availability of computer software capable of synthesizing existing or proposed circuitry has made manual calculations using the methods discussed unnecessary, a thorough understanding of the network theorems is essential to making effective use of the new powerful design and performance analysis tools.

The development of computer–driven monitoring and automatic test apparatus, together with the growth in digital transmission facilities, has reduced the importance of the simple 1000–Hz loss test as an approximation of circuit performance. Nevertheless, the simple test is still a valid service analysis tool, provided its use and limitations are understood. First, the validity of the test results depends on the equipment units and circuit facilities having met all transmission performance specifications, including frequency bandpass, signal, and noise levels, etc., at the time they were placed in service. Second, the test assumes that any changes in circuit performance parameters will cause a change in the 1000–Hz loss. While this is usually the case, there are many documented service problems involving circuits with perfect 1000–Hz loss values that will not satisfactorily pass, for example, a 2200–Hz frequency–shift data signal. The moral—if the 1000–Hz signal does not meet specification limits, the circuit is faulty and simply adjusting the level may cure the problem. A "within–specs" 1000–Hz test result, however, does not necessarily guarantee a trouble–free facility because transmission of a complex wave through a complex network cannot be fully represented by transmission at a single frequency.

4-1 THE BASIC LAWS

In the analysis of the usual electrical networks making up telecommunications circuits, Ohm's and Kirchoff's laws are primary tools. By the aid of these laws, certain theorems have been derived that considerably reduce the effort required for network analysis.

Ohm's Law

The current, I, which flows through an impedance, Z ohms, is equal to the voltage developed across the impedance divided by the value of the impedance, or

$$I = E/Z \quad \text{amperes.} \tag{4-1}$$

This law is illustrated in Figure 4–1 where E and I may be direct or alternating voltages and currents and where Z may be a simple resistor or a complex impedance involving resistance, inductance, and capacitance.

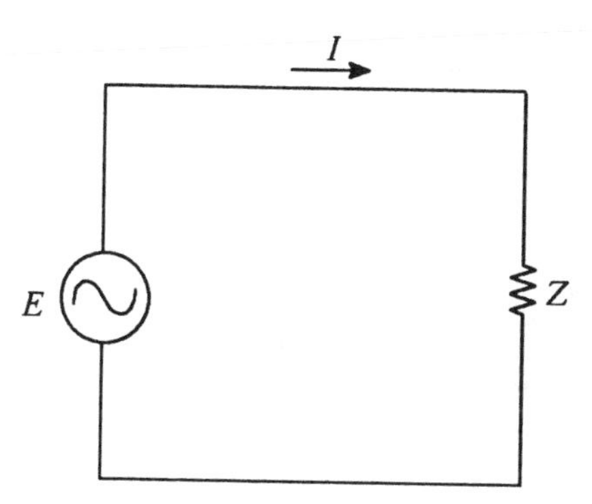

Figure 4–1. Simple series circuit containing an impedance, Z.

Kirchoff's Laws

Law 1: At any point in a circuit, there is as much current flowing into the point as there is flowing away from it. For example, at point x in Figure 4–2,

$$I_1 = I_2 + I_3. \tag{4-2}$$

Law 2: In any closed electrical circuit, the algebraic (or vector) sum of the electromotive forces (emf)s and the potential drops is equal to zero. In Figure 4–2,

$$\left.\begin{aligned} E - I_1 Z_A - I_3 Z_C &= 0, \\ E - I_1 Z_A - I_2 Z_B &= 0, \\ \text{and} \\ I_2 Z_B - I_3 Z_C &= 0. \end{aligned}\right\} \tag{4-3}$$

The arrows in Figure 4–2 indicate the assumed direction of current flow. A battery is assumed to produce a voltage rise from the negative to the positive terminal. A voltage due to current flowing through an impedance is assumed to be in the direction of positive to negative corresponding to the assumed direction of current flow. This accounts for the signs of the terms in Equations 4–3.

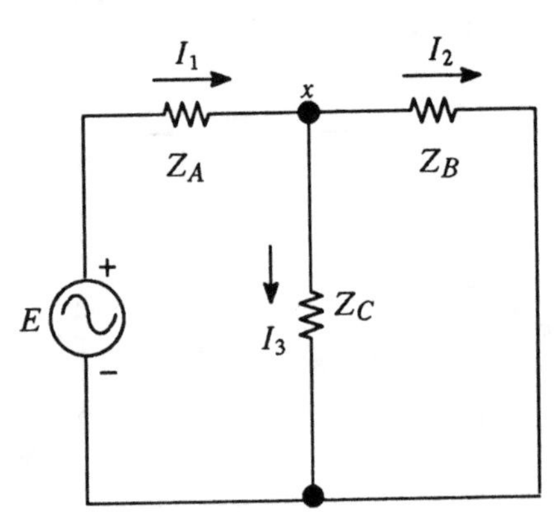

Figure 4–2. Simple series–parallel circuit.

4–2 APPLICATION AND THEOREMS

The application of Ohm's and Kirchoff's laws to more complicated circuits involves setting up simultaneous linear equations for solution. This can be very laborious, and several network theorems have been developed to expedite the process.

Equivalent Networks

From their configurations, two important types of networks are called the T and π electrical networks. A three–element

T structure and a three-element π structure can be interchanged provided certain relations exist between the elements of the two structures and provided the impedances can be realized.

Figure 4-3 represents two forms of a circuit connecting a generator of voltage E and impedance Z_G to a load having impedance Z_L. If the impedances enclosed in the boxes are related by the relationships shown in Figure 4-4, one box may be substituted for the other without affecting the voltages or currents in the circuit outside the boxes.

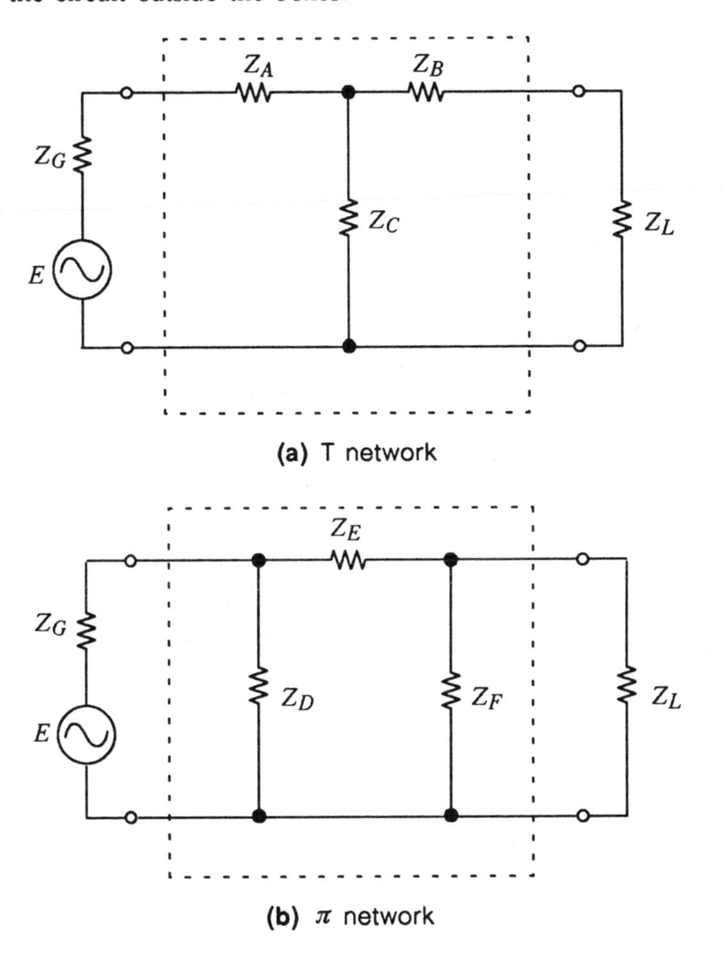

Figure 4-3. Equivalent networks.

π to T	T to π
$Z_A = \dfrac{Z_D Z_E}{Z_D + Z_E + Z_F}$	$Z_D = \dfrac{Z_A Z_B + Z_B Z_C + Z_C Z_A}{Z_B}$
$Z_B = \dfrac{Z_E Z_F}{Z_D + Z_E + Z_F}$	$Z_E = \dfrac{Z_A Z_B + Z_B Z_C + Z_C Z_A}{Z_C}$
$Z_C = \dfrac{Z_F Z_D}{Z_D + Z_E + Z_F}$	$Z_F = \dfrac{Z_A Z_B + Z_B Z_C + Z_C Z_A}{Z_A}$

Figure 4–4. Equivalent network relationships.

This property of networks permits any three–terminal structure, no matter how complex, to be reduced to a simple T. For example, a π – to – T transformation permits converting the circuit in Figure 4–5(a) to that shown in Figure 4–5(b). By combining Z_C with Z_5 and Z_B with Z_6 and making a second π – to – T transformation, Figure 4–5(b) can be reduced to the simple T shown in Figure 4–5(c).

These relationships apply only to networks having three terminals. Similar relations can be developed for four–terminal networks. Figure 4–6 is a typical four–terminal network. If only the voltages measured across terminals 2–2 are significant, the five impedances in Figure 4–6(a) can be replaced by the T structure in Figure 4–6(b).

Thevenin's, or Pollard's, Theorem

To simplify calculations, an arrangement such as that in Figure 4–7(a) may be considered as two networks with one supplying energy to the other. The first of these networks is then replaced by an equivalent simplified circuit consisting of an emf and an impedance in series, as shown in Figure 4–7(b).

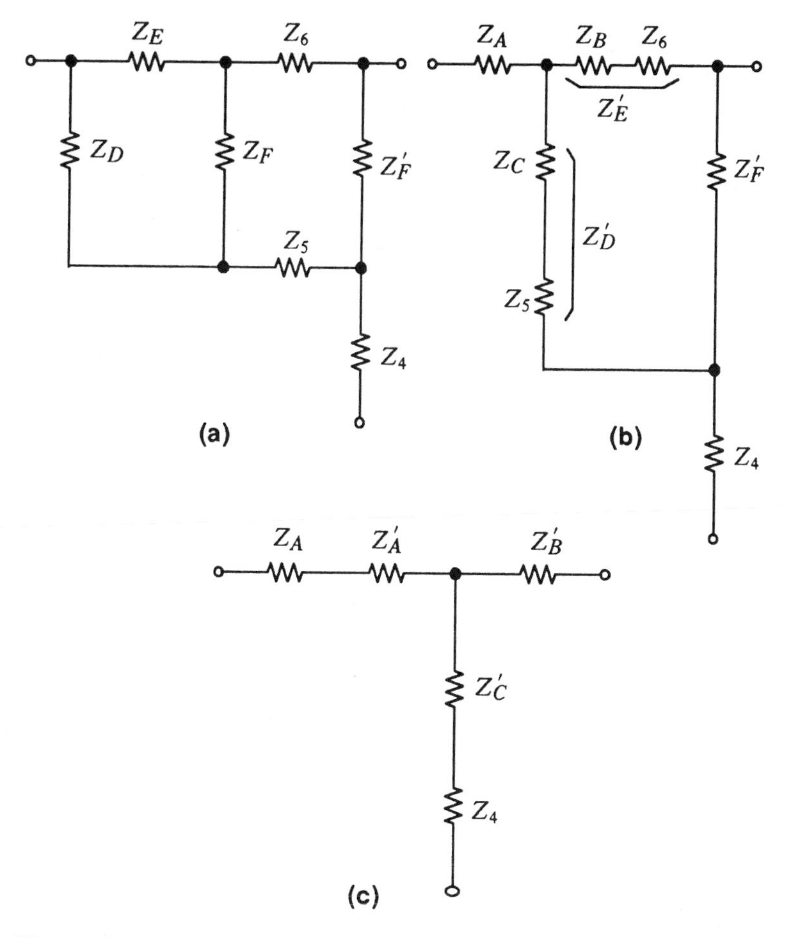

Figure 4–5. Successive simplification of networks by π-to-T transformations.

Thevenin's theorem gives the rules required for this simplification as follows: *The current in any impedance, Z_L, connected to two terminals of a network is the same as that resulting from connecting Z_L to a simple generator whose generated voltage is the open–circuit voltage at the original terminals to which Z_L was connected, and whose internal impedance is the impedance of the network looking back from those terminals with all generators in the original network replaced by their internal impedances.*

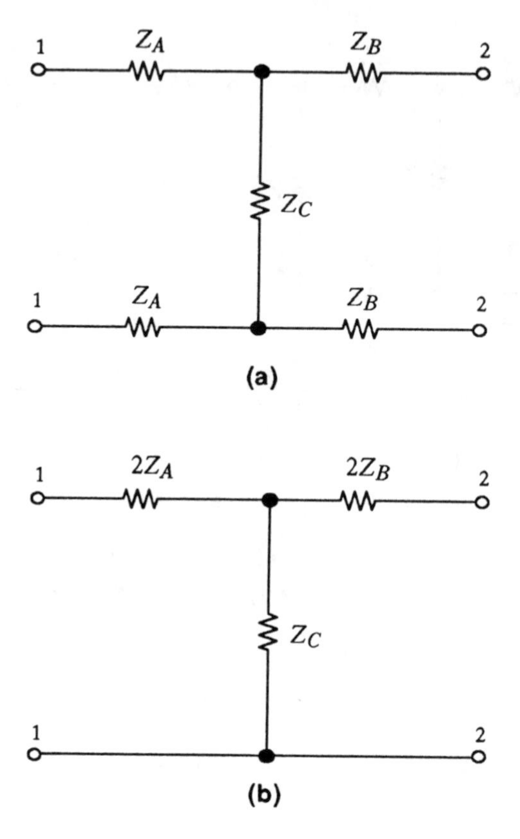

Figure 4–6. Equivalent four–terminal networks.

For example, if the equivalent emf, E' in Figure 4–7(b), is the open–circuit voltage at the terminals of Figure 4–7(a) and if the equivalent impedance, Z' of Figure 4–7(b), is the impedance presented at the terminals of Figure 4–7(a) when E is made zero, the two circuits are equivalent. Another way to compute Z' is to set it equal to the open–circuit voltage at network terminals divided by the short–circuit current at the terminals. Under these conditions the load will draw the same current as in the original connection.

Superposition Theorem

If a network has two or more generators, the current through any component impedance is the sum of the currents obtained by considering the generators one at a time, each of the generators other than the one under consideration being replaced by its internal impedance.

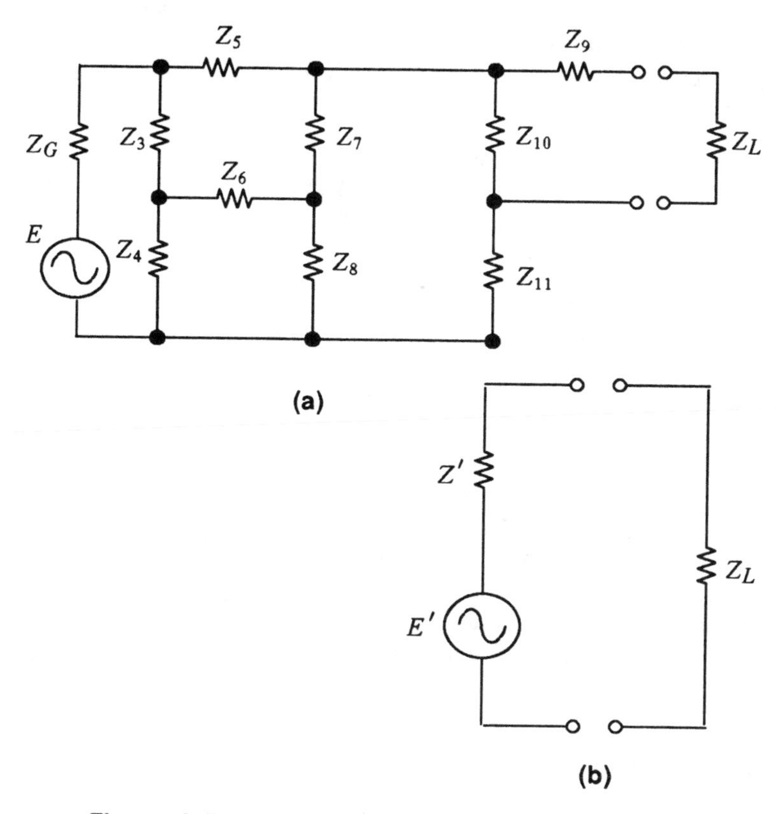

Figure 4–7. Application of Thevenin's theorem.

Multigenerator networks can be solved by Kirchoff's laws, but their solution by superposition requires less complicated mathematics. Perhaps of even greater importance is the fact that this theorem is a useful tool for visualizing the currents in a circuit.

Before an example of the superposition theorem is given, it may be beneficial to review the concept of the internal impedance of a generator. The open–circuit voltage of a real battery will be greater than the voltage across the terminals of the same battery when supplying current to a load. The open–circuit voltage is a fixed value determined by the electrochemical properties of the materials from which the battery is made. Under load, the decrease in terminal voltage is due to the voltage drop across the internal resistance of the battery. If it were possible to construct a battery from materials that had no resistance, the battery would have no internal resistance and no internal voltage drop. Every practical voltage source can be resolved into a voltage in series with an internal resistance or impedance.

Perhaps the superposition theorem can be most easily explained by working out a simple problem. In Figure 4–8(a), which way does the current flow in the 10–ohm resistor?

According to the theorem, the currents caused by each battery should be determined, in turn, with all other batteries replaced by their internal resistances. The currents indicated in Figures 4–8(b) and 4–8(c) are computed by Ohm's law. The currents flowing in the circuit with two batteries will be the sum of these component currents; of course, sum means algebraic sum (or vector sum if the problem is ac). Currents flowing in opposite directions subtract. The resultant currents are shown in Figure 4–8(d), which shows that the 10–ohm resistor carries one ampere in the upward direction. The direction of the current in the 10–ohm resistor could have been estimated by inspection, since the resistances are symmetrical and the 60–volt battery will produce the larger component of current. However, going through the arithmetic illustrates the application of the theorem.

Compensation Theorem

Any linear impedance in a network may be replaced by an ideal generator, one having zero internal impedance, whose generated voltage at every instant is equal in amplitude and phase to the instantaneous voltage drop caused by the current flowing through the replaced impedance.

In Figure 4–9(a), the impedance has been separated from the rest of the network for consideration. The equations of Kirchoff's laws determine the currents and voltages in all parts of the network. According to the compensation theorem, these equations will not be altered if the network is changed to that of Figure 4–9(b) where the generator voltage is the product of current I and impedance Z from Figure 4–9(a).

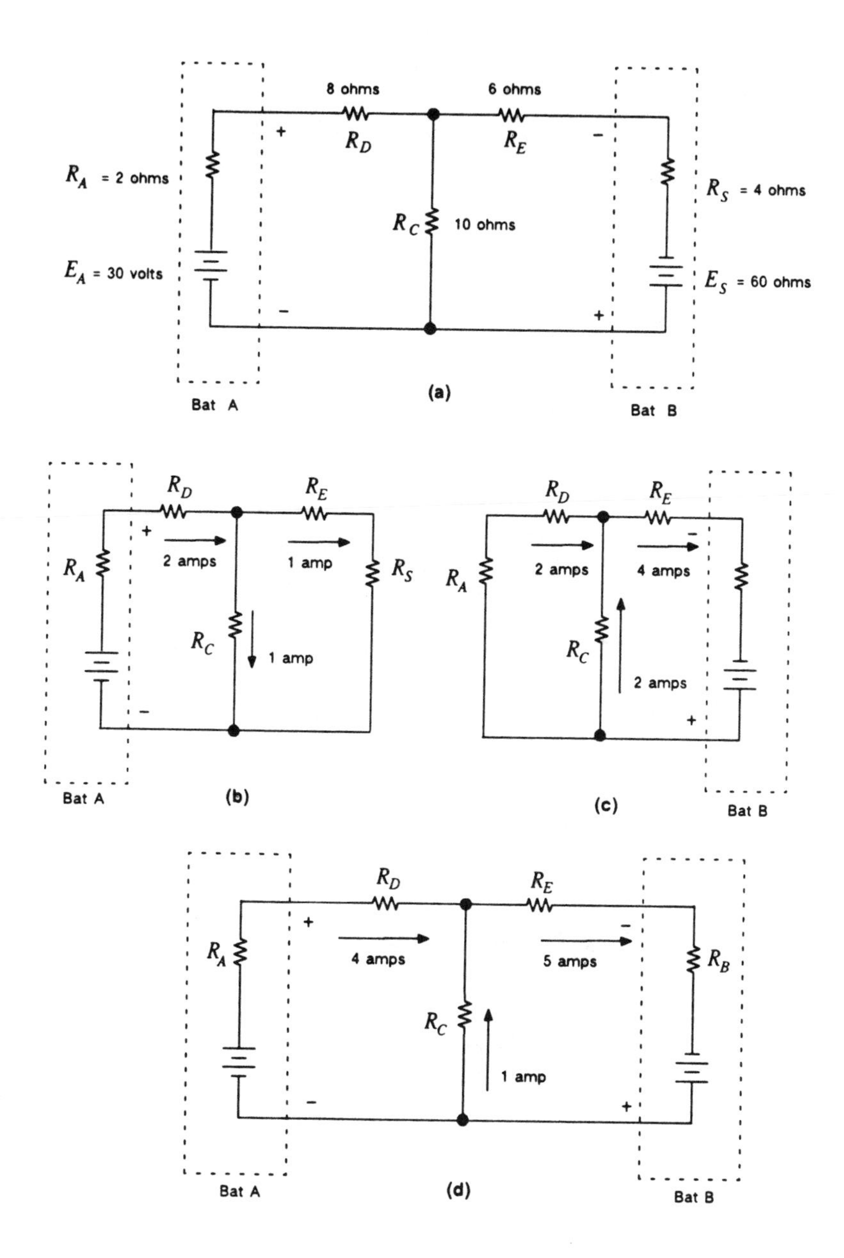

Figure 4-8. Superposition theorem.

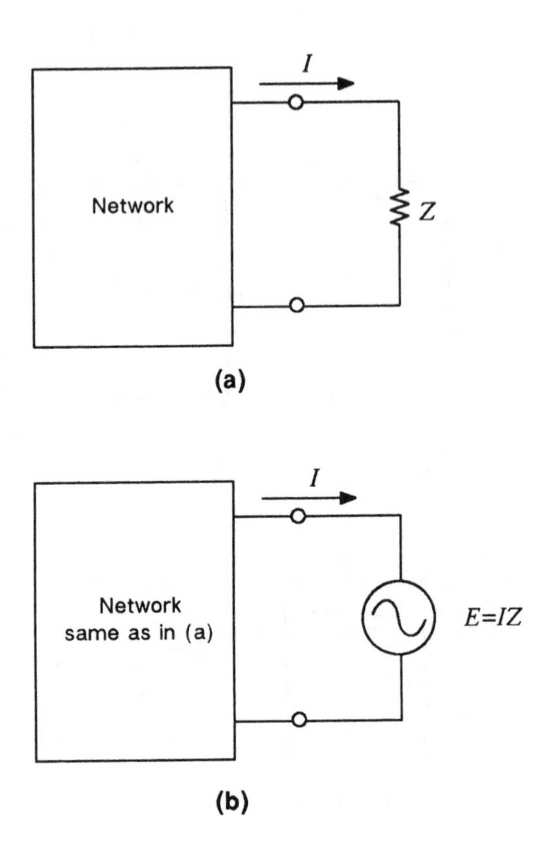

Figure 4–9. Illustration of compensation theorem.

4–3 NETWORK IMPEDANCE RELATIONSHIPS

The analysis of four–terminal networks and their interactions in random connections is primarily related to impedance relationships. The impedances involved are those of the network itself and of its terminations. Relationships are presented to permit calculation of transmission effects (attenuation and phase shift), return loss, echo (magnitude and delay), power transfer, and stability. The networks may be relatively simple configurations made up of discrete components or may be transmission lines, radio circuits, or carrier circuits, any of which may have gain or loss.

Image Impedance

In a four–terminal network, such as that in Figure 4–10, impedances Z_1 and Z_2 may be found such that if a generator of impedance Z_1 is connected between terminals 1–1 and impedance Z_2 is connected as a load between terminals 2–2, the impedances looking in both directions at 1–1 will be equal and the impedances looking in both directions at 2–2 will be equal. Impedances Z_1 and Z_2 are called the *image impedances* of the network.

The values of Z_1 and Z_2 may be determined from Ohm's law and the solution of two simultaneous equations. From inspection of Figure 4–10, the two equations may be written as

$$Z_1 = Z_A + \frac{(Z_B + Z_2)Z_C}{Z_B + Z_C + Z_2}$$

and

$$Z_2 = Z_B + \frac{(Z_A + Z_1)Z_C}{Z_A + Z_C + Z_1},$$

where Z_A, Z_B, and Z_C are the impedances of the T–network equivalent to the four–terminal network.

Solving for Z_1 and Z_2 yields

$$Z_1 = \sqrt{\left(Z_A + Z_C\right)\left(Z_A + \frac{Z_B Z_C}{(Z_B + Z_C)}\right)}$$

and

$$Z_2 = \sqrt{\left(Z_B + Z_C\right)\left(Z_B + \frac{Z_A Z_C}{(Z_A + Z_C)}\right)}.$$

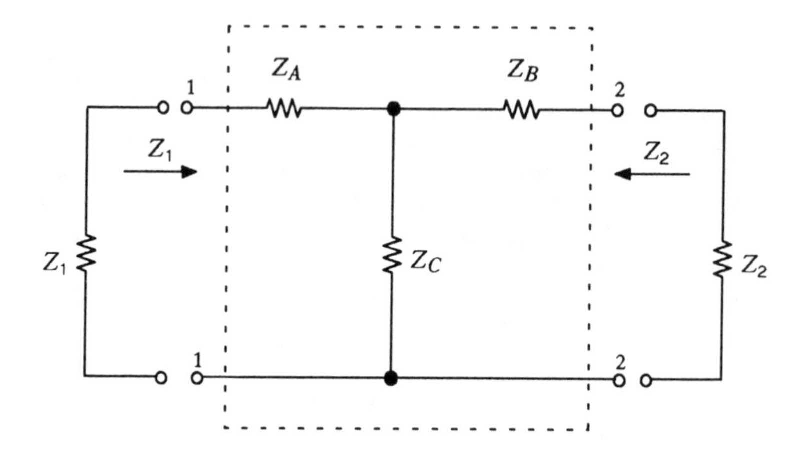

Figure 4-10. Image termination of a four-terminal network.

Examination of the latter equations and Figure 4-10 reveals some interesting relationships. The parenthetical expressions $(Z_A + Z_C)$ and $(Z_B + Z_C)$ are seen to be the impedances of the T network if the impedances are computed, respectively, from terminals 1-1 with terminals 2-2 open and from terminals 2-2 with terminals 1-1 open. Similarly, $Z_A + (Z_B Z_C)/(Z_B + Z_C)$ and $Z_B + (Z_A Z_C)/(Z_A + Z_C)$ are the impedances at terminals 1-1 and 2-2 if the opposite pair of terminals is short-circuited.

Thus, the image impedances of a four-terminal network are most easily determined by measuring the open-circuit and short-circuit impedances as above. Then,

$$Z_1 = \sqrt{Z_{OC}\, Z_{SC}} \tag{4-4}$$

and

$$Z_2 = \sqrt{Z'_{OC}\, Z'_{SC}}\ , \tag{4-5}$$

where

Z_{OC} = impedance at 1-1 with 2-2 open

Z_{SC} = impedance at 1-1 with 2-2 short-circuited

Z'_{OC} = impedance at 2-2 with 1-1 open

Z'_{SC} = impedance at 2-2 with 1-1 short-circuited.

As shown previously, the conversion of any complex network to an equivalent T network can be accomplished for any given frequency. Thus, the processes described allow you to determine the image impedances of a network at any frequency.

Note that if the network is symmetrical, i.e., $Z_A = Z_B$, the image impedances are equal, $Z_1 = Z_2$.

T-Network Equivalent

In the above determination of network image impedances as functions of open-circuit and short-circuit impedances measured (or computed) from the input and output terminals of the network, the assumed impedances of the T network were eliminated mathematically. Sometimes, however, it is also necessary to determine values of Z_A, Z_B, and Z_C of Figure 4–10 in terms of the open-circuit and short-circuit measurements. For a four-terminal network containing only passive components or one in which the gains in the two directions of transmission are equal, this may again be accomplished by solving simultaneous equations.

In the discussion of image impedance, the following relationships among the impedances of Figure 4–10 are shown:

$$Z_{OC} = Z_A + Z_C, \text{ or } Z_A = Z_{OC} - Z_C \tag{4–6a}$$

$$Z'_{OC} = Z_B + Z_C, \text{ or } Z_B = Z'_{OC} - Z_C \tag{4–6b}$$

$$Z_{SC} = Z_A + \frac{Z_B Z_C}{Z_B + Z_C} \tag{4–7a}$$

and

$$Z'_{SC} = Z_B + \frac{Z_A Z_C}{Z_A + Z_C} \quad . \tag{4–7b}$$

In Equations 4–7a and b, substitute the values of Z_A and Z_B from Equations 4–6a and b and solve for Z_C:

$$Z_C = \sqrt{(Z'_{OC} - Z'_{SC})Z_{OC}} \tag{4–8}$$

and also

$$Z_C = \sqrt{(Z_{OC} - Z_{SC})Z'_{OC}}\ . \tag{4-9}$$

The values of Z_C from Equations 4–8 and 4–9 may now be substituted directly in Equations 4–6a and b to give expressions for Z_A and Z_B in terms of input and output open–circuit and short–circuit impedances. Thus, all legs of the equivalent T network may be determined from these measurements, provided the network is bilateral, i.e., contains only passive components or has equal gain in the two directions of transmission.

If the network contains sources of amplification such that the gains in the two directions of transmission are not equal, the circuit cannot be reduced to a simple equivalent T network. Transfer effects, which account for the difference in gain in the two directions, must be taken into account.

Transfer Effects

The determination of image impedances of a four–terminal network and the conversion of such a network to an equivalent T configuration permit establishment of the relationships among input and output current and voltage directly from the application of Ohm's and Kirchoff's laws. However, these relationships may be applied directly only when the four–terminal network is bilateral. When it is not, these relatively simple relationships do not apply directly because of transfer effects that occur as current flows through the network.

Consider the circuit of Figure 4–11. This circuit is similar to Figure 4–10 except that impedance Z_1 is replaced by a voltage generator having an internal impedance equal to Z_1. The circuit arrangements result in voltage V_1 across terminals 1–1 and voltage V_2 across terminals 2–2 when the network is terminated in its image impedances, Z_1 and Z_2. The input current is I_1 and the output current is I_2.

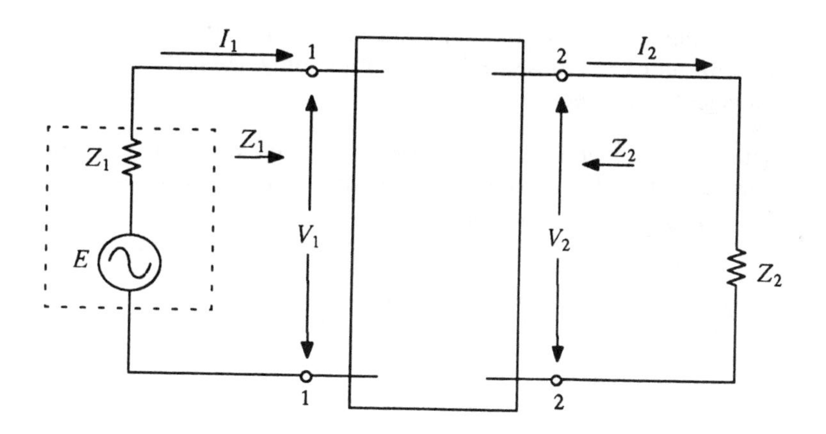

Figure 4-11. Image-terminated network.

If the voltage source is connected at the 2–2 terminals, analogous voltage and current expressions may be written with the symbols V and I changed to V' and I'.

The following relationships may then be written as definitions:

$$G_{1\text{-}2} = \frac{I_2 V_2}{I_1 V_1}, \tag{4–10}$$

$$G_{2\text{-}1} = \frac{I'_1 V'_1}{I'_2 V'_2}, \tag{4–11}$$

and

$$G_I = \sqrt{G_{1\text{-}2}\, G_{2\text{-}1}}, \tag{4–12}$$

where G_I is sometimes called the *image transfer efficiency*. It is a power gain if G is greater than unity, and a loss if G is less than 1.

In these equations, the currents and voltages are complex quantities. The current–voltage products in Equations 4–10 and 4–11 are quantities usually called volt–amperes, or *apparent power*. Thus, Equations 4–10 and 4–11 may be regarded as the gain, in the 1–2 or 2–1 direction, in apparent power resulting from transmission through the network. Equation 4–12 expresses the geometric mean of the apparent power gain in the two directions. In all cases, these definitions apply only when the network is image–terminated.

It will be convenient to express the quantity G_I in terms of open–circuit and short–circuit impedances. It can be shown that G_I may take any of the following forms:

$$G_I = \frac{1 - \sqrt{\frac{Z_{SC}}{Z_{OC}}}}{1 + \sqrt{\frac{Z_{SC}}{Z_{OC}}}} = \frac{1 - \sqrt{\frac{Z'_{SC}}{Z'_{OC}}}}{1 + \sqrt{\frac{Z'_{SC}}{Z'_{OC}}}} ; \qquad (4\text{–}13)$$

$$G_I = \frac{\sqrt{Z_{OC}} - \sqrt{Z_{SC}}}{\sqrt{Z_{OC}} + \sqrt{Z_{SC}}} = \frac{\sqrt{Z'_{OC}} - \sqrt{Z'_{SC}}}{\sqrt{Z'_{OC}} + \sqrt{Z'_{SC}}} ; \qquad (4\text{–}14)$$

and

$$G_I = \frac{Z_1 - Z_{SC}}{Z_1 + Z_{SC}} = \frac{Z_2 - Z'_{SC}}{Z_2 + Z'_{SC}} . \qquad (4\text{–}15)$$

These equations for G_I will be found useful in subsequent discussions of sending–end impedance, echo, and stability.

Sending–End Impedance

The *sending–end impedance* of a four–terminal network is the impedance seen at the input of the network when the output is terminated in any impedance, bZ_2; b is a factor used as a mathematical convenience to modify the terminating image impedance, Z_2. When bZ_2 is equal to the image impedance, Z_2 (i.e., b=1), the sending–end impedance, Z_S, is equal to the image impedance, Z_2. It is important to consider the effects on the value of Z_S of different impedance values for bZ_2, because these effects are related to such phenomena as return loss, singing, and talker echo, any or all of which may be important when a network is terminated in other than its image impedance, as in Figure 4–12.

The development of useful expressions for the analysis of the performance of a four–terminal network terminated in other than its image impedance can be demonstrated conveniently by starting with the image-terminated case as illustrated in Figure 4-13. The voltage V_1 is equal to $E/2$, as shown, because of the assumption of image terminations at both ends of the network. At the receiving terminals 2-2, the network is again assumed to be terminated in its image impedance, Z_2. The voltage appearing across terminals 2-2, V_2, may be defined in terms of Thevenin's theorem. The four-terminal network, which now is the driving point for the load, Z_2, is replaced by a simple impedance (by definition, equal to Z_2) and a generator whose open-circuit voltage is such as to produce V_2, i.e., $E_2 = 2V_2$.

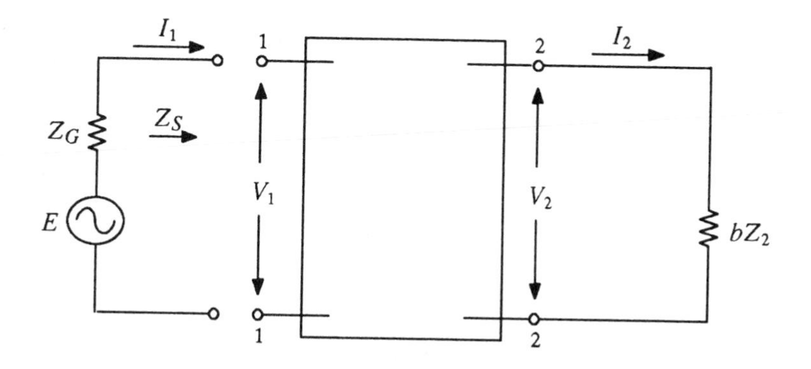

Figure 4-12. Sending-end impedance.

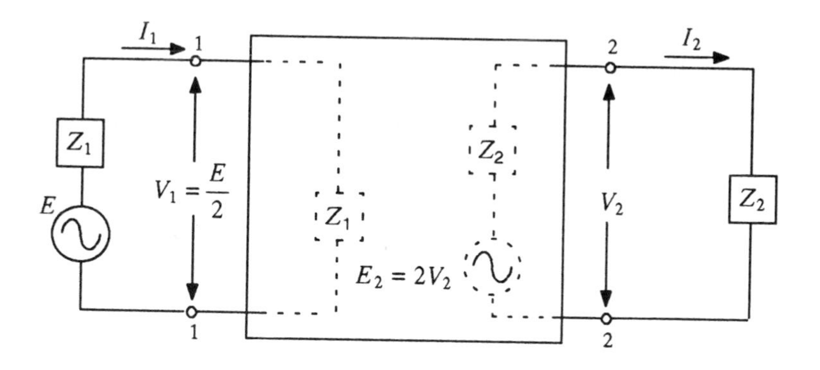

Figure 4-13. Image-terminated network.

By means of Equation 4–10, the input and output portions of the four–terminal network of Figure 4–13 may be related.

Thus,

$$G_{1\text{-}2} = \frac{I_2 V_2}{I_1 V_1} \tag{4–16}$$

The input and output currents may be related to their corresponding voltage drops and impedances by

$$I_1 = V_1/Z_1$$

and

$$I_2 = V_2/Z_2.$$

Substituting these values of current in Equation 4–16,

$$G_{1\text{-}2} = \frac{V_2^2 Z_1}{V_1^2 Z_2}$$

from which

$$V_2 = V_1 \sqrt{G_{1\text{-}2}} \sqrt{Z_2/Z_1} \quad . \tag{4–17}$$

It can also be shown that

$$I_2 = I_1 \sqrt{G_{1\text{-}2}} \sqrt{Z_1/Z_2} \quad . \tag{4–18}$$

Thus, Equations 4–17 and 4–18 may be used to relate input and output voltages and currents in an image–terminated four–terminal network. If the network were driven from the right (generator impedance of Z_2) similar expressions could be derived.

Then,

$$V_1 = V_2 \sqrt{G_{2\text{-}1}} \sqrt{Z_1/Z_2} \tag{4–19}$$

and

$$I_1 = I_2 \sqrt{G_{2\text{-}1}} \sqrt{Z_2/Z_1} \quad . \tag{4–20}$$

Now consider a termination, as shown in Figure 4–14, having a value other than Z_2 at terminals 2–2 of the network. Its value can be expressed in terms of Z_2 and an incremental impedance, Z_r, in series with Z_2. Note that Z_r is a complex impedance whose components may be positive, negative, or zero.

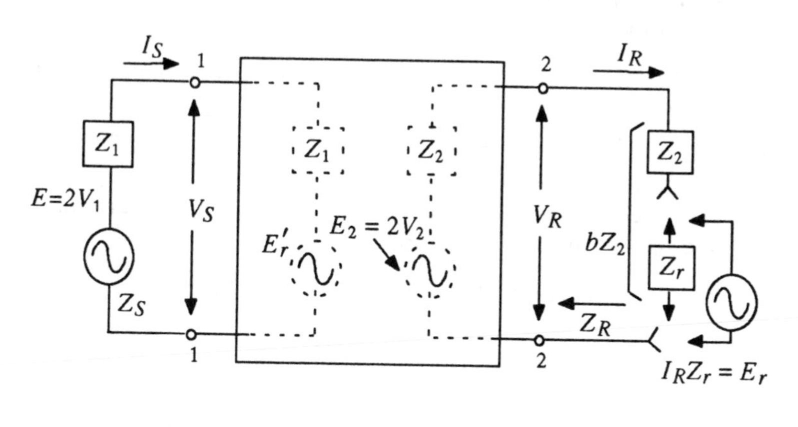

Voltages	Currents
$V_S = V_1 + \frac{E'_r}{2} = \frac{E + E'_r}{2}$	$I_S = I_1 + I'_r$
$V_R = \frac{E_r}{2} + V_2 = \frac{E_r + E_2}{2}$	$I_R = I_2 + I_r$
$E\sqrt{G_{1\text{-}2}}\ \sqrt{Z_2/Z_1} = E_2 = 2V_2$	$I_1\sqrt{G_{1\text{-}2}}\ \sqrt{Z_1/Z_2} = I_2$
$E_r\sqrt{G_{2\text{-}1}}\ \sqrt{Z_1/Z_2} = E'_r$	$I_r\sqrt{G_{2\text{-}1}}\ \sqrt{Z_2/Z_1} = I'_r$

Figure 4–14. Four–terminal network; image matched at input, mismatched at output.

By use of the compensation theorem, Z_r may be replaced by an ideal generator whose internal impedance is zero and whose generated voltage, E_r, is equal to the voltage drop across Z_r caused by the current, I_R, flowing through it.

The circuit may now be analyzed using the superposition theorem. The currents and voltages in the input and output circuits are shown in Figure 4–14. The symbolic form is the same as in Figure 4–13 for voltages and currents analogous to the image–terminated case; other current and voltage components are shown to reflect the presence of the compensating voltage, E_r, substituted for Z_r.

Note that $E'_r/2$ (a component of V_S) and I_r (a component of I_R) may be considered as reflected values of voltage and current at terminals 1–1 that would exist if the compensating voltage, $E_r/2$, at terminals 2–2 acted alone. Similarly, $E_r/2$ and I_r may be considered as the reflected voltage and current at terminals 2–2 due to the compensating voltage.

Now, the sending–end impedance may be written

$$Z_S = \frac{V_S}{I_S} = \frac{(E + E'_r)/2}{I_1 + I_r'}, \qquad (4\text{–}21)$$

where values of V_S and I_S are taken from Figure 4–14.

Equation 4–21 may be developed further. Note that in Figure 4–14 voltage V_R may be written

$$V_R = \frac{E_r + E_2}{2} = I_R b Z_2$$

where b is defined as

$$b = \frac{Z_2 + Z_r}{Z_2} . \qquad (4\text{–}22)$$

Then

$$E_r + E_2 = 2 I_R b Z_2 .$$

Since

$$I_R = \frac{E_2}{Z_2(1 + b)} ,$$

then

$$E_r + E_2 = \frac{2E_2 b Z_2}{Z_2(1 + b)} = \frac{2bE_2}{1 + b} ,$$

and

$$E_r = \frac{2bE_2}{1+b} - E_2$$

$$= \frac{2bE_2 - E_2 - bE_2}{1+b}$$

$$= \frac{E_2(-1+b)}{1+b} .$$

From Figure 4–14,

$$E'_r = E_r \sqrt{G_{2\text{-}1}} \sqrt{Z_1/Z_2} = -E_2 \left(\frac{1 - b}{1 + b}\right) \sqrt{G_{2\text{-}1}} \sqrt{Z_1/Z_2}$$

$$= -E \left(\frac{1 - b}{1 + b}\right) \sqrt{G_{2\text{-}1}} \sqrt{Z_1/Z_2} \sqrt{G_{1\text{-}2}} \sqrt{Z_2/Z_1} .$$

By substituting Equation 4–12,

$$E'_r = -G_I E \left(\frac{1 - b}{1 + b}\right) . \tag{4–23}$$

Then

$$V_S = \frac{E + E'_r}{2} = \frac{E}{2}\left[- G_I \left(\frac{1 - b}{1 + b}\right)\right] . \tag{4–24}$$

The current I_S may be written

$$I_S = I_1 + I'_r = I_1 - \frac{E'_r}{2Z_1} .$$

Substituting Equation 4–23 in the above gives

$$I_S = I_1 + G_I = \frac{E}{2Z_1} \left(\frac{1 - b}{1 + b}\right) = I_1 \left[1 + G_I \left(\frac{1 - b}{1 + b}\right)\right] . \tag{4–25}$$

Equations 4–24 and 4–25 may now be substituted in Equation 4–21 to give

$$Z_S = \frac{V_S}{I_S} = \frac{\frac{E}{2}\left[1 - G_I\left(\frac{1-b}{1+b}\right)\right]}{I_1\left[1 + G_I\left(\frac{1-b}{1+b}\right)\right]} .$$

The image impedance at the input is

$$Z_1 = \frac{E/2}{I_1} .$$

Thus,

$$Z_S = Z_1 \frac{\left[1 - G_I\left(\frac{1-b}{1+b}\right)\right]}{\left[1 + G_I\left(\frac{1-b}{1+b}\right)\right]} . \qquad (4\text{–}26)$$

Equation 4–26 may be used to illustrate the effect on sending–end impedance of providing an image termination (Z_2) at terminals 2–2 of the network. When this is done, the value of Z_r in Equation 4–22 becomes zero. Thus, the value of b becomes unity, the quantity $(1 - b)/(1 + b)$ becomes zero, and Equation 4–26 reduces to $Z_S = Z_1$; i.e., the sending–end impedance equals the image impedance.

An expression similar to Equation 4–26 may be derived for the impedance at terminals 2–2 of Figure 4–14. In this case,

$$Z_R = Z_2 \frac{\left[1 - G_I\left(\frac{1-a}{1+a}\right)\right]}{\left[1 + G_I\left(\frac{1-a}{1+a}\right)\right]} \qquad (4\text{–}27)$$

where a is a measure of the departure of the input terminating impedance from the image impedance. It is written

$$a = \frac{Z_1 + Z_S}{Z_1},$$

where Z_S is the incremental impedance when the terminating impedance is not the image impedance.

All of the quantities in Equations 4–26 and 4–27 are complex, and the labor involved in their evaluation is sometimes considerable. Detailed calculations may be performed on a digital computer. In some cases, tables are available for the evaluation of expressions like those in the two equations above. For ordinary engineering application, however, it is frequently desirable to make quick calculations that need not be extremely accurate. For these purposes, alignment charts have been prepared.

Alignment Charts

The labor of computation arises from the repetitive use of terms in the form of $(1 - b)/(1 + b)$ where b is complex; therefore, it is desirable to reduce this to a single complex quantity in the polar form, $Q \angle \phi$. If b is written in polar form as $X \angle \theta$, the values of X and θ may be written as $X = | \alpha + j\beta |$ or $X = \sqrt{\alpha^2 + \beta^2}$ and $\theta = \tan^{-1}\beta/\alpha$.

Four alignment charts are given here as Figures 4–15, 4–16, 4–17, and 4–18. These charts are used to solve expressions in the form $\frac{1 - X \angle \theta}{1 + X \angle \theta} = Q \angle \phi$. Figures 4–15 and 4–16 give values for Q for various combinations of X and θ, while Figures 4–17 and 4–18 give values for ϕ for various combinations of X and θ. The following example illustrates the use of the charts.

Example 4–1: Use of Alignment Charts

$$\text{Given: } b = 4 \angle 70° = X \angle \theta$$

$$\text{To evaluate: } \frac{1 - b}{1 + b}.$$

First, refer to Figure 4–15. Mark $X = 4$ on the left–hand vertical scale and $\theta = 70°$ on the right–hand vertical scale. Using a straight edge to connect these points, read $Q = 0.848$ on the left side of the Q scale. Next refer to Figure 4–17. Again mark the points $X = 4$ and $\theta = 70°$. With the straight edge, read $\phi = -153.5°$ on the right–hand side of the ϕ scale. Thus, $(1 - b)/(1 + b) = 0.848 \angle -153.5°$.

Note that the charts in Figures 4–15 through 4–18 can be used for a quick evaluation of changes in either the magnitude or the phase angle, or both, of a termination on a network. These evaluations are useful in determining the performance of circuits and in judging what may be done to improve it.

Insertion Loss

Relationships similar to those of Equations 4–26 and 4–27 may be used to determine the insertion loss of a four–terminal network when it is placed between two mismatched impedances. The general case, when the terminating impedances and the input and output image impedances are all different, contains many interaction terms that are difficult to evaluate. Furthermore, this most general situation is usually of only academic interest; since all terminals are accessible, it is easier to measure the insertion loss than to compute it. Therefore, insertion loss (IL) can be simply stated in dB as:

$$\text{IL} = 10 \log \left(\frac{\text{Power delivered to the load with network bypassed}}{\text{Power delivered to the load with network inserted}} \right).$$

Often, the subject network is either symmetrical and has only one value of image impedance, or is a transmission line of characteristic impedance Z_0. Cases of interest frequently apply to the insertion loss of a transmission line as covered in Chapter 5, Part 2.

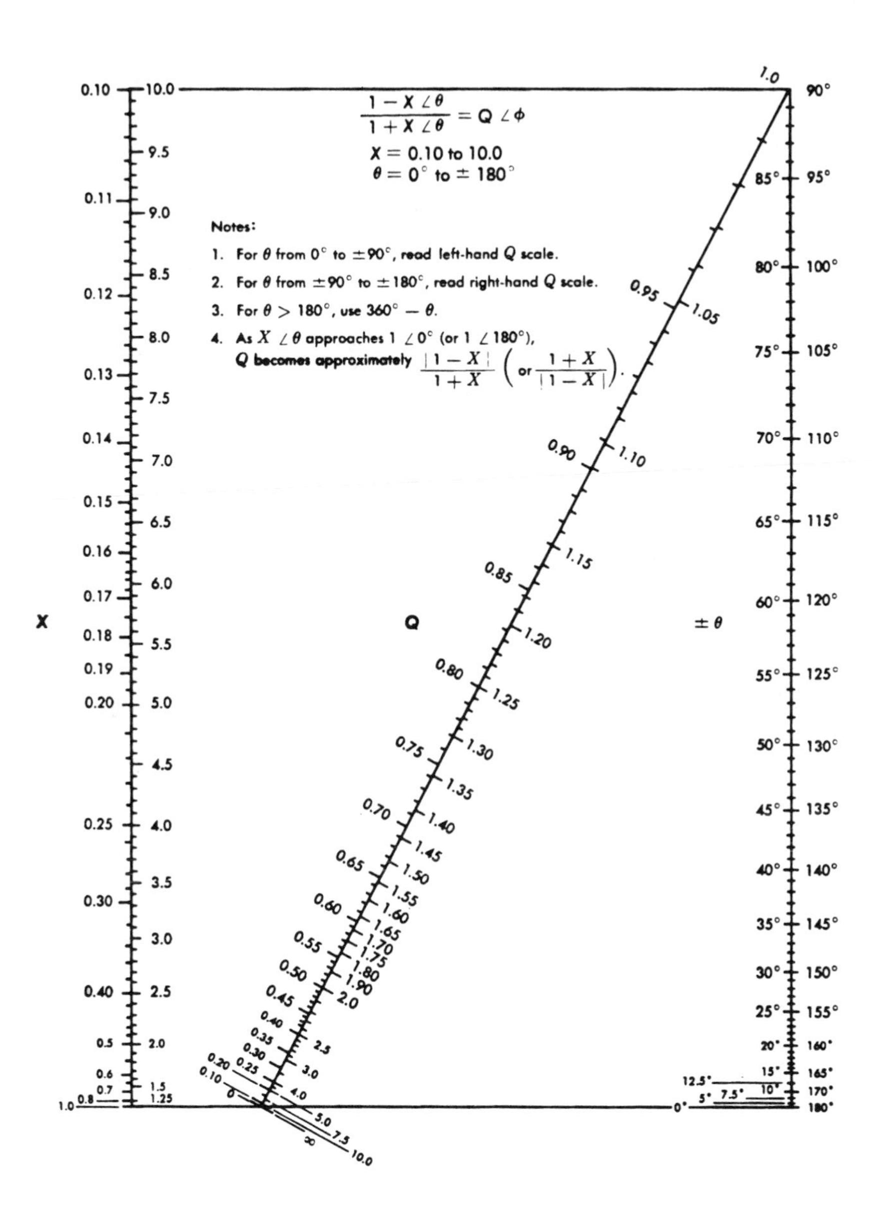

Figure 4–15. Alignment chart Q1.

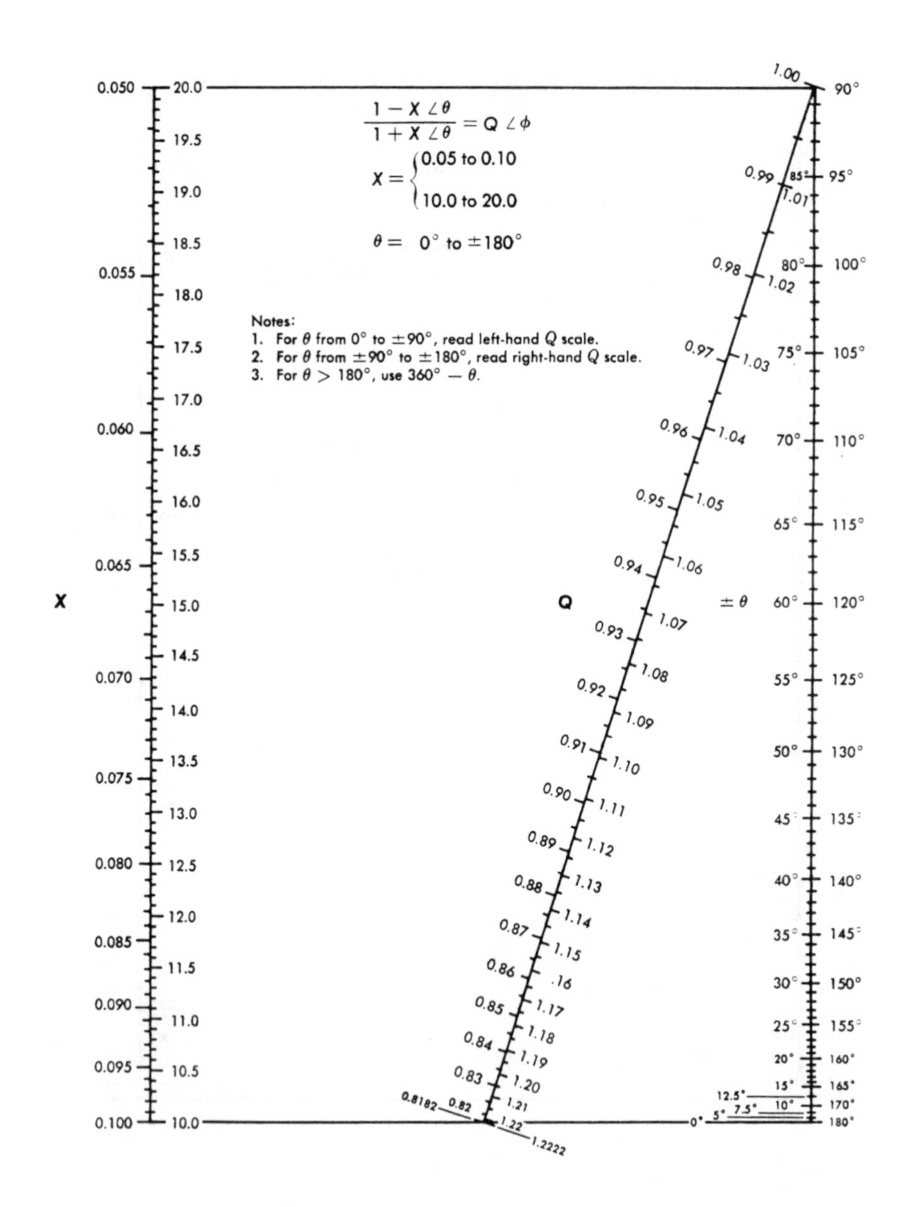

Figure 4–16. Alignment chart Q2.

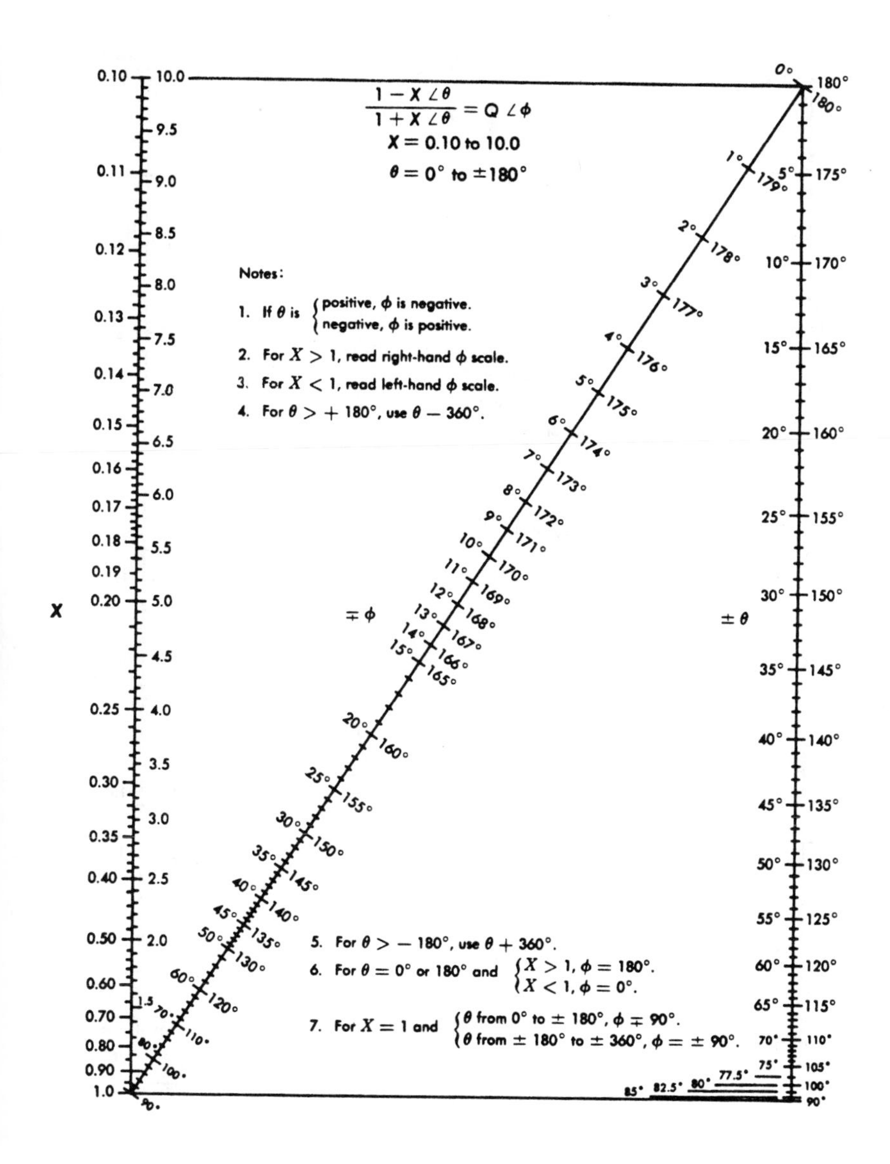

Figure 4–17. Alignment chart $\phi 1$.

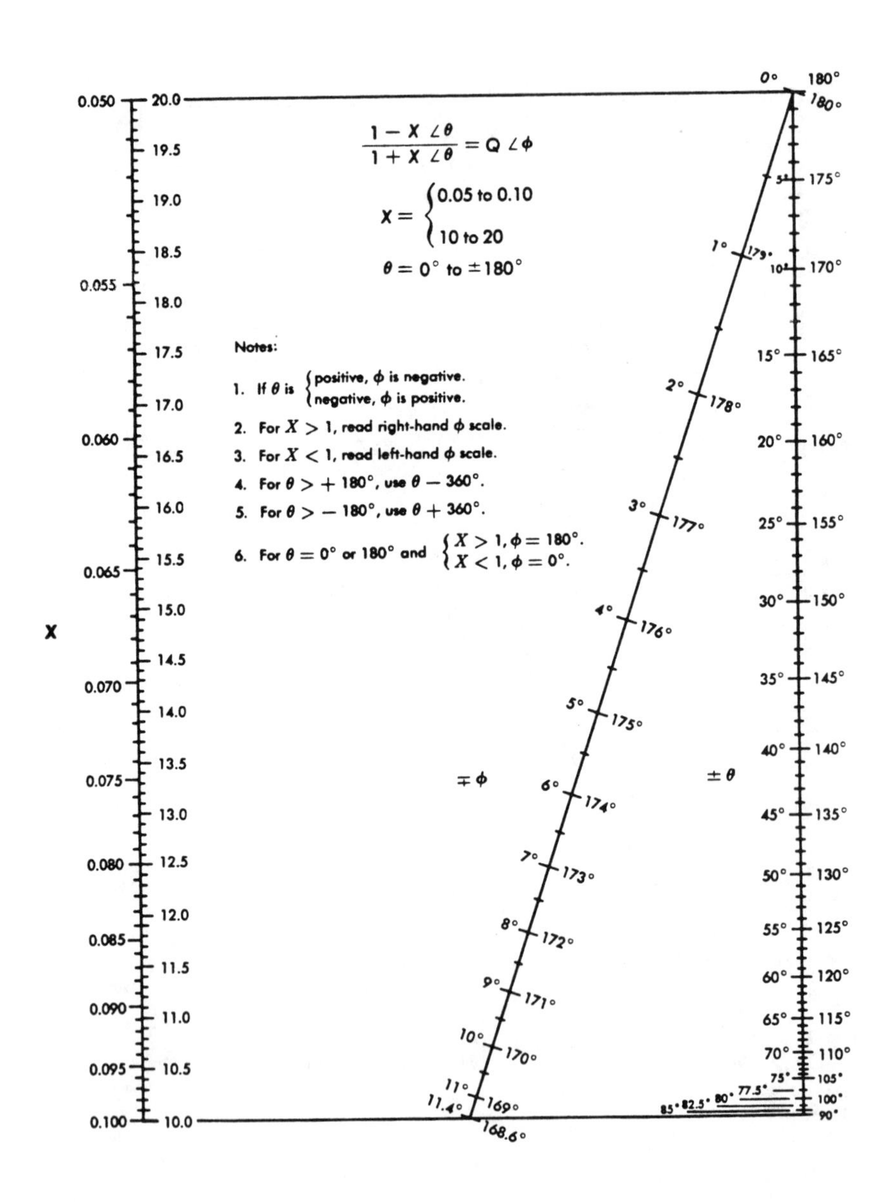

Figure 4–18. Alignment chart $\phi 2$.

Return Loss

In the analysis of transmission circuits, the *return loss* (a measure of the loss in the return path due to an impedance mismatch) is a convenient measure of performance. The return loss is related to the reciprocal of the absolute value of the *reflection coefficient*, a term that relates impedances at a point of connection in such a way as to give a measure of the voltage or current reflected from the mismatch point towards the transmitting end of a circuit.

From Figure 4–19, the reflection coefficient, ϱ, at the terminals of Z_L may be written for voltage,

$$\varrho_v = \frac{Z_L - Z_G}{Z_L + Z_G}, \tag{4–28}$$

or for current,

$$\varrho_i = \frac{Z_G - Z_L}{Z_G + Z_L}. \tag{4–29}$$

The return loss at these terminals is given by 20 log $(1/|\varrho|)$ dB; i.e.,

$$\text{Return loss} = 20 \log \frac{1}{|\varrho|} = 20 \log \left| \frac{Z_G + Z_L}{Z_G - Z_L} \right|. \tag{4–30}$$

The expression for return loss, Equation 4–30, may be written in the form

$$20 \log \frac{1}{|\varrho|} = 20 \log \left| \frac{1}{\left(1 - \dfrac{Z_L}{Z_G}\right) \Big/ \left(1 + \dfrac{Z_L}{Z_G}\right)} \right|.$$

The bracketed expression in this equation may be written in polar form as $(1 - X \angle \theta)/(1 + X \angle \theta)$. Thus, the alignment charts of Figures 4–15 to 4–18 may be used conveniently to determine the return loss at a junction between two impedances whose magnitudes or phase angles, or both, are unequal.

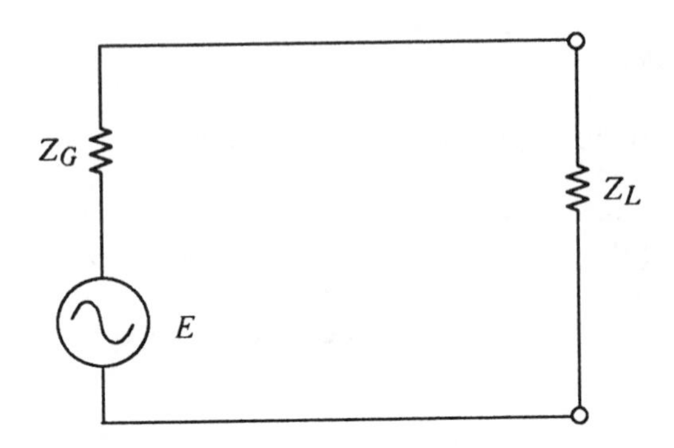

Figure 4–19. Source and load impedances.

The actual voltage across the load, Z_L, is equal to the voltage that would be present across an impedance matched to Z_G plus the reflected voltage. As Z_L approaches zero, the reflection coefficient approaches −1, the measured voltage across Z_L approaches zero, and the return loss approaches 0 dB. As Z_L approaches infinity, the reflection coefficient approaches +1, the voltage across Z_L approaches its open–circuit value (twice the value across Z_L under matched conditions), and the return loss again approaches 0 dB. When Z_L equals Z_G, the reflection coefficient is zero in magnitude and angle, the voltage across Z_L is one–half the open–circuit value, and the return loss is infinite. These stated relationships are oversimplified and must be modified to account for the reactive components of the impedances.

The effects of impedance mismatch on return loss are illustrated in Figure 4–20, which shows that the return loss increases as the angle, θ, decreases and as the ratio of $|Z_L/Z_G|$ or $|Z_G/Z_L|$ approaches unity. The angle θ is that between the load and generator impedances. The values of the parameters of Figure 4–20 may be written $Z_L = |Z_L| \angle \theta_L$, $Z_G = |Z_G| \angle \theta_G$, and $\theta = |\theta_L - \theta_G|$.

Echo—Magnitude and Delay

Return now to Figure 4–14. The condition at terminals 2–2 is one of mismatch; the effect of the mismatch could be evaluated in terms of return loss, as above, simply by using values in Equation 4–30 such that $Z_G = Z_2$, and $Z_L = bZ_2$. However, it is

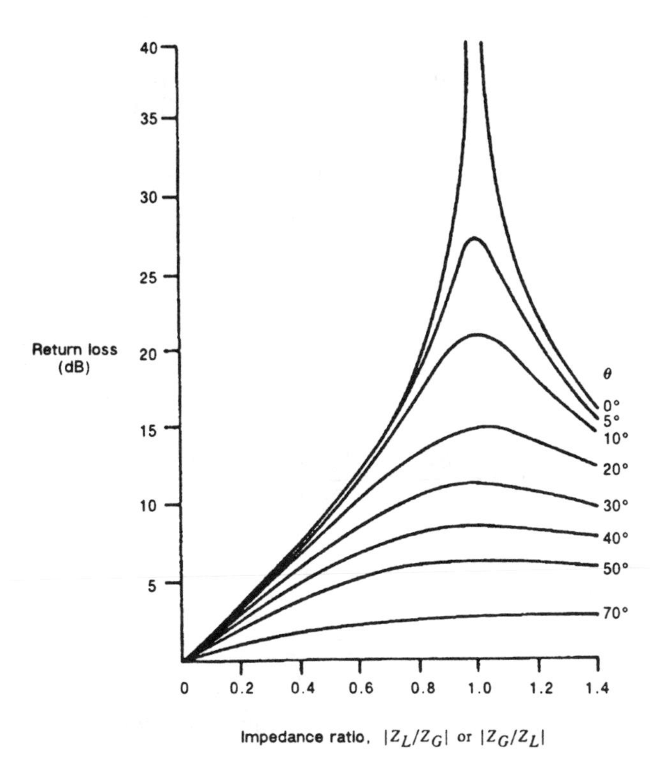

Figure 4–20. Return loss variations.

often desirable to evaluate the magnitude of the reflected voltage or current wave and to determine the delay encountered by the reflected wave in transmission through the network.

Consider first the magnitude of the reflected wave. From Equation 4–22, $b = (Z_2 + Z_r)/Z_2$, or $Z_r = Z_2(b - 1)$. The total current at the output is

$$I_R = \frac{E_2}{2Z_2 + Z_r} = \frac{E_2}{Z_2(b + 1)} . \qquad (4\text{–}31)$$

The voltage across Z_r may be regarded as the reflected voltage due to the mismatch. This is written as

$$E_r = I_R Z_r = I_R Z_2 (b - 1).$$

Substituting Equation 4–31 yields

$$E_r = E_2\left(\frac{b-1}{b+1}\right) = -E_2\left(\frac{1-b}{1+b}\right).$$

The output current, from Figure 4–14, may be written also as

$$I_R = I_2 + I_r .$$

From Ohm's law,

$$I_2 = \frac{E_2}{2Z_2}$$

and

$$I_r = -\frac{E_r}{2Z_2} = \frac{E_2}{2Z_2}\left(\frac{1-b}{1+b}\right).$$

Dividing gives a useful expression for the ratio of reflected to incident current:

$$\frac{I_r}{I_2} = \frac{1-b}{1+b} . \qquad (4\text{–}32)$$

It can be shown that a similar relationship exists for a mismatch at the input; i.e.,

$$\frac{I_S}{I_1} = \frac{1-a}{1+a} . \qquad (4\text{–}33)$$

Equations similar to 4–32 and 4–33 can also be developed to show the ratio of reflected to incident voltages.

Echo evaluations must, of course, take into account the loss encountered in transmission through the network an appropriate number of times. Successive reflections become increasingly attenuated and may eventually be ignored.

The time delay or transmit time for a wave to propagate through a four–terminal network may be shown to be

$$T = \frac{\theta}{2 \times 360° \times f} \qquad (4\text{–}34)$$

where θ is the angle of G_I in degrees and f is the frequency in hertz. Then, the round–trip delay for an echo to be transmitted through a network and back again is

$$T_2 = \frac{\theta}{360° \times f} \cdot \quad (4\text{–}35)$$

Power Transfer

In Figure 4–19, E and Z_G represent a source of power. This source may be a telephone, a repeater amplifier, or the sending side of any point in a telecommunications network connection. The impedance Z_L is the load that receives the power transmitted. It may be another telephone or a radio antenna—the receiving side of any point in a connection. The amount of power transferred from the source to the load will be determined by the relative values of Z_G and Z_L. The power transferred can be shown to be a maximum under three different assumptions as follows:

(1) If Z_G is a fixed impedance and there is no restriction on the selection of Z_L, the power transferred will be a maximum when Z_L is the conjugate of Z_G, i.e., when Z_L and Z_G have equal components of resistance and their reactive components are equal and opposite. This may be written $Z_G = R + jX$, and $Z_L = Z_G^* = R - jX$.

(2) If Z_G is a fixed impedance and the magnitude of Z_L, but not its angle, can be selected, the power transferred will be a maximum when the absolute values of Z_L and Z_G are equal ($|Z_L| = |Z_G|$). That is, the impedances are equal disregarding phase.

(3) If both Z_G and Z_L are pure resistances, the power transferred will be a maximum when the source and load resistances are equal ($R_G = R_L$).

The pure resistance case (3) has been plotted in Figure 4–21 to illustrate the principle. Curves of power and efficiency are drawn over the range $R_L = 0$ to $R_L = 2R_G$. When R_L is zero (a short circuit), the current is at its maximum possible value. The power (I^2R_G) is also at the highest possible value; all the power is dissipated in the source resistance R_G. As R_L is increased, the current and total power ($I^2R_G + I^2R_L$) decreases; however, a portion of the power will be dissipated in the load R_L. Curve A shows that the power dissipated in the external circuit (or load) is a maximum when $R_L/R_G = 1.0$ or $R_L = R_G$. Under this condition, the efficiency is 50 percent. Half of the total power is dissipated in the source and half in the load. This approximates the desirable condition in telecommunications, since in most such applications the primary interest is in receiving all the power possible regardless of the efficiency.

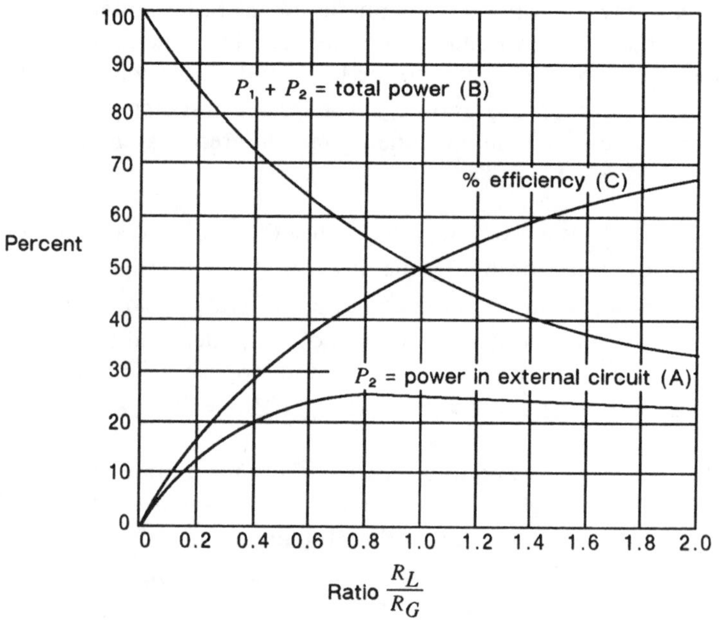

Notes:
Curves A and B = percent of short-circuit power.
Curve C = percent efficiency.

Figure 4–21. Maximum power transfer.

However, in telecommunications, another transmission parameter must be considered—the generation of reflections or echoes. The necessity for compromise between delivering maximum power to a load and maintaining reasonable performance in respect to reflections can best be illustrated by an example.

Example 4-2: Power Transfer and Return Loss

In Figure 4–19, let $Z_G = 900 - j200$ ohms and let $E = 1$ volt rms at 1000 Hz.

(a) What is the return loss at 1000 Hz at the junction between Z_G and Z_L and what is the power delivered to Z_L when $Z_L = 900 + j200$?

(b) What is the return loss at 1000 Hz at the junction between Z_G and Z_L and what is the power delivered to Z_L when $Z_L = 922 + j0$?

Case a:

$$\text{Return loss} = 20 \log \frac{1}{|\varrho|} = 20 \log \left| \frac{1}{\left(1 - \frac{Z_L}{Z_G}\right) \Big/ \left(1 + \frac{Z_L}{Z_G}\right)} \right|$$

$$= 20 \log \left| \frac{Z_G + Z_L}{Z_G - Z_L} \right|$$

$$= 20 \log \left| \frac{1800}{-j400} \right| = \left| \frac{1800 < 0°}{400 < -90°} \right|$$

$$= 20 \log | 4.5 < 90° |$$

$$= 13.1 < 90° \quad \text{dB}$$

where the value of $(1 - 1\angle -25°)/(1 + 1\angle -25°)$ is found from Figures 4–15 and 4–17.

To determine the power delivered to the load, the current may first be determined:

$$I = \frac{E}{Z_G + Z_L} = \frac{1\angle 0}{1800} = 0.000554 \text{ ampere rms.}$$

Then,

$$P_L = I^2R_L = 0.000554^2 \times 900 = 0.000276 \text{ watt.}$$

Case b:

$$Z_G = 900 - j200.$$

$$Z_L = 922 + j0.$$

$$\text{Return loss} = 20 \log \left| \frac{Z_G + Z_L}{Z_G - Z_L} \right|$$

$$= 20 \log \left| \frac{(900 - j200) + (922 + j0)}{(900 - j200) - (922 + j0)} \right|$$

$$= 20 \log \left| \frac{1822 - j200}{-22 - j200} \right|$$

$$= 20 \log \left| \frac{1833 < -6.3°}{201 < -96.3°} \right|$$

$$= 20 \log | 9.1 < 90° |$$

$$= 19.2 \text{ dB.}$$

The power delivered to the load is computed as follows:

$$I = \frac{E}{Z_G + Z_L} = \frac{1 \angle 0}{900 - j200 + 922} = \frac{1 \angle 0}{1822 - j200}$$

$$= \frac{1822 + j200}{3,360,000} = 0.000542 + j0.000060$$

$$= 0.000545 \text{ ampere.}$$

$$P_L = I^2R_L = 0.000545^2 \times 922$$

$$= 0.000274 \text{ watt.}$$

Thus, an improvement of 19.2 − 13.1 = 6.1 dB in return loss is achieved by providing a resistive termination of a value equal to the magnitude of the source impedance, 922 ohms. The penalty paid for this improvement is a reduction from the maximum delivered power of 0.000276 watt to 0.000274 watt, a penalty of 10 log 0.000276/0.000274 = 0.03 dB.

Transducer Loss

Although insertion loss has been used to calculate the expected loss as measured between the input and output of networks such as transmission lines, it is not actually that measured in practice. Factors such as source impedance, load impedance, loss through the network, power delivered to the network, and reflection loss (or reflected power) alter the calculated insertion loss values when the networks (circuits) are actually measured. This is due to the power source impedance and measuring equipment impedance being in variance with the ideal network impedance and assumed terminating impedance. As a result, maximum power transfer rarely takes place when measurements are made across a network. Such a quantitative measurement is called transducer loss. It can be simply stated in dB as,

$$\mathrm{TL} = 10 \log \left(\frac{\text{Maximum power delivered from a source}}{\text{Power delivered to load with network inserted}} \right).$$

It is evident, in practice, that transducer loss factually indicates transmission line or circuit loss whether the impedance at both ends is different or the same. In fact, measurements may vary when the power source and measuring equipment are physically transposed as circuit terminations. Transducer loss is therefore the appropriate measurement for calculating the expected measured loss of a telecommunications transmission line or circuit.

Stability

When a circuit is unstable, it is said to be *singing*; that is, unwanted signal currents and voltages flow in the circuit without an external source of applied signal energy. In Figure 4–22, such conditions of instability would result in signal current flow in aZ_1, bZ_2, and in the network.

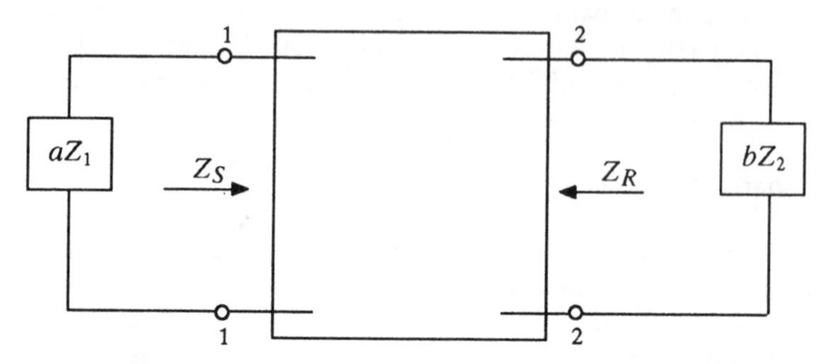

Figure 4–22. Four-terminal network; mismatched at input and output.

The complete development of mathematical criteria for absolute stability or absolute instability is not undertaken here. However, stability criteria are presented with some background to indicate how they are derived.

If currents are circulating in the input circuit of Figure 4–22 without an external source of energy, there must be zero impedance at the frequency at which current is observed. That is, if such a current is circulating through the input, then at that frequency

$$aZ_1 + Z_S = 0.$$

It can be shown that when such a condition exists a similar condition exists at the output; that is,

$$bZ_2 + Z_R = 0.$$

For such a condition to exist, the sending–end impedance, Z_S, must have a negative resistance component equal to the positive resistance component of aZ_1, and the reactive component of impedance Z_S must be equal in magnitude but opposite in sign to the reactive component of aZ_1. Thus, stability is guaranteed if neither aZ_1 nor Z_S has a negative resistance component and at least one has a positive resistance component.

A *stability index* can be derived in terms used earlier in this chapter. It may be written

$$\text{Stability index} = 1 - G_I \left(\frac{1 - a}{1 + a}\right)\left(\frac{1 - b}{1 + b}\right). \tag{4-36}$$

The two parenthetical expressions will be recognized as being of a form that can be evaluated by the alignment charts of Figures 4–15 through 4–18.

Note that if the network is terminated at either end in its image impedance, it cannot be made to sing. Under these conditions, $a = 1$ or $b = 1$, and the stability index = 1, a criterion for absolute stability.

The condition for singing is that the stability index = 0, i.e., that

$$G_I = \frac{1}{\left(\frac{1 - a}{1 + a}\right)\left(\frac{1 - b}{1 + b}\right)}. \tag{4-37}$$

The circuit will not sing provided the magnitude of G_I is slightly less than that given by Equation 4–37. A sample calculation of this type circuit is given in Table 4–1.

Usually, a network is terminated in impedances such that the stability index falls between the extremes of 0 and 1. The margin against singing may be found by

$$\text{Singing margin} = 20 \log \left| G_I = \left(\frac{1 - a}{1 + a}\right)\left(\frac{1 - b}{1 + b}\right)\right|. \tag{4-38}$$

4-4 NETWORK ANALYSIS

The preceding material on the basic network laws and their applications provides the tools for network analysis. Some extensions of these tools and some added sophistication in mathematical manipulations make the analysis job applicable to complex network configurations.

Table 4–1. Computations for Guaranteed Stability

θ_1	θ_{aZ_1}	θ_a	$\frac{1-a}{1+a}$ "Q"	θ_2	θ_{aZ_2}	θ_b	$\frac{1-b}{1+b}$ "Q"	$\frac{1}{\left(\frac{1-a}{1+a}\right)} \times \frac{1}{\left(\frac{1-b}{1+b}\right)} = \lvert G_I \rvert$	Min. Total Loss (dB)
+50°	-90°	-140°	2.70	+30°	-90°	-120°	1.73	(0.370) (0.578) = 0.214	6.7
0°	-90°	-90°	1.00	0°	-90°	-90°	1.00	(1.0) (1.0) = 1.0	0.0
+10°	-90°	-100°	1.19	+10°	-90°	-100°	1.19	(0.841) (0.841) = 0.708	1.5
+20°	-90°	-110°	1.43	+20°	-90°	-110°	1.43	(0.700) (0.700) = 0.490	3.1
+30°	-90°	-120°	1.73	+30°	-90°	-120°	1.73	(0.578) (0.578) = 0.334	4.8
+40°	-90°	-130°	2.15	+40°	-90°	-130°	2.15	(0.465) (0.465) = 0.216	6.7
+50°	-90°	-140°	2.70	+50°	-90°	-140°	2.70	(0.370) (0.370) = 0.137	8.7
+60°	-90°	-150°	3.70	+60°	-90°	-150°	3.70	(0.270) (0.270) = 0.073	11.4
+70°	-90°	-160°	5.50	+70°	-90°	-160°	5.50	(0.182) (0.182) = 0.033	14.8
+80°	-90°	-170°	12.0	+80°	-90°	-170°	12.0	(0.083) (0.083) = 0.0069	21.5
+90°	-90°	-180°	∞	+90°	-90°	-180°	∞	(0) (0) = 0	∞

Notes:

θ_1 = angle of Z_1, the input image impedance.

θ_{aZ_1} = angle of aZ_1, assumed to be -90°.

θ_a = worst angle of a in $\left(\frac{1-a}{1+a}\right)$.

$\lvert a \rvert = 1$.

θ_2 = angle of Z_2, the output image impedance.

θ_{bZ_2} = angle of bZ_2, assumed to be -90°.

θ_b = worst angle of b in $\left(\frac{1-b}{1+b}\right)$.

$\lvert b \rvert = 1$.

Mesh Analysis

A circuit of any complexity may be analyzed by considering each mesh of the circuit independently and writing an equation for the voltage relations in each. To do this, of course, it is first necessary to define a mesh.

In Figure 4–2, for example, *nodes* are defined as those points at which individual series combinations of components are interconnected. The series combinations are called *branches* (each Z in Figure 4–2 may be made up of series–connected elements in any combination). A *mesh* may then be regarded as openings in the network schematic such as those that might be observed in a fish net. The boundary of a mesh, called the mesh contour, is made up of network branches. The least number of independent loops, or closed meshes, is one greater than the difference between the number of branches and the number of nodes. The number of independent loops determines the number of independent mesh equations needed to solve the network problem. Examination of Figure 4–2 shows that there are three branches and two nodes. Application of the rule indicates that there are two independent meshes.

In mesh analysis, the parameters of the branches are expressed as impedances, the independent variables are the voltages and the voltage drops in each of the branches of a mesh, and the dependent variables are the currents in each branch of a mesh. A simple example of mesh analysis of the circuit of Figure 4–2 may be performed by using the rule above regarding the number of independent meshes in the circuit and by applying Kirchoff's laws.

Thus, the equations

$$E - I_1 Z_A - (I_1 - I_2) Z_C = 0$$

and

$$E - I_1 Z_A - I_2 Z_B = 0$$

provide the two independent equations for the two independent meshes. If the values of E, Z_A, Z_B, and Z_C are known, the two mesh currents can be determined from these equations.

Nodal Analysis

In nodal analysis, the branch parameters are most conveniently expressed as admittances (recall that admittance is the reciprocal of impedance; i.e., $Y = 1/Z$), the dependent variables are the voltages at the individual nodes, and the independent variables are the currents entering and leaving each node. Simultaneous equations are written for node currents; their solution is the nodal analysis of the network. The number of independent nodal equations that may be written is one less than the number of nodes.

There is, of course, a direct correspondence between mesh and nodal equations. One approach is often found superior to the other, and the choice, while theoretically a matter of indifference, is often important from the points of view of convenience and flexibility in treating such things as parasitic circuit elements or active device parameters. The more complex circuits are usually more easily analyzed by the nodal approach.

Finding solutions to mesh or nodal circuit equations can become quite complex when all circuit elements are considered. Such equations, except in the simplest cases, are now usually solved by the use of a computer.

Determinants

In simple networks, brute–force solution of simultaneous equations by successive substitution of one equation in another is generally simple and straightforward. Only modest amounts of network complexity, however, make this approach to finding solutions prohibitive in the amount of time consumed. Further, the processes become so involved that the accuracy of the work must always be carefully checked to guard against error.

The coefficients of the dependent variables of the simultaneous equations may be arranged in rows and columns corresponding to the terms of the equations. If the resulting array is square (i.e., if it has the same number of rows and columns), solutions to the simultaneous equations can be found by the methods of determinants [1].

Matrix and Linear Vector Space Analyses

While it is often possible to determine significant but not complete characteristics of a network by means of voltage, current, and impedance measurements made at the terminals, such expressions may not completely define the network. These expressions, however, are often useful in relating the network performance to its interaction with other interconnected networks and in defining certain properties of the subject network. The coefficients of terms in the mathematical expressions derived from such measurements and observations may be arranged in matrix form; the matrix may or may not be square. Manipulation of the matrix expressions provides a convenient method of network analysis. This may be regarded as a "black box" approach to analysis, which ignores the internal structure of the network but permits specification of its external behavior. The application of the concepts of linear vector spaces to matrix analysis adds a significantly greater amount of power to network analysis [2].

4-5 TRANSFORMERS

Many types of transformer are used in telecommunications circuits. In most cases, the applications differ significantly from those applying to alternating current power distribution systems where the principal use is to step voltages up or down. In communications circuits, in addition to voltage transformation, transformers are used to match impedances, to split and combine transmission paths, to separate alternating and direct currents, and to provide dc isolation between circuits. Impedance matching and the splitting and combining of transmission paths are discussed in some detail because of their importance in transmission.

Impedance Matching

Unequal ratio transformers are used to match unequal impedances to permit maximum energy transfer. The currents through any two windings of such a transformer are inversely proportional to the number of turns in the two windings. The voltages across the two windings are directly proportional to the number of turns in the two windings. Thus,

$$\frac{V_S}{V_L} = \frac{N_1}{N_2}, \text{ or } V_L = \left(\frac{N_2}{N_1}\right) V_S \tag{4-39}$$

where N_1 and N_2 are the number of turns on the primary and secondary windings, respectively.

No power is dissipated in an ideal transformer, illustrated in Figure 4–23. In addition, the phase relation between the voltage and current on the two sides of the transformer is exactly the same. Therefore, the product of voltage V_S across the primary winding and current I_S through the primary winding is equal to the corresponding product for the secondary winding; that is,

$$V_S I_S = V_L I_L, \text{ or } V_S/V_L = I_L/I_S .$$

Then, substituting this value of V_S/V_L in Equation 4–39,

$$I_L = \left(\frac{N_1}{N_2}\right) I_S. \tag{4–40}$$

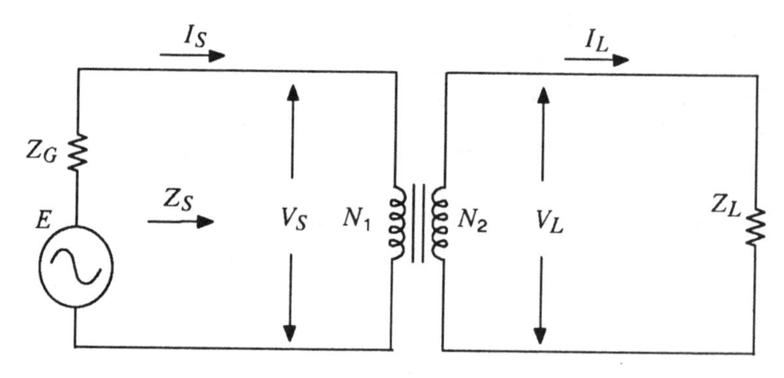

Figure 4–23. Transformer circuit.

From Ohm's law, $V_L = I_L Z_L$. Substituting the values of V_L and I_L from Equations 4–39 and 4–40,

$$V_S\left(\frac{N_2}{N_1}\right) = \left(\frac{I_S N_1}{N_2}\right) Z_L. \tag{4–41}$$

Then,

$$\frac{V_S}{I_S} = Z_S = \left(\frac{N_1}{N_2}\right)^2 Z_L \tag{4–42}$$

and

$$\frac{Z_S}{Z_L} = \left(\frac{N_1}{N_2}\right)^2. \tag{4-43}$$

The relationship shown in Equation 4-43 is used when a transformer is being designed to provide an impedance match, i.e., to provide a design in which $Z_S = Z_G$.

Commercial transformers approach the efficiency of an ideal transformer closely. Small losses are occasioned by currents induced in the core (eddy current losses), by flux in the core (hysteresis losses), and by current flowing in the copper windings.

Splitting and Combining

A common means of splitting and combining transmission paths is by use of a special type of transformer called the *hybrid coil*. In many applications, the resulting circuits are too complex to cover here; therefore, attention is confined to a relatively simple coil structure, a description of its operation, and considerations of some circuit configurations in which the hybrid is used.

The two most common uses for hybrids are illustrated by the block diagram of Figure 4-24(a). First, if Z_d or Z_c, or both, are signal sources, the hybrid circuit will split the energy from these resources between the two loads, Z_a and Z_b. The signal from Z_d (or Z_c) is balanced out in the hybrid so that it does not appear in Z_c (or Z_d). In the second common application, Z_a (or Z_b) may represent a signal source. Energy transmitted through the hybrid divides between Z_d and Z_c but does not appear in Z_b (or Z_a). The two applications, then, are to split and to combine the two signal paths.

In the circuit of Figure 4-24(b), assume that $Z_a = Z_b$, $Z_c = Z_d$, and the number of turns on each of the three windings of the hybrid transformer is the same. With these assumptions, assume a signal source in the branch containing Z_d as shown in Figure 4-24(c). The currents in the right-hand branches of the circuit divide equally between Z_a and Z_b and are cancelled in Z_c. Thus, there is no transmission from Z_d to Z_c. If the signal source were in series with Z_c, the currents would again divide equally between Z_a and Z_b. Their effects would cancel, however, due to the polarity of the magnetic fields in the center–tapped winding of the transformer. Hence, there would be no transmission from Z_c to Z_d.

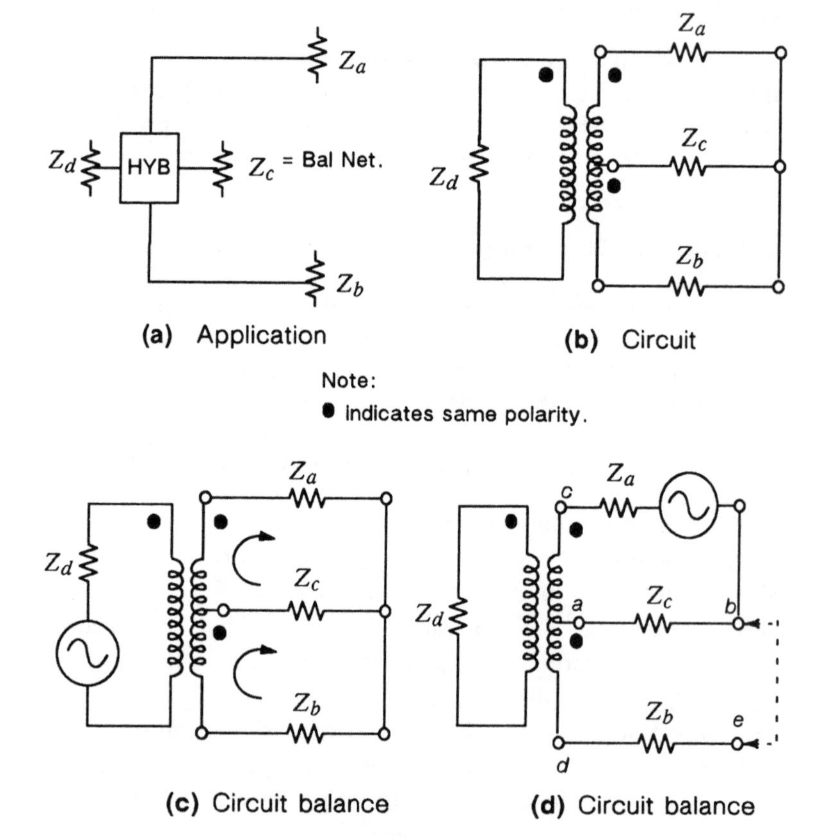

Figure 4–24. Hybrid circuit relationships.

Referring now to Figure 4–24(d), assume the signal source to be in series with Z_a. It is now convenient to imagine the circuit to be opened at the plug shown dashed between points b and e. Under these conditions and with $Z_c = Z_d$, the voltage induced between points a and d is exactly equal to the voltage drop in Z_c so that the voltage at b equals that at e. Then, since points b and e are the same potential, they may be connected without causing current to flow in Z_b. Thus, there is no transmission from Z_a to Z_b.

In each of the above examples, transmission from one impedance to another involves an equal division of energy to two other impedances; each load impedance dissipates half the power from the source, a loss of 3 dB. In addition, core and copper losses are typically about 0.5 dB. Thus, in designing or analyzing transmission circuits in which equal ratio hybrids* are used, 3.5–dB loss is usually assumed for the hybrid.

A common application of hybrid circuits is at the interface between two–wire and four–wire circuits. Such a circuit, illustrated in Figure 4–25, is known as a four–wire terminating set. Transmission is from the amplifier with output impedance Z_a to the two–wire trunk with impedance Z_d, and from the two–wire trunk to the amplifier with impedance Z_b. When transmitting from Z_d to Z_b, half of the energy is dissipated in Z_a (i.e., the output circuit of the amplifier at Z_a), but the signal is not transmitted through amplifier Z_a because of its one–way transmission characteristics. When transmitting from Z_a to Z_d, half the power is absorbed in Z_c. The important thing, however, is that the energy reaching Z_b is reduced to an absolute minimum. If this is not accomplished, the resulting signal will be fed back to the distant end of the four–wire circuit as an echo and, if large enough, may circulate around the four–wire loop resulting in instability or singing as discussed in Part 4–3.

The loss between Z_a and Z_b is known as hybrid balance. In carefully controlled laboratory circuits, a balance of 50 dB is easily achievable. However, in the application described, impedance Z_d represents any of a large number of two–wire trunks and subscriber loops that may be switched into the connection. The

* In some applications, unequal turns ratio hybrids are used. The design of such hybrids involves careful selection of impedances and turns ratios, a process too complex to be covered here.

impedances of these facilities often vary widely; therefore, impedance Z_c must be a compromise value chosen to limit the talker echo originating at the hybrid to an acceptable range.

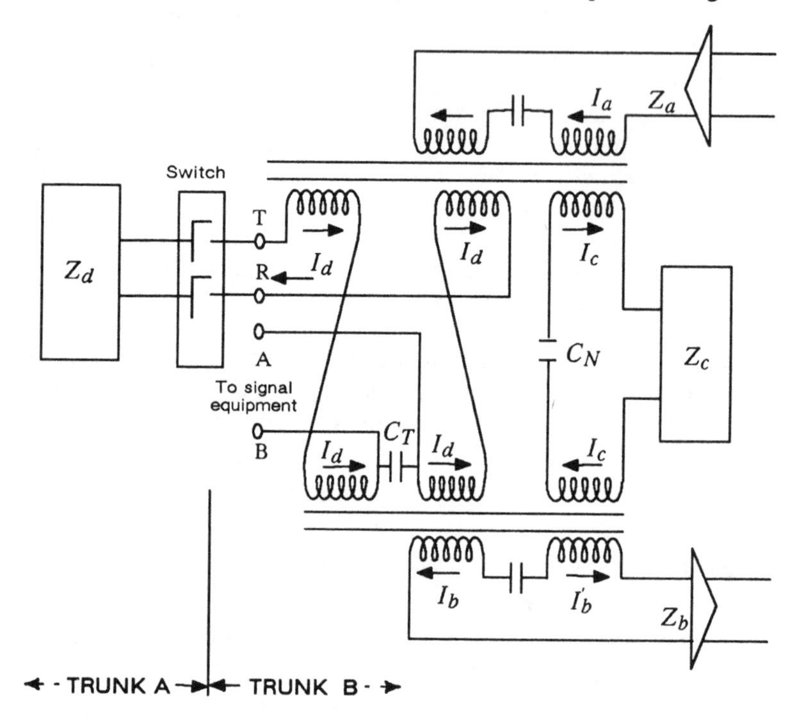

Figure 4–25. Hybrid application—four-wire terminating set.

Connection paths through an analog switching system are not identical. This variance is in addition to that previously mentioned as inherent to the impedance of a group of trunks and loops. In practice, these office wiring differences may be minimized using the following procedure. Replace Z_d with a compromise network chosen to approximate the median impedance of the connecting facilities (trunks and loops). A connection is then established through the switch from a four–wire terminating set to this compromise impedance and the elements of Z_c adjusted to maximize the hybrid balance. This condition is indicated when the loss (called return loss) is greatest between Z_a and Z_b, a situation that occurs when currents Z_c and Z_d are approximately equal in magnitude and phase. As mentioned, hybrid balances of

50 dB or better are attainable in a laboratory environment. In practice, however, return loss values may range from about 15 to 35 dB when hybrids are connected through a switch to two-wire trunks and as low as 5 to 6 dB when the termination is a loop connected to an off-hook telephone.

There are also active electronic hybrids. These have replaced transformer hybrids in some applications as, for example, the line circuits of digital switches.

4-6 RESONANT CIRCUITS

By an appropriate combination of resistors, inductors, and capacitors, circuits may be designed to resonate, i.e., to have extremely high or low loss at a selected frequency. Such circuits, which may be either series or parallel, are often designed as two-terminal networks, which are then used as components of a larger, more complicated four-terminal network. Resonance occurs when the inductive and capacitive components of reactance are equal, i.e., when

$$|X_L| = |X_C| = 2\pi f_r L = \frac{1}{2\pi f_r C} . \qquad (4\text{–}44)$$

The resonant frequency may be found by solving Equation 4–44 for f_r:

$$f_r = \frac{1}{2\pi\sqrt{LC}} . \qquad (4\text{–}45)$$

Selectivity, i.e., the difference in transmission between the resonant frequency and other frequencies, is determined by the amount of resistance in the circuit. Since the resistance is usually concentrated in the inductor, the objective is to have the ratio of the reactance of the inductor to its resistance as high as possible. This ratio is known as the quality factor, or Q, of the inductor and is expressed by

$$Q = \frac{X_L}{R} = \frac{2\pi f L}{R} . \qquad (4\text{–}46)$$

Series Resonance

In a resonant circuit having the inductance and capacitance in series, the circuit reactance is *zero* at the resonant frequency, where the inductive and capacitive reactances are equal as in Equation 4–44; the impedance has a minimum value at this frequency and is equal to the resistance of the circuit. If this resistance is small, a large current will flow as compared to the current flowing at other frequencies, as shown in Figure 4–26. One application of series resonance is in the use of a capacitor of proper value in series with a telephone receiver winding, repeating coil winding, or other inductance, where it is desired to increase the current at specific frequencies.

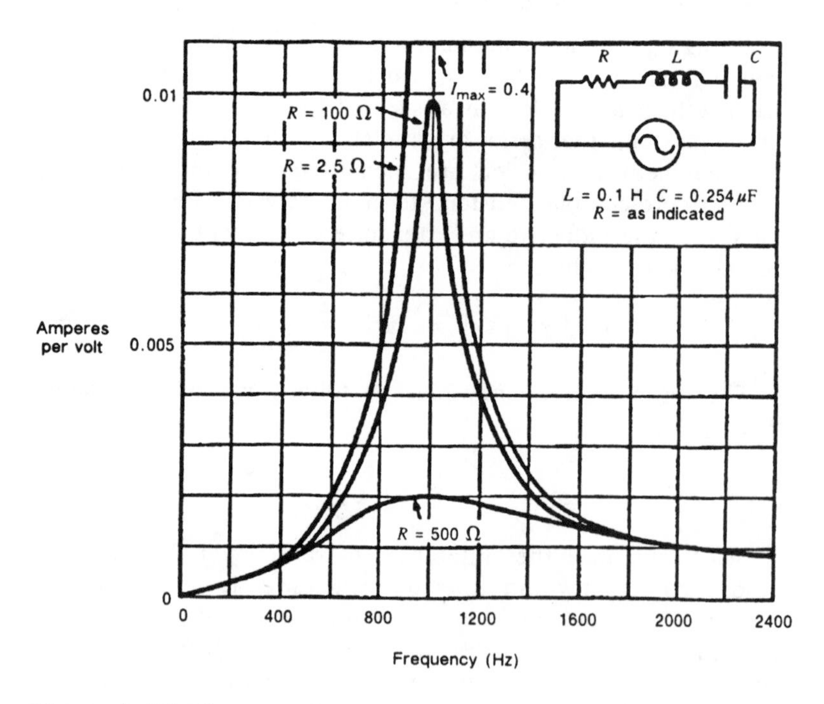

Figure 4–26. Curves of current values in a series resonant circuit.

Parallel Resonance

In a parallel (often called anti-resonant) circuit, which has the inductance and capacitance in parallel, the impedance of the combination is a *maximum* at the resonant frequency, where the inductance and capacitive reactances are equal. Since the impedance is a maximum, the current is a minimum at the resonant frequency. The selectivity of the circuit is decreased as the resistance is increased, reaching a point where the circuit essentially loses its resonant characteristics. This is shown in Figure 4–27. A parallel resonant circuit is often called a tank circuit since it acts as a storage reservoir for electric energy.

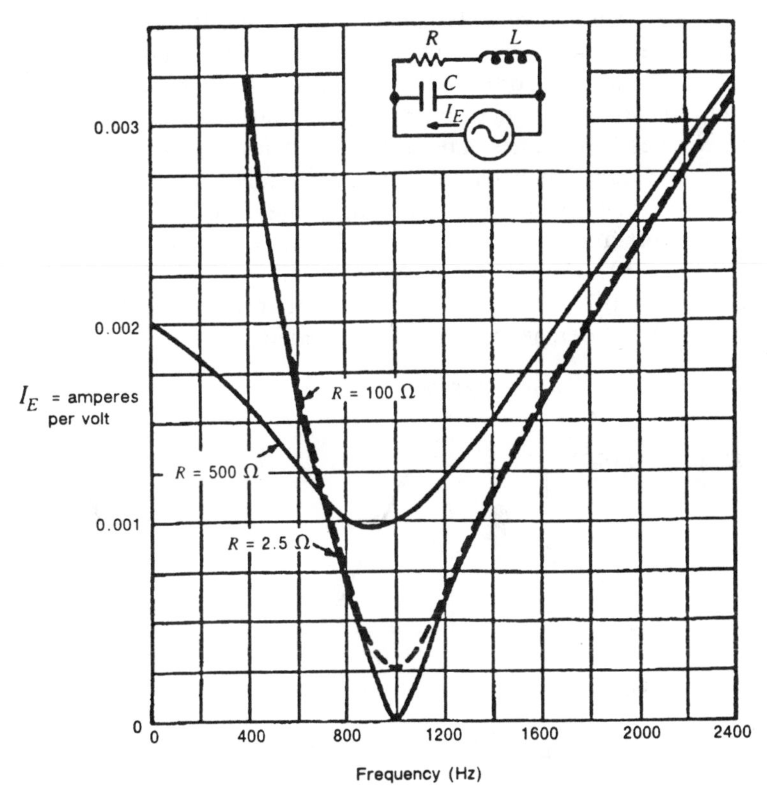

Figure 4–27. Curves of current values in a parallel resonant circuit.

These simplified descriptions of parallel and series resonance are applicable to low-frequency uses. However, as frequencies are increased, it must be recognized that in more complex networks such factors as skin effect in the inductor windings, inductive leakages, and interwinding capacitances may have an appreciable effect on the resonant frequency and the Q of a circuit. Further, as is discussed in Chapter 5, Part 2, the elements making up a tuned resonant circuit may take physical forms (e.g., short sections of transmission line) not usually associated with inductors and capacitors.

4-7 FILTERS

An electrical network designed to permit the flow of current at certain frequencies with little or no attenuation and to present high attenuation at other frequencies is called an electric wave filter. Networks designed to provide very little attenuation of low frequencies and to attenuate high frequencies, as shown in Figure 4–28, are called low–pass or sometimes high–cut filters. Those that pass high frequencies and attenuate frequencies below a given cutoff frequency are called high–pass or low–cut filters (see Figure 4–29). Cutoff frequency is usually defined as that frequency which is attenuated 3 dB more than the insertion loss of the signal frequencies the filter is designed to pass.

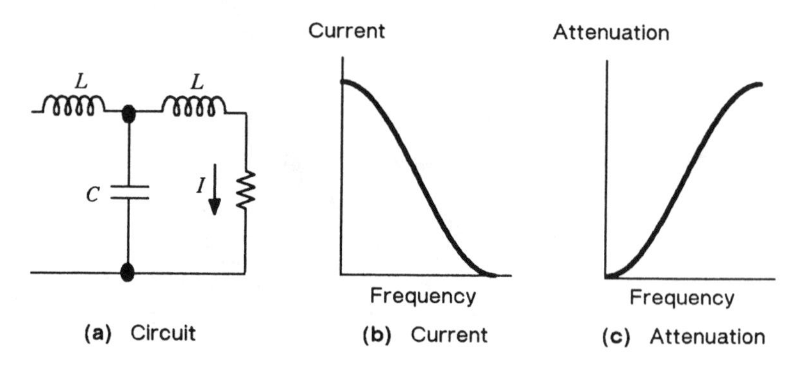

Figure 4–28. Low–pass filter.

Filters may also be designed to pass or greatly attenuate a given band of frequencies. Those which pass a band of frequencies between two cutoff frequencies are called bandpass filters. Those that attenuate a designated frequency spectrum are called band–stop, band–elimination, or band–rejection filters.

Filters may be composed of passive electrical elements, may include mechanical vibrating systems, or may make use of active devices such as operational amplifiers. Filters are used on both analog and digital circuits and may use either analog or digital techniques.

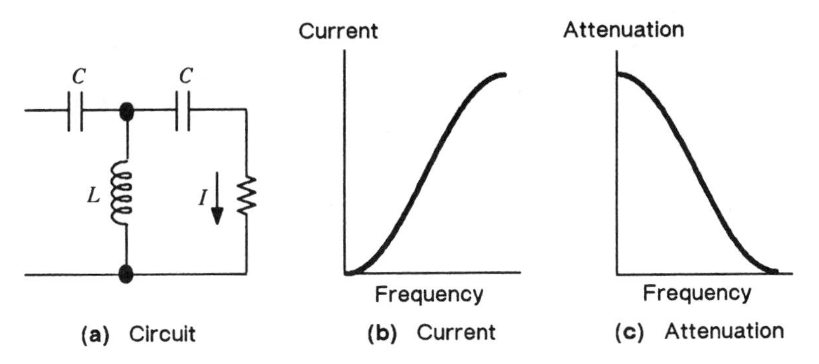

Figure 4–29. High-pass filter.

Passive Filters

Passive filters may be as simple as the high-pass section formed by the single resistor and capacitor in series shown in Figure 4–30. This device makes use of the high impedance of the capacitor at low frequencies and its decreasing impedance with increasing frequency. The more conventional low- and high-pass T-filter sections shown in Figures 4–28 and 4–29, respectively, also make use of this capacitor characteristic and the increase in inductor impedance, which takes place with increasing frequency.

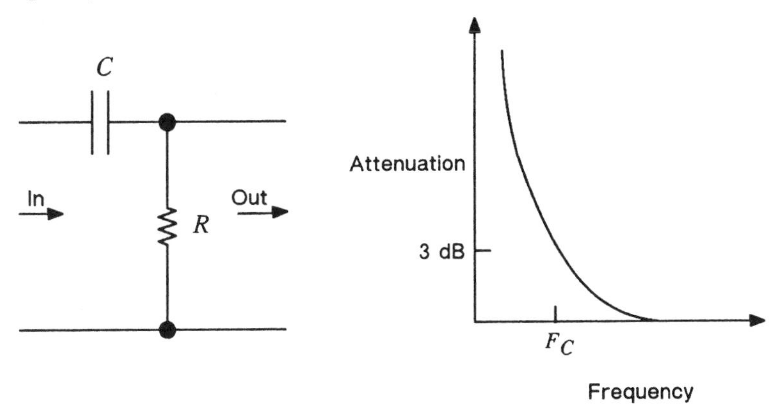

Figure 4–30. RC high-pass filter.

Passive filters designed as bandpass or band–elimination devices usually make use of the resonant properties of series and parallel LC circuits. Figure 4–31 shows the schematic and three attenuation/frequency response curves for a simple bandpass filter. Figure 4–31(a) makes the idealized assumption that all elements of the filter are purely reactive. Under these conditions, if the parallel circuit consisting of L_2 and C_2 is designed to reach resonance at the same frequency at which the two series arms are resonant, the passband will be both narrow and peaked.

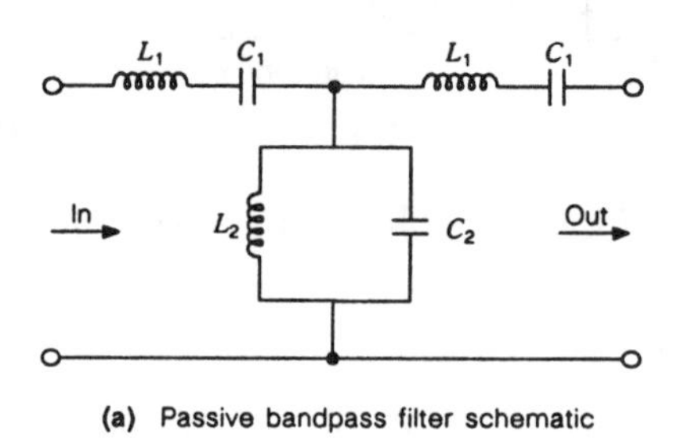

(a) Passive bandpass filter schematic

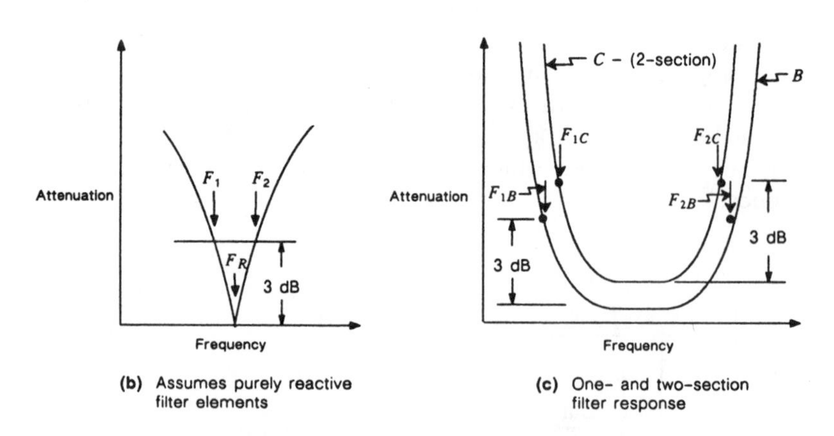

(b) Assumes purely reactive filter elements

(c) One- and two-section filter response

Figure 4–31. Passive bandpass filter.

Of course, the wire used to wind inductors has resistance, as does the wire used to connect the elements of the filter. Further, this resistance changes with frequency due to skin effect. Practical inductors have capacitance between turns of the windings. All capacitors have some leakage resistance. All of these physical properties increase filter insertion losses and reduce the sharpness of frequency cutoff.

Figure 4–31(b) reflects the response of a single-stage filter designed to attain a reasonably flat response between cutoff frequencies F_1 and F_2. Figure 4–31(c) shows the effect of cascading two stages of the filter. Note the increased sharpness of the cutoff, the narrowing of the bandpass and the insertion loss increase resulting from adding the second stage. The additional loss introduced by cascading sections is the penalty paid for obtaining the desired out-of-band frequency rejection characteristics. The bandpass narrowing can be overcome by choosing wider spacing between the cutoff frequencies F_1 and F_2 for the sections making up the cascaded filter.

Some of the more practical ways of obtaining a high ratio of reactance to resistance (i.e, high Q) in a resonant filter circuit are to use a resonant cavity or a mechanical vibrating system, such as the piezoelectric crystal. A crystal resonator is simply a plate of piezoelectric material, usually quartz, with a metal electrode on each side of the plate. If the two electrodes are connected to an alternating current source, the crystal will attempt to vibrate at the signal frequency. If the crystal has been cut or ground to a resonant frequency near that of the driving signal, it will be set into resonance by the signal.

Figure 4–32 shows the equivalent circuit for a single crystal. Note that R_S, L_S, and C_S form a series resonant circuit at a design frequency F_s and that this circuit, in parallel with capacitor C_p, will form a parallel resonant circuit at frequency F_p. Figure 4–33 shows the schematic for a full-lattice crystal network suitable for use as a bandpass filter. This type of filter has found wide application in analog carrier systems.

Simple LC filters still find some application for power supply filtering. Similarly, many devices still make use of passive filter structures such as elliptic-function designs, which are relatively easy to design and build. However, the advent of solid-state circuitry has made the use of active filters practical, and these devices have found high usage in modern telecommunications equipment.

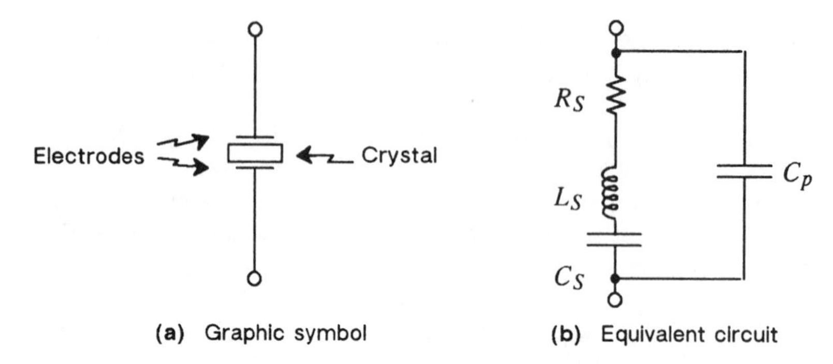

Figure 4–32. Quartz piezoelectric crystal.

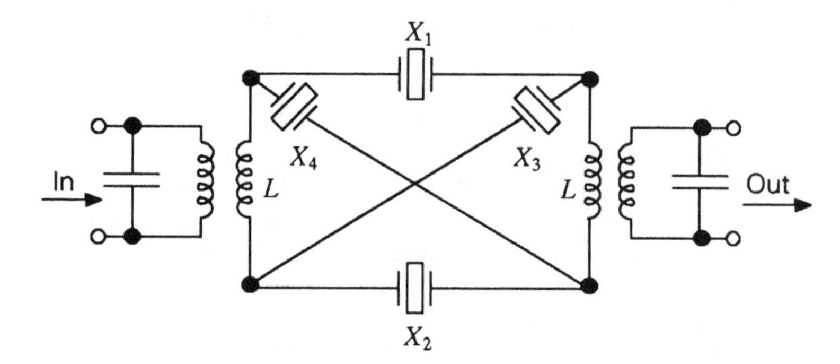

Figure 4–33. Full-lattice crystal filter.

Active Filters

Adding amplifiers with a feedback circuit to a filter design makes it possible to synthesize an RLC filter characteristic without using inductors. These inductorless filters are known as active filters.

Active filters can be used to replace any of the previously noted passive filter designs, including the classical variations whose describing functions were invented by Bessel, Butterworth, and Chebychev. In addition, filters with flat amplitude response, but with a tailored phase/frequency response, can be made. These filters are "all–pass" devices and are used as delay equalizers in telecommunications circuits. A filter having a constant phase shift and tailored amplitude/frequency response is also possible using active techniques.

Active filters have several definite advantages over passive filters. First, they may be made to have unity gain or better, thus offsetting the loss normally associated with the addition of filters to a circuit. Second, active techniques make it possible to obtain the equivalent of inductive reactances without using the heavy, bulky inductors usually required for a typical LC filter designed for low-frequency use. Third, the active circuit is compact, lightweight, and more economical to build and use than its passive counterpart; it can often be fabricated as an integrated circuit.

Figure 4–34 shows the response curves and typical circuit for a single-stage active low-pass filter using discrete components. The circuit makes use of an emitter follower, hence the overall gain of the stage will always be less than one. The amount of gain and the shape of the frequency response are determined by the amount of positive feedback, which in turn is controlled by changing the values of capacitors C_F and C_{IN}. Figure 4–35 shows a similar low-pass filter stage using an IC operational amplifier. Other designs based on switched-capacitor principles are in wide use.

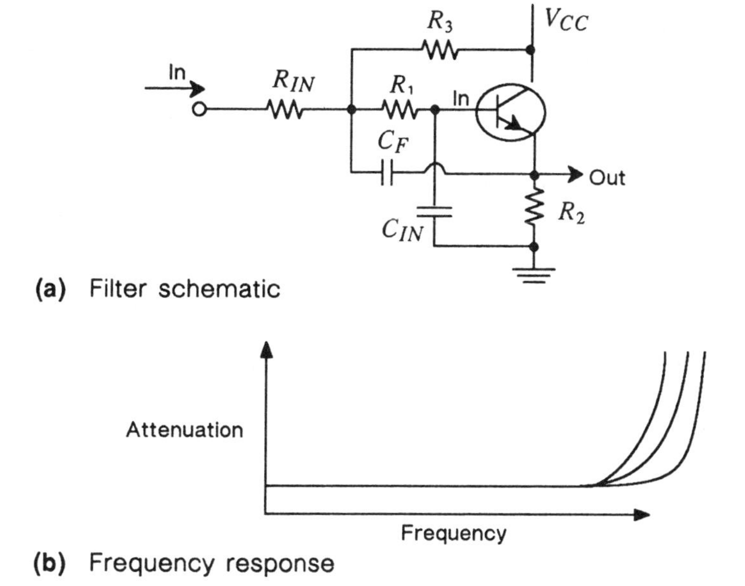

Figure 4–34. Active low-pass filter.

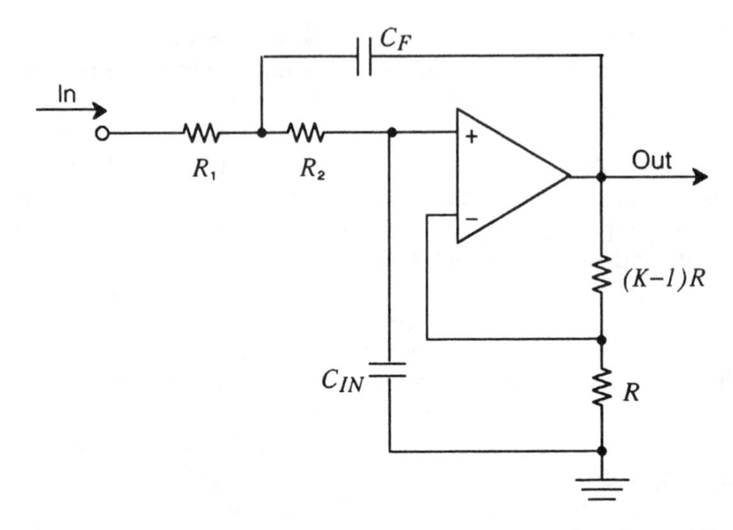

Figure 4-35. Active low-pass filter with operational amplifier.

Digital Filters

It has been recognized for many years that any bandlimited analog signal that is sampled at the Nyquist rate (i.e., twice the highest signal frequency) or higher can be reconstructed from its samples. It follows that filtering performed on a bandlimited analog signal by a linear analog filter can be duplicated by operating on the samples of that signal. Devices that perform such operations are called *sampled–data filters*. If the sample values are represented by code words, the filter is called a *digital filter*.

The output signal produced when an analog signal is passed through a linear filter is given by the product of the original signal and the impulse response of the filter. Similarly, the output of a sampled–data or digital filter is given by the following summation.

$$y(n) = \sum_{k=-\infty}^{\infty} x(k)h(n-k) \qquad (4\text{-}47)$$

where $x(n)$ and $y(n)$ are the filter input and output sample frequencies, respectively, and $h(n)$ is the impulse response of the digital filter, i.e., the output of the filter in response to an isolated input pulse of unit amplitude at time $n = 0$.

From Equation 4–47 it can be shown that a linear filter to be used on encoded digital words can be constructed from digital adders, multipliers, and memory devices.

Digital filters may be designed to give the same time–domain response as their analog counterparts or they may be designed to give similar frequency responses. While it is possible to design a digital filter that has exactly the same impulse response as a sampled version of a corresponding analog filter, it is not possible to make the two filter types have identical frequency responses because the sampling of the time–domain signals introduces spurious signals in the frequency domain. The process, which is called *aliasing*, is illustrated in Figure 4–36.

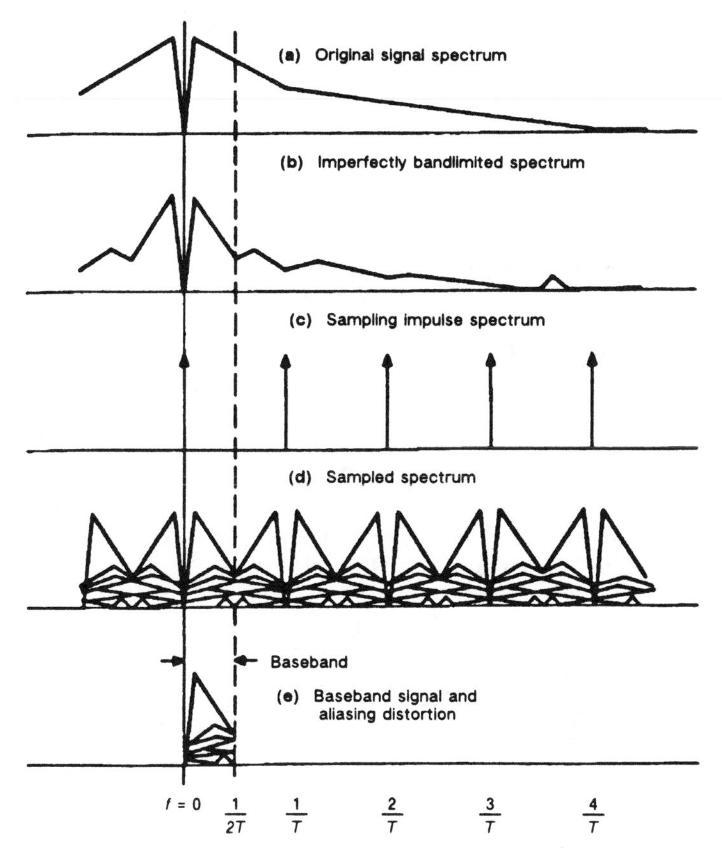

Figure 4–36. Sampling and aliasing.

Figure 4–36(a) shows a message signal that contains considerable energy above the half–sampling frequency and Figure 4–36(b) shows this signal after it has passed through a less–than–perfect filter. Sampling results in the convolution of Figure 4–36(c) with Figure 4–36(b), which in turn results in overlap between the desired baseband signal spectrum and the "tails" from the higher frequency sidebands as shown by Figure 4–36(d). As depicted in Figure 4–36(e), the frequency region from 0 to $1/(2T)$ contains the entire spectrum of Figure 4–36(b), with the higher frequency portions folded over at each boundary. This foldover, or aliasing, is an impairment that can be removed only by using a receiving filter having a bandwidth less than $1/(2T)$. This physical performance limitation makes it necessary to leave a guard band near the half–sampling rate to keep the aliasing distortion within specifications. As a result, with the usual 8–kHz sampling rate, the usable portion of the 4–kHz audio bandwidth is reduced to about 3.5 kHz.

References

1. Bode, H. W. *Network Analysis and Feedback Amplifier Design* (Princeton, NJ: D. Van Nostrand Company, Inc., 1945).

2. Huelsman, L. P. *Circuits, Matrices, and Linear Vector Spaces* (New York: McGraw–Hill Book Company, Inc., 1963).

Additional Reading

Johnson, W. C. *Transmission Lines and Networks* (New York: McGraw–Hill Book Company, Inc., 1950).

Kuo, F. F. *Network Analysis and Synthesis* (New York: John Wiley and Sons, Inc., 1962).

Reference Data for Engineers: Radio, Electronics, Computer, and Communications, Seventh Edition (Indianapolis, IN: Howard W. Sams and Company, Inc., 1985).

Chapter 5

Transmission Line Theory

Transmission line theory has its basis in electromagnetic wave theory from which have been derived relationships among such factors as impedance, impedance matching, loss, velocity of propagation, reflection, and transmission. This chapter reviews many of these important characteristics. Propagation of radio waves, sometimes considered under transmission line theory, is treated in Volume 2.

Chapter 4 deals with networks made up of resistors, capacitors, and inductors. Such components are called lumped constants. A transmission line is an electrical circuit whose constants are not lumped but are uniformly distributed over its length. With care, the theory of lumped constant networks can be applied to transmission lines, but lines exhibit additional characteristics that require consideration.

The detailed characterization of a given type of cable depends on the physical design of the cable. Wire gauge, type of insulation, twisting of the wire pairs, etc., have important effects on attenuation, phase shift, impedance, and other parameters. This chapter contains a general treatment of two–wire transmission lines and a brief introduction to coaxial cables, waveguides, and lightguides. Detailed characteristics of specific designs are found in Volume 2.

5–1 LUMPED CONSTANT EQUIVALENT CIRCUITS

It is convenient to approach the analysis of transmission line characteristics in terms of equivalent lumped constant networks. Since a uniform line appears the same electrically when viewed from either end, the equivalent circuit must be symmetrical. The conductors of an ideal simple transmission line, evenly spaced and extending over a considerable distance, have self–inductance, *L*, and resistance, *R*, which are series–connected elements that must be included in the lumped constant equivalent network. The insulation between the wires is never perfect; there is some leakage between them. The leakage resistance may be very large, as in a dry cable, or it may be fairly small, as in the case of a wet open–wire pair. In any event, the equivalent circuit must contain a conductance, *G*, in shunt between the line conductors. Also, any two conductors separated by an insulator have the properties of a capacitor; therefore, the circuit must have shunt capacitance, *C*.

These electrical components (i.e., resistance, inductance, conductance, and capacitance) are called *primary constants*. They are usually expressed in ohms, millihenries, micromhos, and microfarads per mile. Derived from these are the characteristic impedance and propagation constant; they are the *secondary constants,* both of which are functions of frequency. Although all primary and secondary constants vary with changes in temperature, they are usually expressed as constants at 68°F with correction factors for small changes in temperature. Concepts involving both primary and secondary constants are often used when characterizing transmission lines or equivalent circuits in connection with system performance analysis work.

In the discussion of networks in Chapter 4, Part 3, it is suggested that any circuit could be simulated by a T structure. It is not surprising, then, to find that a useful equivalent circuit for a transmission line is the T network shown in Figure 5–1. For convenience, series constants *R* and *L* can be lumped into impedance Z_A, and shunt constants *C* and *G* into impedance Z_C as in Figure 5–2.

The equivalent circuits in Figure 5–1 or 5–2 are poor approximations of a real transmission line because all of the distributed constants have been concentrated at one point. The approximation is improved by having two T sections in tandem and, in the ultimate, the best representation is an infinite number of tandem–connected T sections, each having the constants of an infinitely short section of the real line.

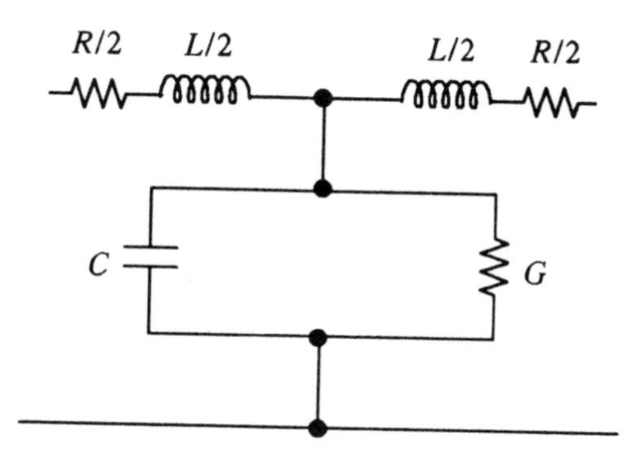

Figure 5-1. Primary constants of a section of uniform line.

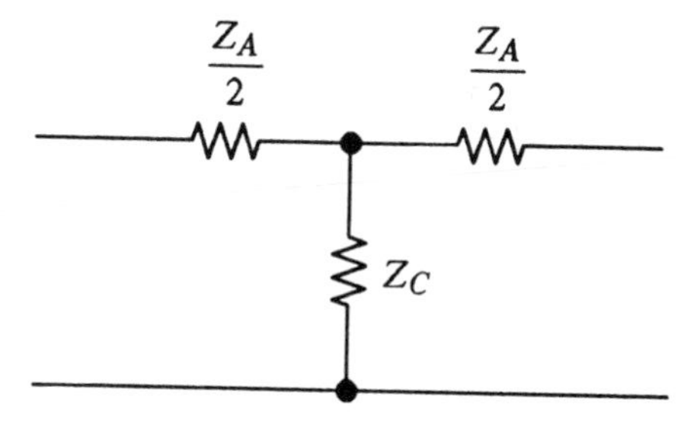

Figure 5-2. Equivalent network of a section of uniform line.

Characteristic Impedance

An example of the simulation of a very long uniform transmission line by an infinite number of identical, recurrent, and symmetrical T networks is shown in Figure 5-3. The impedance, Z_{SA}, at the input to Section A is equal to Z_0. Then, since the line is made up of an infinite number of T sections, if Section A is disconnected, and the input impedance Z_{SB} of the remaining line is measured, the result will be Z_0. Intuitively then, if Section A is terminated in a network of lumped elements equivalent to the impedance Z_{SB} (i.e., Z_0) the input impedance to Section A will equal Z_0 as it did when connected to the infinite line. The value Z_0 is usually called the *characteristic impedance* of the line, although some texts use the term *surge impedance* instead. It is

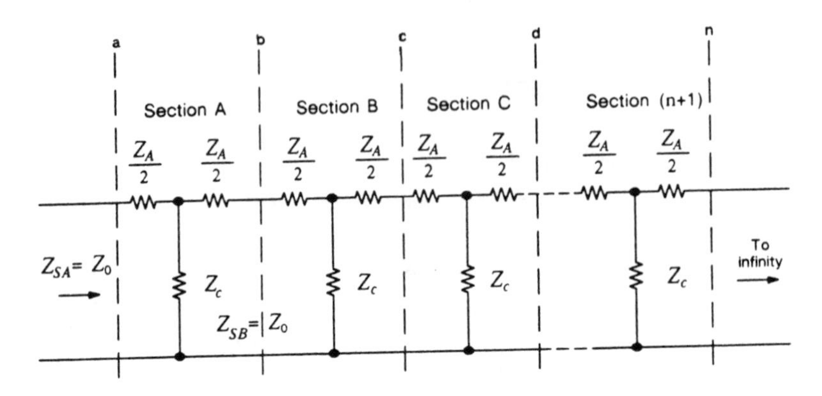

Figure 5-3. Uniform line simulated by an infinite number of identical networks.

related to the equivalent of the T structure of Figure 5–2 by the expression

$$Z_0 = \sqrt{\frac{Z_A{}^2}{4} + Z_A Z_C} \quad \text{ohms.} \tag{5-1}$$

Impedances Z_A and Z_C contain inductance and capacitance, respectively. Since the reactances of inductors and capacitors are functions of frequency, the characteristic impedance of a real or simulated transmission line is also a function of frequency. This property must be recognized when selecting a network that is to terminate a line in its characteristic impedance over a band of frequencies.

It is often more convenient to determine Z_0 by test than by computation. This can be done by measuring the impedance presented by the line when the far end is open–circuited (Z_{OC}) and when it is short–circuited (Z_{SC}). Then, it can be shown that

$$Z_0 = \sqrt{Z_{OC} Z_{SC}} \quad \text{ohms,} \tag{5-2}$$

an equation similar to those given for network image impedances, Equations 4–4 and 4–5.

To summarize, every transmission line has a characteristic impedance, Z_0. It is determined by the materials and physical arrangement used in constructing the line. For any given type of line, Z_0 is by definition independent of the line length but is a function of frequency. The input impedance to a line depends on line length and on the termination at the far end. As the length increases, the value of the input impedance approaches the characteristic impedance irrespective of the far–end termination.

The term characteristic impedance is properly applied only to uniform transmission lines. The corresponding property of a discrete constant network is called image impedance as discussed in Chapter 4, Part 3. If the network is symmetrical, it will have a single image impedance, which is analogous to the characteristic impedance of a uniform line.

Attenuation Factor

If a symmetrical T section is terminated in its image, or characteristic, impedance Z_0 and voltage E_1 is applied to the input terminals, current I_1 will flow at the input. In general the output voltage and current, E_2 and I_2, will be less than E_1 and I_1. Let I_1/I_2 be designated by a; this is the *attenuation factor* for the T *section*. Also, let $\ln \left| I_1/I_2 \right| = \alpha$; this is known as the *attenuation constant* for the T *network*. Then,

$$\left|\frac{I_1}{I_2}\right| = \left|\frac{E_1}{E_2}\right| = a = e^{\alpha} . \qquad (5\text{–}3)$$

If a number of identical symmetrical T sections of image impedance Z_0 are connected in tandem and terminated in Z_0 as shown in Figure 5–4, each T section is terminated in Z_0, and the ratio of its input current to its output current is a. From Figure 5–4, then

$$\left|\frac{I_1}{I_2}\right| = \left|\frac{I_2}{I_3}\right| = \left|\frac{I_3}{I_4}\right| = \left|\frac{I_4}{I_5}\right| = a. \qquad (5\text{–}4)$$

To find the ratio of the input current to the output current for the series, multiply the terms of Equation 5–4. This gives

$$\left|\frac{I_1}{I_5}\right| = \left|\frac{I_1}{I_2}\right| \cdot \left|\frac{I_2}{I_3}\right| \cdot \left|\frac{I_3}{I_4}\right| \cdot \left|\frac{I_4}{I_5}\right| = a^4$$

or, for n sections terminated in Z_0,

$$\frac{I_1}{I_{n+1}} = a^n = e^{na}, \qquad (5\text{–}5)$$

where $a^n = e^{na}$ is the attenuation factor for the tandem–connected T networks.

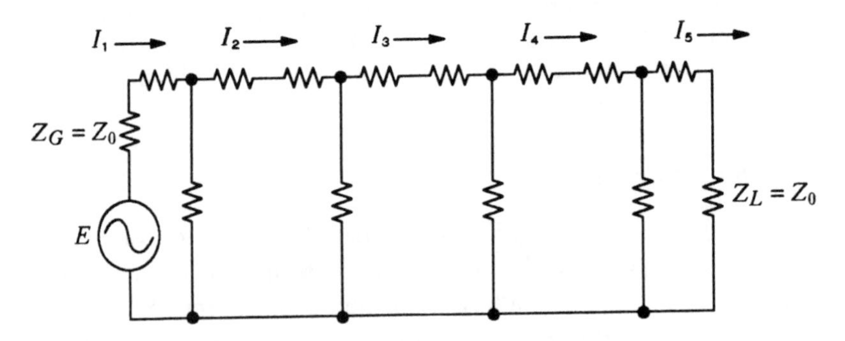

Figure 5–4. Line composed of identical T sections.

Propagation Constant

The ratio of input to output current, I_1/I_2, in a symmetric T section terminated in its characteristic impedance, Z_0, is generally a complex number that may be expressed in a variety of forms to indicate a change of both magnitude and phase. For the arrangement shown in Figure 5–5 where the network is terminated in Z_0 at both the input and the output, the current ratio may be computed as follows:

$$V = I_1\left[\frac{Z_C(Z_A/2+Z_0)}{Z_C+Z_A/2+Z_0}\right] = I_2[Z_A/2+Z_0] \qquad (5\text{–}6a)$$

$$\frac{I_1}{I_2} = \frac{Z_C+Z_A/2+Z_0}{Z_C} = 1 + \frac{Z_A}{2Z_C} + \frac{Z_0}{Z_C} \qquad (5\text{–}6b)$$

where

$$\frac{Z_0}{Z_C} = \sqrt{\frac{Z_A^2}{4} + Z_A Z_C}$$

$$= \frac{1}{Z_C} \sqrt{Z_C^2 \left(\frac{Z_A^2}{4Z_C^2} + \frac{Z_A}{Z_C} \right)}$$

$$= \sqrt{\left(\frac{Z_A}{2Z_C} \right)^2 + \frac{Z_A}{Z_C}} \,.$$

Substituting Z_0 from Equation 5–1,

$$\frac{I_1}{I_2} = 1 + \frac{Z_A}{2Z_C} + \left(\sqrt{\frac{Z_A}{Z_C} + \left(\frac{Z_A}{2Z_C} \right)^2} \right) \tag{5–6c}$$

or, in more general terms,

$$\frac{I_1}{I_2} = e^{\gamma} = e^{(\alpha + j\beta)} \tag{5–6d}$$

where γ is a complex number called the *propagation constant*, or *complex attenuation constant*. Its real and imaginary parts are defined by

$$\gamma = \alpha + j\beta, \tag{5–7}$$

where α is the *attenuation constant,* which represents a change in magnitude, and β is the *wavelength constant*, or *phase constant*, which represents a change in phase. For actual transmission lines with distributed parameters, both constants are usually expressed in terms of units of distance; α is usually expressed as nepers per mile or dB per mile, and β is usually expressed as radians per mile or degrees per mile.

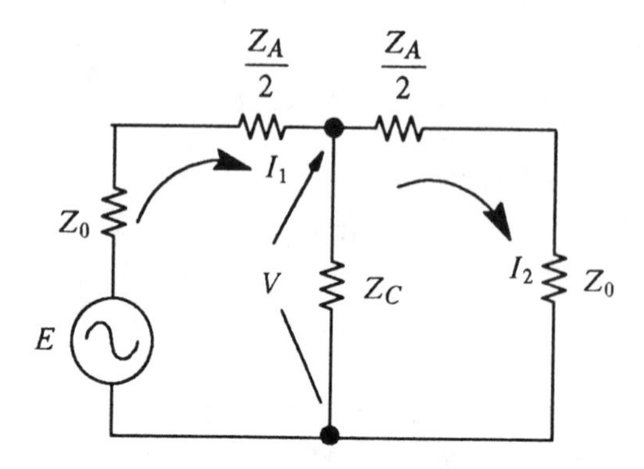

Figure 5-5. Equivalent network of a section of uniform line.

For the network of Figure 5–2, the value of γ may be computed from Equations 5–6a and 5–6b as

$$\gamma = \ln \left[1 + \frac{Z_A}{2Z_C} + \sqrt{\frac{Z_A}{Z_C} + \left(\frac{Z_A}{2Z_C} \right)^2} \right]. \tag{5–8}$$

If the bracketed expression in Equation 5–8 is written in polar form as $A \angle \beta$, where A is the absolute value and β the angle, then

$$\gamma = \ln Ae^{j\beta} = \ln A + j\beta. \tag{5–9}$$

Since the real and imaginary parts of this equation must equal the corresponding parts of Equation 5–7,

$$\alpha = \ln A = \ln \left| \frac{I_1}{I_2} \right| \tag{5–10a}$$

$$= \ln \left| \frac{Z_A/2 + Z_C + Z_0}{Z_C} \right| \text{ nepers/T section}$$

and

$$\beta = arg \frac{Z_A/2 + Z_C + Z_0}{Z_C} \text{ radians/T section} \tag{5–10b}$$

where *arg* stands for *argument* or *angle of*.

Since the attenuation constant can also be expressed in decibels per T section, Equation 5–10a may be written

$$\alpha = \ln A \text{ nepers/section} = 8.686 \ln A \text{ dB/section}$$
$$= 20 \log A \text{ dB/section.}$$

This relationship must not be used indiscriminately; it applies only for an infinite line or one terminated in Z_0.

5–2 LINE WITH DISTRIBUTED PARAMETERS

Characterization of transmission lines involves the same electrical parameters (primary and secondary constants) used in characterizing lumped constant networks. In real lines, however, these constants are not lumped, but rather are distributed uniformly along the line. This part discusses the relationships between primary and secondary constants as well as between these parameters and other important transmission line characteristics including attenuation factor, velocity of propagation, reflections and reflection loss, standing wave ratio, impedance matching, insertion loss, and return loss.

Characteristic Impedance

A single T section may represent a line having distributed elements but at one frequency only. In order to construct an artificial line of lumped elements to simulate a real line, assuming simple T sections are employed, it is necessary to construct many sections of lumped constants to simulate very short sections of line. As the number of sections is increased and the elemental length of line is reduced, the artificial line approaches the actual line in its characteristics over a wide band of frequencies.

Consider an elemental length of line, Δl. Let Z be the impedance per unit length along the line and Y be the admittance per unit length across the line. The T section of Figure 5–6 represents the equivalent network of a line having length Δl.

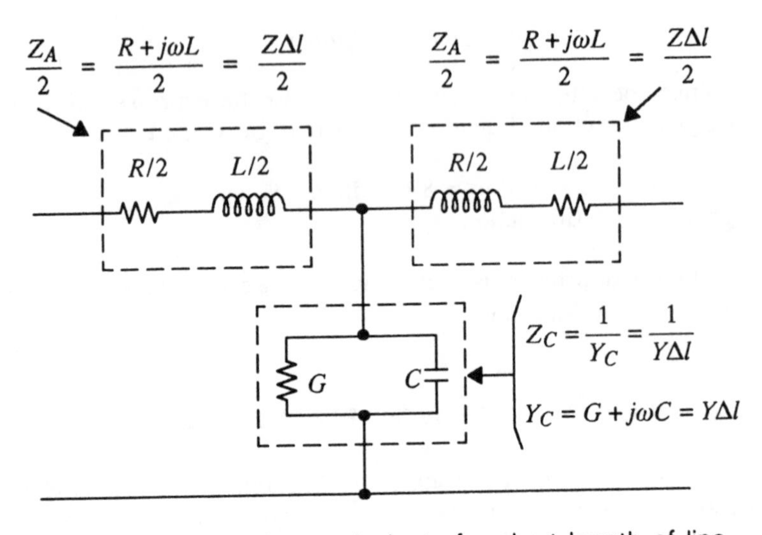

Figure 5–6. T–section equivalent of a short length of line.

The value of Z_A and Z_C of Figure 5–6 may be written

$$Z_A = Z\Delta l$$

and

$$Z_C = \frac{1}{Y\Delta l}\,.$$

Substituting these values in Equation 5–1 gives the characteristic impedance for a lumped constant network as

$$Z_0 = \sqrt{\frac{(Z\Delta l)^2}{4} + \frac{Z}{Y}} \text{ ohms.}$$

For a line with distributed parameters, let Δl approach zero; then, the characteristic impedance is

$$\lim_{\Delta l \to 0} Z_0 = \sqrt{\frac{Z}{Y}} \text{ ohms} \qquad (5\text{–}11\text{a})$$

or, in terms of primary constants R, L, G and C,

$$Z_0 = \sqrt{\frac{R + j\omega L}{G + j\omega C}} \text{ ohms.} \qquad (5\text{–}11\text{b})$$

Propagation Constant

From the derivation of the expression for the propagation constant, Equation 5–8 may be rewritten

$$e^{\gamma} = 1 + \frac{Z_A}{2Z_C} + \sqrt{\frac{Z_A}{Z_C} + \left(\frac{Z_A}{2Z_C}\right)^2} .$$

By expanding the terms under the radical sign by the binominal theorem and rearranging terms, the propagation constant may be written

$$e^{\gamma} = 1 + \left(\frac{Z_A}{Z_C}\right)^{1/2} + \frac{Z_A}{2Z_C} + \cdots \qquad (5\text{–}12)$$

Also, expanding e^{γ} as a power series yields

$$e^{\gamma} = 1 + \gamma + \frac{\gamma^2}{2!} + \cdots \qquad (5\text{–}13)$$

The terms Z_A, Z_C, and γ all place Δl in the numerators of Equations 5–12 and 5–13. As Δl approaches zero, the higher terms of these expansions become insignificant. Combining the two equations and truncating the series yields

$$1 + \gamma + \frac{\gamma^2}{2!} = 1 + \left(\frac{Z_A}{Z_C}\right)^{1/2} + \frac{1}{2}\left(\frac{Z_A}{Z_C}\right) .$$

This equation may be solved algebraically to give

$$\gamma = \sqrt{\frac{Z_A}{Z_C}} = \sqrt{ZY}\ \Delta l$$

for the conditions of Figure 5–6 and for length Δl. For any length, l, made up of $l/\Delta l$ sections,

$$\gamma = \sqrt{ZY}\ \ l,$$

and for a unit length, l = 1, the propagation constant is

$$\gamma = \alpha + j\beta = \sqrt{ZY} = \sqrt{(R + j\omega L)(G + j\omega C)}\,, \qquad (5\text{–}14)$$

where α is in nepers per unit length and β is in radians per unit length. The values of Z and Y are found from the primary constants of the line.

Attenuation Factor

In Part 5–1, the attenuation factor for a number of identical, symmetrical T sections having lumped constants was defined for tandem connections of such sections. Here, where the transmission line is made up of distributed parameters, as the length of an elemental section Δl approaches zero, the attenuation constant becomes α nepers per unit length as discussed. Then, the analogous expression for the attenuation factor is given as

$$\text{Attenuation factor} = e^{\alpha l}. \qquad (5\text{–}15)$$

Velocity of Propagation

The phase shifts represented by the β term in Equation 5–6b express the angular difference in radians between the input and output signals; they imply a time delay between the input signal current (or voltage) and the output signal current (or voltage). This can be used to compute the velocity of propagation through the network or transmission line. The velocity is given by

$$v = \frac{\omega}{\beta}, \qquad (5\text{–}16)$$

typically expressed as miles per second.

Equation 5–16 shows that the velocity of propagation is a function of frequency, since $\omega = 2\pi f$. However, β is also frequency–dependent since it is made up of reactances derived from Z_A and Z_C of Equation 5–8. Thus, while a transmission line having either discrete or distributed elements tends to introduce delay distortion (because the velocity of propagation is different at different frequencies), it is theoretically possible to design a line having no delay distortion by designing β to be directly proportional to ω . This is discussed in Part 5–3.

Reflections

Only lines that are uniform and that are terminated in their characteristic impedance have so far been considered. As long as the signal is presented with the same impedance at all points in a connection, the only loss is attenuation.

If, however, one circuit with impedance Z_G is joined to a second circuit with impedance Z_L, an additional transmission loss is observed. While the signal is traveling in the first circuit, it has a voltage–to–current ratio $E_G/I_G = Z_G$. But before the signal can enter the second circuit, it must adjust to a new voltage–to–current ratio $E_L/I_L = Z_L$. In making this adjustment, a portion of the signal is reflected back toward the sending end of the connection.

It is not surprising that there should be a reflection at an abrupt change in the electrical characteristics of a line. There is a disturbance in any form of wave energy at a discontinuity in the transmission medium. For example, sound is reflected from a cliff; light is reflected by a mirror. These conditions are equivalent to a line terminated in either an open circuit or a short circuit; all of the energy in the incident wave is reflected. A less abrupt change in impedance will cause a partial reflection. For example, a landscape is mirrored in the surface of a pool of water because part of the light falling on the water is reflected. The bottom of the pool is also visible if the pool is not too deep, since part of the light falling on the water passes through the discontinuity of the air–water junction and illuminates the bottom. This type of partial reflection occurs when two circuits with different impedances are joined, as in Figure 5–7. The power, P, in the signal arriving at junction x is divided. A portion of the signal, P_L, is transmitted through the junction to the load Z_L. The remainder of the signal, P_r, is reflected and travels back towards the source.

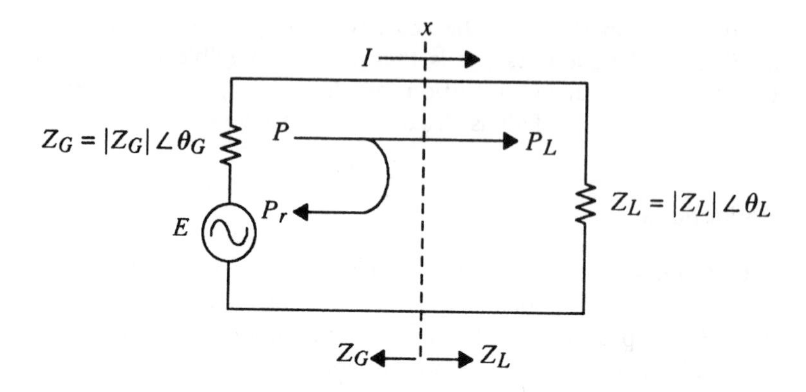

Figure 5-7. Reflection at an impedance discontinuity.

Reflection Loss. A concept frequently used to describe the effect of reflections is that of *reflection loss,* which is defined as the difference in dB between the power that is actually transferred from one circuit to the next and the power that would be transferred if the impedance of the second circuit were identical to that of the first.

Consider the circuit of Figure 5–7 where the impedances do not match. The current in the circuit is $I = E/(Z_G + Z_L)$. The power delivered to Z_L, the load, is

$$P_L = I^2 R_L = \frac{E^2 R_L}{(Z_G + Z_L)^2},$$

where R_L is the resistive component of the load impedance. If impedance Z_L matched Z_G, i.e., $Z_L = Z_G$, the current in the circuit would be $I_m = E/2Z_G$, and the power delivered to the load would be

$$P_m = I_m^2 R_G = \frac{E^2 R_G}{4Z_G^2}.$$

The ratio of the two values of power is

$$\frac{P_m}{P_L} = \frac{(Z_G + Z_L)^2}{4Z_G^2} \times \frac{R_G}{R_L}.$$

Thus, the reflection loss may be written

$$\text{Reflection loss} = 10 \log \frac{P_m}{P_L} = 20 \log \sqrt{\frac{P_m}{P_L}}$$

$$= 20 \log \left| \frac{Z_G + Z_L}{2Z_G} \sqrt{\frac{R_G}{R_L}} \right|$$

$$= 20 \log \left| \frac{Z_G + Z_L}{2\sqrt{Z_G Z_L}} \right| \sqrt{\left|\frac{R_G}{Z_G}\right| \cdot \left|\frac{Z_L}{R_L}\right|}$$

$$= 20 \log \left| \frac{Z_G + Z_L}{2\sqrt{Z_G Z_L}} \right| + 20 \log \sqrt{\frac{\cos \theta_G}{\cos \theta_L}} \text{ dB.} \tag{5-17}$$

Thus, the reflection loss has two components. The first is related to the inverse of the *reflection factor,* defined as K, where

$$K = \left| \frac{2\sqrt{Z_G Z_L}}{Z_G + Z_L} \right| . \tag{5-18}$$

The second depends on the angular relationships between the two mismatched circuits.

It is possible to have negative reflection loss, or reflection gain. This does not mean that power can be generated at an impedance discontinuity. It results from the choice of identical impedances as the reference condition for zero reflection loss. This is not the condition for maximum power transfer, as pointed out in Chapter 4, Part 3.

Standing Wave Ratio. A second concept that is useful in describing the effect of reflections is that of *standing wave ratio*. This concept may be approached from the point of view of a theoretically lossless transmission line.

Equation 5–11 gives the expression for the characteristic impedance of a line as

$$Z_0 = \sqrt{\frac{R + j\omega L}{G + j\omega C}} \quad \text{ohms.}$$

In this equation, the components that produce loss are the resistance and conductance, R and G. When these are negligible relative to $j\omega L$ and $j\omega C$, they may be ignored. This is often true at very high frequencies because of the ω terms in the expression for Z_0. Under these conditions, the characteristic impedance reduces to

$$Z_0 = \sqrt{L/C} \quad \text{ohms.} \tag{5-19}$$

Using Equation 5–14 and again ignoring the R and G terms, the propagation constant for the lossless line is

$$\gamma = j\omega\sqrt{LC}. \tag{5-20}$$

Thus, Equations 5–19 and 5–20 show that for a lossless line the characteristic impedance is a pure real number and the propagation constant is a pure imaginary number. When the propagation constant is expressed as $\gamma = \alpha + j\beta$, the value of attenuation then becomes $\alpha = 0$, and the value of the phase shift constant becomes $\beta = \omega\sqrt{LC}$.

The voltage at any point, x, on a lossless transmission line may be shown to be

$$V_x = V_1\, e^{j\beta x} + V_2\, e^{-j\beta x} \quad \text{volts.} \tag{5-21}$$

Here V_1 and V_2 represent the incident and reflected voltages, respectively, when the line is *not* terminated in its characteristic impedance, Z_0, and where x is the distance from the *load* to the point of measurement.

The velocity of propagation may be found for the lossless line by substituting the value $\beta = \omega\sqrt{LC}$ in Equation 5–16. Thus,

$$v = \frac{\omega}{\beta} = \frac{1}{\sqrt{LC}}.$$

For some structures (such as the coaxial line discussed later in this chapter), the velocity of propagation approaches 186,300 miles per second, which is the speed of light.

It can be seen from Equation 5–21 that voltage V_x is represented as the sum of two traveling waves, the incident and reflected voltage waves. The voltage may also be expressed in terms of trigonometric functions [1]:

$$V_x = V_L \left(\cos \beta x + j\frac{Z_0}{Z_L} \sin \beta x\right)$$

$$= V_L \cos \beta x + jI_L Z_0 \sin \beta x, \qquad (5\text{–}22)$$

where I_L, V_L, and Z_L are, respectively, the current and voltage at the load and the impedance of the load.

For a given frequency and value of x (distance from the load), Equation 5–22 can show $V_x = 0$, e.g., when $V_L = 0$ or when $\cos \beta x + j(Z_0/Z_L) \sin \beta x = 0$. This can occur, for example, when $\beta x = n\pi$ (n, an odd integer) and when Z_L is simultaneously infinite, an open–circuit termination. It can also occur when $Z_L = 0$ (short circuit) and $\beta x = n\pi$ (n, an even integer) simultaneously. These are two cases of special interest that produce standing waves, i.e., waves that do not propagate along the line but that vary between minimum and maximum values at all points except those at which $V_x = 0$.

In the more general case of a line having loss and not terminating in Z_0, standing waves are also produced but not necessarily with null points at which $V_x = 0$. The ratio of maximum to minimum voltages in the envelope along the line is known as the voltage standing wave ratio (VSWR). The VSWR may vary from one to infinity. When a line is terminated in its characteristic impedance, there is no reflected wave and, as a result, the maximum and minimum are the same value to give the ratio of unity. For the boundary conditions of $Z_L = 0$ or ∞ or for $\alpha = 0$, the minima of the envelope are nulls, the value of $V_x = 0$, and the VSWR is infinite.

For cases not involving the boundary conditions of $Z_L = 0$, $Z_L = \infty$, and $\alpha = 0$, the transmission phenomenon can often be analyzed to advantage in terms of a combination of traveling and standing waves; then, the VSWR is used as a measure of the required degree of impedance match. If it is convenient to perform analyses in terms of currents instead of voltages, analogous expressions are used.

Impedance Matching. An impedance discontinuity can often be eliminated by introducing a transformer as an impedance matching device at the junction. In Figure 5–8, the transmission line impedance to the left of x, Z_G, does not match the line impedance to the right of y, Z_L, but the line segments to the left and right do match Z_G and Z_L respectively. By connecting a transformer of turns ratio, $N_x/N_y = \sqrt{Z_G/Z_L}$, into the circuit between x and y, the line to the left of x is made to look into an impedance Z_G, while the line to the right of y looks back into Z_L. In practice, such a transformer will have a loss of a fraction of a dB, but the reflection loss is reduced to near zero. Typical examples of this technique are the transformer in a telephone station set and the input and output transformers in a repeater amplifier.

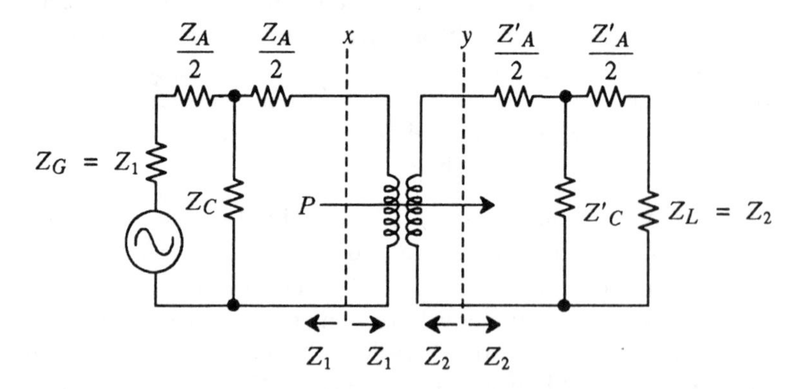

Figure 5–8. Unequal impedances matched by a transformer.

A network (or pad) with image impedances of Z_1 and Z_2 can perform the impedance matching function provided by the transformer in Figure 5–8. These impedance matching pads have limited application, however, because of the loss they introduce into a circuit. For example, the network shown in Figure 5–9 provides a minimum loss pad designed to match two unequal impedances, Z_1 and Z_2. Given $Z_1 = 600$ ohms and $Z_2 = 500$ ohms, R_1 and R_2 calculate to be 244.9 and 1224.7 ohms respectively; the pad loss will be 3.77 dB. If the impedance ratio to be matched is increased to 2 to 1, the pad loss will increase to 7.66 dB [2].

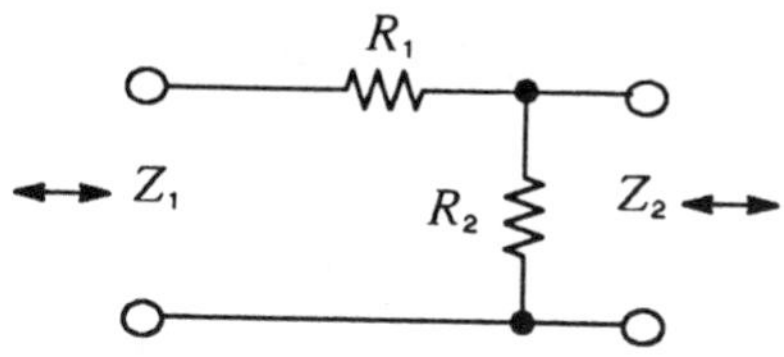

$$Z_1 > Z_2,\ R_1 = Z_1 \sqrt{1 - Z_2/Z_1},$$

$$R_2 = \frac{Z_2}{\sqrt{1 - Z_2/Z_1}}$$

$$\text{Pad loss} = 10 \log \frac{Z_1 (R_2 + Z_2)^2}{R_2^2 Z_2} \text{ dB.}$$

Figure 5–9. Minimum loss matching pad.

At very high frequencies, sections of transmission line with tapered separation may be used as impedance matching devices between two lines with different characteristic impedances or between a feed line and a load. As shown in Figure 5–10, the tapered section of line should be at least one–half wavelength and preferably a full wavelength or longer at the lowest frequency to be transmitted. In general, those frequencies for which the length of the tapered line section is a multiple of a half wavelength will be transmitted with the least reflection.

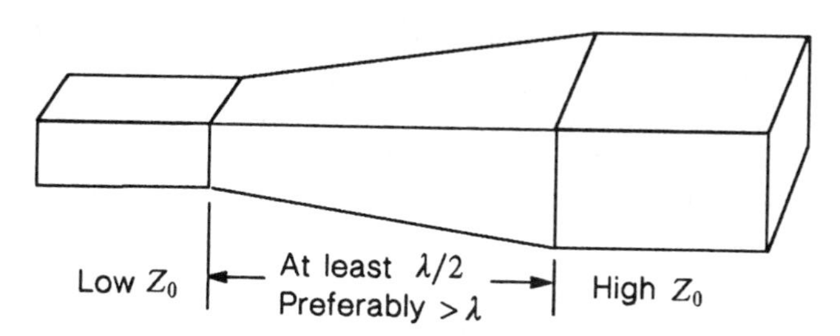

Figure 5–10. Gradual transition of impedance.

Impedance matching between a line and a load may also be accomplished through the use of a line segment whose length is an odd number of quarter wavelengths connected as shown by Figure 5–11. In the usual case for this arrangement, the line segment will be one–quarter wavelength and have very little loss. The characteristic impedance, Z_0, of the line segment may then be closely approximated by the equation,

$$Z_0 = \sqrt{Z_A Z_L}$$

where Z_A and Z_L are, respectively, the impedances of the antenna and the feed line, e.g., if Z_A = 70 ohms and Z_L = 300 ohms, a quarter wavelength section of line having $Z_0 \approx 145$ ohms will minimize the standing wave ratio on the feed line.

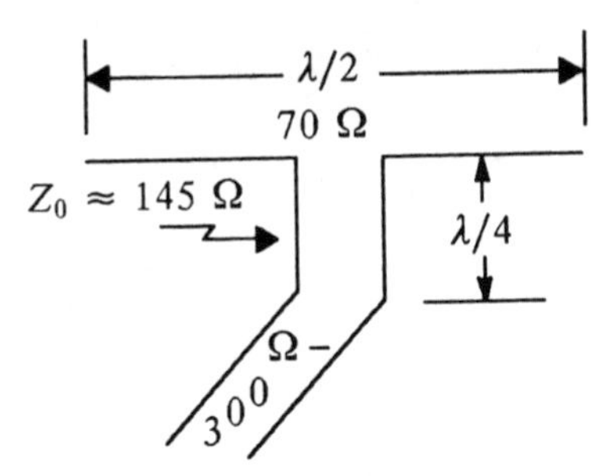

Figure 5–11. Quarter–wavelength section as impedance transition.

The properties of quarter– and half–wavelength sections of transmission line are further discussed later in this chapter under the topic heading, "Lines as Reactive Elements."

Insertion Loss. Insertion loss, first discussed in Chapter 3, Part 1, may be defined as the loss resulting from the insertion of a network between a source and a load. Further consideration of this factor is desirable because important contributions to insertion loss occur as a result of reflections due to impedance mismatches.

Consider the circuits of Figure 5–12. If the four–terminal network or transmission line is not present, as in Figure 5–12(a), the current supplied to Z_L is

$$I_1 = E/(Z_G + Z_L).$$

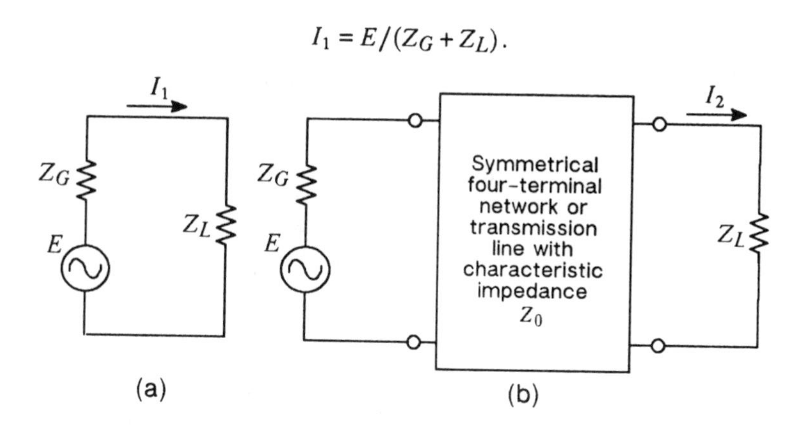

Figure 5–12. Circuits for insertion loss analysis.

When the network is inserted in the circuit, Figure 5–12(b), the current in Z_L may be shown to be

$$I_2 = \frac{2E\ Z_0 e^{-\gamma}}{(Z_0 + Z_G)(Z_0 + Z_L) - (Z_0 - Z_G)(Z_0 - Z_L)e^{-2\gamma}}.$$

Then,

$$\frac{I_1}{I_2} = \frac{(Z_0 + Z_G)(Z_0 + Z_L) - (Z_0 - Z_G)(Z_0 - Z_L)e^{-2\gamma}}{2(Z_G + Z_L)Z_0 e^{-\gamma}}$$

which, when factored, gives

$$\frac{I_1}{I_2} = \left[\frac{(Z_0 + Z_G)(Z_0 + Z_L)}{2(Z_G + Z_L)Z_0 e^{-\gamma}}\right]\left[1 - \frac{(Z_0 - Z_G)(Z_0 - Z_L)e^{-2\gamma}}{(Z_0 + Z_G)(Z_0 + Z_L)}\right]. \quad (5\text{–}23)$$

The second bracketed term is usually neglected. It results from interactions due to network termination impedances not being the proper image impedances for the network. If the network is a transmission line of any significant length, the term $e^{-2\gamma}$ makes the interaction effect negligible. If either Z_G or Z_L is equal to Z_0 or, in the case of a network, is the true image impedance, the interaction effect vanishes completely.

Thus, the insertion loss is usually written,

$$\text{Insertion loss} \approx 20 \log \left| \frac{(Z_0 + Z_G)(Z_0 + Z_L)}{2(Z_G + Z_L)Z_0 e^{-\gamma}} \right|.$$

The right side of this equation may be rearranged to give

$$\text{Insertion loss} \approx 20 \log \left| \frac{(Z_0 + Z_G)}{2\sqrt{Z_0 Z_G}} \right| + 20 \log \left| \frac{(Z_0 + Z_L)}{2\sqrt{Z_0 Z_L}} \right|$$

$$- 20 \log \left| \frac{Z_G + Z_L}{2\sqrt{Z_G Z_L}} \right| + 8.686\, \alpha \quad \text{dB}, \quad (5\text{-}24)$$

where α is the real component of the propagation constant γ.

The approximation to the insertion loss given in Equation 5–24 contains three reflection terms and the attenuation constant, α. The three reflection terms are seen to be related to the reflection factor defined in Equation 5–18. The first two of these increase the insertion loss as a result of the mismatch between either Z_0 and Z_G or Z_0 and Z_L. The third is a negative term representing the reflection loss due to the mismatch between Z_G and Z_L when the intermediate network is not present.

Terms corresponding to the term $20 \log \sqrt{\cos \theta_G / \cos \theta_L}$ of Equation 5–17 do not appear in the equation for insertion loss because, in the insertion loss definition, the ratio of currents (or, more precisely, powers) is taken with respect to the same impedance, Z_L. Thus, the power factor term reduces to unity.

If the transmission line is short or for any other reason has very low loss, the insertion loss expression of Equation 5–24 must be modified by the interaction (second bracketed) term in Equation 5–23. The evaluation of this equation is usually quite laborious. Computer programs, tables, and nomographs are available to simplify computations.

Return Loss. In studies of echo and in the design of two-wire repeatered circuits, the amount of signal reflected at a junction is of greater interest than the loss to the signal transmitted through the discontinuity. As discussed in Chapter 4, Part 3, the difference in dB between the incident current or voltage and the reflected current or voltage at an impedance discontinuity is called *return loss*. If the two impedances at a junction are matched, the return loss will be infinite, since there is no energy reflected. The greater the difference between impedances on each side of a junction, the lower the return loss. At the junction x in Figure 5–7,

$$\text{Return loss} = 20 \log \frac{1}{|\varrho|} = 20 \log \left| \frac{Z_G + Z_L}{Z_G - Z_L} \right| \text{ dB,} \tag{5–25}$$

where ϱ is the reflection coefficient.

In a telecommunications system, a serious source of low return loss occurs in the four–wire terminating set whose compromise balancing network cannot perfectly match the impedance of the office wiring or other connected two–wire circuits at all frequencies within the voice passband. The more serious problems occur at local switching systems (end offices) where tandem-connecting trunks are switched to a wide variety of loops.

Structural Return Loss. As previously discussed, any difference between the impedances of a transmission line and a connected load will reduce the amount of energy transferred to the load and cause a quantity of energy to be reflected back toward the source. In practice, impedance discontinuities may also exist at various points along a paired cable transmission line as the result of manufacturing variances (e.g., conductor gauge, insulation thickness, etc.), field handling and placing operations, splicing, load coil spacing, and gas plug installations. The total effect of these irregularities may be obtained by terminating the transmission line (i.e., cable pair) with its nominal impedance as shown by Figure 5–13 and measuring the return loss.

The line termination and hybrid balance networks consist of the nominal line impedance and an adjustable matching network. In the measuring process, the networks are alternately adjusted to minimize the effect of impedance mismatch at the terminals and obtain the maximum return loss of the cable pair. The structural return loss of a facility directly impacts voice frequency circuit design by defining the upper gain limits for a repeater connected to the facility. However, its effects have been minimized with the use of electronic hybrid and canceller circuitry.

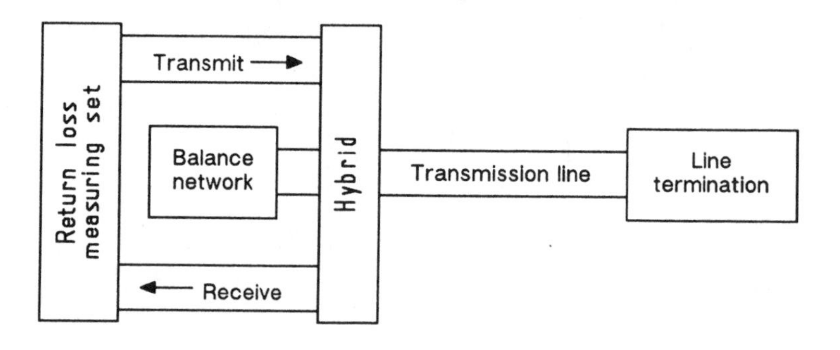

Figure 5–13. Structural return loss measurements.

Lines as Resonant Elements. The concept of voltage and current standing waves on transmission lines was introduced earlier in this chapter. Figure 5–14 shows these voltages and currents on a quarter–wavelength section of essentially lossless line with the load end open–circuited. Obviously there can be no current flowing at the load end of the line as the impedance is infinite so voltage will be at maximum. At the input end of the line, current is maximum and voltage minimum, which indicates an input impedance near zero.

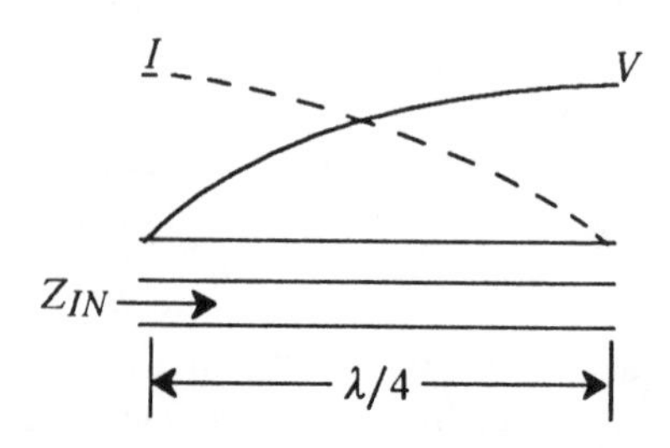

Figure 5–14. Quarter–wavelength open–circuited stub.

The frequency, f_R, necessary for this condition is called the resonant frequency. The resulting input impedance, Z_{IN}, is resistive. At frequencies above the resonant frequency, Z_{IN} becomes inductive and increases with increases in frequency. Conversely, at frequencies below f_R the input impedance is inductive and increases as frequencies are reduced. These characteristics are similar to those of a resonant circuit consisting of an inductor and capacitor in series.

As tabulated in Table 5–1, for frequencies above f_R the physical length, x, is greater than $\lambda/4$ and at frequencies less than f_R the line length is less than $\lambda/4$. Unlike the lumped element series circuit which has only one resonant frequency, the transmission line characteristics are repeated at each odd harmonic of f_R as illustrated in Figure 5–15(a). At even harmonics, the line length will become multiples of $\lambda/2$ and the input impedance will behave in the manner described below for a short–circuited quarter–wavelength line section.

Table 5–1. Transmission Characteristics of Resonant Quarter–Wavelength Lossless Line

Line Section	Equivalent Circuit	Line Input Impedance		
		$f = f_R$ $x = \lambda/4$	$f < f_R$ $x < \lambda/4$	$f > f_R$ $\lambda/4 < x < \lambda/2$
Short circuit 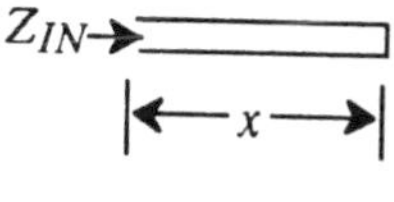 Z_{IN}→ ← x →	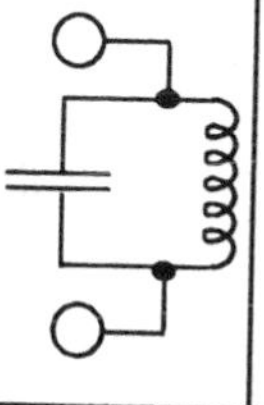	High resistive	Inductive	Capacitive
Open circuit 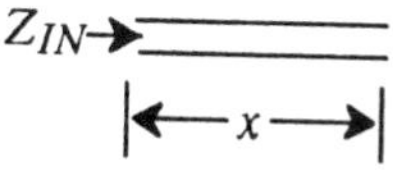 Z_{IN}→ ← x →	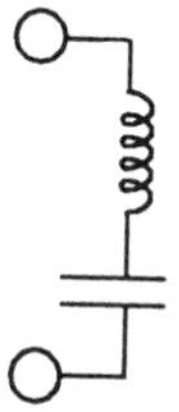	Low resistive	Capacitive	Inductive

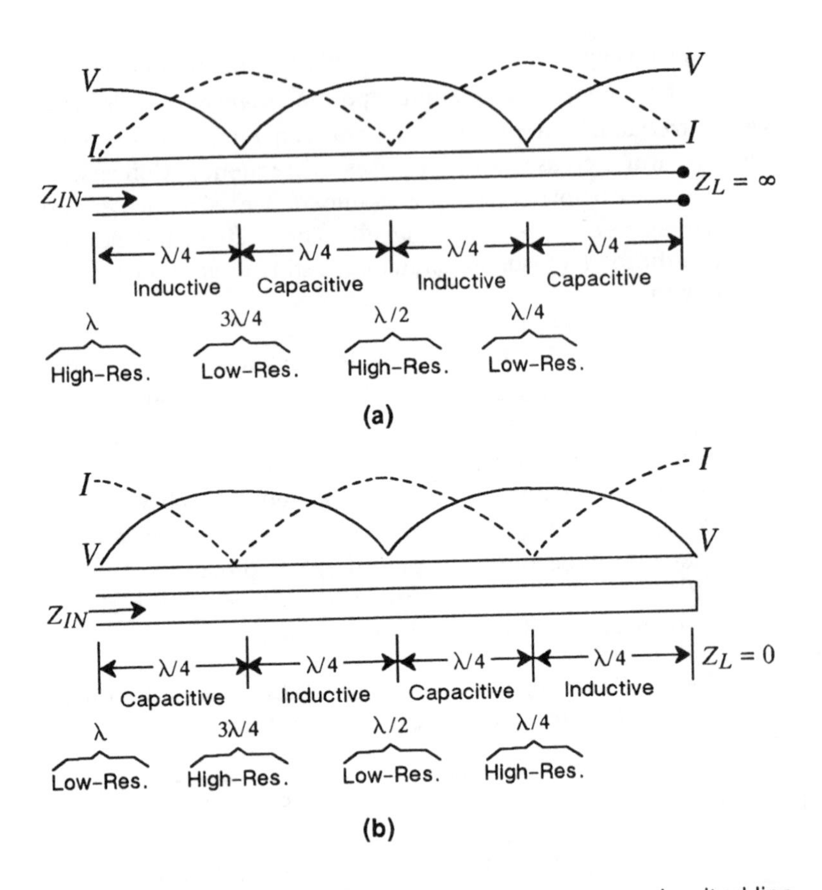

Figure 5–15. Input impedance along (a) an open-circuited line and (b) a short-circuited line.

Figure 5–16 shows a quarter-wavelength line segment with the load end short circuited. At the short, the current is high and the impedance and voltage approach zero. At resonant frequency the input current is low, voltage is high, and the impedance high and resistive. Then, as tabulated in Table 5–1, at frequencies greater than f_R, Z_{IN} becomes capacitive. At frequencies less than f_R, Z_{IN} is inductive. The input impedance at all intermediate frequencies is less than the impedance at resonance. These characteristics are similar to those of a parallel resonant circuit except that they will recur for each odd harmonic of f_R, as shown in Figure 5–15(b). For even harmonics of f_R, the input impedance will be as described above for an open–circuited quarter–wavelength of line.

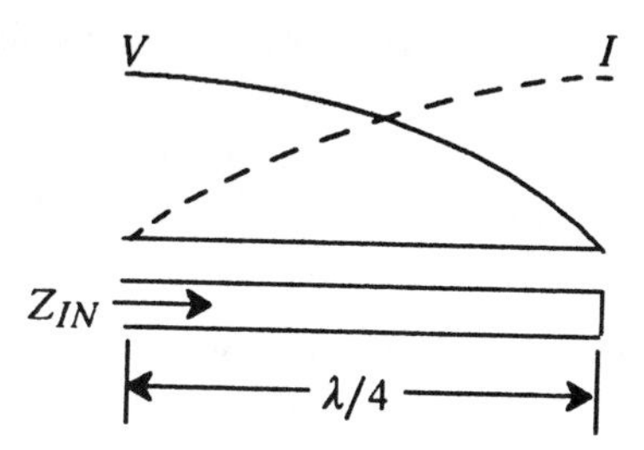

Figure 5–16. Quarter–wavelength short–circuited stub.

Figure 5–17 illustrates a design that takes advantage of the high input impedance of a shorted quarter–wavelength section of line. As shown, the shorted stub is used to support the inner conductor of a rigid coaxial line.

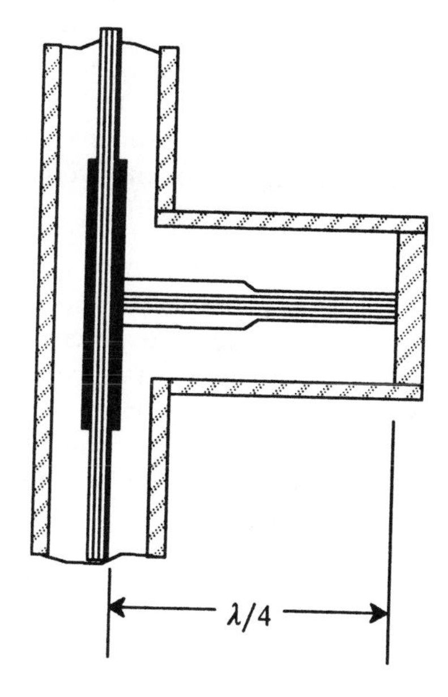

Figure 5–17. Shorted stub support for coaxial inner conductor.

The tunable shorted quarter–wavelength stub may be used to add selectivity at the input to a receiver. A half–wavelength section may be used as a slot filter to attenuate a narrow band of frequencies. This capability is useful when a nearby transmitter overloads a receiver, preventing reception from a weaker station operating on a frequency only slightly removed from that used by the more powerful station. Open–circuited quarter– and half–wavelength line segments may also be used in filters but they are usually more difficult to tune than the equivalent short–circuited sections.

Lines as Reactive Elements. At wavelengths measured in centimeters, ordinary inductors are unsatisfactory as reactive elements. However, as discussed in the previous section, short segments of either short–circuited or open–circuited lines may provide inductive or capacitive reactance. These line segments may be used as inductance or capacitance in a filter or in an impedance matching network.

Although at exact resonance the input impedance to either an open– or short–circuited line is resistive, the impedance is almost purely reactive at intermediate frequencies. If the length of the line is less than a quarter wavelength, the reactance of an open–circuited line is capacitive (negative); reactance of a short–circuited line is inductive (positive). For line segments one–quarter to one–half wavelength long, an open–circuited line is inductive and a short–circuited line is capacitive.

The exact impedance and velocity of propagation of an open–circuited stub vary with the physical parameters of the conductors and the dielectric of the insulation. In practice, the values for these parameters are rarely precisely known, therefore laborious calculations of stub lengths to gain accuracy cannot be justified. With this in mind, the following equations will give good approximations for the reactance of stub lines *unless* the length is near a multiple of a quarter wavelength.

For an open–circuited line:

$$X = -Z_0 \cot \beta x = -Z_0 \cot \left(2\pi \frac{x}{\lambda}\right),$$

and for a short–circuited line:

$$X = Z_0 \tan \beta x = Z_0 \tan \left(2\pi \frac{x}{\lambda}\right)$$

where Z_0 is the line characteristic impedance, x is the line length, and λ is the wavelength based on the velocity of propagation on the line.

5-3 LOADED LINES

In the previous discussion of velocity of propagation, it was pointed out that transmission lines normally introduce delay distortion. It was further suggested that delay distortion can be theoretically eliminated by design. This section considers how such a design can be approached.

Analysis

To achieve the theoretically distortionless line, it is necessary that the series impedance, Z, and the shunt admittance, Y, have the same angle. This may be expressed

$$\frac{\omega L}{R} = \frac{\omega C}{G},$$

or

$$LG = RC. \tag{5-26}$$

The impedance and admittance may be written $Z = R + j\omega L$ and $Y = G + j\omega C$, respectively. Thus,

$$Y = \frac{RC}{L} + \frac{j\omega LG}{R} = \frac{G}{R}Z. \tag{5-27}$$

Substituting in Equation 5-14,

$$\gamma = \sqrt{ZY} = Z\sqrt{G/R} = Z\sqrt{C/L}. \tag{5-28}$$

Then,

$$\gamma = \alpha + j\beta = \sqrt{G/R}\,(R + j\omega L)$$

and

$$\alpha = \sqrt{RG} = R\sqrt{C/L}, \tag{5-29}$$

$$\beta = \omega L\sqrt{G/R} = \omega\sqrt{LC}, \tag{5-30}$$

$$v = \omega/\beta = 1/\sqrt{LC}. \tag{5-31}$$

Substituting Equations 5–26 and 5–27 in Equation 5–11a,

$$Z_0 = \sqrt{Z/Y} = \sqrt{R/G} = \sqrt{L/C} \text{ ohms.} \tag{5-32}$$

Thus, when $\omega L/R = \omega C/G$, the attenuation (α), the velocity (v), and the characteristic impedance (Z_0) are independent of frequency, and Z_0 is a pure resistance. Such a line, terminated in its characteristic impedance, will have no loss distortion or delay distortion. Note, however, that the attenuation and velocity both decrease, and the characteristic impedance increases.

Unfortunately, a transmission line having such optimum characteristics is not readily attainable in practice. In transmission lines made up of pairs in well–maintained cables, the value of the conductance, G, is very small. It is not desirable to increase it artificially because that increases the attenuation correspondingly, as indicated in Equation 5–29. The value of capacitance cannot be changed appreciably because of practical considerations of spacing between conductors. To approach the optimum condition where $LG = RC$ as indicated by Equation 5–26, it would be necessary either to increase the inductance (L) or to reduce the resistance (R). The latter is not economical (it may be accomplished by increasing the size of the wire), and so the only remaining alternative is to increase the inductance. This practice, known as *inductive loading*, is used widely to reduce the attenuation of voice–frequency cables.

Inductive Loading

In considering the effect of inductive loading, note that the configuration of the T section of Figure 5–5 is basically that of a low–pass filter as illustrated in Figure 4–28. The cutoff frequency of such a structure is the frequency below which there is very little attenuation (ideally none) and above which the attenuation increases very rapidly. For the structure of Figure 5–5 in which $G = 0$, this frequency is

$$f_C = \frac{1}{\pi\sqrt{LC}} \text{ Hz.} \tag{5–33}$$

When applied to lines loaded with discrete lumped elements, the value of L is the load coil inductance. In theory, the inductive component (primary constant) of the medium for the length of the load section should be added; however, it is usually negligible. Similarly, the value of C to be used is the primary constant value of capacitance for the medium, multiplied by the length of the load section. In theory, the result should be increased by the incremental capacitance of the load coil, but this also is usually negligible.

As previously mentioned, series inductance may be added to reduce the attenuation in a cable pair. Below the cutoff frequency, f_C, the attenuation is reduced as indicated by Equation 5–29. Unfortunately, line inductance cannot be increased by loading without increasing resistance by virtue of the wire used in the load coils. Because of the resistance increase and other frequency–dependent limitations in the application of inductive loading, the attenuation is, in practice, more nearly $a = (R\sqrt{C/L})/2$ than the value of Equation 5–29. Above the cutoff frequency, the attenuation increases rapidly. The effect of an increase in L is illustrated by Figure 5–18.

Loading Methods. Loading may be accomplished by either of two practical methods. The first, called continuous loading, involves winding magnetic material (e.g., permalloy tape) around the copper conductors. This method is expensive and has been employed in only a few cases on submarine cable installations. The second method uses discrete, lumped inductive components, called *load coils,* introduced along the line at regular intervals.

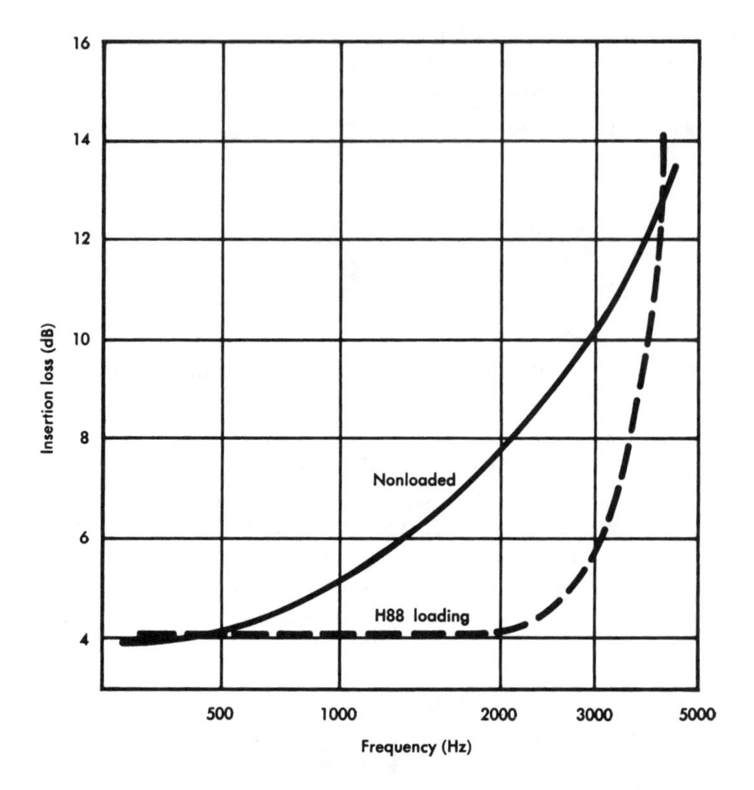

Figure 5–18. Insertion loss of 12,000 feet of 26-gauge cable measured between 900 ohms and 2 μF.

Below the cutoff frequency, the transmission line analysis for such an arrangement is reasonably accurate on the assumption that such lumped elements are uniformly distributed along the line. This assumption does not hold near or above the cutoff frequency.

Load Coil Spacing. Load coils introduce impedance discontinuities into an otherwise smooth or uniform line. This effect is minimized by making the spacing between load coils short compared to the wavelength of the transmitted signal (i.e., the coil spacing must not appreciably exceed a quarter wavelength of the highest frequency to be transmitted), and by spacing the coils precisely along the line. Nonuniform coil spacing, the additional line resistance contributed by the coil, and coil inductance variations introduce transmission irregularities in the passband and result in a deterioration of the structural return loss of the medium. Manufacturing specifications cover the allowable coil resistance and inductance variations as well as control the range of cable pair parameters.

In practice, geographical and other considerations make it impossible to space load coils precisely. For example, it is usually impractical to place a coil in the middle of a busy street intersection or a flowing stream. Also, the distance between equipment office buildings is rarely an exact multiple of the standard load coil spacing. To overcome these problems various schemes have been devised for building out short load coil sections, simulating added cable length, to more nearly match the average load section of a line. Load sections adjacent to repeater or office terminals (called end sections) may also be adjusted to match the impedance of repeaters or other terminal equipment.

Design rules have been devised for deciding when it is appropriate to build out individual load sections so that the repeater section facility will meet minimum structural return loss values. These rules generally specify three limitations that, if met, assure satisfactory performance of the line. These rules are as follows:

(1) The deviation of the average coil spacing from the nominal or standard value shall not exceed a specified percentage of the nominal spacing, usually two percent.

(2) No individual load section length shall vary from the average spacing by more than a given percentage of the average section spacing—the usual allowance varies from two to five percent, with the specified value dependent on the service requirements.

(3) The deviation of each section length from the average length is determined without regard to sign. The average of these deviations must not exceed a specified percentage of the average section length. The usual allowance ranges between 0.5 and 1.2 percent and is determined by service requirements.

Load coil spacing specifications are most stringent for trunks and least stringent for subscriber loop service.

A number of loading arrangements have been used in the United States through the years, each with a designated nominal coil spacing and uniform value of inductance. Each of the standard coil spacings has been assigned a code letter that, when coupled with the load inductance value and cable wire gauge, gives a complete description of the loading system. For example, the cable of Figure 5–18 is designated as 26H88, indicating 26–gauge wire, 6000–foot spacing, and load coils having 88 millihenries of inductance. H88 is the most common loading system. A less popular arrangement is D66, which is used by telephone companies associated with the Rural Electrification Administration (REA) program for rural telephone development. D66 loading consists of 66 millihenry coils placed at 4500–foot intervals.

The principal uses of various types of loading and their effects on attenuation, phase shift, and line impedance are covered in Volume 2.

5-4 COAXIAL CABLE

Consideration of transmission lines thus far has been confined to lines made of two parallel wire conductors. However, a coaxial configuration as shown in Figure 5–19 offers advantages at high and very high frequencies. These include low loss relative to paired wire, freedom from interference from external sources, and minimal crosstalk susceptibility.

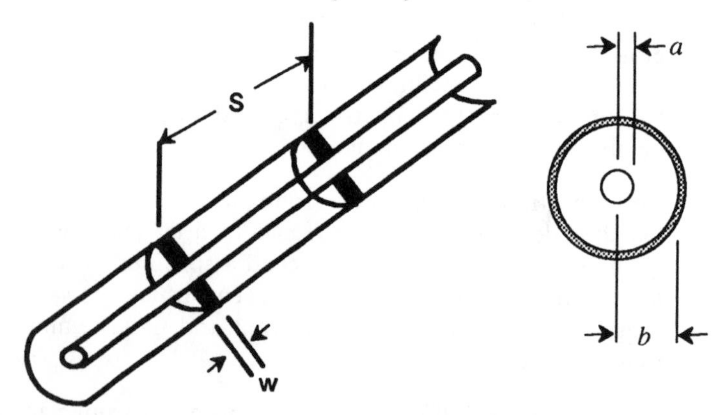

Figure 5–19. Coaxial cable.

The conductors are usually made wholly of copper or of copper center wire with an aluminum shield. The diameter of the outer conductor may range from about 0.25 inch to several inches and may be a solid tube or, in the case of small flexible cables, braided construction. The inner conductor may be a single solid wire, a stranded wire, or a hollow tube. In practice, the central wire may be held in place accurately by insulating material that may be in the form of a plastic or solid foam core, a spirally wrapped plastic string or, as shown by Figure 5–19, a series of beads or discs spaced along the inner conductor.

If the outside radius of the coaxial central conductor is designated as a and the internal radius of the outer conductor is b, the characteristic impedance, Z_0, at high frequencies, neglecting leakage, and assuming that air is the medium separating the conductors, is approximately

$$Z_0 = \sqrt{L/C} = 138 \log (b/a) \text{ ohms} \qquad (5\text{–}34a)$$

where L and C are the inductance, in henries per unit length, and capacitance, in farads per unit length, of the coaxial tube.

For disc–insulated lines, a more exact approximation that reflects the effect of the material separating the conductors is given by the following

$$Z_0 = \frac{138 \log(\frac{b}{a})}{\sqrt{1 + (k-1)(\frac{w}{s})}} \qquad (5\text{–}34b)$$

where k is the relative dielectric constant of the insulation between the conductors (e.g., 1 for air and 2.16 for polyethylene),
s is the insulation disc spacing,
w is the width of the inner conductor supporting discs.

When a uniform insulation is used, the denominator of Equation 5–34b reduces to k. Thus, where the insulation is air, $k = 1$ and the equation reverts to the form of Equation 5–34a.

The attenuation constant for a coaxial cable line at high frequencies varies with the square root of the frequency. For the case where both conductors are made of copper and leakage is near zero, the attenuation constant may be approximated by the following equation:

$$a = \frac{R}{2Z_0} = \left(\frac{K\sqrt{f}}{276}\right)\left(\frac{\sqrt{1 + (k-1)(\frac{w}{s})}}{b}\right)\left(\frac{1 + \frac{b}{a}}{\log \frac{b}{a}}\right)(6.336 \times 10^4)(8.686)\text{dB/mi} \qquad (5\text{–}35)$$

where a and b are the inner and outer radii, in inches,
f is frequency, in Hz,
k, w, and s are as shown above under Equation 5–34b,

and

$$K = \frac{\sqrt{\varrho}}{1006\,\pi} = 41.6 \times 10^{-9}$$

where ϱ is the resistivity of copper.

The last two terms of Equation 5–35 merely serve to convert the answer from nepers/inch to dB/mile. If these terms are omitted and conductor radii a and b are dimensioned in centimeters, the attenuation constant will be given in nepers/cm.

It can be shown from Equation 5–35 that coaxial attenuation will be minimized by making the b/a ratio equal to 3.6. This ratio is not critical, as the attenuation will not exceed the minimum more than ten percent for ratios between 2.3 and 6.0. As a matter of interest for the condition of air dielectric and a 3.6 ratio, Z_0 will be approximately 77 ohms.

The coaxial cable used for intercity communications service in the United States employs a copper tube having a 0.369–inch inside diameter and a copper center wire 0.1003 inch in diameter. The b/a ratio for this cable at 3.68 is only slightly above the optimum. From Equation 5–34a, Z_0 for this cable computes to be 78.1 ohms, while using Equation 5–34b to compensate for the polyethylene disc construction puts Z_0 at 73.6 ohms for the case where w = 0.1 inch and s = 1 inch. The loss figures at 1 MHz from Equation 5–35 with and without compensation for the disc construction are 3.95 and 3.72 dB/mi, respectively. The manufacturer lists Z_0 and α as 75.5 ohms and 3.85 dB/mi respectively for this cable at 20°C. Thus, it can be seen that Equations 5–34a and 5–35 provide valid performance approximations for a typical coaxial cable.

Theoretically, a coaxial tube could be used to transmit any frequency from direct current to infinity; however, in practice its applications are generally limited to frequencies between 1 MHz and 3 GHz. Although coaxial lines operated on some early inter-city routes in the United States at frequencies as low as 60 kHz, coaxial lines provide poor shielding at these low frequencies where the skin depth is comparable to the thickness of the outer conductor. For this and economic reasons, coaxial installations that prove viable at frequencies above 1 MHz usually lose out to paired cable at lower frequencies.

At very high frequencies (approaching 150 MHz), impedance irregularities introduced by cable placing operations and cable manufacturing processes, together with discontinuities inherent in disc center conductor construction, limit the capability of coaxial lines. These irregularities cause reflections that cause frequency–response problems on long lines. On a disc line, if the disc spacing approximates a half wavelength, reflections may accumulate to an excessive level.

In order that modes other than the ordinary transverse electromagnetic (TEM) mode will not be propagated on a coaxial line, the free space wavelength of the highest transmitted frequency must exceed the average circumference of the inner and outer conductors. Thus, as signal frequencies move higher the coaxial tube must get smaller. The result is an increase in signal attenuation and a decrease in the maximum power that may be transmitted by the cable. Fortunately, as will be discussed later in this chapter, the use of waveguides for antenna–to–transmitter connections and lightguide for longer haul low–power services becomes feasible before the physical limits of coaxial facilities are reached.

5–5 WAVEGUIDES

The purpose of transmission lines is to guide the energy of electromagnetic waves from place to place. Thus, any transmission line can be called a waveguide. However, in common usage the term is used to designate a line consisting of a hollow metal tube or duct with a rectangular, circular, or elliptical cross section.

Research on the electromagnetic energy propagation properties of waveguides began in the early 1930s. Their first practical applications were on centimeter wavelength radar equipment developed early in World War II. Research directed toward developing a high–capacity intercity telecommunications system using waveguide transmission line continued following the end of the war. Workable systems were designed. However, the considerable costs associated with manufacturing, placing, and maintaining the hollow tube free of transmission–affecting deformities made the systems uneconomical unless they could be operated near their very large facility capacity (200,000 voice circuits). No commercial intercity route was ever built.

The development of optical fiber cables has made foreseeable use of long–haul waveguide systems uneconomical. Waveguides, however, continue to fill the need for a relatively low–loss medium connecting high–powered radio transmitters to antennas where the frequency to be transmitted is between about 1 GHz and the upper end of the radio frequency spectrum.

Modes of Propagation

In free space, the electric and magnetic fields that make up an electromagnetic wave are always perpendicular to each other and to the direction of wave propagation at any instant in time. When a wave travels through a waveguide, however, the guide confines the forces of one of the fields, but not both, so that it has a component that is parallel to the direction of propagation. There is, in general, an infinite number of ways in which the energy fields can arrange themselves in a waveguide as long as there is no upper limit to the frequency to be transmitted. Each field configuration is called a propagation mode. Each mode generally falls into one of two fundamental classes.

In one class, the *transverse electric* (TE) mode, the electric field is everywhere perpendicular, or transverse, to the direction of propagation and the magnetic field has a longitudinal element. In the other class, it is the electric field that has a longitudinal element; this is called the *transverse magnetic* (TM) mode.

Each propagation mode is identified by its class and by a set of two subscripts (e.g., TE_{10}). For rectangular waveguides, the subscripts indicate the number of half–wave field variations occurring across the lateral dimensions of the duct. The first subscript is associated with the larger dimension and the second subscript with the short side. Figure 5–20 illustrates the use of the subscripts and shows the longitudinal field element in the TE_{11} and TM_{11} modes.

The rectangular TE_{10} mode shown in Figure 5–21 is the most widely used waveguide propagation mode. The arrows in the sketch indicate the electric field direction; their grouping and the graph indicate the relative field strength across the width of the guide. Note that there is no field strength variation parallel to the short side of the guide, hence the zero for the second subscript.

In circular waveguide notation, the first subscript indicates the number of full–wave electric field variations around the circumference of the guide. The second subscript denotes the number of radial variations. Figure 5–22 shows the use of this notation for the circular TE_{11} mode. This figure also shows an overlay of the TE_{10} rectangular mode on the TE_{11} circular mode, which illustrates the similarity between these two popular operating modes. This meshing of the two modes partially explains why they are frequently used when it is desired to interconnect circular and rectangular guide sections.

Subscript notation for elliptical waveguides is similar to that used for circular guides with the exception of an additional preceding subscript indicating whether the wave is "even" (*e*) or "odd" (*o*). Thus, an elliptical waveguide with the electric field parallel to the major axis of the ellipse would be described as operating in a ${}_oTE_{11}$ (odd) mode if there were one full–wave electric field variation around the circumference of the guide and one variation across the ellipse at its minor axis. An elliptical waveguide operating in an ${}_eTE_{11}$ (even) mode will have the electric field parallel to the minor axis of the ellipse, one full–wave variation around the circumference and a half–wave variation along the major axis of the ellipse.

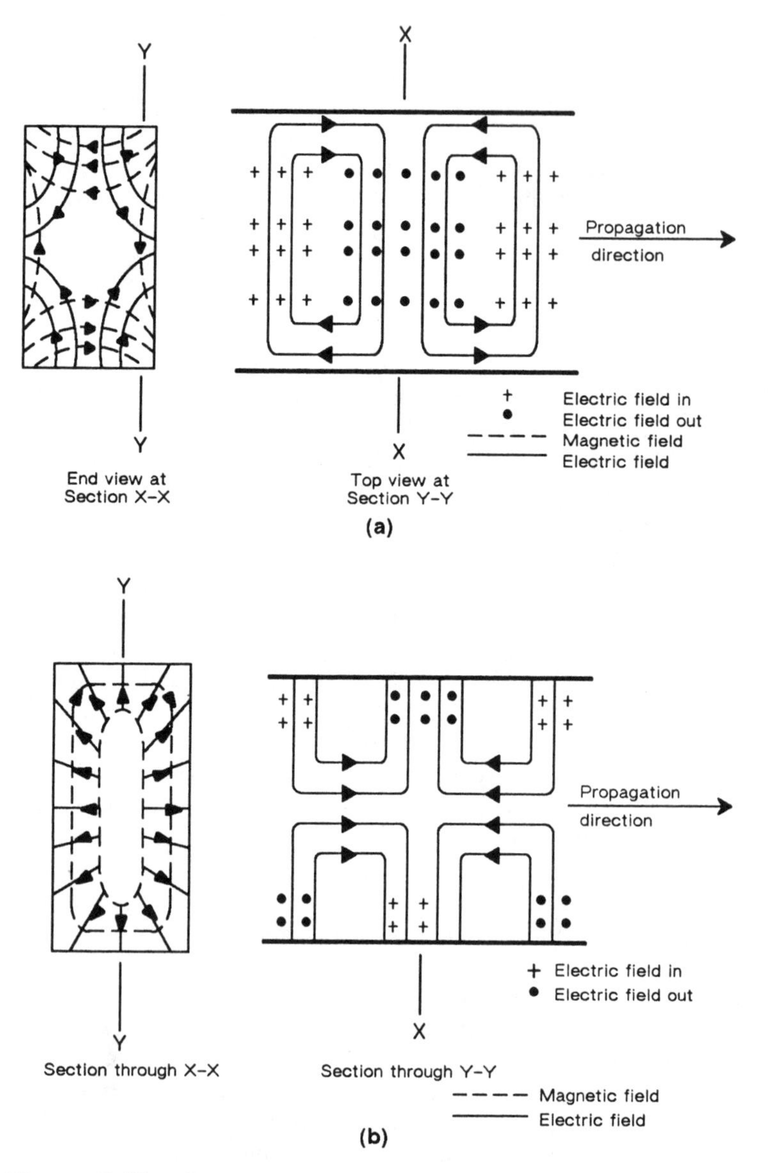

Figure 5–20. Rectangular waveguide in (a) the TE_{11} mode and (b) the TM_{11} mode.

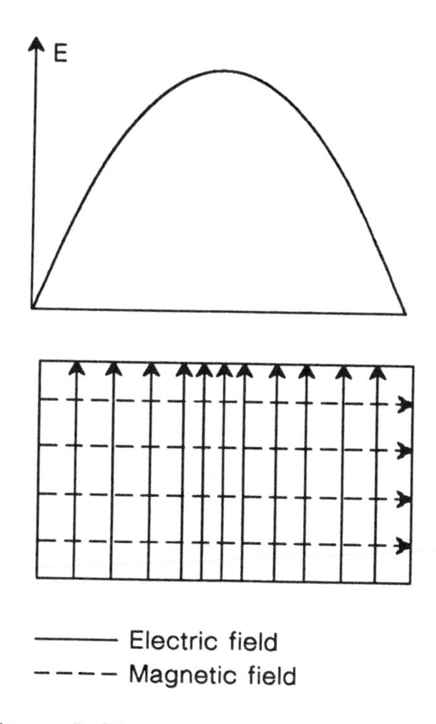

Figure 5-21. TE_{10} rectangular mode.

Waveguide Dimensions

Propagation of a specific mode is possible only when the wavelength of the applied energy is less than the "cutoff" wavelength of the waveguide. Longer wavelengths (i.e., lower frequencies) are not transmitted. A given guide, however, may transmit longer wavelengths (i.e., lower frequencies) in some other mode. The mode having the longest cutoff wavelength is called the *dominant* mode. The dominant mode is usually the preferred operational mode because, for a given frequency, it permits use of a physically smaller waveguide than other modes and minimizes transmission losses.

Many waveguides presently in use have a rectangular cross section with a two to one dimensional ratio and operate in the TE_{10} mode. The cutoff wavelength of such a design is twice the dimension of the longer side of the rectangle. Any other modes that might be generated in a waveguide with these dimensions will have cutoff wavelengths no more than half that of the dominant mode. Therefore, the theoretical band of frequencies that may be propagated in the dominant mode alone ranges from the frequency corresponding to the cutoff wavelength of that mode to twice the cutoff frequency of the mode; all other modes are suppressed.

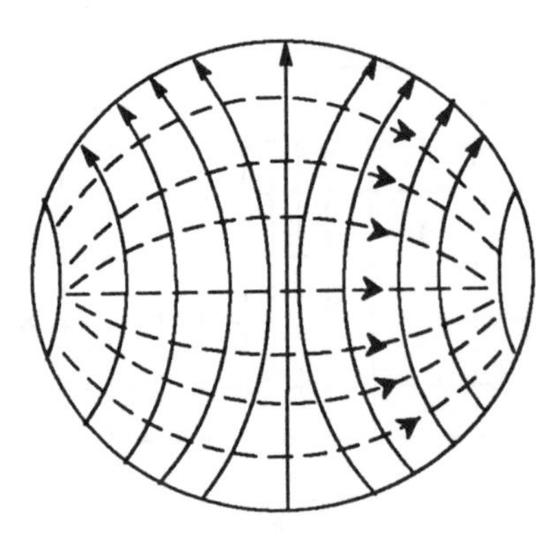

(a) TE_{11} circular mode

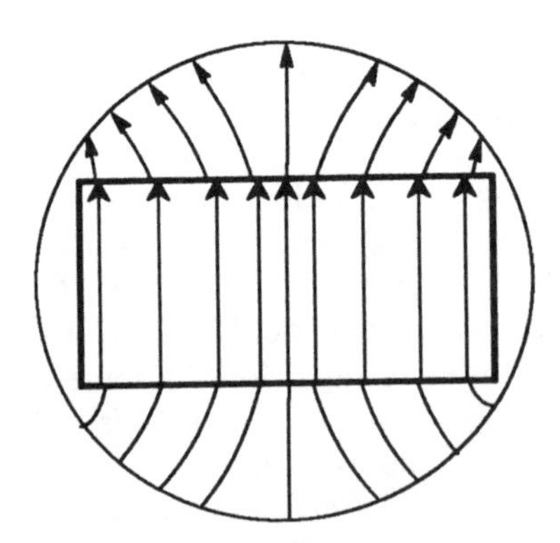

(b) TE_{10} rectangular overlaid on TE_{11} circular mode

Figure 5-22. Circular propagation modes.

In practice, the usable frequency bandpass is somewhat less than twice the cutoff frequency. Therefore, systems are usually designed for an operating frequency band extending from about 1.3 to 1.9 times the dominant mode cutoff frequency. Figure 5–23 indicates the recommended operating range and the cutoff frequency of the WR 137 (RG–50) rectangular waveguide operated in the TE_{10} mode. WR 137 has inside cross-sectional dimensions of 1.372 × 0.622 inches, which gives a height/base ratio of 0.45. The figure shows the calculated loss-versus-frequency response of the guide when made of copper. Also shown are curves reflecting the expected changes in the loss/frequency curves if the height/base ratios are increased to 0.7 or decreased to 0.3.

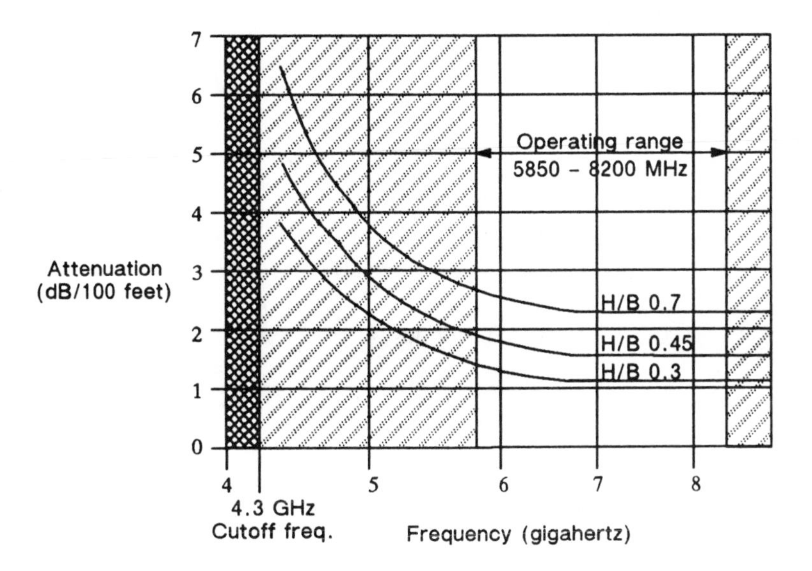

Figure 5–23. Recommended operating range and cutoff frequency of WR 137 (RG–50) rectangular waveguide.

The lower height/base ratio (0.3) provides a lighter, more easily installed waveguide at the expense of additional loss and a reduction in the amount of power that may be transferred from the transmitter to the antenna. The larger ratio (0.7) reduces the theoretical unit loss, increases bulk and, by increasing the cutoff wavelength of the TE_{10} mode, introduces the possibility that spurious propagation modes can exist at frequencies above about

6 GHz, which is within the proposed operating range of the standard WR 137 waveguide [3].

An in–depth study of waveguide characteristics with its attendant derivation of attenuation formulas is beyond the scope of this text. Those readers wishing to verify the information shown in Figure 5–23 may use the following equation, which is applicable to the TE_{10} mode operating in a rectangular guide.

$$a = \frac{(R_S)(8.686)(30.48)}{(b\sqrt{\mu/\epsilon})\sqrt{1-\left(\frac{fc}{fo}\right)^2}}\left[1+\frac{2b}{a}\left(\frac{fc}{fo}\right)^2\right] \text{ dB/100 ft.} \quad (5\text{–}36)$$

where R_S = (for copper) $2.61 \times 10^{-7} fo$
(for brass) $5.01 \times 10^{-7} fo$
(for silver) $2.52 \times 10^{-7} fo$

fc = Cutoff frequency = $\frac{3 \times 10^8}{2a}$

fo = Operating frequency

μ = $4\pi \times 10^{-7}\ H/m$ } In free space

ϵ = $8.854 \times 10^{-12}\ F/m$ } In free space

λ_C = Cutoff wavelength

a = Long dimension, in meters

b = Short dimension, in meters.

The corresponding equation for attenuation of the TE_{11} mode in a circular waveguide is

$$a = \frac{R_S \times 8.686 \times 30.48}{r\sqrt{\mu/\epsilon}\sqrt{1-\left(\frac{fc}{fo}\right)^2}}\left[\left(\frac{fc}{fo}\right)^2 + 0.42\right] \text{ dB/100ft} \tag{5-37}$$

where r = Radius of waveguide, in meters

$fc = \dfrac{0.293}{r\sqrt{\mu\,\epsilon}}$.

See Equation 5–36 above for definitions of the other symbols. Other propagation modes and the derivation of the above equations are covered in Reference 4.

The selection of waveguide configurations and propagation modes is directly associated with the antenna types to be fed, the polarization of the electromagnetic fields to be propagated by the antennas, and the transmission systems to be served. For example, antennas are usually designed to propagate an electric field vector either parallel or perpendicular to the earth's surface. The parallel arrangement is called *horizontal polarization* and the perpendicular vector results in the term *vertical polarization*. Some antennas are designed to transmit horizontally and vertically polarized signals simultaneously. In the latter case, a waveguide feeding the antenna must be able to maintain the relative 90–degree positioning of the signals as well as present each signal to the antenna in its designated orientation.

In theory, vertically and horizontally polarized waves can exist in the same medium without mutual coupling. In practice, as energy is radiated in one polarization, small portions are converted to the other polarization by imperfections in the antenna systems. Nevertheless, cross–polarization discriminations of over 30 dB can be obtained in the 4– and 6–GHz radio bands with commonly used antenna systems. In these bands, adjacent channel spectrum allocations are alternately assigned to operate with vertical and horizontal polarizations in order to use the frequency spectrum more efficiently.

Rectangular waveguides are often preferred over circular configurations because they provide simple solutions to the problem of maintaining field polarization for antenna feeds. This may be an important selection consideration where the physical layout requires numerous turns or twists in the waveguide run.

Circular waveguides are always used where swivel joints are required, such as when feeding a rotating radar antenna. Circular guides with the approximate physical sizes of rectangular guides can transmit the same frequencies with appreciably lower losses and can handle both polarizations simultaneously. For this reason they are often substituted for rectangular guides on long antenna feed sections, particularly in long–haul microwave systems where the signal–to–noise ratio may be critical. Circular configurations are also the favorite choice when frequency allocations from two or more radio bands (e.g., 4, 6, and 11 GHz) are to be transmitted in the same antenna feed.

Coupling

Any method used to couple energy into a waveguide may be used to remove energy from the guide. The coupling may be by way of either the guide's electric or magnetic fields. Regardless of the coupling arrangement, it must provide for matching the waveguide impedance to that of the interconnecting transmission line or other device.

The most popular coupling choice is a probe or antenna placed near a maximum point in the electric field of the mode to be propagated. Impedance matching is achieved by proper placement of the probe in the waveguide electric field and adjusting the shape and extension of the probe into the guide. Figure 5–24(a) shows a coaxial cable/waveguide junction and an example of a probe placed to excite the TE_{10} mode in a rectangular waveguide. The probe consists of an extension of the coaxial inner conductor into the guide to form an antenna. The waveguide forms a ground plane which must be solidly connected to the coaxial outer conductor. Figure 5–24(b) illustrates a method of exciting the TE_{10} mode in a rectangular waveguide by coupling between the magnetic field in the guide and a loop consisting of the inner coaxial conductor looped back and connected to the outer coaxial conductor and the waveguide wall.

While the above coupling methods are the ones most frequently used, they are by no means the only ones possible. For example, two pieces of waveguide may be coupled by cutting holes in each section and clamping them together so as to match the locations of the holes physically. The hole locations are chosen so that the two guide propagation modes have some common field component over the extent of the holes. This coupling method is frequently used to connect waveguides to cavity resonators such as those used on some microwave radio transmitters.

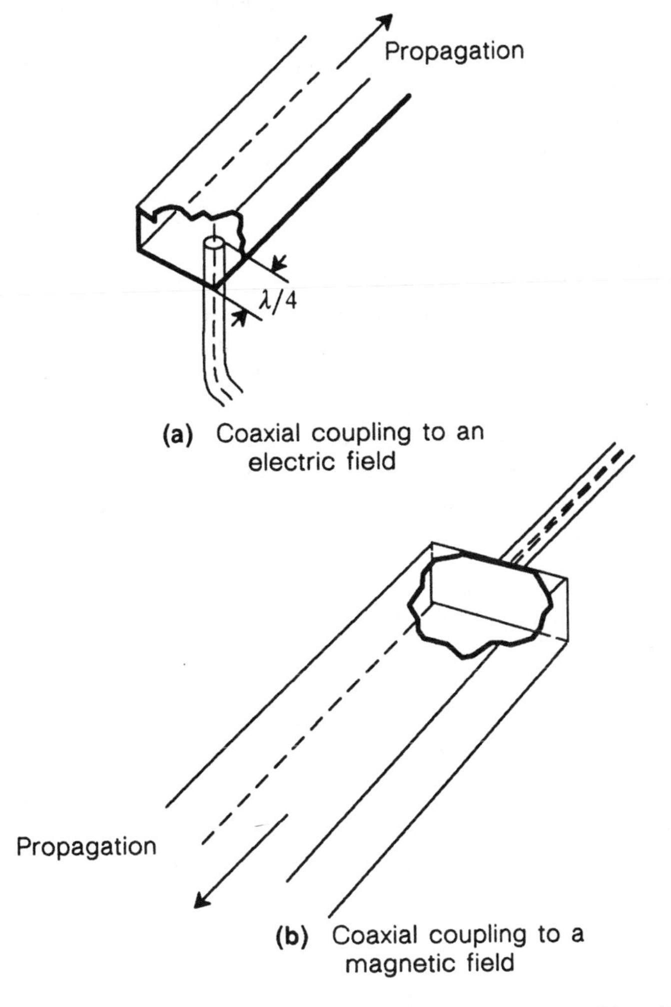

Figure 5–24. Coaxial cable coupling to waveguide TE_{10} mode.

5-6 OPTICAL FIBER CABLE

The use of optical fibers as a transmission medium was determined to be practical about 1966. Early experiments indicated that fiber, as a transmission medium using the visible light down to the near-infrared spectrum, would have losses of only a few dB per kilometer. Also, the invention of light-emitting diodes and solid-state lasers made fiber interesting and practical. With the coincidental development of light-emitting devices and glass fiber technology, work commenced in the 1960s identifying lightguide systems as a viable transmission medium. In 1970, a revolutionary discovery produced fiber having an attenuation of "only" 20 dB/km. Advances in fabrication technology in the 1980s have produced optical cable having an attenuation of 0.15 dB/km at a wavelength of 1.55 μm (micrometers).

The principal advantage of fiber is its wide bandwidth (at least 30,000 GHz) which permits thousands of simultaneous conversations over a fiber pair. Fiber optic bandwidth generally is specified as a function of distance, e.g., 1000 MHz/km. A one-quarter-inch diameter cable containing two optical waveguides can carry the same volume of communications traffic as five three-inch diameter cables, each containing 2400 pairs of copper wire. It is apparent that another advantage of fiber is its small size and resultant low weight.

Fiber waveguides have a distinct advantage over metallic facilities due to immunity to induced noise because they can be constructed entirely of dielectric material. This feature eliminates electromagnetic and radio interference, lightning surge, corona discharge, and other interference sources from affecting transmission. Crosstalk is not a problem in fiber optic systems.

As shown in Figure 5-25, the optical fiber transmission medium is constructed of two coaxial transparent tubular glass strands surrounded by a light-absorbing jacket [5]. Generally, the optical fiber consists of a core in the center made of inorganic glass or a combination of material (usually silicon dioxide) through which most of the information is transmitted. It is surrounded by a cladding material made also of glass with a refractive index slightly lower than the core. As a result, the optical rays will be continuously reflected off the cladding and guided in a forward direction along the length of the fiber, as illustrated in Figure 5–26 [6]. The outer protective coating attached to the cladding is a jacket usually made of organic polymer. It does not degrade optical performance.

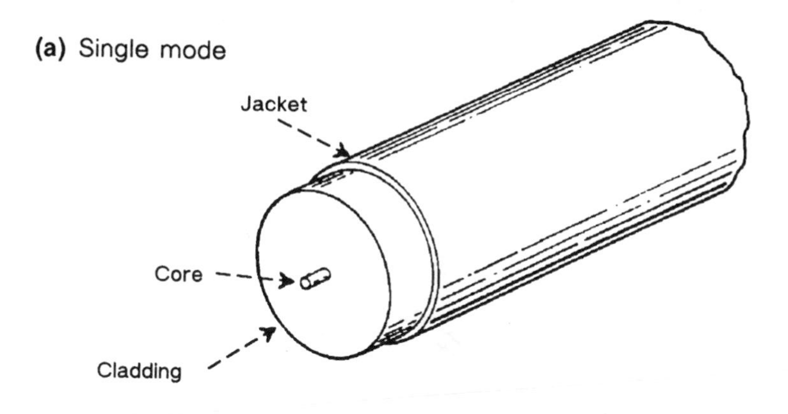

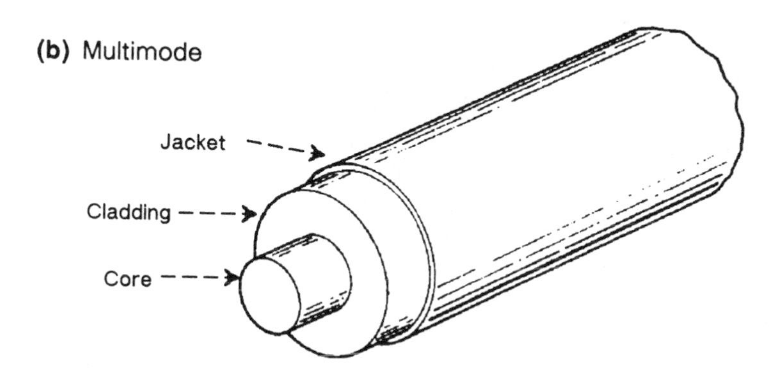

Figure 5–25. Optical fiber construction.

The core diameter of the single–mode fiber in Figure 5–25(a) is only a few microns, but that of the multimode fiber illustrated in Figure 5–25(b) is a few mils. The thickness of the cladding in both modes is less than 10 mils. Laser transmission works well with both modes of fiber, but light–emitting diodes generally match multimode fiber better because of low coupling losses.

There are three basic types of optical fiber as distinguished by their modal and physical properties. Single–mode fiber (step–index monomode), Figure 5–27(a), is constructed with a core diameter so small that essentially only one axial mode can propagate. Single–mode fiber is preferred if a large bandwidth is required. The transmission bandwidth is limited only by material dispersion in connection with the emission bandwidth of the light source.

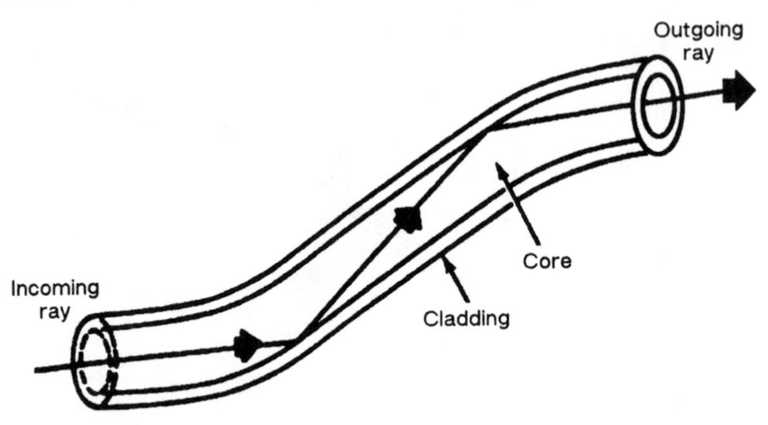

Figure 5–26. Optical ray transmission in fiber waveguide.

Multimode fiber is constructed with a larger core (typically 50 × 62.5 μm) than single–mode fiber, which results in a reduction of the transit time between partial rays. This is accomplished by a gradation of the refractive index, Figure 5–27(b), from the core center to the cladding. The radial decrease of the index away from the fiber axis results in an increase in the velocity of the partial light rays outside the core center, thus compensating for larger pair length. Therefore, almost the same velocity can be achieved for all modes by suitable choice of the index profile.

The earliest form of fiber is the step–index multimode, Figure 5–27(c), which consists of a homogeneous core with a refractive index, n_1, surrounded by cladding with a slightly lower refractive index, n_2. Light rays are guided by total reflection within the core and propagate at different angles; they therefore exhibit different transmit times, resulting in relatively limited bandwidth. This type of fiber is generally not used in telecommunications plant due to its inherently high modal distortion.

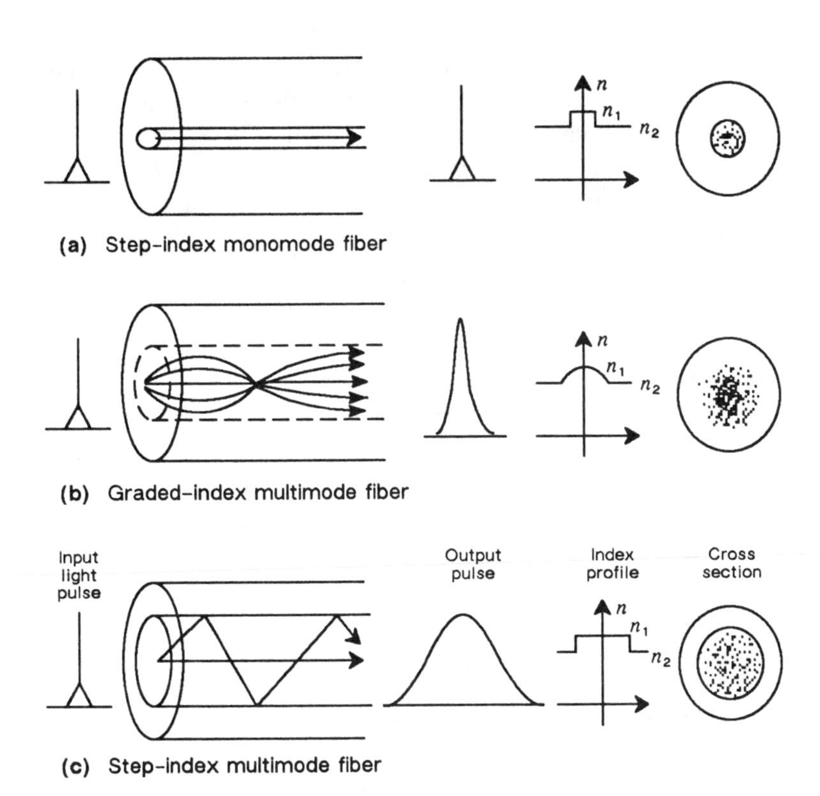

Figure 5–27. Basic types of optical fiber.

Optical Ray Propagation

The index of refraction of the core, n_1, is larger by design than that of the cladding, n_2. This difference in the step–index modes can produce total internal reflection of propagated rays. Rays that strike the core–cladding interface at an angle greater than the critical angle, θ_C, Figure 5–28(a), are totally reflected [7]. The angle θ_C is defined as

$$\theta_C = \sin^{-1} \frac{n_2}{n_1} . \qquad (5\text{–}38)$$

In a perfectly straight optical fiber having no attenuation, guided light rays having an angle of incidence greater than θ_C would theoretically propagate indefinitely. Rays that strike the interface at angles less than θ_C are only partially reflected, and their intensity diminishes rapidly with distance. Figure 5–28(b) shows that the angle θ_C defines the maximum launch angle, ϕ_C, at which guided rays can be injected into the fiber. This launch angle is usually given in terms of the numerical aperture (NA) which is defined as

$$NA = \sin \phi_C = \sqrt{n_1^2 - n_2^2}. \tag{5–39}$$

Since the energy launched at angles greater than ϕ_C is rapidly attenuated, the coupling efficiency of a signal source to a fiber depends on NA.

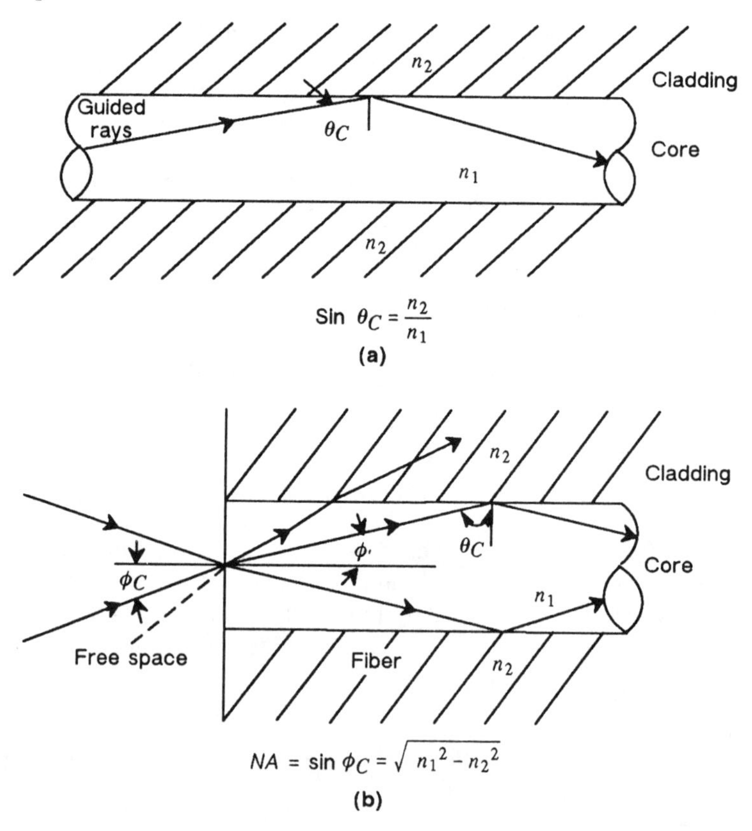

Figure 5–28. Propagation and launching angles for optical fibers.

Fiber guides are designed to be single mode or multimode, with the number of modes increasing with *NA* and core radius, provided that the wavelength of the signal source remains constant. With multimode fibers separated into two groups, step index and graded index, the difference between the two finds n constant along the core radius with step index, and the index of the graded–index core varies along the radius r according to the relationship

$$n(r) = n_1 \left[1 - 2\Delta \left(\frac{r}{r_1} \right)^g \right]^{1/2} r < r_1 \qquad (5\text{–}40)$$

where $\Delta = \dfrac{n_1^2 - n_2^2}{2\, n_1^2}$, r_1 is the core radius, and g is the index profile parameter. The value of Δ for the waveguides is typically a few percent or less.

The cladding in all cases has a constant index, n_2. The material properties of the core and cladding are circularly symmetrical with respect to the fiber axis and do not vary appreciably along its length.

Dispersion

Light ray transmission can be associated with modes, as Figure 5–28 illustrates. The rays (modes) that travel parallel to the fiber axis will arrive at their destination sooner than those making many reflections along the way. Each light pulse injected into the fiber core is made up of more than one ray so that they arrive at the fiber's terminus at different times, resulting in a broadened pulse. This is called modal dispersion (or distortion) and is shown in Figure 5–27(b) and (c). This type of dispersion is eliminated in single–mode fibers since only one mode can propagate. However, in a single–mode fiber, parallel rays of a different wavelength will be dispersed because the velocity of light varies with wavelength in a dielectric. This is called intramodal, or chromatic, dispersion (distortion).

Total dispersion in an optical fiber medium is the result of both intermodal and intramodal dispersion. It causes the pulses of light traveling through the fiber to broaden as a function of the length of the fiber. This pulse spreading determines the bandwidth and hence the information–carrying capacity of the fiber. It is limited to a point where pulses, broadened by dispersion, may spread into each other causing the original pulse to be indistinguishable from the others in the bit stream.

The total dispersion can be determined by the width of the output pulse when the input pulse is very narrow in time (i.e., approaching a delta function). The output pulse is then called the impulse response, $h(t)$. The characteristics of $h(t)$ can be described in terms of one parameter, the rms pulse width, σ. Its value is defined by

$$\sigma^2 = \frac{1}{\varrho}\int_{-\infty}^{\infty} (t - T)^2 \, h(t)dt \qquad (5\text{–}41)$$

where

$$T = \frac{1}{\varrho}\int_{-\infty}^{\infty} th(t)dt$$

is the pulse delay and

$$\varrho = \int_{-\infty}^{\infty} h(t)dt \ .$$

Intramodal Dispersion. A generated optical signal pulse will spread, even in a single–mode fiber, with the amount of spreading dependent on the nonzero spectral width of the light source and the propagation constant of light in the fiber as a function of wavelength. Part of this variation is caused by composition of the fiber material resulting in a change of the index of refraction with wavelength, λ, and is called material dispersion. Its magnitude, σ_M, can be expressed by

$$\sigma_m = L(\sigma_M)_0 \qquad (5\text{–}42)$$

where

$$(\sigma_M)_0 = \sigma_S \frac{\lambda_S}{c}\left[\frac{d^2n}{d\lambda^2}\right]_{\lambda_S} ,$$

σ_S is the standard deviation of the spectral distribution of the source, L is the fiber length, c is the speed of light, and λ_S is the mean wavelength of the signal source. The quantity

$$\frac{\lambda_S}{c}\left[\frac{d^2n}{d\lambda^2}\right]\lambda_S$$

varies with wavelength and passes through zero at a wavelength that depends on the core material.

The variation in the propagation constant depends on the ratio of the core radius to the wavelength of the light ray. This is called waveguide dispersion. It is small over the spectral range of interest and is important only in single–mode fibers when material dispersion is at its minimum. In the absence of material dispersion, waveguide dispersion, σ_ω, can be approximated by

$$\sigma_\omega = \frac{L\sigma_S}{2\pi c} V^2 \left[\frac{\delta^2\beta}{\delta V^2}\right]\lambda_S \tag{5–43}$$

where

$$V = \frac{2\pi r_1}{\lambda_S}\sqrt{n_1^2 - n_2^2}$$

and β is the propagation constant of the fundamental mode of the fiber. The derivation and value of β can be found in other documents [8]. The parameter V determines the number of modes which can propagate.

The pulse spreading (S_M) caused by intramodal dispersion can be quantitatively expressed by

$$S_M = (L\lambda/c)\frac{d^2n}{d\lambda^2}\,\delta\lambda \tag{5–44}$$

which indicates that pulse spreading due to intramodal dispersion increases with the spectral width of the light source $(\delta\lambda)$, with the emitted wavelength (λ), with the amount of material dispersion $(d^2n/d\lambda^2)$, and with the length of the fiber (L). Material dispersion is a fundamental property of the fiber and is not changeable for a given type. The only way that intramodal pulse spreading can be reduced for a given wavelength and system length is to reduce the spectral width of the light source, e.g., by using laser sources of improved design. A narrower input pulse width implies more pulses per unit of time, a higher pulse frequency, and an increase in the bandwidth.

Intermodal Dispersion. The existence of several hundred modes, each having its own propagation constant, causes intermodal dispersion that does not exist in single–mode fibers. In a step–index fiber, there is a maximum delay difference between the mode that takes the longest path and a mode that is exactly parallel to its axis in the shortest path. The longest path occurs for a ray that enters the fiber just at the critical angle, θ_C. As a result, pulse spreading increases linearly with fiber length, and bandwidth is inversely proportional to distance. The product of bandwidth (B) and length (L) can be used as a figure–of–merit for the information–carrying capacity of optical fiber cable.

For the axial mode of propagation, the time (t) that a ray takes to traverse a fiber of a particular length (L) is expressed by

$$t = \frac{L}{c} n_1 \qquad (5\text{–}45)$$

where n is the refractive index of the core and c is the speed of light. However, the ray entering at the critical angle (θ_C) takes the longest path through the fiber. Its travel time is expressed as the time (t') required to transverse the fiber medium where

$$t' = \frac{d/\sin a}{c} n_1 \qquad (5\text{–}46)$$

and where a is the angle of extinction.

Using Snell's law, the delay difference (δT) is calculated as

$$\delta T = t' - t = \frac{L}{c} n_1 (1/\sin a - 1) = \frac{L}{c} n_1 (n_1/n_2 - 1) . \qquad (5\text{–}47)$$

Differential delay between modes is greatly reduced when the core profile is changed from the step–index variety to the graded–index type described in Equation 5–40.

Glass fibers have attenuation losses caused principally by light absorption in the core, by scattering due to imperfections at the core–cladding interface, and by dielectric differences and variations molded into the fiber during manufacture.

Scattering

The phenomenon of (Rayleigh) scattering arises from density and composition fluctuations in optical fiber. It is the scattering of light rays in the fiber due to microirregularities in the dielectric medium through which light propagates. The magnitude of scattering is inversely proportional to the fourth power of the wavelength ($1/\lambda^4$), so that it becomes rapidly smaller at the longer wavelengths. The density and compositional variations in the uniformity of the core material are fundamental, cannot be eliminated, and, as a result, set the lower limit on fiber loss.

Transmission Losses

The most important parameters in evaluating a transmission medium are attenuation and bandwidth. They affect the physical range of transmission and the speed of data that can be sent over the facility.

Fiber loss is expressed in decibels per kilometer (dB/km)

$$\text{Loss} = 10 \log \frac{\text{power in}}{\text{power out}} \text{ dB.} \qquad (5\text{–}48)$$

Factors contributing to loss in an optical fiber path are absorption, scattering, bending loss, splicing and connector loss, and coupling losses at terminal locations.

As with dispersion, there are two sources of attenuation. The first is associated with the waveguide properties of fiber and has several components, the most important of which is microbending loss. Commercial fibers have minute deviations of the fiber axis, called microbends, that cause rays that strike the core–cladding interface at angles close to the critical angle to be partially, rather than totally, reflected. If energy is transferred between rays, as is usually the case, then energy is transferred into these loss rays and escapes into the cladding. As a result, microbending loss can add substantially to the overall attenuation.

Optical fiber cables are manufactured in limited lengths, and are available with individual fibers (stranded) or multiple fibers assembled into a ribbon. Therefore, optical fiber sections must be connected together at various points between terminations. Fibers are joined using various methods such as fusion or mechanical connectors. Some cables, with up to 144 fibers, use array connectors that permit splicing multiple fibers at a time. Rotary splicing of fibers provides efficient fiber alignment without "tuning." Using a tab lineup procedure, splices can be obtained with 0.2 dB of loss, or one can tune the splice to the lowest signal loss on the spot and achieve a loss down to 0.05 dB. In controlled environments, fiber optic cross connections can be effected using optical fiber patch bays, permitting up to 576 cross connections per bay.

In addition to waveguide losses, there are losses due to the material properties of the core that depend on wavelength. Material attenuation generally decreases rapidly with increasing wavelength. At longer wavelengths, absorption associated with atomic vibrations takes over and the core becomes opaque. For low–loss fibers, material attenuation is dominated by scattering and ion absorption.

Ion absorption is caused by the introduction of the hydroxyl (–OH) ion from moisture introduced during manufacture. The fundamental vibration of the ion, illustrated in Figure 5–29 [8], is centered around 2.8 microns, but overtones at about 0.9 and 1.4 μm are in the usable range of the fiber optic transport system. Consequently, fiber optic systems operate efficiently between the "water peaks." Therefore, the total loss, a, for fiber can be expressed as

$$a = \frac{A}{\lambda^4} + B(\lambda) + C \text{ dB} \qquad (5\text{–}49)$$

where A/λ^4 is the amount of Rayleigh scattering, $B(\lambda)$ is the loss associated with all other wavelength–dependent absorption processes, and C is the loss due to wavelength–dependent effects (such as microbending loss).

Optical Fiber Strength

Due to the brittle nature of glass, fiber strength is of particular interest. The tensile strength of the glass fiber is determined by

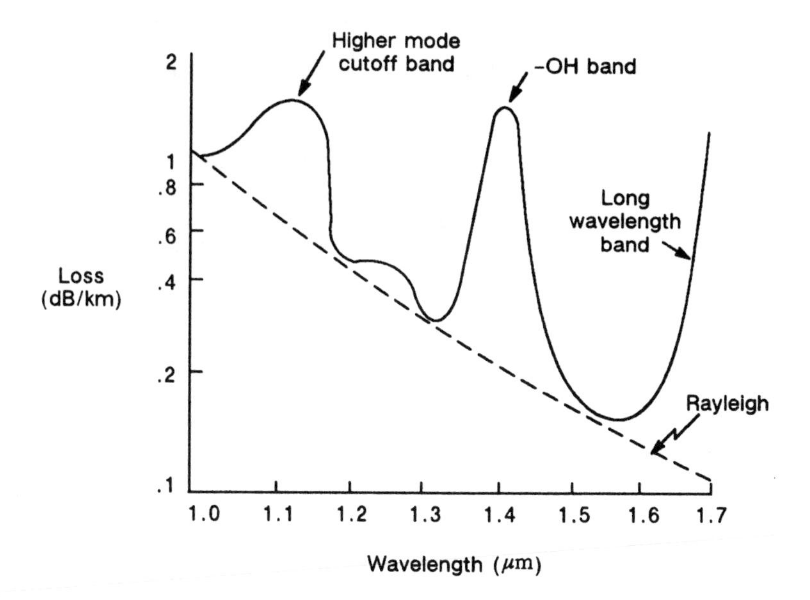

Figure 5–29. Rayleigh scattering limit at long wavelengths, and loss versus wavelength of an excellent optical fiber.

the deepest surface flaw. Advances in optical fiber cable manufacture, installation, and use are rapidly increasing the reliability of fiber optics. Optical fiber cable can be preassembled and buried as a cable–in–a–conduit installation, pulled through the conduit, placed in the air, or installed under water. The properties of strength member materials, such as steel, polyester monofilament, etc., produce variations in cable strength. As a result of technological advances in optical fiber cable construction, users must be aware of constant cable data updates.

References

1. Skilling, H. H. *Electric Transmission Lines* (New York: McGraw–Hill Book Company, Inc., 1951).

2. *Reference Data for Engineers: Radio, Electronics, Computer, and Communications,* Seventh Edition (Indianapolis, IN: Howard W. Sams and Company, Inc., 1985).

3. Freeman, R. L. *Reference Manual for Telecommunication Engineering* (New York: John Wiley and Sons, Inc., 1985), p. 933.

4. Ramo, S., T. Van Duzer, and J. R. Whinnery. *Fields and Waves in Communication Electronics* (New York: John Wiley and Sons, Inc., 1984).

5. Members of Technical Staff. *Engineering and Operations in the Bell System*, Second Edition (Murray Hill, NJ: AT&T Bell Laboratories, Inc., 1983), pp. 132–134.

6. *Optical Fiber Communication* (Gaithersburg, MD: Future Systems Incorporated, July 1982), pp. 32–42.

7. Personick, S. D. *Fiber Optics—Technology and Applications* (New York: Plenum Press, 1985).

8. Members of Technical Staff. *Transmission Systems for Communications*, Fifth Edition (Murray Hill, NJ: AT&T Bell Laboratories, Inc., 1982), pp. 821–829.

Additional Reading

Anner, G. E. and W. L. Everitt. *Communication Engineering* (New York: McGraw–Hill Book Company, Inc., 1956).

Creamer, W. J. *Communication Networks and Lines* (New York: Harper and Brothers, 1951).

Johnson, W. C. *Transmission Lines and Networks* (New York: McGraw–Hill Book Company, Inc., 1950).

Reed, H. R. and L. A. Ware. *Communication Circuits* (New York: John Wiley and Sons, Inc., 1955).

Schelkunoff, S. A. *Electromagnetic Waves* (Princeton, NJ: D. Van Nostrand Company, Inc., 1943).

Smith, P. H. "Transmission Line Calculator," *Electronics* (Jan. 1939 and Jan. 1944).

Chapter 6

Wave Analysis

A signal may be represented by a function whose value is specified at every instant of time. In transmission work, however, this characterization is not always the most convenient to use because information about transmission lines and networks is usually presented as a function of frequency rather than of time. Therefore, a method is needed for translating between time–domain and frequency–domain expressions of signal and network characteristics. It is possible to pass from one domain to the other by using mathematical transformations. This ability is useful in providing answers to questions that arise when a pulse signal expressed as a time–domain function is applied to a transmission line whose characteristics are known in terms of frequency–domain functions. These questions might be: How does the pulse look at the output of the line? What is the frequency spectrum of a pulse signal as a function of pulse repetition rate and duty cycle? What is the resultant energy distribution? What are the bandwidth considerations?

The duality between frequency and time domains in describing signals and linear networks is a concept that can become so familiar that a person may unconsciously transfer from one domain to the other. For instance, a sine wave of frequency f_0 might be pictured in the time domain as a curve that crosses the time axis $2f_0$ times per second, or in the frequency domain as a narrow spike located at a point $f = f_0$ on the frequency axis and characterized by two numbers giving amplitude and phase. In this simple case, a method of passing from one of these representations to the other is simple to visualize and to formulate. The frequency, amplitude, and phase can be obtained from a time–domain representation of a sinusoidal wave by merely counting and measuring appropriate dimensions; on the other hand, if the frequency, amplitude, and phase are given, the time–domain waveform can be constructed. However, for more complicated waveforms, this transformation is not so simple, and a well–defined mathematical procedure must be employed to pass from one domain to the other.

The *Fourier transform pair* is the mathematical formulation of this useful concept; as such, it is indispensable for dealing with signals and linear networks. A review of some of the important properties of the Fourier transform and illustrations of its uses can give a qualitative understanding of its meaning and application. The reader is referred to standard mathematics texts for more rigorous treatment of the subject.

6–1 INSTRUMENTATION

To complement the mathematical procedures of defining signals, interferences, or networks by means of the Fourier transform pair, field or laboratory observations are often needed to study engineering problems of maintenance, design, or performance evaluation. Two types of instrumentation are commonly used to display signals and interferences in the time or frequency domain for visual study. These are the oscilloscope, which displays a signal or an interference in the time domain, and the spectrum analyzer, which displays signal or interference components in a frequency band. For either oscilloscope or spectrum–analyzer functions, microprocessors can be used that measure and calculate parameters, assemble the data, and present the results either as a cathode ray tube display or a printer output.

Figure 6–1 illustrates, in a simple block diagram, the operation of an oscilloscope. The position of the cathode ray spot on the tube face is a function of the voltages impressed on the horizontal and vertical deflection plates. The output signal of the sawtooth generator is impressed on the horizontal deflection plates and causes the spot to move repetitively from left to right. The signal under study, usually a periodic time function, is applied to the vertical deflection plates. This causes the spot to move vertically in accordance with the voltage applied. When the two signals are properly synchronized, a time–domain representation of the test signal waveform is traced on the tube face.

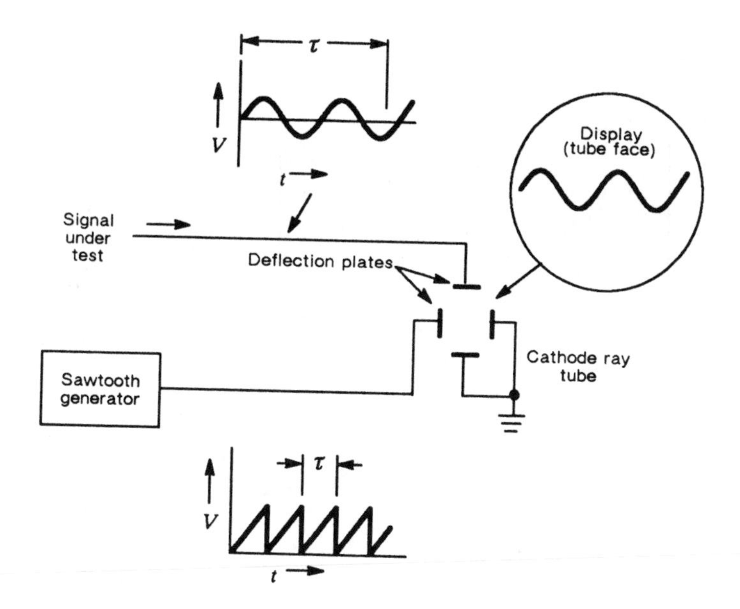

Figure 6–1. Oscilloscope operation (time domain).

The operation of a spectrum analyzer is somewhat more complicated. A sawtooth signal is used to deflect the spot from left to right on the tube face as in the oscilloscope. However, the signal applied to the vertical plates must be subjected to a number of transformations before it can be so used. The signal consists of a broad band of frequencies illustrated by the band from f_B to f_T in Figure 6–2. This signal is impressed at the input to the analyzer and mixed with the output of the tunable oscillator. This oscillator is driven by the sawtooth generator. Its output signal varies in frequency to sweep across the band from f_B to f_T in a repetitive fashion as the output voltage of the sawtooth generator increases from its minimum to its maximum value. This process converts the signal, in effect, from a voltage–frequency function to a voltage–time function. The output of the mixer, however, has many unwanted components. A bandpass filter is used to select the desired component, which is then impressed on the vertical deflection plates of the cathode ray tube. The filtering is usually accomplished through several stages of intermediate frequencies where the bandlimiting is more easily carried out.

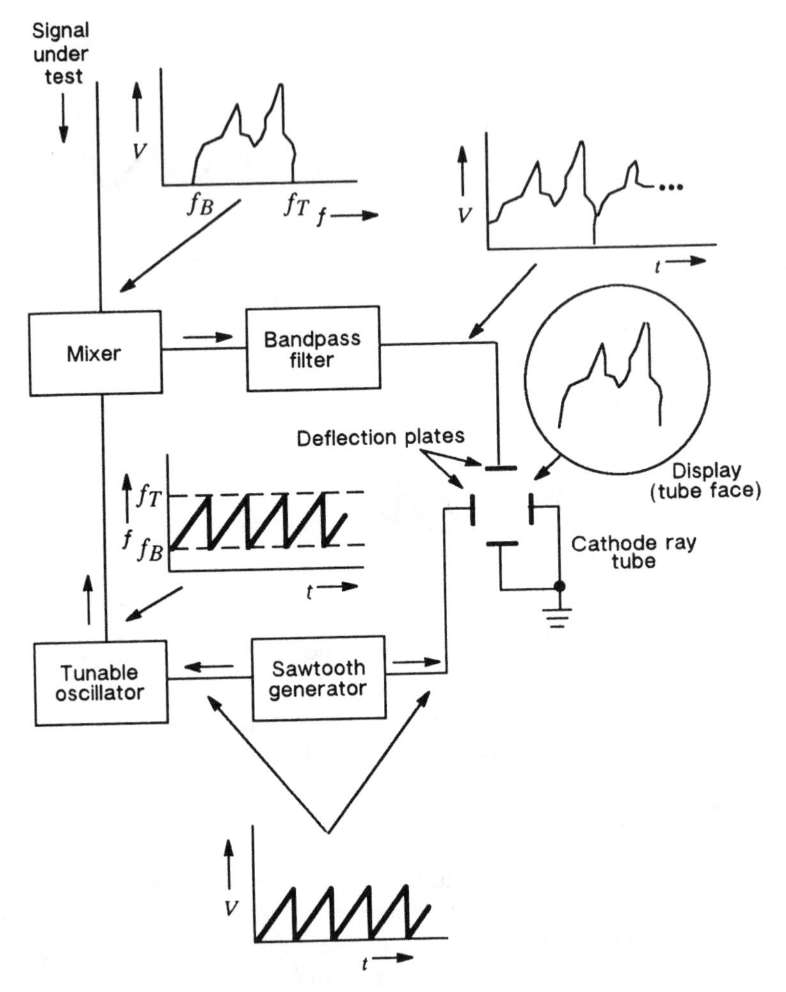

Figure 6–2. Spectrum analyzer operation (frequency domain).

6–2 PERIODIC SIGNALS

Alternate descriptions of periodic signals in the time and frequency domains are based on the fact that when sine waves of appropriate frequencies, phases, and amplitudes are combined, their sum can be made to approximate any periodic signal.

Similarly, any periodic signal can be decomposed into its component sine waves.

Fourier Series Representation

A good starting point for discussing wave analysis is the familiar Fourier series representation of periodic functions given by

$$f(t) = \frac{A_0}{2} + \sum_{n=1}^{\infty} (A_n \cos n\omega_0 t + B_n \sin n\omega_0 t), \qquad (6\text{–}1)$$

where A_0, A_n, and B_n are constants that may be computed by the following equation [1]:

$$A_0 = \frac{1}{\pi} \int_0^{2\pi} f(t)dt, \qquad (6\text{–}2)$$

$$A_n = \frac{1}{\pi} \int_0^{2\pi} f(t) \cos nt \, dt, \qquad (6\text{–}3)$$

and

$$B_n = \frac{1}{\pi} \int_0^{2\pi} f(t) \sin nt \, dt. \qquad (6\text{–}4)$$

The interval over which the integration is performed, 0 to 2π, is the fundamental period of the function $f(t)$.

The validity of the Fourier series may be demonstrated by simultaneously displaying a square wave on an oscilloscope and on a spectrum analyzer. The spectral components of such a signal may be filtered and displayed on the two instruments in various combinations. To illustrate such a demonstration, Equation 6–1 may be rewritten

$$f(t) = \frac{C_0}{2} + \sum_{n=1}^{\infty} C_n \cos (n\omega_0 t + \phi_n), \qquad (6\text{–}5)$$

where

$$C_0 = A_0,$$

$$C_n = (A_n^2 + B_n^2)^{1/2},$$

$$\cos\phi_n = A_n/C_n,$$

$$\sin\phi_n = -B_n/C_n.$$

Figure 6–3 illustrates various displays that might be observed; sketches of oscilloscope patterns are at the left and spectrum analyzer displays are at the right. In Figure 6–3(a), the output of a square–wave generator is shown at the left. The period of the wave is $1/\omega_0 = T$ seconds. The wave is shown as having an amplitude of unity ($A = 1$) and a pulse width of $T/2$. The corresponding spectrum analyzer display shows components at as many odd harmonics of the fundamental as the analyzer will accommodate. In the illustration, harmonics are shown up to the eleventh.

Figures 6–3(b), 6–3(c), and 6–3(d) illustrate the displays when the inputs to the measuring sets are limited to the fundamental, third, and fifth harmonics, respectively. It is interesting to note how quickly the oscilloscope display approaches the original square wave.

Consideration of Figure 6–3 and Equation 6–5 shows how the Fourier expansion for the square wave, $f(t)$, may be used to determine certain requirements on a channel that is to be used to transmit the square wave. The extent to which pulse distortion can be tolerated determines the number of signal components that must be transmitted and, therefore, the bandwidth that must be provided. Further detailed study would also show how much distortion (gain and phase) can be tolerated. The idealized sketches of Figure 6–3 indicate no distortion. Gain distortion would change the relationships among the amplitudes of the signal components. Phase distortion would cause relative shifts of the components along the time axis. Such shifts would also cause distortion of the pulse.

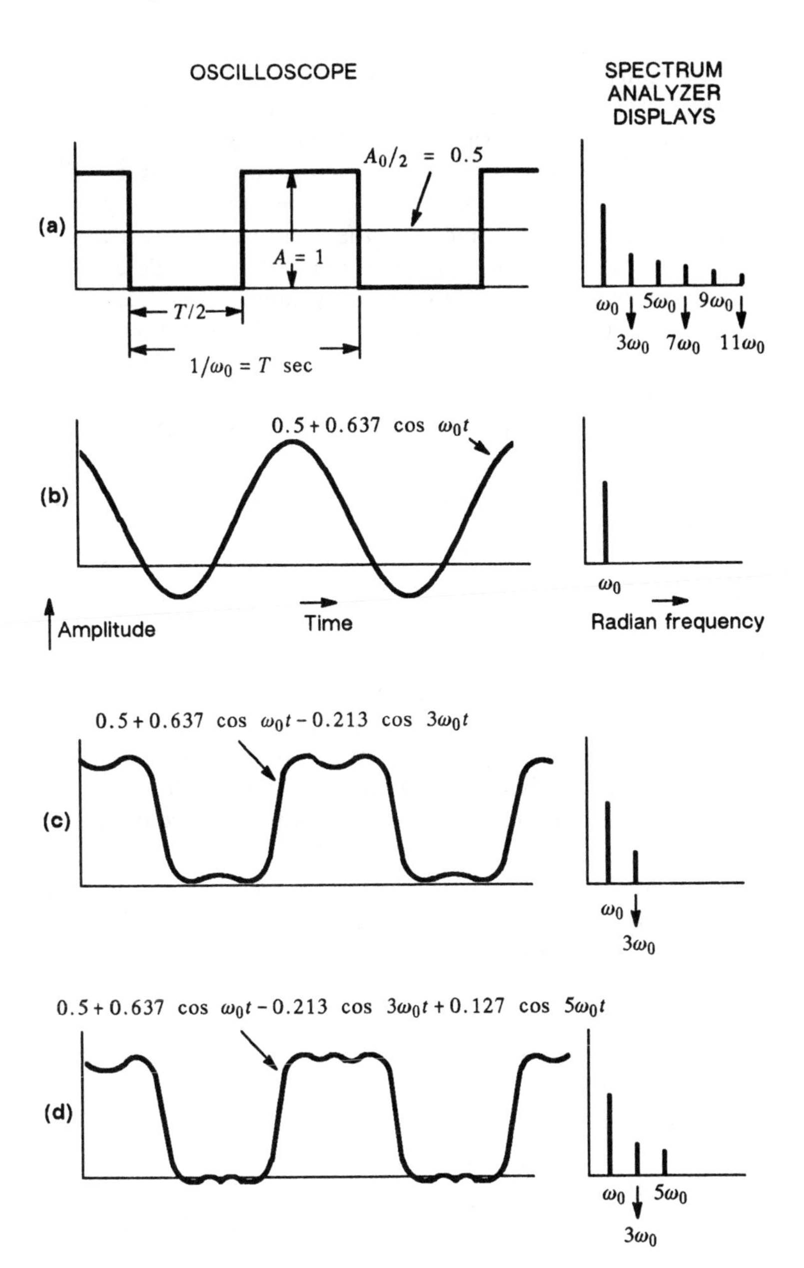

Figure 6–3. Fourier components of a square wave.

Symmetry. The Fourier analysis of certain periodic waveforms can frequently be simplified by observing properties of symmetry in the waveform and by selecting the coordinates about which the waveform varies to take maximum advantage of the observed properties of symmetry. It can be shown that, by proper choice of axes, one or more of the coefficients can always be made zero; however, if more than one is to be made zero, the waveform *must* exhibit odd or even symmetry. It is desirable to define these properties of symmetry and to illustrate them mathematically and graphically because, by taking advantage of such properties, it is possible to greatly reduce the mathematics necessary to evaluate the coefficients A_n and B_n of Equation 6–1.

Periodic functions exhibiting *odd symmetry* have the mathematical property that

$$f(-t) = -f(t). \qquad (6\text{–}6)$$

That is, the shape of the function, when plotted, is identical for positive and negative values of time, but there is a reversal of sign for corresponding values of positive or negative time. A familiar function exhibiting this property is the sine function. Figure 6–4(a) also illustrates a function having odd symmetry.

A function having odd symmetry contains no cosine terms and, in addition, contains no dc component. Thus, in Equation 6–1, since $A_n = 0$ and $A_0 = 0$, the Fourier series is written

$$f(t)_{\text{odd}} = \sum_{n=1}^{\infty} B_n \sin n\,\omega_0 t. \qquad (6\text{–}7)$$

Graphically, the function can be seen to have odd symmetry by folding the right side of the time axis over upon the left side and then rotating the folded half 180 degrees about the abscissa, which must be selected as the dc component of the waveform. When the function is folded and rotated as indicated, the folded portion is superimposed directly on the unfolded function for negative time.

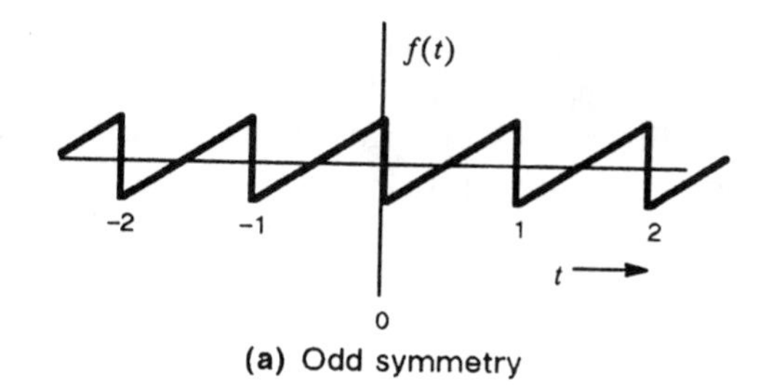

(a) Odd symmetry

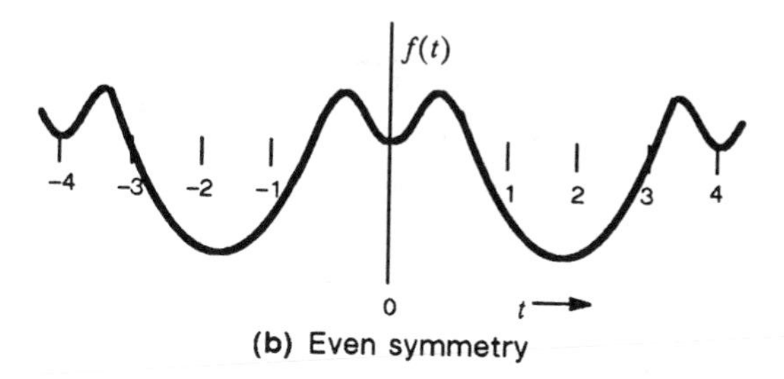

(b) Even symmetry

Figure 6–4. Symmetrical functions.

A function having *even symmetry* contains no sine terms. That is

$$B_n = 0$$

in Equation 6–1. In this case, the Fourier series is written

$$f(t)_{\text{even}} = \frac{A_0}{2} + \sum_{n=1}^{\infty} A_n \cos n\,\omega_0 t. \tag{6–8}$$

The mathematical property that such a function exhibits is that

$$f(t) = f(-t). \tag{6–9}$$

Graphically, this function may be seen to be even if the portion to the right of the vertical axis (positive time) is folded about the axis to fall upon the left portion (negative time). If the function is even, the folded portion would fall directly upon the unfolded left portion. Such a function is illustrated in Figure 6–4(b).

Some functions can be adapted to have either odd or even symmetry by the appropriate selection of axes. One such example is given in Figure 6–5. In Figure 6–5(a), the function exhibits odd symmetry. By shifting the vertical axis to the right by one-half a unit time interval, the function is translated into one having even symmetry. This is shown in Figure 6–5(b) and is also illustrated in Figure 6–3.

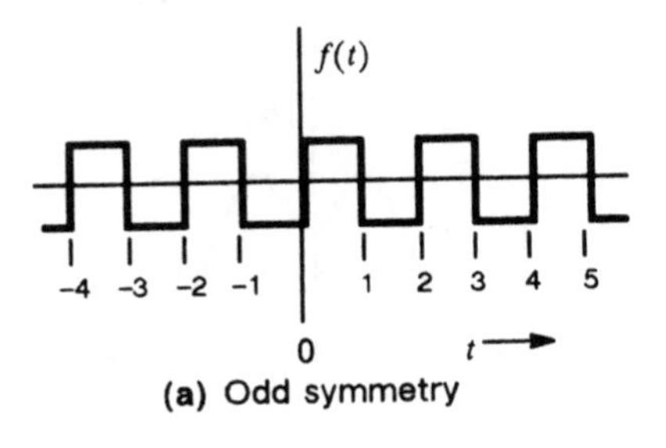

(a) Odd symmetry

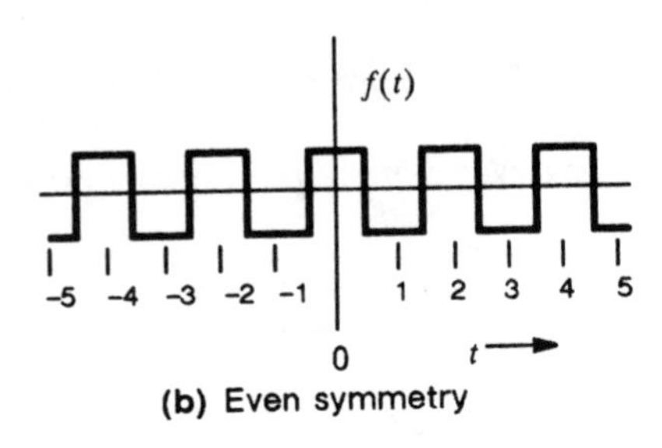

(b) Even symmetry

Figure 6–5. Symmetry by choice of axes.

Example 6–1: A Fourier Series Application

It has been shown how the Fourier analysis of a square wave can be used to illustrate the way such a wave can be decomposed into its harmonically related components (Figure 6–3). Similarly, a square wave was used to illustrate how a proper choice of coordinates can simplify a problem by taking advantage of symmetry properties in the wave to be analyzed (Figure 6–5).

This example of Fourier analysis demonstrates the effect on frequency content, harmonic amplitudes, and required relative bandwidth of changing the period of a rectangular wave. The waveforms are illustrated in Figure 6–6. In each of the waveforms illustrated, the pulse amplitude is unity and the pulse duration is τ seconds. In Figure 6–6(a) the repetition period is $T_a = 2\tau$ seconds; in Figure 6–6(b) the period is $T_b = 2T_a$ seconds; in Figure 6–6(c) the period is $T_c = 2T_b$ seconds. In each, the vertical axis is chosen so that the function exhibits even symmetry. Thus, there are no sine terms in the Fourier series. For each, then, the Fourier series may be written as in Equation 6–8.

$$f(t)_{\text{even}} = \frac{A_0}{2} + \sum_{n=1}^{\infty} A_n \cos n\,\omega_0 t.$$

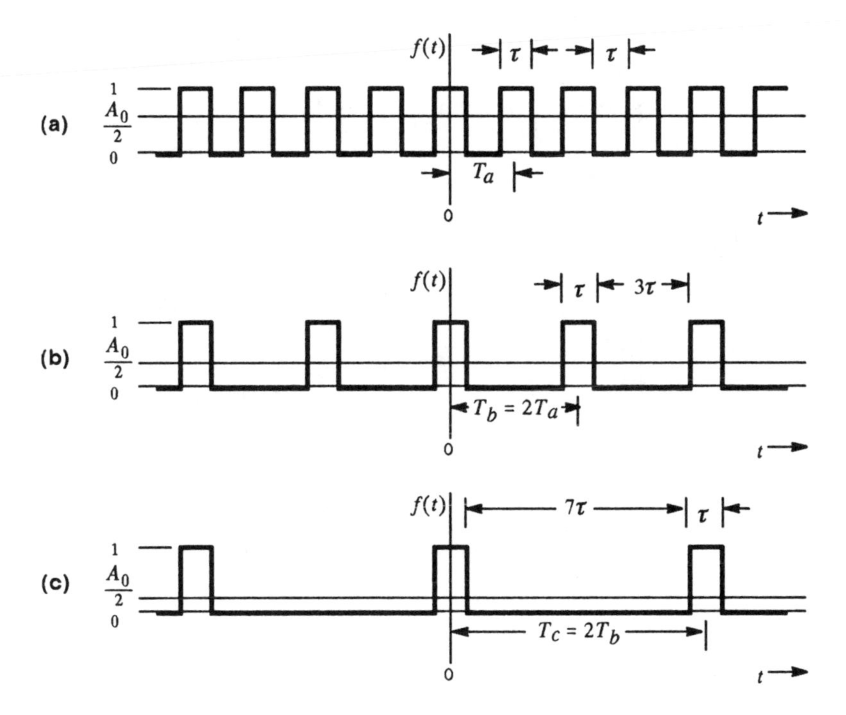

Figure 6–6. Periodic rectangular pulses with different periods.

For the three cases in Figure 6–6, the periodic functions may be written, respectively, as follows:

Figure 6–6(a)

$$\left.\begin{array}{l} f(t) = 1 \\ \\ f(t) = 0 \end{array}\right\}\left(m - \frac{1}{2}\right)\tau \le t \le \left(m + \frac{1}{2}\right)\tau \left\{\begin{array}{l} \frac{m}{2} = 0 \text{ or integer} \\ \\ \frac{m}{2} = \text{fraction} \end{array}\right.$$

In this case, $\tau = \frac{T_a}{2} = \frac{2\pi}{2} = \pi.$

Therefore,

$$\left.\begin{array}{l} f(t) = 1 \\ \\ f(t) = 0 \end{array}\right\}\left(m - \frac{1}{2}\right)\pi \le t \le \left(m + \frac{1}{2}\right)\pi \left\{\begin{array}{l} \frac{m}{2} = 0 \text{ or integer} \\ \\ \frac{m}{2} = \text{fraction} \end{array}\right.$$

Figure 6–6(b)

$$\left.\begin{array}{l} f(t) = 1 \\ \\ f(t) = 0 \end{array}\right\}\left(m - \frac{1}{2}\right)\tau \le t \le \left(m + \frac{1}{2}\right)\tau \left\{\begin{array}{l} \frac{m}{2} = 0 \text{ or integer} \\ \\ \frac{m}{4} = \text{fraction} \end{array}\right.$$

Now, $\tau = \frac{T_b}{4} = \frac{2\pi}{4} = \frac{\pi}{2},$

and

$$\left.\begin{array}{l} f(t) = 1 \\ \\ f(t) = 0 \end{array}\right\}\left(m - \frac{1}{2}\right)\frac{\pi}{2} \le t \le \left(m + \frac{1}{2}\right)\frac{\pi}{2} \left\{\begin{array}{l} \frac{m}{4} = 0 \text{ or integer} \\ \\ \frac{m}{4} = \text{fraction} \end{array}\right.$$

Figure 6–6(c)

$$\left.\begin{array}{l} f(t)=1 \\ \\ f(t)=0 \end{array}\right\}\left(m-\frac{1}{2}\right)\tau \le t \le \left(m+\frac{1}{2}\right)\tau\left\{\begin{array}{l} \frac{m}{8}=0 \text{ or integer} \\ \\ \frac{m}{8}=\text{fraction} \end{array}\right.$$

In this instance, $\tau = \frac{T_c}{8} = \frac{2\pi}{8} = \frac{\pi}{4}$.

Thus,

$$\left.\begin{array}{l} f(t)=1 \\ \\ f(t)=0 \end{array}\right\}\left(m-\frac{1}{2}\right)\frac{\pi}{4} \le t \le \left(m+\frac{1}{2}\right)\frac{\pi}{4}\left\{\begin{array}{l} \frac{m}{8}=0 \text{ or integer} \\ \\ \frac{m}{8}=\text{fraction} \end{array}\right.$$

In the above equations, m is an integer from $-\infty$ to $+\infty$. The value of the dc component, $A_0/2$, may be determined for each case by means of Equation 6–2. Thus,

(a) $A_0/2 = \frac{1}{2}$; (b) $A_0/2 = \frac{1}{4}$; (c) $A_0/2 = \frac{1}{8}$.

Note that the value of $A_0/2$ for a periodic function may be determined as the value of $f(t)$ averaged over one period. Where the function represents rectangular pulses, the value of $A_0/2$ is $A\tau/T$ where A is the amplitude of the pulse.

Now, to further illustrate, consider the frequencies of the fundamentals and third and fifth harmonics for the three waveforms of Figure 6–6. The frequency of the fundamental, f_1, is the reciprocal of the fundamental period, T_a, T_b, or T_c. For the three cases of interest, the frequencies are

(a) $f_1 = \frac{1}{T_a} = \frac{1}{2\tau}$; $f_3 = 3f_1 = \frac{3}{2\tau}$; $f_5 = 5f_1 = \frac{5}{2\tau}$.

(b) $f_1 = \frac{1}{T_b} + \frac{1}{4\tau}$; $f_3 = 3f_1 = \frac{3}{4\tau}$; $f_5 = 5f_1 = \frac{5}{4\tau}$.

$$\text{(c)}\quad f_1 = \frac{1}{T_c} = \frac{1}{8\tau};\ f_3 = 3f_1 = \frac{3}{8\tau};\ f_5 = 5f_1 = \frac{5}{8\tau}.$$

Thus, the frequencies of the fundamentals and their harmonics are seen to decrease as the period, T, of the fundamental increases.

Finally, the amplitudes of these signal components may be determined from Equation 6–3:

$$\text{(a)}\quad A_1 = \frac{1}{\pi}\int_0^{2\pi} f(t)\cos t\,dt$$

$$= \frac{1}{\pi}\int_0^{\pi/2} f(t)\cos t\,dt + \frac{1}{\pi}\int_{\pi/2}^{3\pi/2} f(t)\cos t\,dt$$

$$+ \frac{1}{\pi}\int_{3\pi/2}^{2\pi} f(t)\cos t\,dt$$

$$= \frac{1}{\pi}\sin t\Big]_0^{\pi/2} + 0 + \frac{1}{\pi}\sin t\Big]_{3\pi/2}^{2\pi}$$

$$= \frac{1}{\pi}(1+1) = \frac{2}{\pi} = +0.637,$$

$$A_3 = \frac{1}{\pi}\int_0^{2\pi} f(t)\cos 3t\,dt$$

$$= \frac{1}{3\pi}\sin 3t\Big]_0^{\pi/2} + 0 + \frac{1}{3\pi}\sin 3t\Big]_{3\pi/2}^{2\pi}$$

$$= \frac{1}{3\pi}(-1-1) = -\frac{2}{3\pi} = -0.213,$$

and

$$A_5 = +0.127.$$

(b) $$A_1 = \frac{1}{\pi}\int_0^{2\pi} f(t)\ \cos\ t\ dt$$

$$= \frac{1}{\pi}\int_0^{\pi/4} f(t)\ \cos\ t\ dt + \frac{1}{\pi}\int_{\pi/4}^{7\pi/4} f(t)\ \cos\ t\ dt$$

$$+ \frac{1}{\pi}\int_{7\pi/4}^{2\pi} f(t)\ \cos\ t\ dt$$

$$= \frac{1}{\pi}\ \sin\ t\ \Big]_0^{\pi/4} + 0 + \frac{1}{\pi}\ \sin\ t\ \Big]_{7\pi/4}^{2\pi}$$

$$= \frac{1}{\pi}\ (0.707 + 0.707) = 0.450,$$

$$A_3 = \frac{1}{\pi}\int_0^{2\pi} f(t)\ \cos\ 3t\ dt$$

$$= \frac{1}{\pi}\ \sin\ 3t\ \Big]_0^{\pi/4} + 0 + \frac{1}{3\pi}\ \sin\ 3t\ \Big]_{7\pi/4}^{2\pi}$$

$$= \frac{1}{3\pi}\ (+0.707 + 0.707) = +0.151,$$

and

$$A_5 = -0.899.$$

(c) $$A_1 = +0.244,$$

$$A_3 = +0.196,$$

and

$$A_5 = +0.118.$$

While the amplitudes of A_n can be seen to decrease with increasing n for each of the three cases, observe that there is no obvious, simple relationship among the values of A_n from case to case in the example. The value of A_1 logically decreases as T/τ increases, but A_3 and A_5 change with regard to amplitude and sign.

The (sin x)/x Function

The lengthy and laborious calculations of Example 6–1 are given to illustrate in detail how the coefficients of a periodic function expressed as a Fourier series can be determined; however, for a number of commonly found waveforms, these coefficients have already been calculated [2]. Many of the expressions for such coefficients contain a term in the form (sin x)/x. This function is so commonly found that a plot of the function on a normalized scale is given in Figure 6–7.

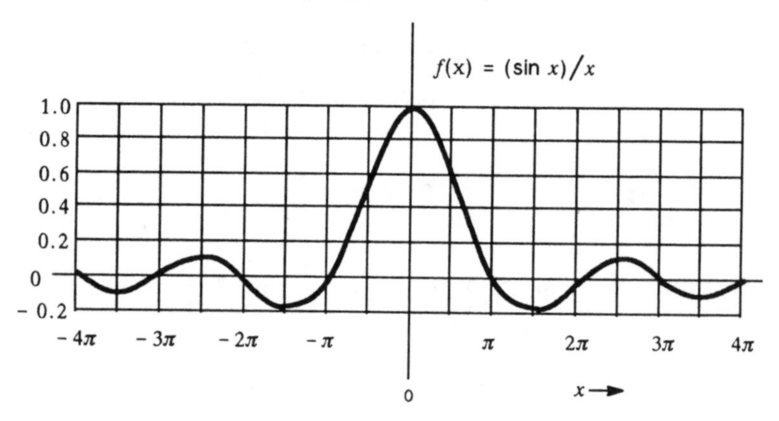

Figure 6–7. The (sin x)/x function.

In Example 6–1, the amplitude coefficient may be computed for each harmonic component by

$$A_n = A_0 \left(\frac{\sin \dfrac{n\pi\tau}{T}}{\dfrac{n\pi\tau}{T}} \right) \qquad (6\text{–}10)$$

Values for $n\pi\tau/T = x$ may be found from Figure 6–7 for values of n, τ, and T defined as in Example 6–1. Recall also that for rectangular pulses, $A_0 = A\tau/T$.

6–3 NONPERIODIC SIGNALS

Although the Fourier series is a satisfactory and accurate method of representing a periodic function as a sum of sine and cosine waves as illustrated by Equation 6–1, somewhat broader mathematical expressions, known as the Fourier transform pair, must be used to represent nonperiodic signals as functions of time or as functions of frequency. Although these are most useful in characterizing nonperiodic signals, they may also be applied to the analysis or synthesis of periodic signals. Similar mathematical representations may be used to describe the transmission response of a network or transmission line by combining expressions representing signals with those representing network characteristics.

The Fourier Transform Form

The determination of the components of a signal can be accomplished by the methods of Fourier analysis. If the signal is periodic, the analysis is relatively simple and can be carried out, as previously described, by a Fourier series representation. If the signal is nonperiodic, the *Fourier transform* may sometimes be used.* It is written

$$g(\omega) = \int_{-\infty}^{\infty} f(t)e^{-j\omega t}dt. \tag{6–11}$$

This equation may be used to determine the function of frequency, $g(\omega)$, given a function of time, $f(t)$, that is single–valued, has only a finite number of discontinuities, possesses a finite number of maxima and minima in any finite interval, and whose integral converges.

The inverse function, written

$$f(t) = \frac{1}{2\pi} \int_{-\infty}^{\infty} g(\omega)e^{j\omega t}d\omega, \tag{6–12}$$

is known as the *Fourier integral*, or the inverse Fourier transform. This expression is used for Fourier synthesis. Given the

* Many signals cannot be expressed in terms of Fourier components because the function $f(t)$ is not deterministic. Methods of analyzing these functions depend on expressing them in probabilistic forms, usually in terms of the spectral density function [3]. Much work has been done to analyze such signals with digital computers. This procedure has been made more efficient by use of the *fast Fourier transform*.

function of frequency, $g(\omega)$, of a signal, the signal may be synthesized as a function of time by Equation 6–12. Together, Equations 6–11 and 6–12 are the Fourier transform pair.

Most signals transmitted over the telecommunications network are random in parameters such as probability of occurrence, amplitude, or phase. Such signals usually cannot be expressed in terms of Fourier components because the function $f(t)$ is not deterministic. Much can be learned, however, by examining some random signals, such as the random data signals depicted in Figure 6–8 (also see Figure 14–3), in terms of the characteristics of one pulse [for which $f(t)$ is deterministic], provided the interaction among pulses is not neglected.

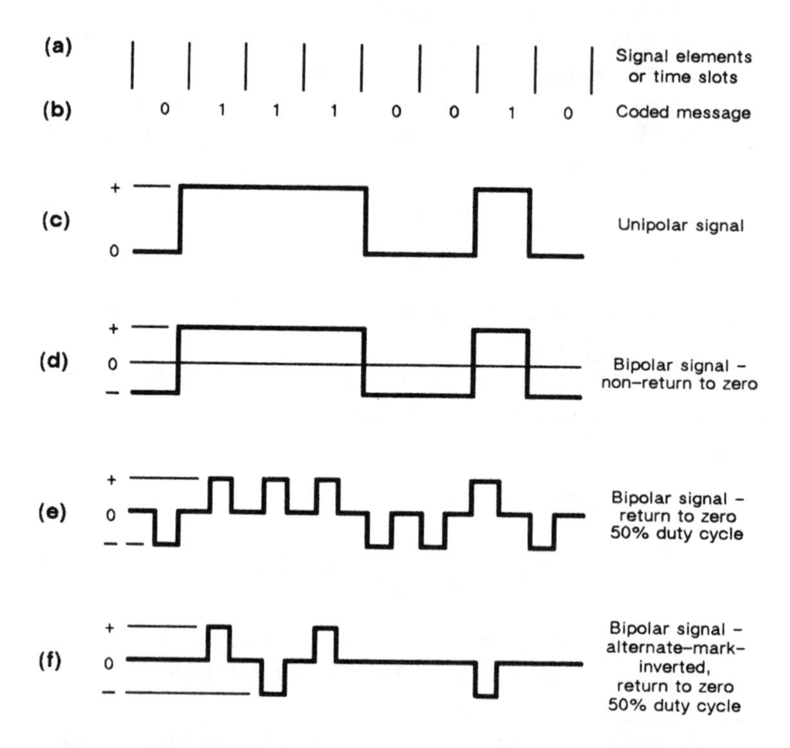

Figure 6–8. Some signal formats for a random signal.

The Single Rectangular Pulse. Consider the single rectangular pulse of Figure 6–9. From Equation 6–11 and from examination of the pulse [$f(t) = A$ from $-\tau/2$ to $+\tau/2$ and zero elsewhere], the Fourier transform may be written

$$g(\omega) = A\int_{-\infty}^{\infty} f(t)e^{-j\omega t}\,dt = A\int_{-\tau/2}^{\tau/2} e^{-j\omega t}\,dt.$$

Observation of the nature of the function $f(t)$ and subsequent substitution of the limits of integration make this equation tractable. Integrated, the equation becomes

$$g(\omega) = \frac{2A\sin\frac{\omega\tau}{2}}{\omega} = \frac{A\tau\sin\frac{\omega\tau}{2}}{\frac{\omega\tau}{2}}, \qquad (6\text{–}13)$$

the familiar $(\sin x)/x$ form. Note that the expression has a continuous distribution of energy at all frequencies, rather than at discrete frequencies as indicated for the components of the Fourier series for the periodic function represented by Equation 6–10. The function of Equation 6–13 is a pure real and, therefore, the components of the signal in the time domain are all cosine functions, in phase at $t = 0$. Values for $g(\omega)$ in Equation 6–13 may be found by using appropriate values for A and τ and, substituting $x = \omega\tau/2$, by use of the plotted values of $(\sin x)/x$ in Figure 6–7.

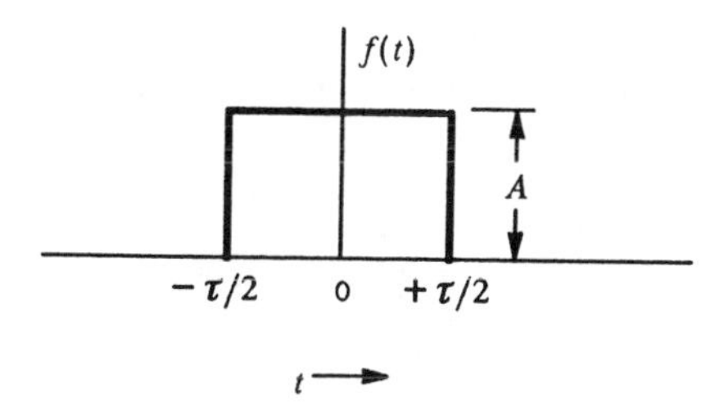

Figure 6–9. A single pulse.

The Impulse. An impulse is approximated when a rectangular pulse is narrowed without limit while keeping its area ($A\tau$ in Figure 6–9) unchanged. To simplify the treatment, the area may be assumed to be equal to unity. Thus, in the time domain an impulse is a signal having energy but infinitesimal duration.

The corresponding frequency spectrum may be found from Equation 6–13 by noting the assumption that $A\tau = 1$ and that $(\sin \omega\tau/2)/(\omega\tau/2) = 1$ when $\omega\tau/2 = 0$ (see Figure 6–7). Thus, the resulting spectrum contains all frequencies from $-\infty$ to $+\infty$ of equal phase at $t = 0$ and each having an amplitude of unity. This description of an impulse is useful in discussing the impulse response of a network.

Transmission Response

Transmission of nonperiodic signals through a network or transmission line may be studied by Fourier transform methods in either the frequency or the time domain.

Frequency Response. The complex frequency spectrum can often be used to simplify complicated problems. The advantages to be had by operating in the frequency domain arise from the relatively simple relationship between input and output signals transmitted through linear networks or transmission lines when the relationship is specified in that dimension. In a typical problem, the input signal has a spectrum $g_i(\omega)$ and the output $g_o(\omega)$. The transmission path can be described by a frequency function that is its transfer impedance (transfer voltage or current ratio), or what is commonly called its frequency response. This function, $H(\omega)$, can be established by computation from the known circuit constants of the system or network. It can also be found experimentally by applying a sine–wave test signal of known characteristics at the input and measuring amplitude and relative phase at the output.

The relationship between the input and output spectra of a signal applied to a network is particularly simple;

$$g_o(\omega) = H(\omega)\ g_i(\omega), \tag{6–14}$$

where g_o, g_i, and H are, in general, complex functions of the radian frequency, ω. In polar form, the amplitude and phase relationships are respectively,

$$|g_o(\omega)| = |H(\omega)|\ |g_i(\omega)| \tag{6–15}$$

$$\theta_o(\omega) = \theta_h(\omega) + \theta_i(\omega)\ . \tag{6–16}$$

The validity of these relationships rests upon the superposition principle since $g_O(\omega)$ is computed by assuming that it is a linear combination of the responses of the network to each frequency component (taken individually) in the input wave. This observation implies that if the response of a linear system to the gamut of sine–wave excitations is known, then its response to any other waveform can be found uniquely by decomposing that wave into its Fourier components and computing the response to each individual component. The output waveform $f_O(t)$, can be found by evaluating the Fourier integral of $g_O(\omega)$. The principle outlined here is the basis for all sine–wave testing techniques used in practice. It should be noted, however, that it is useful only for *linear systems* since it is only in such systems that superposition is generally valid. In the case of a nonlinear device, such as a rectifier, the response to each input waveform must be computed separately; the complex frequency response of the network does not allow generalization to include other functions.

Impulse Response. Transmission through a network can also be completely described in terms of its impulse response, which is defined as the function $h(t)$ that would be found at the output as a result of applying an impulse (previously defined) to the input terminals. Since the time function applied to the input has a flat frequency spectrum, it would be expected that $h(t)$ will have a spectrum that differs from flatness by the frequency characteristic of the network. In other words, $H(\omega)$ gives the frequency and phase spectra of $h(t)$. Expressed analytically, a unit impulse input to a network $H(\omega)$ produces an output $h(t)$ given by

$$\mathrm{F}[h(t)] = H(\omega) \tag{6–17}$$

where the F denotes the Fourier transform, from which it follows that

$$H(\omega) = \int_{-\infty}^{\infty} h(t)\ e^{-j\omega t} dt \tag{6–18}$$

and also

$$h(t) = \frac{1}{2\pi} \int_{-\infty}^{\infty} H(\omega)\ e^{\,j\omega t} d\omega. \tag{6–19}$$

The impulse response is, of course, a real function of time. Certain relationships between $H(\omega)$, $H(-\omega)$, and the conjugate of $H(\omega)$, written $H^*(\omega)$, can be shown [4]. These lead to the following:

$$\begin{aligned} H(-\omega) &= H^*(\omega) \\ H_R(\omega) &= H_R(-\omega) \\ -H_I(\omega) &= H_I(-\omega) \\ |H(\omega)| &= |H(-\omega)| \end{aligned} \qquad (6\text{–}20)$$

where H_R and H_I are the real and imaginary parts, respectively, of $H(\omega)$.

These are extremely important mathematical properties of any physical transmission path—network or transmission line. The first expression in the series of equations numbered 6–20 shows that the transfer impedance of the network, $H(\omega)$, expressed for negative frequencies, $H(-\omega)$, is equal to its conjugate expressed for positive frequencies, $H^*(\omega)$. From this fact, the second expression is derived directly to show that the real part of the impedance function, $H_R(\omega)$, has even symmetry about zero frequency. The third expression shows that the imaginary (phase) component of $H(\omega)$, H_I, has odd symmetry about zero frequency. The last expression, showing the relation between absolute values of H, follows from the first.

Bandwidth. It was previously shown, in the discussion of the single rectangular pulse, that the ability to establish limits of integration led to a useful expression for a frequency–domain description of the pulse. In a similar manner, the recognition of the finite bandwidth of a channel makes practical the impulse response analysis of transmission through a network.

An examination of the Fourier integral of Equation 6–19 indicates that, in order to determine the function of time corresponding to a particular frequency spectrum, it is necessary to know that spectrum from $-\infty$ to $+\infty$. However, in the application of Fourier synthesis to any real situation, the signal under study is always generated by a source capable of producing only a finite range of frequencies.

Similarly, the signal is carried on a channel capable of transmitting only a finite bandwidth. Hence, it is necessary to examine the spectrum only in this region, and the signal can be assumed to be zero outside this region. Such a finite bandwidth would restrict the number of time functions that can be synthesized to those whose fastest time rate of change is of the same order as the rate of the highest frequency component that may be present.

In practice, limits are used that depend on the characteristics of the physical system or circuit being dealt with, rather than using the infinite limits given in Equation 6–19. This equation may be modified to account for the finite bandwidth of any real system, and the Fourier integral can be written

$$h(t) = \frac{1}{2\pi} \int_{-\omega_2}^{-\omega_1} H(\omega)e^{j\omega t}d\omega + \frac{1}{2\pi} \int_{\omega_1}^{\omega_2} H(\omega)e^{j\omega t}d\omega. \qquad (6\text{–}21)$$

Example 6–2: Impulse Response of an Ideal Low–Pass Filter

As an example of the usefulness of the Fourier transform pair, consider a problem in pulse transmission, where information is being transmitted in digital form. At the transmitting terminal, a pulse is either sent or not sent at times t_1, t_2, etc. The problem is to tell, after the signal has been sent through the transmission medium (represented here by a low–pass filter), whether a pulse is present for each signal element or time slot at the receiver.

For this example, assume that the difference between two successive coded signals, illustrated by S_1 and S_2 in Figure 6–10, lies in the fact that S_1 has a pulse in position 5, whereas S_2 does not. Further assume that these signals are passed through a low–pass filter that has an idealized transmission characteristic shown in Figure 6–11. This idealized characteristic has a constant finite value of attenuation (assumed to be 0 dB for this problem) from zero frequency to f_1 and has infinite attenuation above f_1. It has no delay distortion for frequencies from zero to f_1; delay distortion above f_1 is of no consequence since there is no signal transmission above f_1. (This is an easy case to analyze; such characteristics are impossible to achieve but can be approximated.) The example, then, illustrates how bandwidth limitation alone can cause energy in the fourth position of S_2 to spill over into pulse position 5.

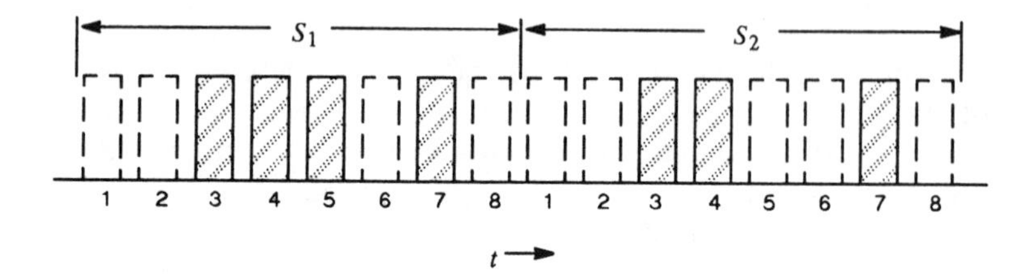

Figure 6–10. Successive code signals.

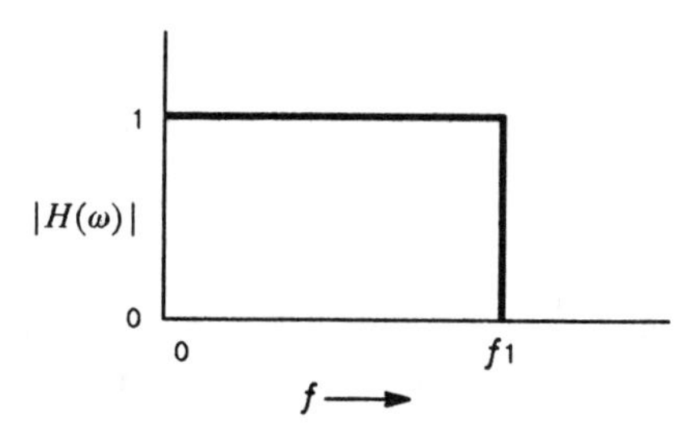

Figure 6–11. Idealized low-pass transmission characteristic.

If the transmission characteristic of the network is known, it is next necessary to assume a spectrum for the input pulse at position 4 and, in turn, determine its effect on the pulse or lack of pulse in position 5. Although the first inclination would probably be to assume a rectangular pulse like that of Figure 6–9 (even though real pulses are never exactly rectangular), the problem can be simplified by assuming an impulse. Compare the spectrum of an impulse (flat versus frequency, with no phase reversals) with the spectrum of a rectangular pulse in the region of $\omega = 0$ (almost flat for very low frequencies). It is seen that, if the transmitted bandwidth is small enough compared to the first frequency at which $(\sin x)/x$ becomes zero, the output will be the same whether the input is taken to be a narrow rectangular pulse or an impulse. Since the spectrum of an impulse is easier to handle analytically, the input is assumed to be an impulse. If it is desired to refine the results later, the input spectrum may be modified to have the (sin x)/x shape, or the frequency response, $H(\omega)$, may be modified.

Moreover, if the input signal is assumed to be an impulse, the task is to determine the signal (as a function of time) at the output of a path having the transmission characteristic shown in Figure 6–11. First notice that $|H(\omega)|$ can be plotted for negative as well as positive frequencies. By the relations of Equations 6–20, the plot would look like Figure 6–12, where $\omega_1 = 2\pi f_1$ has been substituted for f_1.

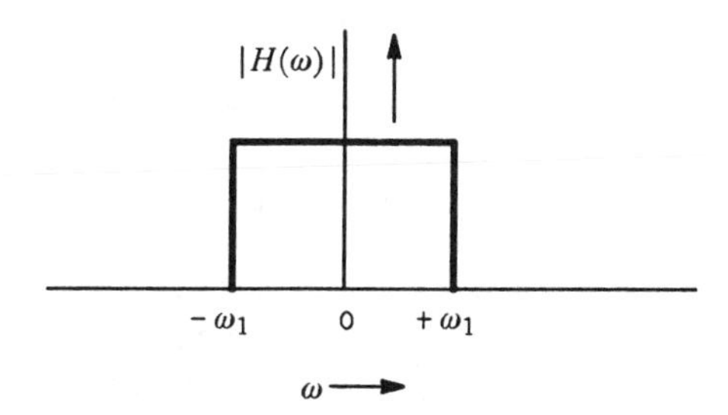

Figure 6–12. Idealized low-pass transmission characteristic (positive and negative frequencies).

If Equation 6–19 is applied to Figure 6–12 and constant delay is ignored, the output pulse may be represented as

$$h(t) = \frac{1}{2\pi} \int_{-\omega_1}^{\omega_1} e^{j\omega t} d\omega \ .$$

The term $H(\omega)$ in Equation 6–19 is shown in Figure 6–12 to be equal to unity in the interval from $-\omega_1$ to $+\omega_1$, and so does not appear in the above expression for $h(t)$.

This equation may be integrated to yield

$$h(t) = \frac{\omega_1}{\pi} \times \frac{\sin \omega_1 t}{\omega_1 t}.$$

This is a (sin x)/x function of time plotted in Figure 6–13. On this plot, $t = 0$ is arbitrary; for a physical network that approximates the characteristic of Figures 6–11 and 6–12, the zero time point represents the absolute delay of the transmission path. The optimum time for the next pulse is at $t = 1/(2f_1)$ because $h(t)$ goes through zero at that point, and interpulse (or intersymbol) interference is minimized.

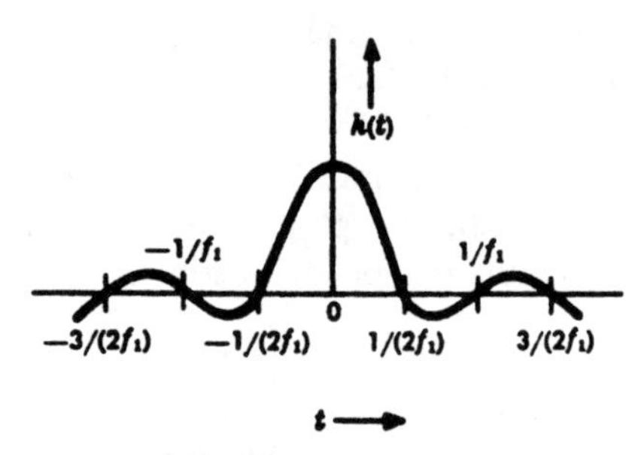

Figure 6–13. $h(t)$ at output of low-pass transmission path.

If the cutoff of the transmission path is at 500 kHz, then the interval between impulses should be 1 microsecond (repetition rate, 1 MHz). A shorter interval would tend to make the receiver think a pulse is present when in fact it is not; a longer interval would result in some cancellation when the following pulse is present. The spacing of pulses to avoid intersymbol interference is one of the fundamental requirements in digital transmission.

The necessity for distinguishing between the presence or absence of a pulse in position 5 of S_1 and S_2 in Figure 6–10 depends on a design that minimizes the effect of the presence of an unwanted signal in position 5 due to the pulse in position 4. This is accomplished by relating, in the design, the system transmission characteristic and pulse repetition rate so that the next pulse position (position 5) corresponds to the crossover of pulse number 4 at time $1/(2f_1)$ as illustrated in Figure 6–13.

This example illustrates the way in which the Fourier transform pair can be used. If an input signal, which is a given function of time, is assumed, the signal (as a function of time) at the output of a network can be found if the transmission characteristic of the network is known. The results may be expressed in very general functional terms in order to display the nature of a problem, or specific formulas may be used to obtain specific numerical results. In any particular case, finding the solution may be easy (as in Example 6–2) or may involve laborious or sophisticated mathematical manipulation of the specific functions involved in the problem. The basic idea remains the same.

Another class of transmission problems involves circuits having bandpass characteristics. Such problems are often difficult to solve directly, but are amenable to solution by the methods of Fourier analysis using an equivalent low–pass circuit arrangement such as that in Example 6–2 [5,6].

References

1. Scott, R. E. *Linear Circuits* (Reading, MA: Addison–Wesley Publishing Company, Inc., 1960).

2. *Reference Data for Engineers: Radio, Electronics, Computer and Communications,* Seventh Edition (Indianapolis, IN: Howard W. Sams and Company, Inc., 1985), pp. 7–10 to 7–12.

3. Franks, L. E. *Signal Theory* (Englewood Cliffs, NJ: Prentice–Hall, Inc., 1969).

4. Bogert, B. P. et al. *IEEE Transactions on Audio and Electroacoustics* (Special issue on the fast Fourier transform), Vol. AU–15 (June 1967).

5. Panter, P. F. *Modulation, Noise, and Spectral Analysis* (New York: McGraw–Hill Book Company, Inc., 1965).

6. Sunde, E. D. "Theoretical Fundamentals of Pulse Transmission," I and II, *Bell System Tech. J.*, Vol. 33 (May 1954), pp. 721–788 and (July 1954), pp. 987–1010.

Additional Reading

Anner, G. E. and W. L. Everitt. *Communication Engineering* (New York: McGraw–Hill Book Company, Inc., 1956).

Van Valkenburg, M. E. *Network Analysis* (Englewood Cliffs, NJ: Prentice–Hall, Inc., 1964).

Wylie, C. R., Jr. *Advanced Engineering Mathematics* (New York: McGraw–Hill Book Company, Inc., 1951).

Chapter 7

Negative Feedback Amplifiers

Detailed knowledge of feedback principles is needed only by those involved in the design and development of active transmission circuits. However, the high performance of modern transmission equipment so depends on the use of negative feedback that it appears desirable to provide some appreciation of why feedback is used, what it accomplishes, how it operates in electronic circuits, and what limitations exist in its application. With the design of feedback amplifiers used as the basis for discussion, feedback mechanisms and the interactions among them may be covered as background for an understanding of the interdependence of system and amplifier, or repeater, performance.

Negative feedback is commonly used in transmission systems because it suppresses unwanted changes in amplifier gain and substantially reduces harmonic distortion and interchannel modulation noise. It facilitates the design of amplifiers having much better broadband return loss characteristics than can be achieved without feedback. It is also widely used in voltage regulators, constant–current sources, and temperature regulators.

7–1 THE PRINCIPLE OF NEGATIVE FEEDBACK

In its simplest form, a negative feedback amplifier can be regarded as a combination of an ordinary amplifier (the μ circuit) and a passive network (the β circuit); by means of the latter, a portion of the output signal of the amplifier is combined out of phase with its input signal as illustrated in Figure 7–1. Ideally, this phase difference is 180 degrees and hence the term *negative feedback*.

The gain of a feedback amplifier may be written

$$\frac{e_2}{e_1} = \frac{\mu}{1 - \mu\beta} \quad . \tag{7-1}$$

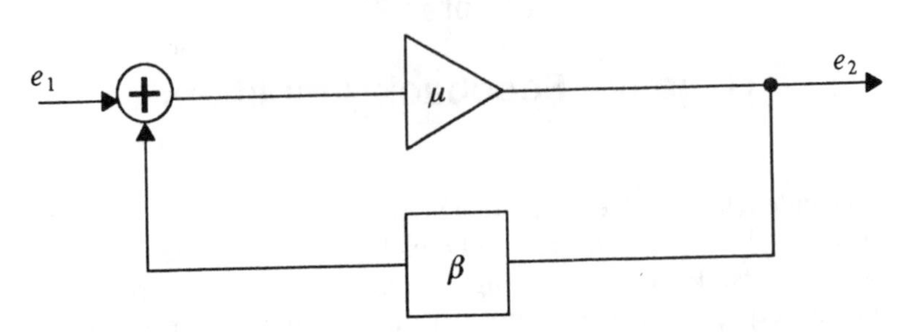

Figure 7-1. Feedback amplifier configuration.

Without feedback ($\beta = 0$), the gain would be simply $e_2/e_1 = \mu$. Thus, one effect of feedback is the reduction of gain by the term $1/(1 - \mu\beta)$.

In general, the μ gain is much larger than unity. As a result, an approximation may be derived from Equation 7-1 as follows:

$$\frac{e_2}{e_1} = \frac{1}{1 - \mu\beta} = \frac{\mu}{(1/\mu) - \beta} \approx -\frac{1}{\beta} \; ; \tag{7-2}$$

that is, the gain of a feedback amplifier is approximately proportional to β-circuit loss and is independent of μ-circuit gain.

These characteristics result in feedback amplifiers having attributes that far outweigh the disadvantage of reduced amplifier gain; consequently, in modern design, negative feedback is used in nearly all electronic amplifiers. It is especially valuable in amplifiers used in analog transmission systems where many amplifiers are connected in tandem. Here, without feedback, the cumulative effect of small imperfections in individual amplifiers would be intolerable.

7-2 APPLICATIONS OF FEEDBACK

The design of analog transmission systems involves finding simultaneous solutions to problems of bandwidth, repeater spacing, and signal–to–noise performance. These in turn are related to channel capacity, line loss and the achievability of compensating gain, the cumulation of interferences such as thermal and intermodulation noise, and the provision of adequate signal load–carrying capacity. The design of amplifiers to meet such requirements is made possible by feedback. It is incorporated in amplifiers of line repeaters used in analog and digital cable systems as well as in the amplifiers that are found in all types of terminal and station equipment. Feedback principles also apply to the design of preamplifier circuits in regenerators for metallic and fiber digital systems, in digital echo cancellers, and in adaptive hybrids.

One other important transmission system application is the use of feedback in dynamic backward–acting regulator and equalizer circuits. Such circuits use one or more single–frequency signals, called pilots, which are applied to a transmission system at the transmitting terminal at precise and carefully controlled frequencies and amplitudes. Immediately following a point of regulation, the pilot signal is picked off the line, rectified, and compared with a reference voltage. The error signal, i.e., the difference between the rectified pilot and the reference, is fed back to the input of a regulating amplifier through a network. The response of this network to the error signal changes the transmission gain in a direction and by an amount to correct the pilot amplitude at the output of the regulator. By the use of several pilots appropriately positioned in the signal spectrum, complex gain/frequency corrections are made across the entire signal band, resulting in dynamic equalization of the high–frequency line. Similar regulation principles apply to the intermediate–frequency portions of digital microwave receivers.

7–3 BENEFITS OF FEEDBACK

Once a system design is chosen, any departure from the ideal represents a penalty in performance. Departures in system gain result in increases in thermal noise if the gain is greater than desired. Furthermore, in the latter situation the system may become overloaded.

Equation 7–2 shows that gain of a feedback amplifier is nearly independent of the μ circuit. Thus, departures from the ideal gain/frequency characteristic (i.e., departures from design values) that are caused by changes in the μ circuit are effectively reduced by feedback. These changes may be caused by manufacturing, aging, and temperature–induced variations in μ–circuit components, which include the active devices, and by power supply fluctuations.

The nonlinear input/output characteristics of all active devices are another source of impairment in broadband electronic circuits. This type of impairment, often referred to as harmonic distortion or intermodulation noise, is also reduced by the use of negative feedback. If no other benefits accrued from using feedback, this alone would justify application in analog cable transmission systems and in FM terminals of microwave radio systems.

Additional feedback benefits accrue in the control of amplifier input and output impedances. Usually it is required that these impedances, or at least their absolute values, match the impedances of the circuits to which they connect. In nonfeedback amplifiers it is difficult to meet this requirement because the desired impedances are incompatible with the impedances of the devices used in the amplifiers. Circuit compromises often must be made to achieve an acceptable impedance match. In feedback amplifiers, however, the provision of feedback increases the design choices that can be made. It is usually possible to achieve a better impedance match over a wide bandwidth by using a feedback amplifier than otherwise.

Example 7–1: Feedback Effects

This simple example illustrates how a μ–gain change of about 0.8 dB may be suppressed by feedback to an amplifier gain change of approximately 0.1 dB.

Let the overall gain of an amplifier be 10 dB; that is,

$$20 \log \frac{e_2}{e_1} = 10; \frac{e_2}{e_1} \approx 3.16.$$

From Equation 7–2,

$$\frac{e_2}{e_1} = \frac{\mu}{1 - \mu\beta} \approx 3.16.$$

Assume the μ gain (without feedback) is $20 \log \mu = 30$ dB; then

$$\mu = 31.6$$

and, by substitution,

$$\beta = 0.248.$$

Now, let the μ gain increase from 30 dB to 30.8 dB; that is, μ increases by about 10 percent from 31.6 to 34.8.

Then the overall amplifier gain is

$$\frac{e_2}{e_1} = \frac{34.8}{1 - 34.8\,(-0.284)} = 3.2$$

and

$$20 \log \frac{e_2}{e_1} = 20 \log 3.2 = 10.1 \text{ dB}.$$

Thus a 10–percent change in μ–circuit gain is held to about 1.2–percent change in overall gain (0.1 dB).

That the amplifier gain increased as the μ gain increased is due to the phase relationships implied by the simple substitutions made. In complex feedback structures, the amplifier gain might increase or decrease over limited portions of the band and within a limited range of the μ–gain change.

7–4 CIRCUIT CONFIGURATIONS

The principal circuit configurations useful in feedback circuits can be classified most easily in terms of the way in which the μ and β circuits are connected to each other and to the external interconnections at amplifier input and output. The variety of connections that can be made cannot be clearly demonstrated by a simple drawing such as that of Figure 7–1. The actual situation is that shown broadly by Figure 7–2 in which the μ, β, input, and output circuits are interconnected by means of six–terminal networks. The classification of feedback circuits then depends on the forms that these six–terminal networks assume.

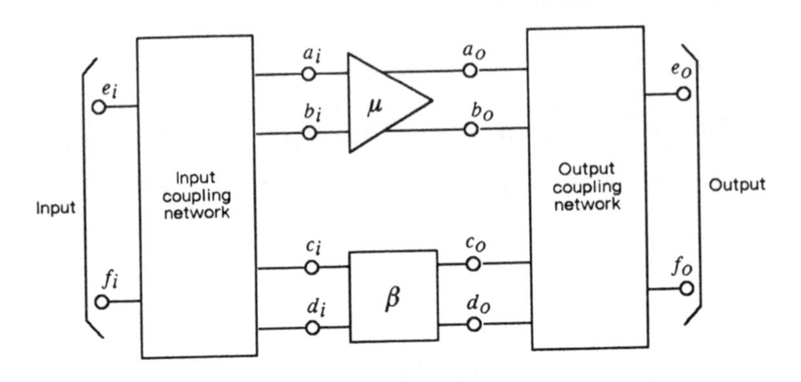

Figure 7–2. Feedback amplifier representation.

Illustrations of some of the more common feedback amplifier structures are given in Figures 7–3 through 7–6. Where appropriate, the network terminals are identified in accordance with the notation used in Figure 7–2. The μ circuits commonly have one, two, or three stages of gain; an unlimited number of network configurations may be found in the passive networks shown in the figures. To avoid complexity here, the internal network configurations are generally omitted in the figures.

Series and Shunt Feedback

The configuration of Figure 7–3 is called series feedback because, as seen from the input and output terminals, the μ and β circuits are in series. The β circuit, shown here as the π arrangement of three impedances (A, B, and C) may be much simpler or much more complex than that illustrated. The effective line terminals (e_i, f_i, e_o, and f_o) are shown at the high sides of the transformers since the transformer characteristics in this case may be added directly to those of the connecting circuits.

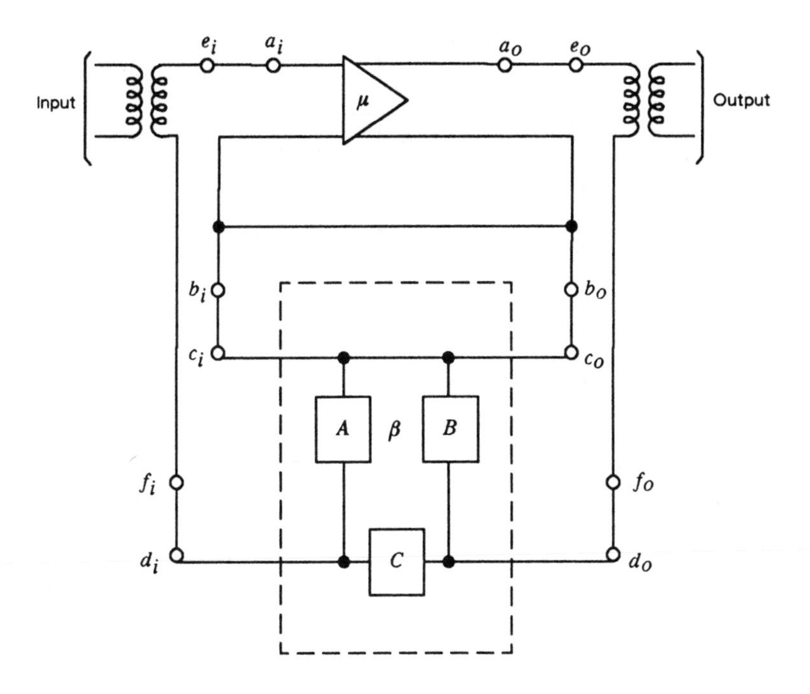

Figure 7–3. Series feedback amplifier.

Figure 7–4 shows how feedback may be provided by means of shunt connections. The β circuit, here represented as a T network of the three impedances, may again take on many configurations. Note that the connecting terminals (input and output), β network, and μ network are all in parallel.

Series and shunt feedback designs are simple and convenient for many applications. Feedback tends to make the effective input and output impedances of the amplifier very high or very low. As a result, it is possible to build out these impedances conveniently by the use of discrete components to achieve a good impedance match to the connecting network or transmission line. A disadvantage is that the connecting impedances form a part of the $\mu\beta$ loop. As a result, variations in the line impedance, sometimes large and impossible to control, affect the $\mu\beta$ characteristics; in some cases, the effect may be great enough to cause amplifier instability.

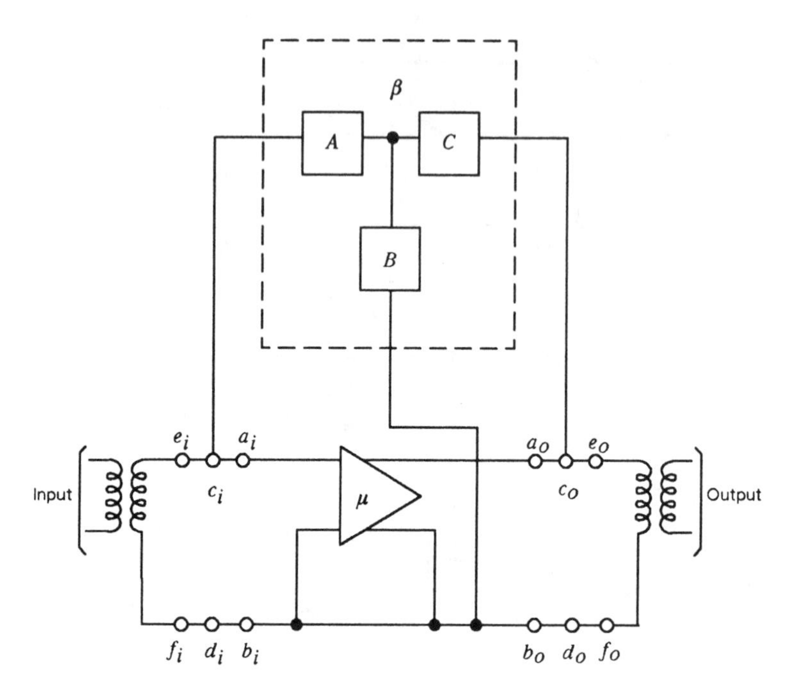

Figure 7-4. Shunt feedback amplifier.

Bridge-Type Feedback

These difficulties may be mitigated by using bridge-type feedback circuits. The configuration that is most commonly used, especially for broadband repeaters in analog cable systems, is the high-side hybrid feedback arrangement illustrated in Figure 7-5. Several network branches must be added in this configuration to provide hybrid balance and input and output impedance control. These branches are designated Z_n and Z_1 in Figure 7-5. The advantages of this circuit include the achievement of minimum noise and improved intermodulation performance while controlling both the input and output impedances.

Figures 7-3 through 7-5 show symmetrical arrangements at each end of the amplifier. This has been done only to simplify the illustrations. The number of configurations is increased greatly by combining different types of connections at input and output. Furthermore, circuit advantages can sometimes be realized by providing multiple loop configurations. An example of such a configuration is given in Figure 7–6. Here, a feedback amplifier with a series feedback network $Z_{\beta 1}$, similar to that of Figure 7–3, is shown with local shunt feedback $Z_{\beta 2}$ around the last stage of a three-stage configuration in the μ path. The impedances Z_{i1} and Z_{i2} are interstage networks in the μ path.

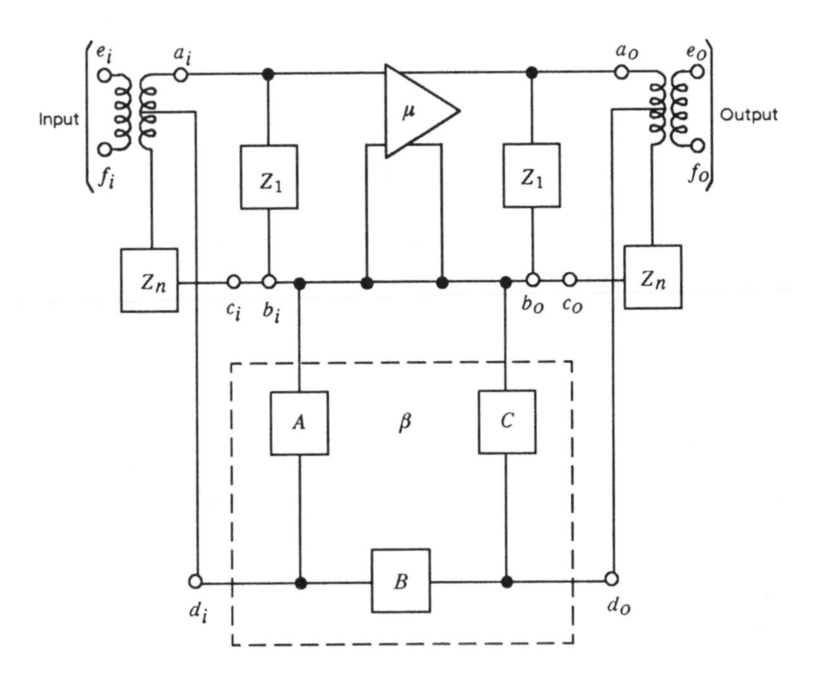

Figure 7–5. Amplifier with high-side hybrid feedback.

7–5 DESIGN CONSIDERATIONS

It is not possible or desirable to review here the entire procedure followed in designing a feedback amplifier. However, some important relationships and design limitations are discussed in order to provide an improved understanding of how transmission systems operate and how system performance is related to the design of the individual amplifier.

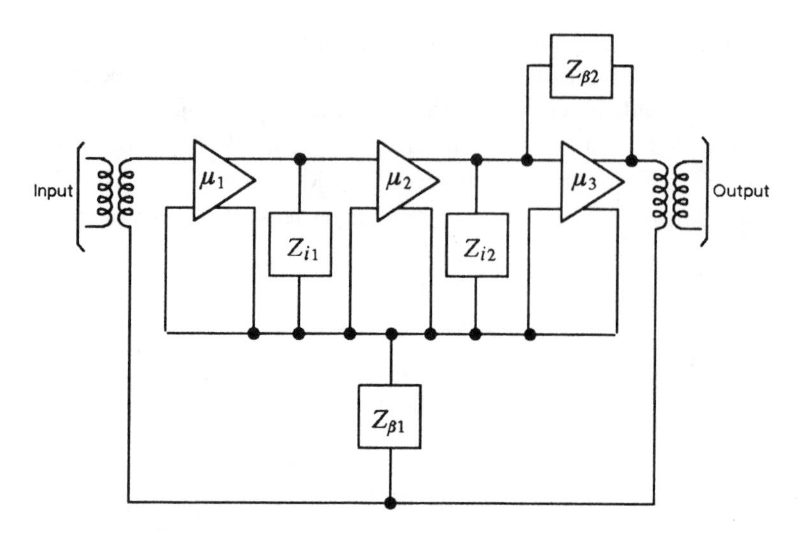

Figure 7-6. Three-stage series feedback amplifier with local shunt feedback on last stage.

Gain and Feedback

The shape and magnitude of the gain/frequency characteristic are basic design considerations. The closeness of the gain/frequency characteristic to the desired characteristic may be determined by the degree of circuit complexity that can be tolerated; however, the better the match, the better will be the ultimate transmission characteristic of the system.

Characteristic Shaping. The characteristics of feedback amplifiers are all complex functions of frequency that are importantly related to the transmission characteristics of all the networks making up the complete amplifier and its external terminations.

In many applications, it is desirable to design the amplifier to a flat gain, one that is equal over the entire transmitted band. In the case of line repeaters for analog cable systems, it is usually desirable to have the gain of the amplifier sections of the repeaters match the loss of the cable section over the band of interest. In either case, the desired flat or shaped gain/frequency characteristic is produced primarily by proper design of the β-circuit network since the gain is approximately equal to $-1/\beta$ as shown in Equation 7–2. Some gain shaping may also be provided in those networks that are outside the $\mu\beta$ loop, such as the coupling networks shown in many of the figures as simple transformers.

To achieve optimum signal–to–noise performance, it is also desirable in many cases to shape the feedback/frequency characteristics of an amplifier. For example, it is possible to increase low–frequency feedback at the expense of high–frequency feedback. This can be accomplished by careful design of all networks in the $\mu\beta$ loop, using frequency–dependent reactive components, since the feedback is, by definition, proportional to $1/(1-\mu\beta)$.

Gain and Phase Margins. The selection of a circuit configuration and the amount of feedback to be provided depend on the magnitudes of the gain and bandwidth required and on the characteristics of available active devices. These considerations include the linearity of the device input/output characteristics, the noise figure of the input device, and the need for minimizing variations in circuit parameters due to device aging and ambient temperature changes.

As shown in Equation 7–2, the insertion gain of a feedback amplifier is

$$\frac{e_2}{e_1} = \frac{\mu}{1-\mu\beta} \approx -\frac{1}{\beta} \; .$$

The total gain around the feedback loop is defined as $\mu\beta$, where μ is the total gain provided by the active devices (and their related μ–circuit networks) and β is the loss of the network that connects the output back to the input. From these relationships, the loop gain in dB is

$$20 \log \mu\beta = 20 \log \mu + 20 \log \beta \approx 20 \log \mu - g_R \quad (7\text{–}3)$$

where g_R is the insertion gain of the complete closed–loop amplifier in dB. It is approximately equal to $-20 \log \beta$. Thus,

$$20 \log \mu \approx 20 \log \mu\beta + g_R \; . \quad (7\text{–}4)$$

That is, the sum of the loop gain and insertion gain cannot exceed the total gain available in the μ circuit. It is therefore impossible to get loop gain in excess of the difference between the μ gain and the desired insertion gain. When the desired loop gain is greater, the design is said to be *gain limited.*

Most broadband amplifiers, however, are not gain limited; the need for adequate stability margins is usually controlling. In the gain expression $\mu/(1-\mu\beta)$, the denominator may become zero, depending on phase relationships, when $\mu\beta = 1$. If $\mu\beta$ is equal to unity at any frequency, inband or out-of-band, the amplifier may become unstable and break into spontaneous oscillation at that frequency if the phase of $\mu\beta$ is unfavorable. If it were possible to hold $|\mu\beta| >> 1$ for all frequencies, this would not be a problem, but every active device has some frequency above which its gain decreases monotonically. The rate of decrease may be increased by stray inductance or capacitance. Thus, there is always a frequency at which $|\mu\beta| = 1$.

Two criteria must be satisfied to guarantee a stable amplifier; the phase must be greater than 0 degrees where $|\mu\beta|$ passes through 0 dB, and $|\mu\beta|$ must represent several dB of loss where the phase passes through 0 degrees. These criteria are known as the *phase and gain margins* in an amplifier design. If an amplifier has such margins, it is said to meet the Nyquist stability criteria. Such margins are illustrated in Figure 7–7 where the characteristics are plotted on an arbitrary, normalized frequency scale. A phase margin of about 30 degrees and gain margin of about 10 dB, as illustrated, allow for variations in device characteristics that result from manufacturing processes, aging, and temperature.

The achievement of adequate phase and gain margins sets an upper limit on the achievable inband feedback. When this limit is lower than that set solely by gain considerations, the design is said to be *stability limited.*

Ideally, maximum stability margins would result if the phase of the $\mu\beta$ characteristics could be held at 180 degrees. Then, the gain expression could be written as $1/(1+|\mu\beta|)$. Within the transmission band, the phase is often controlled to approach this condition. However, out-of-band phase can change due to phase shifts inherently associated with any gain/frequency characteristic, such as the gain cutoff mentioned earlier. Furthermore, for very high frequencies the propagation time around the feedback loop contributes additional phase shift that can be minimized, but not eliminated, by careful design.

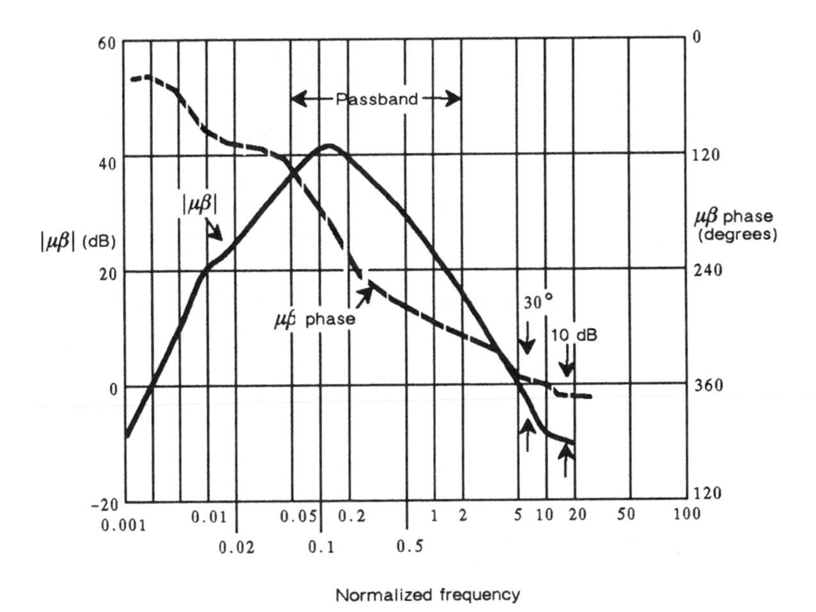

Figure 7–7. Typical feedback amplifier characteristics.

Nonlinear Distortion and Overload

In addition to the related considerations of gain and achievable feedback, the combination of overload, gain, and nonlinear distortion must be considered in feedback amplifier design. These can be studied by first examining the phenomenon of nonlinear distortion and its reduction by feedback and then relating these to the problems of gain and overload.

Nonlinear Distortion. The generation of intermodulation products by nonlinear input/output characteristics of transistors is a very complex phenomenon. The analysis here is oversimplified in order to illustrate how products are generated, how feedback tends to suppress them, and how gain and overload are affected.

The nonlinear input/output voltage relationships of an amplifier may be represented by the expression

$$e_O = a_0 e_i^0 + a_1 e_i^1 + a_2 e_i^2 + a_3 e_i^3 + \ldots \quad , \qquad (7\text{–}5)$$

where e_O and e_i are the output and input signal voltages, and the a coefficients provide magnitude values of various wanted and unwanted components in the output signal. If the input signal has many frequency components, Equation 7–5 may be used to study intermodulation by assuming

$$e_i = A \cos at + B \cos \beta t + C \cos \gamma t.$$

When this value of e_i is substituted in Equation 7–5, the expression can be expanded by trigonometric identities. The output voltage then contains an infinite number of terms consisting of various combinations of input signal components; the magnitudes are represented by the coefficients A, B, and C of the input signal and a_0, a_1, etc., of the input/output expression. Fortunately, in most applications the magnitudes of terms in Equation 7–5 having exponents of the fourth power and higher are so small that they usually may be ignored.

To demonstrate the nonlinear phenomenon and the effects of feedback, a few specific terms of the output voltage, extracted from expansion of Equation 7–5 after substituting the expression for e_i, may be examined. The terms of interest are

$$e_1 = a_1 A \cos at \qquad (7\text{–}6)$$

$$e_2 = a_2 AB \cos (a+\beta)t \qquad (7\text{–}7)$$

$$e_3 = \frac{3}{2} a_3 ABC \cos (a+\beta-\gamma)t. \qquad (7\text{–}8)$$

The first term, Equation 7–6, is a component of the output that corresponds exactly with the first term of the input signal ($A \cos at$) except for the coefficient a_1. This coefficient may be regarded as a measure of gain of the amplifier, g_R. As shown in Equation 7–4, the value of g_R, and therefore the value of a_1, is a function of the feedback, $\mu\beta$.

Equation 7–7 represents an intermodulation distortion component derived from the second–order term of Equation 7–5. The coefficient of this term involves magnitudes A and B of the feedback, $\mu\beta$; to a first approximation, the value of a_2 is reduced in direct proportion to the amount of feedback provided.

Equation 7–8 represents an intermodulation distortion component derived from the expansion of the third–order term of Equation 7–5. The coefficient involves the magnitudes A, B, and C of the three intermodulating input signal components and the coefficient a_3 of Equation 7–5. The value of a_3 is also reduced by feedback but not by as simple a relationship as a_1 and a_2. Second–order modulation components, fed back to the input, mix with fundamental signal components to produce products that appear at the output as third–order products. The result is that the reduction of third–order intermodulation is not quite as effective as the reduction of second–order intermodulation.

Overload. The coefficients 20 log M_2 and 20 log M_3 are essentially constant over most of the signal amplitude range of interest (though they may be functions of frequency). However, as overload is approached, departures from constant values of 20 log M_2 and 20 log M_3 are observed, as are departures from normally constant gain. These observations lead to a number of definitions of overload in a feedback amplifier. Typical characteristics are plotted in Figure 7–8 for departures of 20 log M_3 and gain from their nominal values as functions of the signal power at the output of a repeater. Three definitions of overload are discussed briefly below; two are related to departure of 20 log M_3 from a constant value, and one is related to the departure of gain from constant value.

Definition 1: By this definition, the overload point is that value of output signal power at which 20 log M_3, the third–order modulation coefficient, increases by 0.5 dB relative to its nominal constant value. This is identified as point P_{R1} at 20 dBm in Figure 7–8. This definition, appropriate for use in systems limited by intermodulation, is conservative in the sense that only a slight performance impairment results from exceeding the limit by a small amount. A relatively small amount of overload margin would be allowed in a design based on this definition.

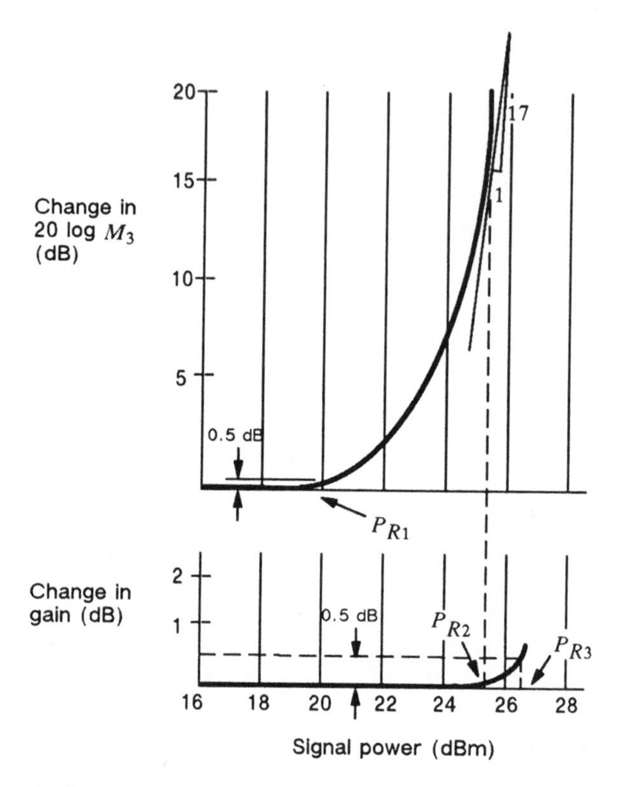

Figure 7–8. Overload point definitions as applied to a typical amplifier.

Definition 2: In this case, the overload point is defined as that value of output signal power at which the third–harmonic power increased by 20 dB for a 1–dB increase in signal power; this corresponds to a 17–dB increase in 20 log M_3. Since under these conditions very serious transmission impairment may result, a more generous overload margin must be provided. This definition of overload is recommended by the International Telegraph and Telephone Consultative Committee (CCITT) [1]. Its use is justified by the statistics of system performance interactions in long analog cable systems and by the amplitude/frequency statistics of a broadband signal; together, these statistics are used to show a very low probability of overload. The overload point is illustrated by point P_{R2} in Figure 7–8 at a signal power of about 25.5 dBm.

Definition 3: The overload phenomenon may be related to changes in amplifier gain, whereby the overload point is defined as the signal power at the output at which the amplifier gain departs from its nominal value of 0.5 dB, as illustrated by point P_{R3} on the lower portion of Figure 7–8. For this illustration, the overload point is about 26.5 dBm. The use of this definition may be appropriate when intermodulation distortion is not a major consideration.

The range of values of defined amplifier overload points is fairly wide; for example, it is 6.5 dB in Figure 7–8. However, in the event that the overload point is exceeded under definition 2 or 3, performance degradation is so severe that wider system margins must be provided in most cases. Thus, the actual operating value of load might well be approximately the same no matter which definition is used.

Noise and Terminations

It is desirable to introduce the subject of thermal noise generation in networks and systems here in order to relate the phenomenon to amplifier design and, thus, to overall system performance [2].

It can be shown that the available noise power of a thermal noise source is directly proportional to the product of the bandwidth of the system or detector and the absolute temperature of the source. This relation can be expressed as

$$p_a = kTB \text{ watts}, \tag{7–9}$$

where k is Boltzmann's constant (1.3805×10^{-23} joule per Kelvin), T is the absolute temperature in degrees Kelvin (290°K is taken as room temperature), and B is the bandwidth in hertz. Available noise power may also be expressed as

$$P_a = -174 + 10 \log B \quad \text{dBm}. \tag{7–10}$$

The noise figure for a two–port network is defined as follows: "The noise figure at a specified input frequency is the ratio of (1) the total noise power per unit bandwidth at a corresponding output frequency available at the output when the noise temperature of the input source is standard (290°K) to (2) that portion of this output power engendered at the input frequency by the input source" [3]. The noise figure, when applied according to this definition to a narrow band, ΔB, is called a *spot noise figure*. The spot noise figure may vary as a function of frequency.

Alternately, the spot noise figure, n_F, can be expressed in terms of signal-to-noise ratio. This may be expressed as

$$n_F = \frac{p_{si}/p_{ni}}{p_{so}/p_{no}} \qquad (7\text{-}11)$$

where p represents power and the subscripts are s for signal, n for noise, i for input, and o for output. Here, the noise figure is defined as the ratio (p_{si}/p_{ni}) of the available signal-to-noise power ratio at the input of the two-port network to the available signal-to-noise power ratio (p_{so}/p_{no}) at the output of the two-port when the temperature of the noise source is standard (T = 290°K).

The value of p_{ni} can be determined, by substitution in Equation 7-9, as $p_{ni} = kT\Delta B$. The ratio p_{si}/p_{so} is the gain, $g_a(f)$, of the network. The substitution of these values in Equation 7-11 yields

$$n_F = \frac{p_{no}}{g_a(f) \cdot kT\Delta B} \;. \qquad (7\text{-}12)$$

Examination of Equation 7-12 shows that the noise figure of an amplifier is importantly related to the thermal noise generated at the input (where the signal is at its lowest amplitude), to the gain of the amplifier, and to any sources of noise greater than the input noise amplified by $g_a(f)$. These internal noise sources are to some extent subject to control by circuit design techniques. The dominant source, however, is usually at the amplifier input. Here, the noise source is outside the $\mu\beta$ loop and, as a result, the noise figure is not improved by feedback.

The selection of components and the design of the input circuits of amplifiers for minimum noise figure is important in transmission system design (e.g., in preamplifiers for fiber optic regenerators). The cumulation of noise in tandem–connected analog amplifiers is directly related to the number of amplifiers in tandem. Thus, when the number of amplifiers has been set by repeater spacing, gain, and bandwidth considerations, the noise performance is controlled by the individual noise figures of the amplifiers.

As mentioned, the design of feedback amplifiers and their classification into a variety of types depend on the forms that the six–terminal coupling networks take and the manner in which β and μ circuits and external circuit connections are made. At the input, the design must simultaneously (1) satisfy return loss requirements by providing a termination to properly match the amplifier input impedance to the line impedance, (2) minimize the noise figure of the first–stage device by suitably matching its input impedance to the driving point impedance, and (3) meet feedback and gain–shaping requirements. At the output, the design must again satisfy impedance matching and feedback requirements and, in addition, must minimize penalties in nonlinear and overload performance that might result from improper last–stage terminations. In general, these combinations of requirements can best be met by the use of hybrid feedback connections, described previously and illustrated in Figure 7–5.

7–6 OTHER TECHNIQUES

While this chapter focuses on negative feedback as the principal means of obtaining ultralinear amplifiers, it is appropriate to mention that two other techniques are available for VHF–UHF–SHF applications. At such high frequencies, the phase shift in the loop of a feedback amplifier becomes intolerable. One potential design method without this difficulty is feed–forward, in which the intermodulation products generated in an amplifier are partially cancelled by use of suitable delay lines and summation circuits [4]. Another is the use of a predistorter circuit, which produces out–of–phase intermodulation products that subtract from equivalent products generated in the main amplifier. This method has been applied commercially in 6000–channel single–sideband microwave radio systems operating in the 6–GHz region [5].

References

1. CCITT, IInd Plenary Assembly, Rec. G.222 (New Delhi, 1960).

2. Members of Technical Staff. *Transmission Systems for Communications,* Fifth Edition (Murray Hill, NJ: AT&T Bell Laboratories, Inc., 1982), Chapter 4.

3. *Definition of Electrical Terms,* ASA–C42.65, American Standards Association (1957).

4. Seidel, H. "A Microwave Feed–Forward Experiment," *Bell System Tech. J.*, Vol. 50, No. 9 (Nov. 1971).

5. Hecken, R. P., R. C. Heidt, and D. E. Sanford. "The AR6A Single–Sideband Microwave Radio System—Predistortion for the Traveling–Wave–Tube Amplifier," *Bell System Tech. J.*, Vol. 62, No. 10, Part 3 (Dec. 1983).

Additional Reading

Blecher, F. H. "Design Principles for Single Loop Feedback Amplifiers," *Transactions of the IRE,* Vol. CT–4 (Sept. 1957).

Bode, H. W. *Network Analysis and Feedback Amplifier Design* (Princeton, NJ: D. Van Nostrand Company, Inc., 1945).

Hakim, S. S. *Junction Transistor Circuit Analysis* (New York: John Wiley and Sons, Inc., 1962).

Thomas, D. E. "High–Frequency Transistor Amplifiers," *Bell System Tech. J.*, Vol. 38 (Nov. 1959).

Chapter 8

Modulation

Communications signals must usually be transmitted via a medium separating the transmitter from the receiver. Since the information to be sent is rarely in the best form for direct transmission, transmission efficiency requires that it be processed before being transmitted. *Modulation may be defined as that process whereby a signal is converted from its original form into one more suitable for transmission over the medium between the transmitter and receiver* [1]. The process may shift the signal frequencies to facilitate transmission or to change the bandwidth occupancy, or may materially alter the form of the signal to optimize noise or distortion performance. At the receiver, this process is reversed by demodulation.

Satisfactory transmission and recovery of modulated signals depend on the introduction by the medium of no more than a specified amount of distortion. The effects of distortion in the medium may be quite different for different modulation modes. If maximum distortion values are exceeded, signal impairments at the receiver are excessive. Distortions that must be considered are of many types. They include amplitude distortion, which results from the variation of transmission loss with frequency, and phase distortion (often expressed as delay distortion), which results from the phase/frequency characteristic of the channel departing from linear. Other forms of signal impairment that may result in imperfect signal demodulation include nonlinear channel input/output characteristics, frequency offset, amplitude and phase jump, echoes, and noise. These impairments are treated in later chapters. They are discussed in this chapter only where they result directly from the modulation or demodulation process.

The modulation process can be represented mathematically by an equation that, in its most general form, can be used to express any of several forms of modulation. The several forms include amplitude modulation (AM), angle modulation (frequency or phase), and pulse modulation. While other expressions are more representative of the various forms of pulse modulation, there is one form of the equation that lends itself particularly to studies of amplitude and angle modulation,

$$M(t) = a(t)\cos[\omega_c t + \phi(t)]. \tag{8-1}$$

Here $a(t)$ represents the amplitude of the sinusoidal carrier, and $\cos[\omega_c t + \phi(t)]$ is the carrier and its instantaneous phase angle. An amplitude-modulated system is one in which $\phi(t)$ is a constant, and $a(t)$ is functionally related to the modulating signal. An angle-modulated system results when $a(t)$ is held constant and $\phi(t)$ is made to bear a functional relationship to the modulating signal. It is appropriate to discuss each of these two types separately and in some detail.

All three general types of modulation (amplitude, angle, and pulse) are used extensively in telecommunications equipment. For example, broadband multiplex equipment and N-type carrier systems employ several forms of amplitude modulation, most microwave radio systems employ angle modulation or quadrature amplitude modulation (QAM) for the high-frequency signal transmitted between transmitting and receiving antennas, and digital carrier systems employ pulse modulation.

8-1 PROPERTIES OF AMPLITUDE-MODULATED SIGNALS

Equation 8-1 can be modified to represent amplitude modulation by making $\phi(t)$ a constant. For convenience, let $\phi(t) = 0$ to obtain

$$M(t) = a(t)\cos\omega_c t, \tag{8-2}$$

where the carrier is at the frequency $f_c = \omega_c/2\pi$ and where $a(t)$ is the modulation signal which is a function of time. Since the modulated wave, $M(t)$, is the product of $a(t)$ and a carrier wave, the process can be called product modulation.

A general expression for $a(t)$ may be written as

$$a(t) = a_0 + mv(t). \qquad (8\text{–}3)$$

If Equation 8–3 is normalized by letting the dc component, a_0, equal 1, the coefficient m is defined as the modulation index. It is equal to unity for 100–percent modulation.

Now, let $v(t)$ represent a signal containing two components at different frequencies, f_m and f_n, having amplitudes of a_m and a_n, respectively. Then,

$$mv(t) = m(a_m \cos\omega_m t + a_n \cos\omega_n t)$$

and, by substitution in Equation 8–3,

$$a(t) = a_0 + m(a_m \cos\omega_m t + a_n \cos\omega_n t). \qquad (8\text{–}4)$$

By substitution in Equation 8–2 and by trigonometric expansion, the modulated signal becomes

$$\begin{aligned} M(t) &= [a_0 + m(a_m \cos\omega_m t + a_n \cos\omega_n t)] \cos\omega_c t \\ &= a_0 \cos\omega_c t \\ &\quad + \frac{m}{2} [a_m \cos(\omega_c - \omega_m)t + a_n \cos(\omega_c - \omega_n)t]. \\ &\quad + \frac{m}{2} [a_m \cos(\omega_c + \omega_m)t + a_n \cos(\omega_c + \omega_n)t]. \qquad (8\text{–}5) \end{aligned}$$

If in Equation 8–5 the coefficients a_0 and a_n are zero, and in addition a_m and m both equal unity, the resulting modulated wave expressed by Equation 8–5 reduces to

$$M(t) = \frac{1}{2} \cos(\omega_c - \omega_m)t + \frac{1}{2} \cos(\omega_c + \omega_m)t. \qquad (8\text{–}6)$$

Equation 8–6 contains no component at the original carrier frequency, f_c, but only a side frequency on either side of the carrier and spaced f_m hertz from the carrier frequency, as shown in Figure 8–1. The terms in Equation 8–6 containing $(\omega_c - \omega_m)$ and $(\omega_c + \omega_m)$ are known as the lower and upper sidebands (LSB and

USB), respectively. The resultant wave of Equation 8–6 represents a form of modulation known as double–sideband suppressed carrier (DSBSC). This form is characterized by a zero–amplitude dc component in the modulating signal and, as a result, a modulated signal having no component at the carrier frequency.

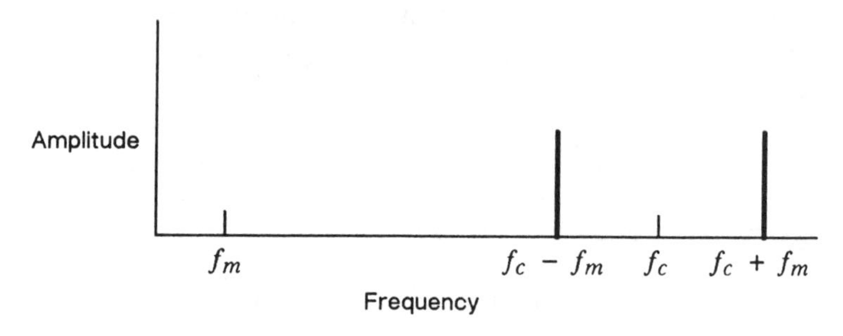

Figure 8–1. Product modulator—single–frequency modulating signal.

Consider next the resultant form of Equation 8–5 if $a_0 = 0$ and m, a_n, and a_m are all unity. Then the modulated wave is

$$M(t) = \frac{1}{2}[\cos(\omega_c - \omega_m)t + \cos(\omega_c - \omega_n)t] + \frac{1}{2}[\cos(\omega_c + \omega_m)t + \cos(\omega_c + \omega_n)t]. \qquad (8\text{–}7)$$

The result is as if the two modulating frequency components at f_m and f_n were modulated independently and then added linearly. Thus, superposition holds, the product modulation process is quasi–linear, and it may be inferred that product modulation translates the baseband signal in frequency and reflects it symmetrically about the carrier frequency without distortion.* The result is illustrated in Figure 8–2(a), which shows the two–frequency case, and in Figure 8–2(b), which shows the more general case of a modulating wave having a

*Note that while the mathematical analysis for product modulation is linear, the physical realization of the process often involves the use of non-linear devices. The mode of operation in these cases still results in a quasi-linear process output.

spectrum from f_a to f_b where $f_b < f_c/2$. Note that if $f_b > f_c/2$, the baseband and LSB signals overlap. Ambiguity or distortion, which can occur in the recovered signal, may be avoided in design by choosing frequencies to make $f_b < f_c/2$.

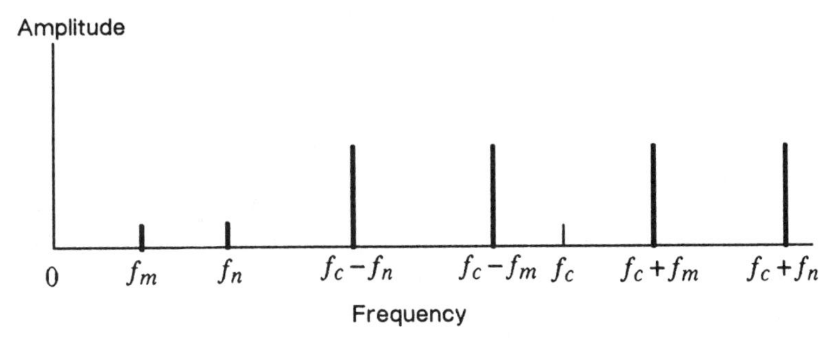

(a) Modulated spectrum of two modulating frequencies

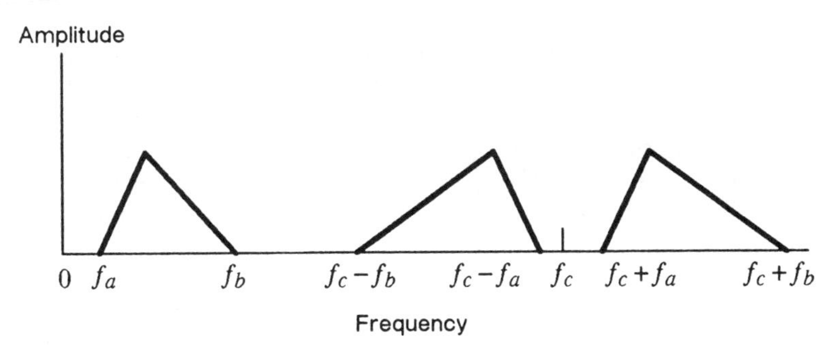

(b) Modulated spectrum of modulating band of frequencies

Figure 8–2. Product modulator frequency spectrum—complex modulating signal.

If $a(t)$ is given a strong dc component, i.e., $a_0 \neq 0$, the function $a(t)$ may be restricted to values of one sign only (for example, positive values only). Then, a carrier component in the output wave will result, as shown by the first term in Equation 8–5. The resultant wave is known as a double sideband with transmitted carrier (DSBTC) signal.

If either sideband in the DSBSC spectrum, Figure 8–2(b), is rejected by a filter or other means, the result is a single–sideband (SSB) wave. Basically, SSB modulation is simply frequency translation, with or without the inversion obtainable by selecting the LSB rather than the USB. Sideband suppression by filtering is the most common method. When this is done, the carrier component is usually effectively suppressed with the unwanted sideband.

Up to this point, three types of amplitude–modulated signal have been mentioned: double sideband with transmitted carrier, double sideband suppressed carrier, and single sideband. Subsequently, the properties of these three signals are examined, and finally a fourth type, known as vestigial sideband (VSB), is considered.

Double Sideband with Transmitted Carrier

DSBTC provides a basis for discussing various forms of amplitude modulation. Consider a baseband signal (e.g., a complex wave with a continuous but bandlimited frequency spectrum) with a time function represented by $v(t)$ and, for simplicity, a maximum amplitude of unity. The modulating function, $a(t)$, can be forced positive at all times by letting $a_0 \geq 1$ in Equation 8–3. This ensures that there are no phase reversals in the carrier component.

For a single–frequency modulating wave, a_n equals zero in Equation 8–5 and, letting $a_0 = 1$ and $a_m = 1$, the modulated wave is

$$M(t) = \cos \omega_c t + \frac{m}{2} \cos(\omega_c - \omega_m)t + \frac{m}{2} \cos(\omega_c + \omega_m)t. \qquad (8\text{–}8)$$

In many instances the use of exponential notation for periodic functions has advantages over the trigonometric notation that has been used thus far in this chapter. A particularly useful application is in the phasor representation of modulated waves as an aid in understanding the various modulation processes. A sinusoidal carrier, $\cos \omega_c t$, can be written

$$\mathrm{Re}\ [e^{j\omega_c t}] \quad ,$$

where Re represents the real part of the complex quantity and

$$e^{j\omega_c t} = \cos \omega_c t + j \sin \omega_c t.$$

The exponential $e^{j\omega_c t}$ is a counterclockwise rotating phasor of unit length in the complex plane, and its real part is its projection on the real axis. This phasor is shown for three values of time in Figure 8–3.

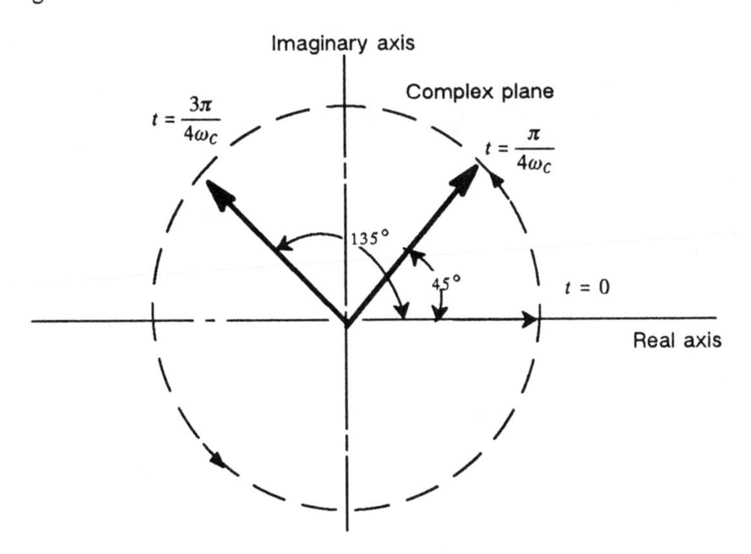

Figure 8–3. Phasor diagram of $e^{j\omega_c t}$.

Now consider the amplitude-modulated wave of Equation 8–8. This can be written in exponential notation as

$$M(t) = \mathrm{Re}\left[e^{j\omega_c t} + \frac{m}{2} e^{j(\omega_c - \omega_m)t} + \frac{m}{2} e^{j(\omega_c + \omega_m)t}\right]$$

$$= \mathrm{Re}\left[e^{j\omega_c t}\left(1 + \frac{m}{2} e^{j\omega_m t} + \frac{m}{2} e^{-j\omega_m t}\right)\right].$$

In this form the carrier phasor is multiplied by the sum of a stationary vector and two rotating vectors of equal size that rotate in opposite directions. As may be seen in Figure 8–4, the sum of these three vectors is always real and, consequently, acts only to modify the length of the real part of the rotating carrier phasor. This produces amplitude modulation as expected.

At this point, the average power in the carrier and in the sideband frequencies should be considered. For a unit amplitude carrier and a circuit impedance such that average carrier power is 1 watt, the power in each side frequency is $m^2/4$ watts; thus, the total sideband power is $m^2/2$ watts. Thus, for 100–percent modulation, only one–third of the total power is in the information–bearing sidebands. The sidebands get an even smaller share of the total power when the modulating function is a speech signal that has a higher peak–to–rms ratio than a sinusoid. The sideband power must be reduced to a few percent of the total power to prevent occasional peaks from overmodulating the carrier.

While the DSBTC signal is sensitive to certain types of transmission phase distortion, it is not impaired by a transmission phase characteristic that is linear with the frequency. The basic requirement for no impairment is that the transmission characteristic have odd symmetry of phase about the carrier frequency.

An interesting degradation occurs under certain extreme transmission phase conditions. Suppose that the LSB frequency vector in Figure 8–4 is shifted clockwise by θ degrees, and the USB frequency is shifted clockwise by $180 - \theta$ degrees. The resulting signal, Figure 8–5, consists of a carrier phasor with the sideband frequency vectors adding at right angles. The resultant vector represents a phase–modulated wave whose amplitude modulation has been largely cancelled, or washed out. A low–index DSBTC signal so distorted is indistinguishable from a low–index phase–modulated signal.

The condition of an LSB vector shifted by θ degrees and the USB vector shifted by $(180 - \theta)$ degrees represents a worst case. Any change in phase relationship between the two sideband vectors and the carrier, other than in odd symmetry, causes partial washout and some phase modulation. Among other things, the effective index of modulation is reduced.

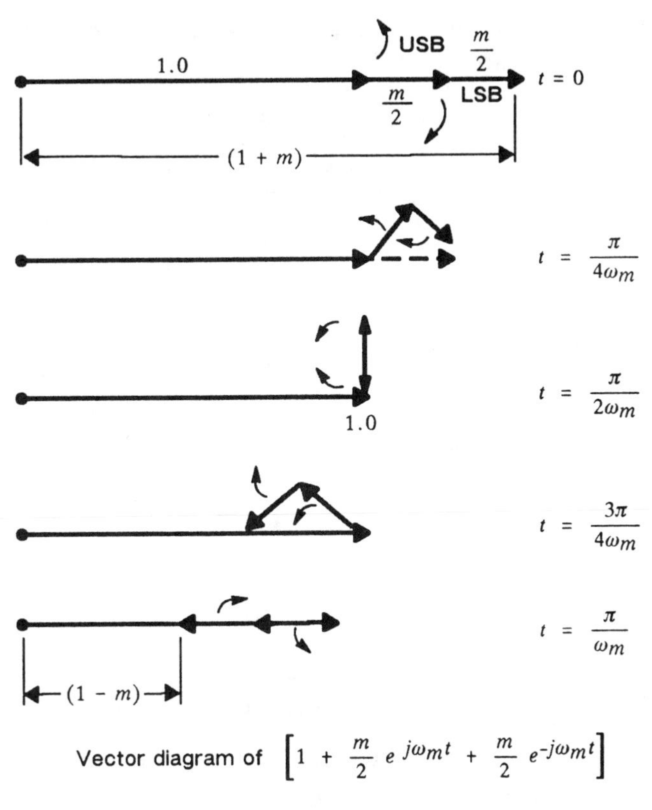

Figure 8–4. Amplitude modulation—index of modulation = *m*.

Double-Sideband Suppressed Carrier

The DSBSC signal requires the same transmission bandwidth as DSBTC, but the power efficiency is improved by suppression of the carrier. This requires reintroduction of a carrier at the receiving terminal, which must be done with extreme phase accuracy to avoid the type of washout distortion just discussed. Examination of Figure 8–4 shows that a θ–degree phase error of the inserted carrier results in the effective amplitude modulation being reduced by the factor cos θ. In the extreme, this effect can be seen by shifting only the stationary unit phasor (the carrier) of Figure 8–4 by 90 degrees to obtain the washout result of Figure 8–5. If the phase error θ is $\Delta\omega_e t$ radians and the baseband signal is a single–frequency sinusoid, the demodulated signal consists of two sinusoids separated by twice the error frequency of the inserted carrier, Δf_e hertz.

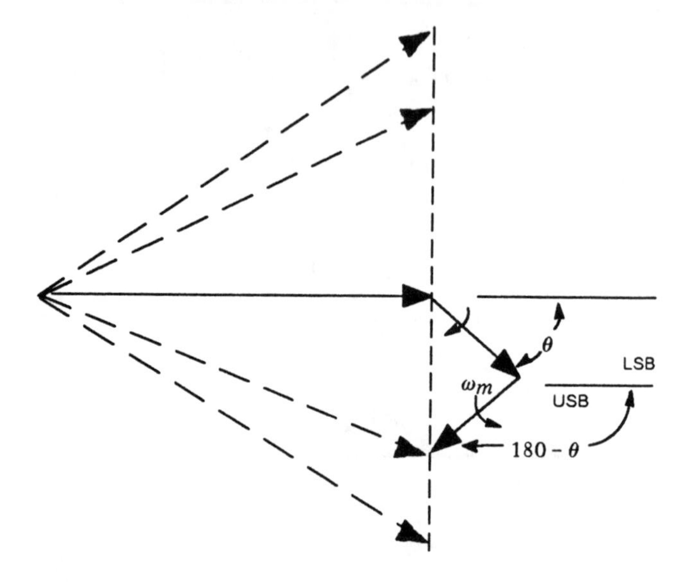

Figure 8–5. Result of certain extreme phase distortion of DSBTC signal to produce phase modulation.

The difficulty of accurately reinserting the carrier is the greatest disadvantage of DSBSC and is probably the reason this form has not seen more use. However, the transmitted sidebands contain the information required to establish the exact frequency and, except for a 180–degree ambiguity, the phase of the required demodulating carrier. This is so by virtue of symmetry about the carrier frequency, even with a random modulating wave. One means of establishing the carrier at f_C is to square the DSBSC wave, filter the component present at frequency $2f_C$, and electrically divide the frequency in half [2]. It should be noted that a carrier thus derived disappears in the absence of modulation.

Single Sideband

The SSB signal is not subject to the demodulation washout effect discussed in connection with the DSB signals. In fact, the local carrier at the receiving terminal can have a slight frequency error. This produces a frequency shift in each demodulated baseband component. Even with errors of 2 to 5 Hz, the system is adequate for high–quality telecommunications circuits. However, the SSB method of transmission with a fixed or rotating phase error in demodulation does not preserve the baseband waveform at all. This may be seen in Figure 8–6 by considering the phasor representing the USB signal as arising from a single baseband frequency component at f_m. The dashed line represents the reference carrier phasor about which the sideband rotates with a relative angular velocity, ω_m.

Figure 8–6. USB and reference carrier phasors for SSB signal.

If a strong carrier of reference phase is added to the received sideband (as can be done in the receiving terminal just ahead of an envelope detector), the envelope of the resultant wave is sinusoidal and peaks when the sideband phasor aligns itself with the carrier. An envelope detector would produce, in the proper phase, a sinusoidal wave of frequency f_m.

If the phase of the added carrier is advanced 90 degrees, the peaks in the demodulated wave occur 90 degrees later; as a result, the baseband signal is retarded by 90 degrees. Although this does not distort the waveform of the single–frequency wave considered, each frequency component in a complex baseband wave would be retarded 90 degrees causing gross waveform distortion as illustrated in Figure 8–7 where the baseband fundamental and the third harmonic are both shifted 90 degrees. Although an envelope detector is assumed here, similar results would follow from analyzing product detection of the SSB signal if the demodulating carrier were shifted relative to the required value (i.e., relative to the real or virtual carrier of the transmitted signal).

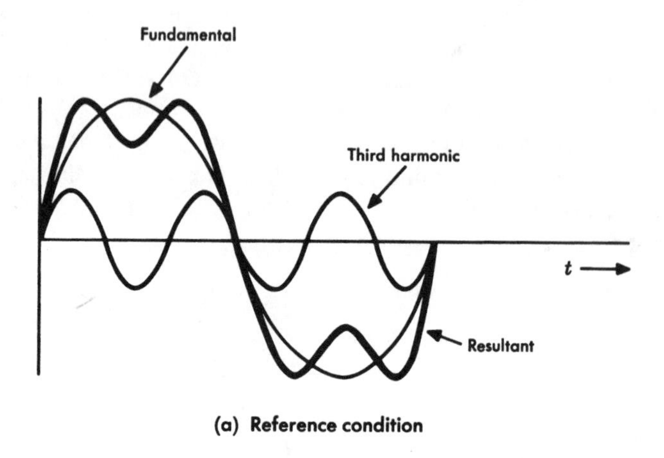

(a) Reference condition

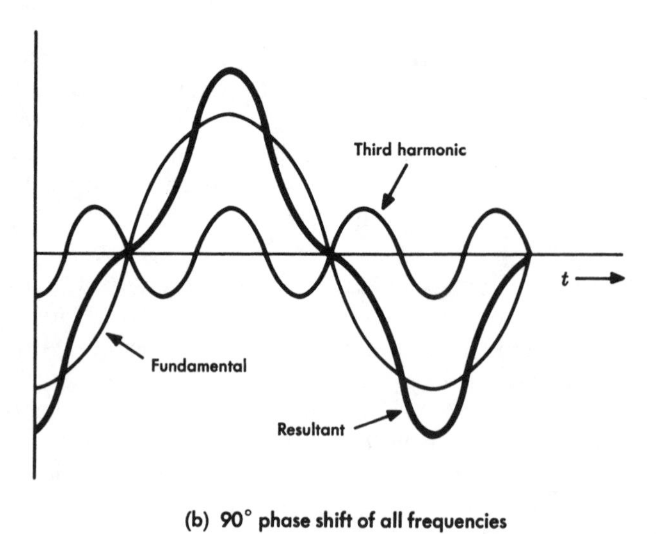

(b) 90° phase shift of all frequencies

Figure 8-7. Waveform distortion due to 90° reference carrier phase error causing 90° lag of all frequencies.

SSB signals inherently contain quadrature components, a source of distortion that can cause serious impairment where faithful recovery of the (time–domain) baseband waveform is necessary for satisfactory transmission quality. An SSB signal can be represented as two DSB signal pairs superimposed, as in Figure 8–8. One DSB pair has its resultant at right angles, or in quadrature. The inherent quadrature components and their related desired components are sometimes further shifted by a form of channel distortion called intercept distortion. Whether the distortion is inherent (quadrature distortion) or added (intercept distortion) its reduction or elimination from the demodulated signal depends on the signal format and on the design of the demodulator. The desired condition can be approached by adding a strong, or exalted, local carrier to the signal and then using an envelope detector. This approach, illustrated in Figure 8–9, shows that the angle θ (a measure of unwanted phase modulation) is reduced with exalted carrier as in Figure 8–9(b) relative to its value in Figure 8–9(a). However, the index of modulation is seen to be reduced also. When it is possible to establish the correct phase of the transmitted or virtual (suppressed) carrier, a more effective way to eliminate quadrature distortion is to use product detection.

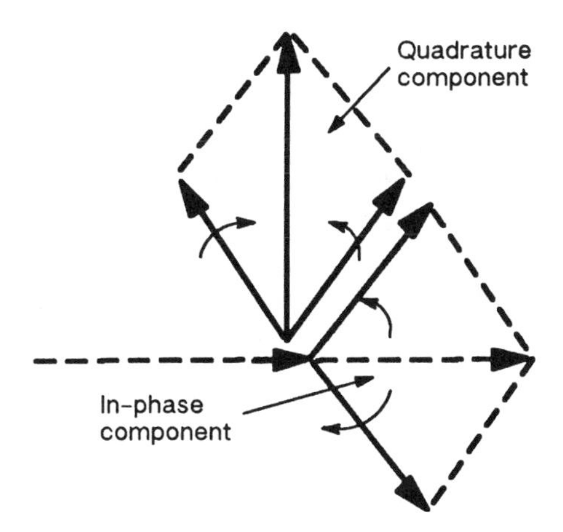

Figure 8–8. Analysis of SSB signal into in-phase and quadrature components.

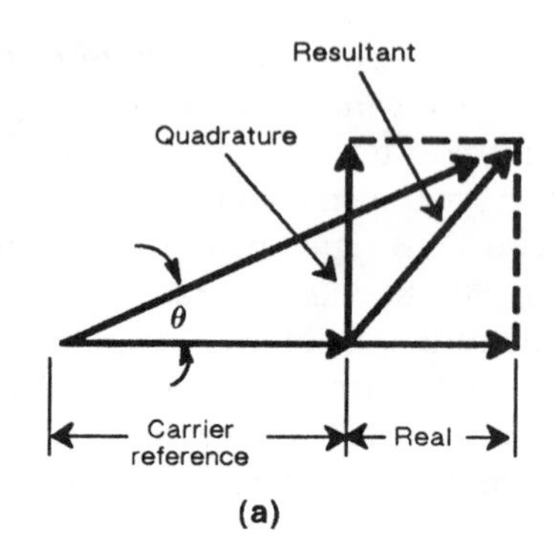

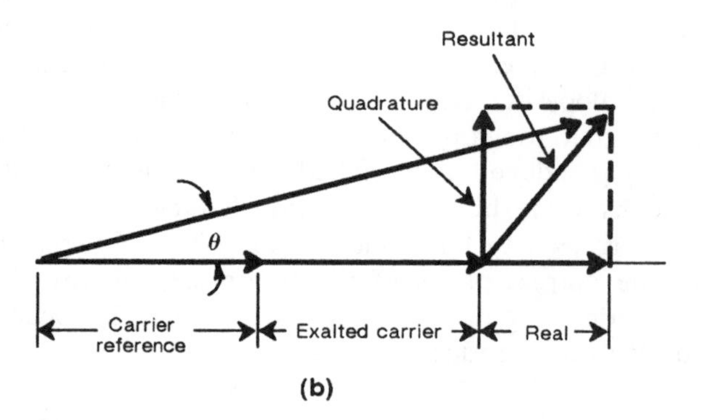

Figure 8-9. Quadrature distortion and reduction of phase modulation by exalted carrier.

Since voice transmission is very tolerant of quadrature distortion, the design of early carrier systems allowed reintroduction of the carrier with a frequency error. The resulting severe quadrature distortion renders these systems unsuitable for transmission of accurate baseband waveforms and makes these systems theoretically unfit for data pulse transmission. Also, some data signals contain very–low–frequency components. An SSB system will not transmit these components since practical filters cannot be built to suppress all of the unwanted sideband without cutting into the carrier frequency and the equivalent low frequencies of the wanted sideband.

A common technique used in carrying data traffic on SSB channels is to modulate a subcarrier in the data terminal using angle modulation or types of amplitude modulation that permit transmission of dc components. This also solves the quadrature distortion problem, since the subcarrier is transmitted and used in the ultimate demodulation in the receiving data terminal. Since the data subcarrier and the data sidebands travel the same path, the former provides the proper reference information for modulating the latter, even in the presence of frequency shift. Of course, the baseband channel must be adequately equalized for delay and attenuation.

SSB is the modulation technique usually used for the frequency–division multiplexing of multiple–message channels prior to transmission over broadband facilities. Actually, SSB techniques are often used for interim frequency translations in the multiplex terminal for convenient filtering of the channel [3]. The bandwidth of the signal, measured in octaves, may be increased by such translations.

Vestigial Sideband

VSB modulation is a modification of DSB in which part of one sideband is removed. It is produced by passing a DSB wave through a filter to suppress part of the spectrum as shown in Figure 8–10, leaving one full sideband and a vestige of the other. Demodulation of such a wave adds the LSB and USB components to recover the baseband signal. To preserve the baseband frequency spectrum without accentuating or depressing parts of it, the amplitude cutoff characteristic of the filter must be made symmetric about the carrier frequency as indicated in the figure. This causes the VSB to complement the attenuated part of the full sideband. For the same reason, and to avoid quadrature distortion, the varying phase shift through the filter must possess odd symmetry about the carrier frequency. As long as the cutoff is symmetrical, it can be gradual (approaching full DSB), sharp (approaching SSB), or intermediate (between these extremes).

The VSB signal is similar to DSBTC for low baseband frequencies and to SSB for high baseband frequencies. In the cutoff region, the behavior is as shown in Figure 8–11. The USB and LSB component vectors add to unity when they peak along the reference carrier line. If properly demodulated, they produce the same baseband signal as an SSB signal of unit amplitude.

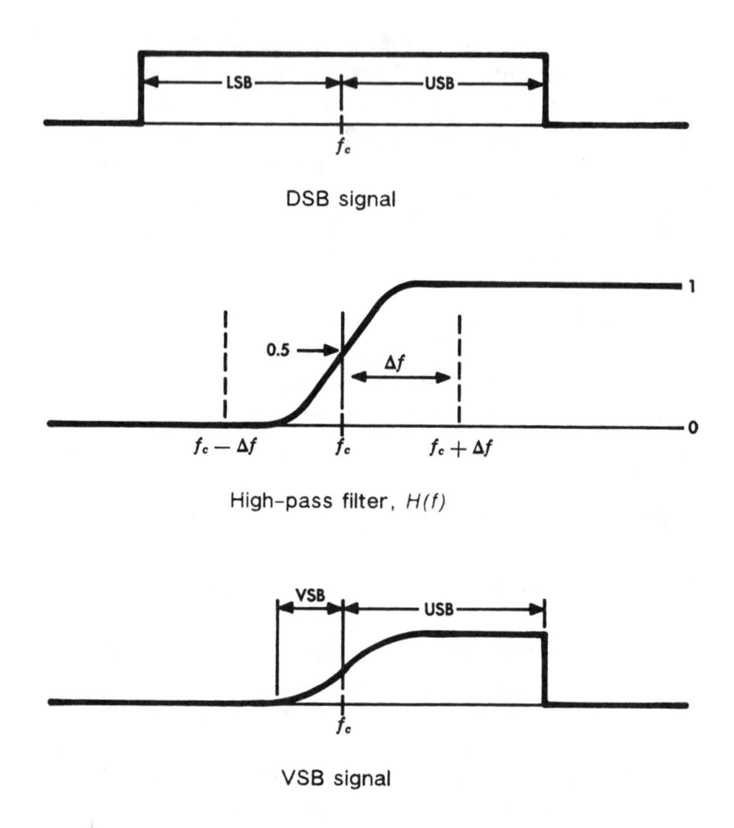

Figure 8–10. Generation of VSB wave; for no distortion, $1 - H(f_c + \Delta f) = H(f_c - \Delta f)$.

The desired filtering characteristics may be shared among the transmitting and receiving terminals and the transmission medium. The apportioning of the characteristic is determined by economics and signal–to–noise considerations.

Transmission by VSB conserves bandwidth almost as efficiently as SSB while retaining the excellent low–frequency baseband characteristics of DSB. Although the ideal SSB signal should allow the sideband spectrum to extend all the way to the carrier frequency, practical limitations on filters and phase distortion make it impractical. Thus, VSB has become standard for television and similar signals where good phase characteristics and transmission of low–frequency components are important, but the bandwidth required for DSB transmission is unavailable or uneconomical. It requires somewhat more bandwidth than SSB and has the additional disadvantage that the transmitted carrier, only partially suppressed, may add significantly to signal loading.

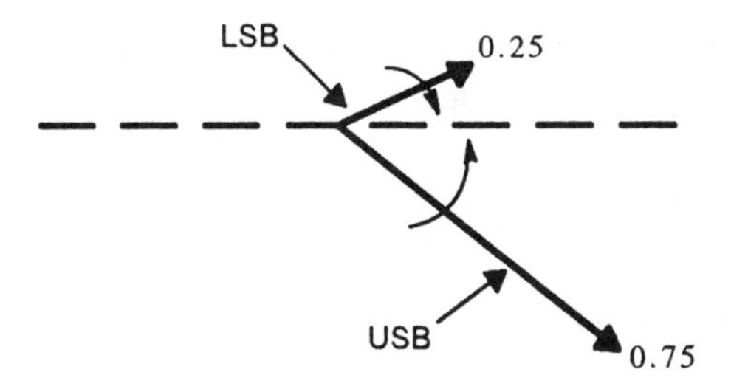

Figure 8-11. VSB phasors for intermediate modulating frequency.

Quadrature Amplitude Modulation

An application of amplitude modulation of considerable importance is quadrature operation. It is possible to AM-modulate a carrier wave with a set of signals while simultaneously AM-modulating the same carrier, shifted by 90 degrees of phase (or in quadrature), with a second set of signals. After sending the two independent signals through a common transmission medium, it is theoretically possible to demodulate them with complete independence, even though they occupy the same frequency spectrum.

QAM is usable with analog signals, as in the most widely used system for AM stereophonic broadcasting. It is also important for sending digital signals with high bandwidth efficiency. By giving the carrier eight possible levels and giving the quadrature carrier eight levels of its own, it is common to send 9600 b/s or higher data speeds on a voiceband channel. In this "64 QAM" operation, the two eight-level signals form a 64-point two-dimensional graph or "constellation." QAM is also used in digital radio systems for bandwidth conservation (e.g., with 64 QAM to transmit 135 Mb/s through 30 MHz of spectrum). Other numbers of levels (16, and even 256) may be used.

8-2 PROPERTIES OF ANGLE-MODULATED SIGNALS

Equation 8–1, with $a(t)$ held constant, may be rewritten

$$M(t) = A_c \cos [\omega_c t + \phi(t)] \tag{8-9}$$

where $\phi(t)$ is the angle modulation in radians. If angle modulation is used to transmit information, it is necessary that $\phi(t)$ be a prescribed function of the modulating signal. For example, if $v(t)$ is the modulating signal, the angle modulation $\phi(t)$ can be expressed as some function of $v(t)$.

Many varieties of angle modulation are possible depending on the selection of the functional relationship between the angle and the modulating wave. The most important of these are phase modulation (PM) and frequency modulation (FM).

Phase Modulation and Frequency Modulation

The difference between phase and frequency modulation can be understood by first defining four terms with reference to Equation 8–9:

$$\text{Instantaneous phase} = \omega_c t + \phi(t) \quad \text{rad,} \tag{8-10}$$

$$\text{Instantaneous phase deviation} = \phi(t) \quad \text{rad,} \tag{8-11}$$

$$\text{Instantaneous frequency}^{*} = \frac{d}{dt}[\omega_c t + \phi(t)]$$

$$= \omega_c + \phi'(t) \quad \text{rad/sec,} \tag{8-12}$$

$$\text{Instantaneous frequency deviation} = \phi'(t) \quad \text{rad/sec.} \tag{8-13}$$

Phase modulation can then be defined as angle modulation in which the instantaneous phase deviation, $\phi(t)$, is proportional to the modulating signal voltage, $v(t)$. Similarly, *frequency modulation* is angle modulation in which the instantaneous frequency

---

*The instantaneous frequency of an angle-modulated carrier is defined as the first time derivative of the instantaneous phase.

deviation, $\phi'(t)$, is proportional to the modulating signal voltage $v(t)$. Mathematically, these statements become, for PM,

$$\phi(t) = kv(t) \quad \text{rad} \tag{8-14}$$

and, for FM,

$$\phi'(t) = k_1 v(t) \quad \text{rad/sec} \tag{8-15}$$

from which

$$\phi(t) = k_1 \int v(t)dt \quad \text{rad} \tag{8-16}$$

where k and k_1 are constants.

These results are summarized in Table 8–1, which also shows phase–modulated and frequency–modulated waves that occur when the modulating wave is a single sinusoid.

Table 8–1. Equations for Phase– and Frequency–Modulated Carriers

Type of Modulation	Modulating Signal	Angle-Modulated Carrier
(a) Phase	$v(t)$	$M(t) = A_C \cos [\omega_C t + kv(t)]$
(b) Frequency	$v(t)$	$M(t) = A_C \cos [\omega_C t + k_1 \int v(t)dt]$
(c) Phase	$A_m \cos \omega_m t$	$M(t) = A_C \cos (\omega_C t + kA_m \cos \omega_m t)$
(d) Frequency	$-A_m \sin \omega_m t$	$M(t) = A_C \cos\left(\omega_C t + \frac{k_1 A_m}{\omega_m} \cos \omega_m t\right)$
(e) Frequency	$A_m \cos \omega_m t$	$M(t) = A_C \cos\left(\omega_C t + \frac{k_1 A_m}{\omega_m} \sin \omega_m t\right)$

Figure 8–12 illustrates amplitude, phase, and frequency modulation of a carrier by a single sinusoid. The similarity of waveforms of the PM and FM waves shows that for angle–modulated waves it is necessary to know the modulation function; that is, the waveform alone cannot be used to distinguish between PM and FM. Similarly, it is not apparent from Equation 8–9 whether an FM or a PM wave is represented; it could be either. A knowledge of the modulation function, however, permits correct identification. If $\phi(t) = kv(t)$, it is PM, and if $\phi'(t) = k_1 v(t)$, it is FM.

Comparison of (c) and (d) in Table 8–1 shows that the expression for a carrier which is phase– or frequency–modulated by a sinusoidal–type signal can be written in the general form of

$$M(t) = A_c \cos(\omega_c t + X \cos \omega_m t) \qquad (8\text{–}17)$$

where

$$X = kA_m \quad \text{rad for PM} \qquad (8\text{–}18)$$

and

$$X = \frac{k_1 A_m}{\omega_m} \quad \text{rad for FM.} \qquad (8\text{–}19)$$

Here X is the peak phase deviation in radians; it is called the index of modulation. For PM, the index of modulation is a constant, independent of the frequency of the modulating wave; for FM, it is inversely proportional to the frequency of the modulating wave. Note that in the FM case the modulation index can also be expressed as the peak frequency deviation, $k_1 A_m$, divided by the modulating signal frequency, ω_m. The terms *high index* and *low index* of modulation are often used. It is difficult to define a sharp division; however, in general, low index is used when the peak phase deviation is less than one radian. It is shown later that the frequency spectrum of the modulated wave depends on the index of modulation.

When the modulation function consists of a single sinusoid, it is evident from Equation 8–17 that the phase angle of the carrier varies from its unmodulated value in a simple sinusoidal fashion,

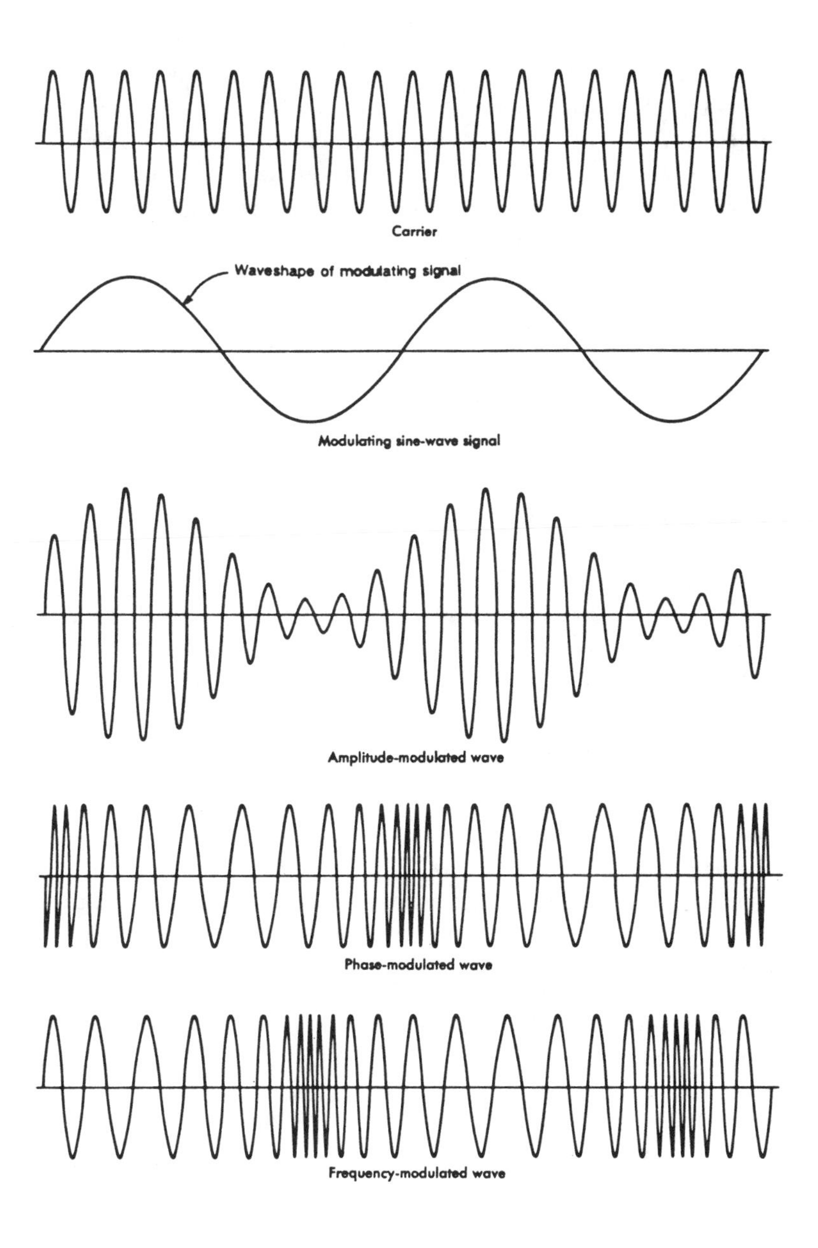

Figure 8–12. Amplitude, phase, and frequency modulation of a sine-wave carrier by a sine-wave signal.

with the peak phase deviation being equal to X. The phase deviation can also be expressed in terms of the mean square phase deviation, D_ϕ, which for this case is $X^2/2$. Similarly, the frequency deviation of a sinusoidally modulated carrier can be expressed either in terms of the peak frequency deviation, $k_1 A_m$ rad/sec = $k_1 A_m / 2\pi$ Hz, or the mean square frequency deviation, D_f, which is $k_1{}^2 A_m{}^2 / 8\pi^2$ Hz^2.

Where a large number of speech signals comprise the complex modulating function, the modulated signal closely approximates a random signal having a Gaussian spectral density function. Hence, from the statistics of the modulated signal, it is possible to define the value of instantaneous voltage that will be exceeded only a specified percentage of the time. Since instantaneous frequency deviation is proportional to instantaneous voltage, it follows that this voltage defines the value of instantaneous frequency deviation that is exceeded only the specified percentage of the time. It is customary to define the peak frequency deviation produced by the complex message load as the deviation exceeded 0.001 percent of the time. The peak deviation determines the required bandwidth.

Phasor Representation

A wave angle–modulated by sinusoids can be represented by phasors as is done for the AM waves. Generally, the angle–modulated case is more complex, as can be seen by expanding Equation 8–17 into a Bessel series of sinusoids. In the special case of very low index (X less than 1/2 radian), all terms after the first can be ignored, and the phasor diagram is similar to that for an AM wave except for the phase relationship of the sidebands relative to the carrier. In the PM case, the sidebands are phased to change the angle, rather than the amplitude, of the carrier as illustrated in Figure 8–13. A close examination of the phasor diagrams shows that one sideband of the PM wave is 180 degrees out of phase with the corresponding sideband in the AM wave. This can be seen by comparing Figure 8–4 at $t = 0$, for example, with Figure 8–13 at $t = \pi/2\omega_m$. In fact, it was pointed out in the AM discussion that if the inserted carrier of a DSBSC signal has a phase error of 90 degrees, severe washout occurs and the previously amplitude–modulated wave has very little amplitude modulation but considerable phase (or angle) modulation. The approximate phasor diagram for a low–index angle–modulated system modulated by a single–frequency sinusoid at f_m is shown in Figure 8–13 for several values of time. The resultant vector has an amplitude close to unity at all times and an index, or maximum phase deviation, of X radians. A true angle–modulated wave would include higher order terms and would have no amplitude variation. If X is small enough, these terms are often ignored.

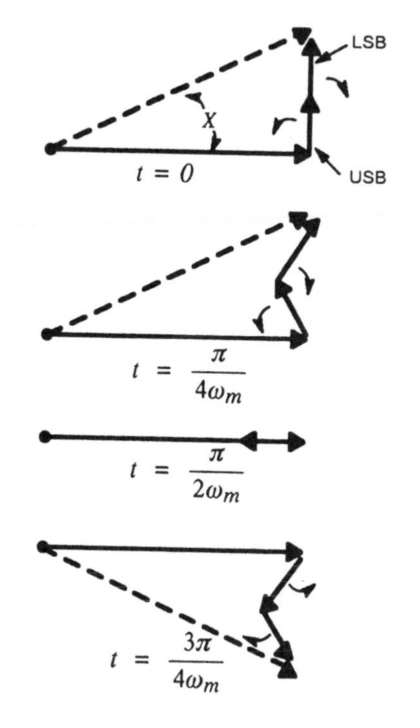

Figure 8–13. Phase modulation—low index.

Several interesting conclusions may be observed by comparing the low–index angle–modulated wave with the AM signal shown in Figure 8–4. Both types of modulation are similar in the sense that they contain the carrier and the same first–order sideband frequency components. In fact, for the low–index case, the amplitudes of the first–order sidebands are approximately the same when the indices are equal ($X = m$). The important difference is the phase of the sideband components. It may be expected, therefore, that in the transmission of an FM or PM wave the phase characteristic of the transmission path is extremely important. Certain phase irregularities can easily convert PM components into AM components.

Average Power of an Angle–Modulated Wave

The average power of an FM or PM wave is independent of the modulating signal and is equal to the average power of the carrier when the modulation is zero. Hence, the modulation process takes power from the carrier and distributes it among the many sidebands but does not alter the average power. This may be demonstrated by assuming a voltage of the form of Equation 8–9, squaring, and dividing by a resistance, R, to obtain the instantaneous power,

$$\begin{aligned} P(t) &= \frac{M^2(t)}{R} \\ &= \frac{A_c^{\,2}}{R} \cos^2\left[\omega_c t + \phi(t)\right] \\ &= \frac{A_c^{\,2}}{R} \left\{ \frac{1}{2} + \frac{1}{2} \cos\left[2\omega_c t + 2\phi(t)\right] \right\} . \end{aligned} \quad (8\text{–}20)$$

The second term can be assumed to consist of a large number of sinusoidal sideband components about a carrier frequency of $2f_c$ Hz; therefore, the average value of the second term of Equation 8–20 is zero. Thus, the average power is given by the zero frequency term

$$P_{avg} = \frac{A_c^{\,2}}{2R} . \quad (8\text{–}21)$$

This, of course, is the same as the average power in the absence of modulation.

Bandwidth Required for Angle–Modulated Waves

For the low–index case, where the peak phase deviation is less than one radian, most of the signal information of an angle–modulated wave is carried by the first–order sidebands. It follows that the bandwidth required is at least twice the frequency of the highest frequency component of interest in the modulating signal. This would permit the transmission of the entire first–order sideband.

For the high–index signal, a different method called the quasi–stationary approach must be used [4]. In this approach, the assumption is made that the modulating waveform is changing very slowly so that static response can be used. For example, assume that a 1–volt baseband signal causes a 1–MHz frequency deviation of the carrier. This corresponds to $k_1 = 2\pi \times 10^6$ radians per volt–second. Then, if the modulating signal has a 1–volt peak, the peak frequency deviation is 1 MHz. Thus, it is obvious that *if the rate of change of frequency is very small* the bandwidth is determined by the peak–to–peak frequency deviation. It was mathematically proven by J. R. Carson in 1922 that FM could not be accommodated in a narrower band than AM, but might actually require a wider band [5]. The quasi–stationary approach for large index indicates that the minimum bandwidth required is equal to the peak–to–peak (or twice the peak) frequency deviation.

Thus, for low–index systems ($X < 1$) the minimum bandwidth is given by $2f_T$, where f_T is the highest frequency in the modulating signal. For high–index systems ($X > 10$), the minimum bandwidth is given by $2\Delta F$, where ΔF is the peak frequency deviation. It would be desirable to have an estimate of the bandwidth for all angle–modulated systems regardless of index. A general rule (first stated by Carson in 1939) is that the minimum bandwidth required for the transmission of an angle–modulated signal is equal to two times the sum of the peak frequency deviation and the highest modulating frequency to be transmitted. Thus,

$$Bw = 2(f_T + \Delta F) \quad \text{Hz.} \tag{8–22}$$

Carson's rule gives results that agree quite well with the bandwidths actually used in the telecommunications industry. It should be realized, however, that this is only an approximation. The actual bandwidth required is, to some extent, a function of the waveform of the modulating signal and the quality of transmission desired.

8-3 PROPERTIES OF PULSE MODULATION

In pulse–modulation systems, the unmodulated carrier is usually a series of regularly recurrent pulses. Modulation results from varying some parameter of the transmitted pulses, such as the amplitude, duration, or timing. If the baseband signal is a continuous waveform, it is broken up by the discrete nature of the pulses. In considering the feasibility of pulse modulation, it must be recognized that the continuous transmission of information describing the modulating function is unnecessary, provided the modulating function is bandlimited and the pulses occur often enough. The necessary conditions are expressed by the sampling principle, as discussed below.

It is usually convenient to specify the signalling speed or pulse rate in *bauds*. A baud (named in honor of a French telegraph engineer named Baudot) is defined as the unit of modulation rate corresponding to a rate of one unit interval per second; i.e., baud $= 1/T$ where T is the minimum signalling interval in seconds. When the duration of signalling elements in a pulse stream is constant, the baud rate is equal to the number of signalling elements or symbols per second. Thus, the baud rate denotes pulses per second in a manner analogous to hertz denoting cycles per second. Note that all possible pulses are counted whether or not a pulse is sent, since no pulse is usually also a valid symbol. Since there is no restriction on the allowed amplitudes of the pulses, a baud can contain any arbitrary information rate in bits per second. Unfortunately, *bits per second* is often used incorrectly to specify a digital transmission rate in bauds. For binary symbols of equal time duration, the information rate in bits per second is equal to the signalling speed in bauds if there is no redundancy. In general, the relation between information rate and signalling rate depends on the coding scheme employed.

Sampling

In any physically realizable transmission system, the message or modulating function is limited to a finite frequency band. Such a bandlimited function is continuous with time and limited in its possible range of excursions in a small time interval. Thus, it is necessary only to specify the amplitude of the function at discrete time intervals in order to specify it exactly. The basic principle discussed here is called the sampling theorem, which in a restricted form states [6]:

> *If a message that is a magnitude–time function is sampled instantaneously at regular intervals and at a rate at least twice the highest significant message frequency, then the samples contain all of the information of the original message.*

The application of the sampling theorem reduces the problem of transmitting a continuously varying message to one of transmitting information representing a discrete number of amplitude samples per given time interval. For example, a message bandlimited to f_T hertz is completely specified by the amplitudes at any set of points in time spaced T seconds apart, where $T = 1/2f_T$ [7]. Hence, to transmit a bandlimited message, it is necessary only to transmit $2f_T$ independent values per second. The time interval, T, is often referred to as the Nyquist interval.

The process of sampling can be thought of as the product modulation of a message function and a set of impulses, as shown in Figure 8–14. The message function of time, $v(t)$, is multiplied by a train of impulses, $c(t)$, to produce a series of AM pulses, $s(t)$. If the spectrum (i.e., the Fourier transform) of $v(t)$ is given by $F(f)$, as shown in Figure 8–14, the spectrum of the sampled wave, $s(t)$, is then shown by $S(f)$ in the figure. The output spectrum, $S(f)$, is periodic on the frequency scale with period f_S, the sampling frequency. It is important to note that a pair of sidebands has been produced around f_S, $2f_S$, and so on through each harmonic of the sampling frequency. This figure also shows the need for $f_S > 2f_T$, so that the sidebands do not overlap. Note also that all sidebands around all harmonics of the sampling frequency have the same amplitude. This results because the frequency spectrum of an impulse is flat with frequency. In a practical case, of course, finite–width pulses would have to be used for the sampling function, and the spectrum of the sampled signal would fall off with frequency as the spectrum of the sampling function does.

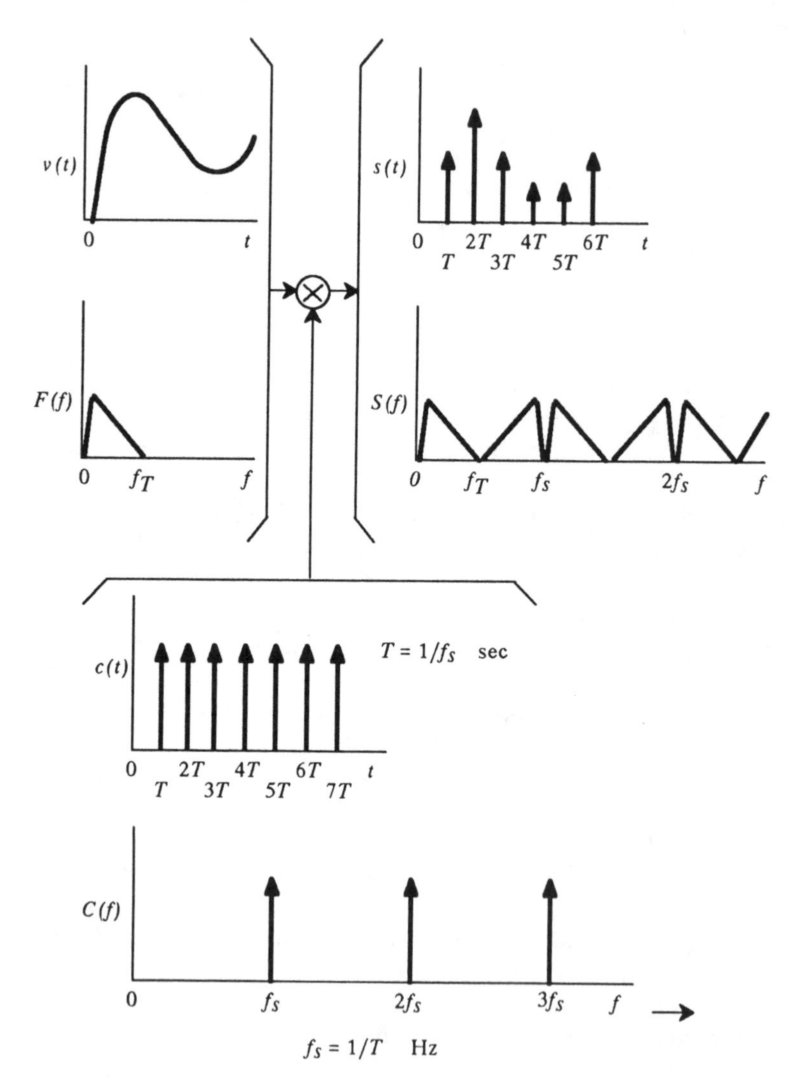

Figure 8–14. Sampling with an impulse modulator.

The AM pulse signal that results from sampling the input message may be transmitted to the receiver in any form that is convenient or desirable from a transmission standpoint. At the receiver, the incoming signal, which may no longer resemble the impulse train, must be operated on to recreate the original pulse–amplitude–modulated sample values in their original time sequence at a rate of $2f_T$ samples per second. To reconstruct the message, it is necessary to generate from each sample a proportional impulse and to pass this regularly spaced series of impulses through an ideal low–pass filter having a cutoff frequency f_T. Examination of the spectrum of $S(f)$ in Figure 8–14 makes the feasibility of this obvious. Except for an overall time delay and possibly a constant of proportionality, the output of this filter would then be identical to the original message. Ideally, then, it is possible to transmit information exactly, given the instantaneous amplitude of the message at intervals spaced not farther than $1/2f_T$ seconds apart.

Pulse Amplitude Modulation

In pulse amplitude modulation (PAM), the amplitude of a pulse carrier is varied in accordance with the value of the modulating wave as shown in Figure 8–15(c). It is convenient to look upon PAM as modulation in which the value of each instantaneous sample of the modulating wave is caused to modulate the amplitude of a pulse. Signal processing in time–division multiplex terminals often begins with PAM, although further processing usually takes place before the signal is launched onto a transmission system.

Pulse Duration Modulation

Pulse duration modulation (PDM), sometimes referred to as pulse length modulation or pulse width modulation, is a particular form of pulse time modulation. It is modulation of a pulse carrier in which the value of each instantaneous sample of a continuously varying modulating wave is caused to produce a pulse of proportional duration, as shown in Figure 8–15(d). The modulating wave may vary the time of occurrence of the leading edge of the pulse, the trailing edge, or both. In any case, the message to be transmitted is composed of sample values at discrete times, and each value must be uniquely defined by the duration of a modulated pulse.

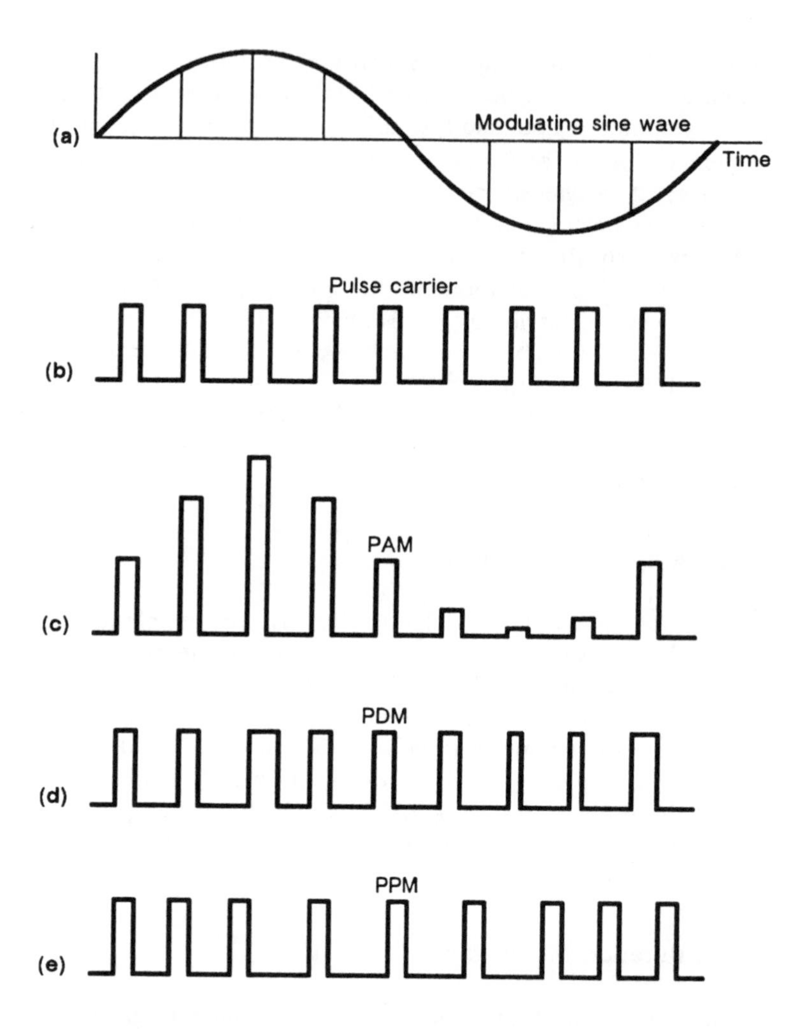

Figure 8–15. Examples of pulse-modulation systems.

Pulse Position Modulation

A particular form of pulse time modulation, in which the value of each instantaneous sample of a modulating wave varies the position of a pulse relative to its unmodulated time of occurrence, is pulse position modulation (PPM). This is illustrated in Figure 8–15(e). The variation in relative position may be related to the modulating wave in any predetermined unique manner.

Practical applications of PPM systems have been on a modest scale, even though their instrumentation can be extremely simple.

If either PDM or PPM is used to time–division multiplex several channels, the maximum modulating signal must not cause a pulse to enter adjacent allotted time intervals. In telecommunications systems with high peak–to–rms ratios, this requirement leads to a wasteful use of time space. In fact, almost all of the time available for modulation is wasted because many of the busy channels may be expected to be inactive and most of the rest will be carrying small signal power. Consequently, although PPM is more efficient than PDM, both fall short of the theoretical ideal when used for multiplexing ordinary telephone channels.

Pulse Code Modulation

A favored form of pulse modulation is known as pulse code modulation (PCM). This mode of signal processing may take any of several forms; each requires the successive steps of sampling, quantizing, and coding. If the input signal is analog in nature, the sampling is usually a sampling of the signal amplitude; if the signal is digital, the sampling process may take the form of time–sampling to determine the presence of transitions from one signal state to another. The process of sampling, common to pulse modulation in general, was described previously. Following is a discussion of quantizing and several forms of coding.

Quantizing. Instead of attempting the impossible task of transmitting the exact amplitude of a sampled signal, suppose only certain discrete amplitudes of sample size are allowed. Then, when the message is sampled in a PAM system, the discrete amplitude nearest the true amplitude is sent. When received and decoded, this signal sample has an amplitude slightly different from the original PAM sample. But a new signal can be created that has the amplitude originally sent, plus or minus a tolerable degree of error.

Representing the message by a discrete and therefore limited number of signal amplitudes is called *quantizing*. It inherently introduces an initial error in the amplitude of the samples, giving rise to quantization noise. But once the message information is in a quantized state, it can be relayed for any distance without further loss in quality, provided only that the added noise in the signal received at each repeater is not too great to prevent correct recognition of the presence or absence of a line pulse. If the received signal lies between *a* and *b* and is closer to *b*, it is surmised that *b* was sent. If the noise is small enough, there are no errors. Note, therefore, that in quantized signal transmission the maximum received noise is determined by the number of bits in the code, while in analog signal transmission it is controlled by the repeater spacing, the characteristics of the medium, and the amplitude of the transmitted signal.

Coding. A quantized sample can be sent as a single pulse having certain possible discrete amplitudes or certain discrete positions with respect to a reference position. If, however, many discrete sample amplitudes are required (100, for example), it is difficult to design circuits that can distinguish between amplitudes. It is much less difficult to design a circuit that can determine whether a pulse is present. If several pulses are used as a code group to describe the amplitude of a single sample, each pulse can be present (1) or absent (0). For instance, if three pulse positions are used, then a code can be devised to represent the eight different amplitudes shown in Table 8–2. These codes are, in fact, just the numbers (amplitudes) at the left written in binary notation. In general, a code group of n on–off pulses can be used to represent 2^n amplitudes. For example, 7 binary pulses yield 128 sample levels.

Table 8–2. Binary Code Representation of Sample Amplitudes

Amplitude Represented	Code
0	000
1	001
2	010
3	011
4	100
5	101
6	110
7	111

It is possible, of course, to code the amplitude in terms of a number of pulses that have discrete amplitudes of 0, 1, and 2 (ternary, or base 3) or 0, 1, 2, and 3 (quaternary, or base 4), etc., instead of the pulses with amplitudes 0 and 1 (binary, or base 2). If ten levels are allowed for each pulse, then each pulse in a code group is simply a digit or an ordinary decimal number expressing the amplitude of the sample. If n is the number of pulses and b is the base, the number of quantizing levels the code can express is b^n. To decode this code group, it is necessary to generate a pulse that is the linear sum of all pulses in the group, each pulse of which is multiplied by its place value ($1, b, b^2, b^3 \ldots$) in the code.

Differential Pulse Code Modulation. This form of pulse modulation has two major potential advantages that can sometimes be used advantageously in particular design situations. First, it can sometimes result in a lower digital rate than straight PCM coding and yet give equivalent transmission performance. Second, the sampling, quantizing, and coding of a signal can be accomplished without the use of large amounts of common equipment. Thus, in situations where large numbers of signals need not be processed simultaneously, it may be more economical than conventional PCM.

Many forms of differential PCM exist [8]. One, known as delta modulation, samples the analog signal at a high rate and codes the samples in terms of the *change* of signal amplitude from sample to sample. The digital rate must be higher than the sampling rate given previously (sampling at a rate at least twice the highest message frequency) because of distortion that might be introduced when the rate of change of signal amplitude is high. The combined sampling and coding process, however, may still result in a lower net digital rate.

References

1. Panter, P. F. *Modulation, Noise, and Spectral Analysis* (New York: McGraw–Hill Book Company, Inc., 1965), p. 1.

2. Graham, R. S. and J. W. Rieke. "The L3 Coaxial System—Television Terminals," *Bell System Tech. J.*, Vol. 32 (July 1953), pp. 915–942.

3. Blecher, F. H. and F. J. Hallenbeck. "The Transistorized A5 Channel Bank for Broadband Systems," *Bell System Tech. J.*, Vol. 41 (Jan. 1952), pp. 321–359.

4. Rowe, H. E. *Signals and Noise in Communication Systems* (Princeton, NJ: D. Van Nostrand Company, Inc., 1965), pp. 103 and 119–124.

5. Carson, J. R. "Notes on the Theory of Modulation," *Proceedings of the IRE* (Feb. 1922).

6. Black, H. S. *Modulation Theory* (Princeton, NJ: D. Van Nostrand Company, Inc., 1953), p. 37.

7. Oliver, B. M., J. R. Pierce, and C. E. Shannon. "The Philosophy of PCM," *Proceedings of the IRE,* Vol. 36 (Nov. 1948), pp. 1324–1331.

8. Members of Technical Staff. *Transmission Systems for Communications,* Fifth Edition (Murray Hill, NJ: AT&T Bell Laboratories, Inc., 1982), pp. 643–659.

Chapter 9

Probability and Statistics

The parameters in most engineering problems are not unique or deterministic; that is, they can assume a range of values. If extreme or worst-case values of the important parameters are used, solutions to such problems are seldom economical, frequently inaccurate, and sometimes not even realizable. Probabilistic solutions must be sought; that is, the nature of the distribution of parameters must be studied and understood, appropriate values must be found to represent the parameters in question, and answers must be found that adequately represent the range of values that the solutions can take as a result of the range of values of the important parameters. The tools for finding economic solutions to such problems are provided by the related subjects, *probability* and *statistics*.

While the use of extreme values of parameters often leads to impractical solutions to problems, the use of other parameter values (nominal, mean, or average) may also lead to impractical solutions. It is important to consider overall distributions of values; in some cases only the average is important, but in other cases extreme values (the tails of the distributions) may have to be taken into account. Sometimes, the extreme cases are solved by legislating against them. For example, telephone loops could be laid out by assuming the use of a single gauge of wire in the cables used in the loop plant. If this were done, the losses of loops longer than some specific value would exceed the loss that can give satisfactory service. Loops having such excess loss are avoided by applying loop design rules that require the addition of gain devices, the use of loading coils, or the use of heavier gauge wire when the loop length exceeds the limit. Losses, however, are still functions of all the parameters mentioned (wire gauge, distance, loading, and gain). If the rules were written so that no possible connection could have excessive loss, the solution would not be economical; if too many connections have excess loss, grade of service suffers. Thus, the problem is to find an economical compromise that can provide an overall satisfactory grade of service.

Since a telecommunications system may be large and complex, it is impractical to measure the values of all similar parameters (noise and loss on all trunks, for example) in order to determine the performance or to describe the characteristics of any part of the plant. Instead, the plant is described on the basis of statistical parameters using only a few key numbers, such as one to represent some central or average value and one to represent the dispersion or spread of the data. Estimates of such numbers can be determined by measuring only a properly chosen sample of the total universe of values.

Probability theory provides a mathematical basis for the evaluation and manipulation of statistical data. The theory treats events that may occur singly or in combination as a result of interacting phenomena which also may be occurring sequentially or simultaneously.

Following a classical process of deductive reasoning, the theory of probability [1] evolved from a number of postulates that were based on experimental observations. The postulates were tested and, where necessary, modified to fit observed data. Finally, clearly defined axioms evolved, and the entire theory was built upon these axioms. Probability theory provides the means for expressing or describing a set of observations more efficiently than by enumerating all numbers in the set. The unknowns are expressed as functions of a random variable; these functions, which describe the domain and range of the unknown, are derived by a mapping process. This process, together with some of the terminology and symbology that are unique to probability theory, must be described.

The *mean* (or *expected value*), the *standard deviation*, and the *variance* are the principal parameters used in expressions for discrete and continuous functions of a random variable. Methods of summing random variables are available and a number of different types of distribution may be used to represent communications phenomena of various characteristics. Each is represented by a different distribution function. Where functional relationships are not known, statistical analyses are often used.

9–1 ELEMENTS OF PROBABILITY THEORY

Probability theory is applied to the study of and relationships among *sets* of observations or data. The largest set, consisting of all the observations or all the data, is known as the *universe*, the *domain*, or the *sample space*. *Subsets*, which are made up of certain interrelated elements defined according to specified criteria, are all contained in the sample space. The interrelations among subsets, often referred to simply as sets, are conveniently displayed for study in a sketch called a *Venn diagram*, in which the sample space is displayed as a square. Subsets are depicted as geometrical figures within the square; within each figure are located all the elements of that subset.

Figure 9–1 is an example of a Venn diagram illustrating the relationships among sets A, B, and C and the sample space, S, of which they are parts. Examination of Figure 9–1 shows that C is a subset of B, B is a subset of A, and A is a subset of S. It follows that C is a subset of A and that B and C are subsets of S. The above statements regarding subsets may be written as follows:

$$C \subset B,\ B \subset A,\ A \subset S,\ C \subset A,\ B \subset S,\ C \subset S,$$

where the symbol $\subset$ is used to indicate that every element of the subset shown at the closed end of the symbol is also an element of the larger set shown at the open end of the symbol. Thus, $C \subset B$ (C is contained in B) may also be written $B \supset C$ (B contains C).

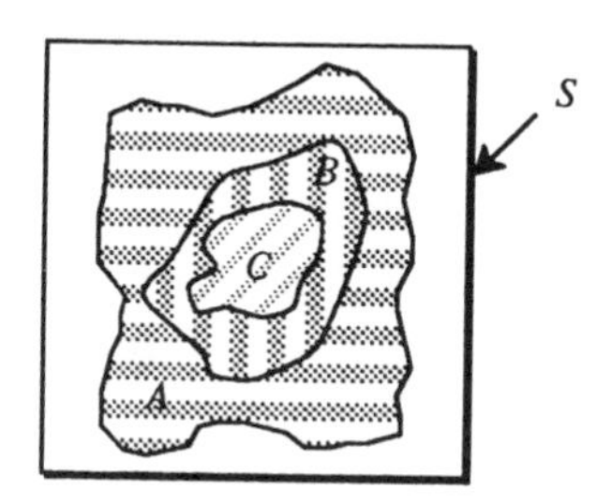

Figure 9–1. Venn diagram of three subsets.

Axioms

A number of axioms form the basis of probability theory. These are:

(1) *The probability of an event, A, is the ratio of the outcomes favorable to A to the total number of outcomes, n,* where it is assumed that all n outcomes are equally likely. Here the total number of events represents the sample space, and the event A represents the subset of the sample space which satisfies some specific criterion. The axiom may be expressed $P(A) = n_A/n$.

(2) *Probability is a positive real number between 0 and 1 inclusive;* i.e.,

$$0 \leq P \leq 1.$$

In the physical world, negative probability has no meaning and nothing can occur more than 100 percent of the time.

(3) *The probability of an impossible event is zero.* Note that the rule does not imply the converse; i.e., a probability of zero does not mean that an event is impossible. (The impossible event is sometimes called the *empty set*, or *null set*, one that contains no elements.)

(4) *The probability of a certain event is unity.* By certain event is meant one that is certain to occur at every trial. It is the set represented by the sample space. Again, the converse is not necessarily true; i.e., a probability of unity does not necessarily mean that the event is certain.

(5) *The probability that at least one of two events occurs is the sum of the individual probabilities of each event minus the probability of their simultaneous occurrence.*

(6) *The probability of the simultaneous occurrence of two events is the product of the probability of one event and the conditional probability of the second event given the first.*

Set Operations

Many relationships among the sets (or subsets) of a sample space may be established for the purpose of performing mathematical operations. Such set operations include those of union, intersection, and complement, each of which requires the introduction of additional commonly used symbology.

(1) The *union* of two sets, written $A \cup B$, is defined as the set whose elements are all the elements either in A or in B or in both. The union of A and B is illustrated in the Venn diagram of Figure 9–2(a).

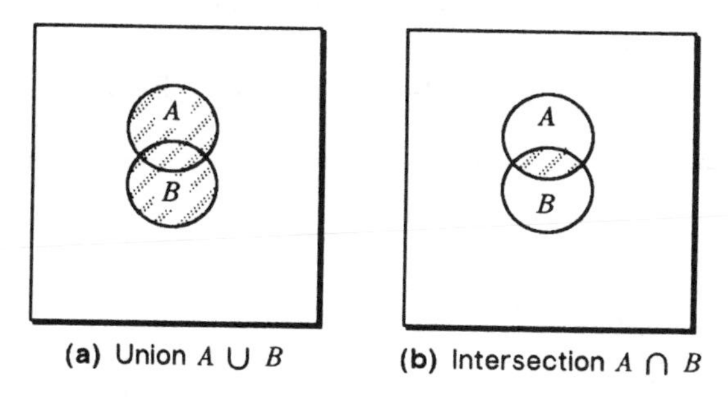

(a) Union $A \cup B$ (b) Intersection $A \cap B$

Figure 9–2. Union and intersection of sets.

(2) The *intersection* of two sets, written $A \cap B$, is defined as the set whose elements are common to set A and set B, as illustrated in Figure 9–2(b).

(3) The *complement* of set A is the set consisting of all the elements of the sample space that are not in A. The complement is identified by the use of the prime symbol. It is illustrated in Figure 9–3 as A'.

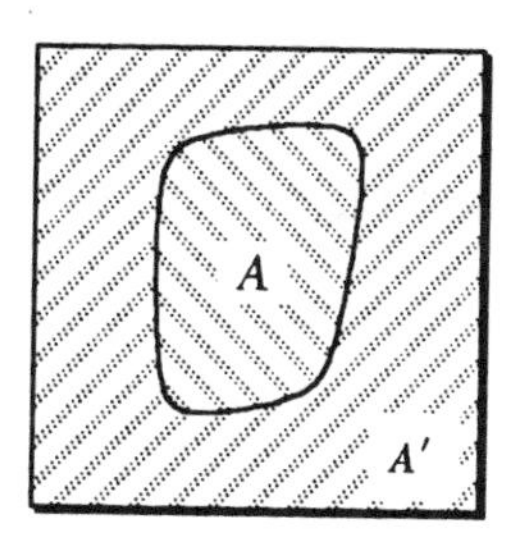

Figure 9–3. Complement sets.

Consider now a hypothetical experiment where the totality of results makes up a sample space, S, and involves two events, A and B. In developing the probabilities associated with these two events, it is convenient to use the above symbology to indicate various compound events, i.e., those involving union or intersection. The total number of possible outcomes (elements of the sample space) is taken as n. Any of the n outcomes is assumed to be equally probable. Compound events involving A and B can be summarized as in Table 9–1 and as illustrated by the Venn diagram of Figure 9–4. Each area in Figure 9–4 illustrates the events shown in the first column of Table 9–1, one of which occurred after each performance of the experiment.

Table 9–1. Compound Events

Event	No. of Outcomes	Probability
$A \cap B'$	n_1	n_1/n
$A' \cap B$	n_2	n_2/n
$A \cap B$	n_3	n_3/n
$A' \cap B'$	n_4	n_4/n

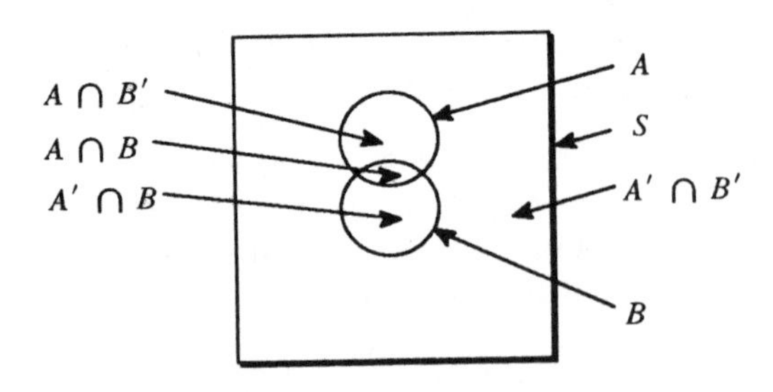

Figure 9–4. Venn diagram of compound events.

A number of probability relations can be defined and related, by observation, to Table 9–1 and Figure 9–4. The probability of A, without regard for the occurrence of another event, is

$$P(A) = P(A \cap B') + P(A \cap B) = (n_1 + n_3)/n. \qquad (9\text{–}1)$$

Similarly, the probability of event B is

$$P(B) = P(B \cap A') + P(B \cap A) = (n_2+n_3)/n. \tag{9-2}$$

The probability of either A or B or both is

$$P(A \cup B) = P(A) + P(B) - P(A \cap B) = (n_1+n_2+n_3)/n. \tag{9-3}$$

The probability of the event which is the intersection of A and B may be written

$$P(A \cap B) = P(A)\ P(B \mid A) \tag{9-4}$$

or

$$P(A \cap B) = P(B)\ P(A \mid B). \tag{9-5}$$

The expressions $P(B \mid A)$ and $P(A \mid B)$ are known as conditional probabilities. These may be read "the conditional probability of B, given A" and "the conditional probability of A, given B," respectively. These conditional probabilities may then be determined as

$$P(B \mid A) = \frac{P(A \cap B)}{P(A)} = n_3/(n_1+n_3) \tag{9-6}$$

and

$$P(A \mid B) = \frac{P(A \cap B)}{P(B)} = n_3/(n_2+n_3). \tag{9-7}$$

Note also that the probability of A and B occurring simultaneously may then be determined

$$P(A \cap B) = n_3/n. \tag{9-8}$$

Some additional definitions and conclusions may now be presented. If events A and B cannot occur simultaneously, they are mutually exclusive, or *disjoint*; the probability of their simultaneous occurrence is the probability of the empty set, ϕ; that is,

$$P(A \cap B)_{\text{disjoint}} = P(\phi) = 0.$$

If this conclusion is combined with Axiom 5, it may be stated that if two events are mutually exclusive, the probability of at least one of them is the sum of their individual probabilities; that is,

$$P(A \cup B)_{\text{disjoint}} = P(A) + P(B).$$

If the occurrence of an event in no way depends on the occurrence of a second event, the two are *independent*. Mathematically, A and B are independent if

$$P(A \mid B) = P(A)$$

or if

$$P(B \mid A) = P(B).$$

Then from Equation 9–4 or 9–5, it is seen that $P(A \cap B) = P(A)\,P(B)$. Note that this does not mean that $P(A \cap B) = 0$. The fact that two events are independent means that there is no functional relationship between their probabilities of occurrence. The expression $P(A \cap B) = 0$ says that A can never occur when B does, a functional relationship of mutual exclusion.

If events A and B can occur simultaneously, then a certain fraction of events B have event A associated with them. If this fraction is the same as the fraction of all possible events that have event A associated with them, then the events A and B are independent. Symbolically, independence implies that

$$P(A \mid B) = \frac{P(A \cap B)}{P(B)} = \frac{P(A)\,P(B)}{P(B)} = P(A). \qquad (9\text{–}9)$$

This can be demonstrated by combining the definition of independence with Axiom 6. If A and B are statistically independent, the probability of their simultaneous occurrence is the product of their individual probabilities; that is,

$$P(A \cap B)_{\text{independent}} = P(A)\, P(B). \qquad (9\text{–}10)$$

Note that Equation 9–10 is symmetric in A and B. This implies that if A is independent of B, then B is independent of A. This need be true only in the statistical sense. It is important to recognize the difference between statistical dependence and causal dependence. From the causal viewpoint, subscriber complaints are dependent on noisy trunks, but noisy trunks are not dependent on subscriber complaints. Statistical analysis would merely show a dependence or correlation between the two without any indication as to which is the cause and which is the effect.

Much statistical work is simplified if it can be assumed that events are either mutually exclusive or independent. Where events are mutually exclusive, the probability of at least one of the events is the simple sum of the probabilities of the mutually exclusive events. The probability of the simultaneous occurrence of independent events may be found as the product of the probabilities of the independent events.

Example 9–1:

This example concerns a group of 1000 trunks between two cities. All of these trunks are measured for loss and noise. It is found that 925 trunks meet the noise objective, 875 trunks meet the loss objective, and 850 trunks meet both objectives. If a connection is established between the two cities and if there is an equal probability that any trunk may be used, what relationships can be evaluated from the foregoing set operations in regard to calls between the two cities?

Various events and their probabilities may now be tabulated, as in Table 9–1; both symbolic and numerical values are given in Table 9–2. A Venn diagram of the relationships among the subsets of trunks is given in Figure 9–5.

Table 9–2. Tabulation of Events in Example 9–1

Event	Note	No. of Occurrences	Probability or Relative Frequency
S	1	$n = 1000$	$n/n = 1000/1000 = 1$
A	2	$n_1 = 925$	$n_1/n = 925/1000 = 0.925$
B	2	$n_2 = 875$	$n_2/n = 875/1000 = 0.875$
A'	3	$n_3 = 75$	$n_3/n = 75/1000 = 0.075$
B'	3	$n_4 = 125$	$n_4/n = 125/1000 = 0.125$
$A \cap B$	4	$n_5 = 850$	$n_5/n = 850/1000 = 0.85$
$A \cap B'$	4	$n_6 = 75$	$n_6/n = 75/1000 = 0.075$
$A' \cap B$	4	$n_7 = 25$	$n_7/n = 25/1000 = 0.025$
$A' \cap B'$	4	$n_8 = 50$	$n_8/n = 50/1000 = 0.05$
$A \mid B$	5	$n_9 = 850$	$n_5/(n_7 + n_5) = 850/875 = 0.971$
$B \mid A$	6	$n_{10} = 850$	$n_5/(n_6 + n_5) = 850/925 = 0.919$
$A \cup B$	7	$n_5 + n_6 + n_7 = 950$	$(n_5 + n_6 + n_7)/n = 950/1000 = 0.95$

Notes:

1. S is the sample space, 1000 trunks.
2. A and B are two subsets, the trunks that meet the noise objective and the loss objective, respectively.
3. A' and B' are the complements of A and B.
4. These are the four mutually exclusive events that make up the sample space, S.
5. $A \mid B$, the event A given B, consists of those trunks meeting the noise objective among the trunks that meet the loss objective.
6. $B \mid A$, the event B given A, contains the trunks meeting the loss objective among those that meet the noise objective.
7. $A \cup B$ represents all of the trunks that meet the noise objective, the loss objective, or both.

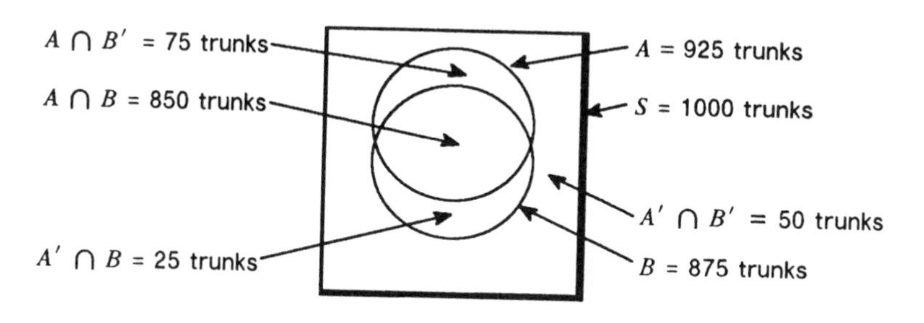

Figure 9–5. Venn diagram for a set of trunks.

9–2 DISCRETE AND CONTINUOUS FUNCTIONS

An important objective in working with statistics and with probability theory is a more efficient way of describing a set of observations than by enumerating all the numbers in the set. A common problem is that of characterizing a set of measurements that are supposed to be similar or identical but are not. The random variable is a function that may be discrete, as the trunks in Example 9–1, where the trunks either met objectives or did not. The random variable may also be continuous. In Example 9–1, the data may have related to actual measurements of loss and noise, and the random variable might have represented the distribution of these measurements, i.e., the number of trunks showing noise or loss values in some recognizable measurement system such as dB of loss or dBrnc0 of noise.

Mapping

Consider a sample space made up of elements designated as ϱ_i. By a process called mapping, the elements of the space (or domain) can be expressed in terms of a random variable, X, which is plotted along an axis. The mapping process is illustrated in Figure 9–6, where $X(\varrho_i) = x_i$. By virtue of the rule of correspondence, each element, ϱ_i, maps into one and only one value, x_i, although it is possible for more than one ϱ_i to map into the same x_i. While every element ϱ_i must map into some value x_i, it is not necessary that every x_i be an image of an element ϱ_i.

Theoretically, the variable $X(\varrho_i)$ may take any value from $-\infty$ to $+\infty$, as indicated in Figure 9–6. It is generally true, however, that the mapping process establishes a restricted range of x between minimum and maximum values. This is also illustrated in Figure 9–6.

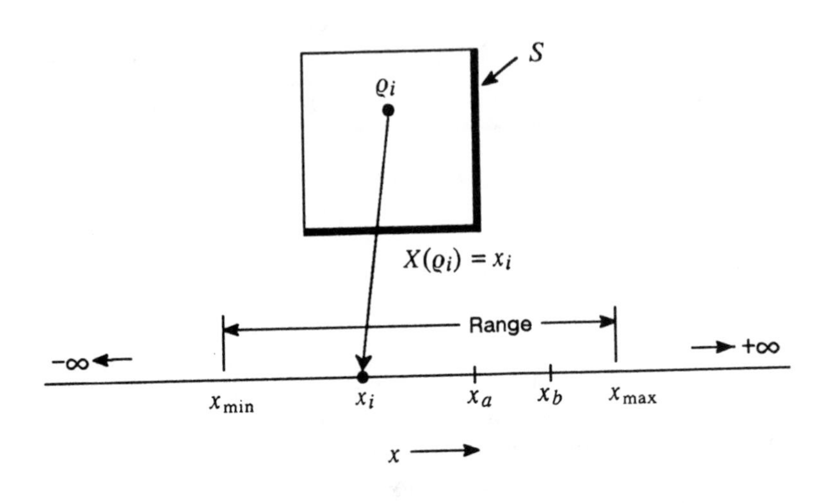

Figure 9–6. Mapping.

If the elements of a sample space exhibit characteristics that involve two parameters, the mapping becomes a two–dimensional process as illustrated by Figure 9–7, where the trunks of Example 9–1 are mapped onto the x–x and y–y axes. The various events then map into areas in the x–y plan of Figure 9–7.

As mentioned previously, the random variable, X or Y, may be continuous or discrete. In either case, the treatment and manipulation of data depend on the ability to express these variables by suitable functional relationships, such as the cumulative distribution function (c.d.f.) or the probability density function (p.d.f.).

Cumulative Distribution Function

A real random variable is a real function whose domain is the sample space, S, and whose range is the real line (the x axis). The random variable also satisfies the conditions (1) that the set

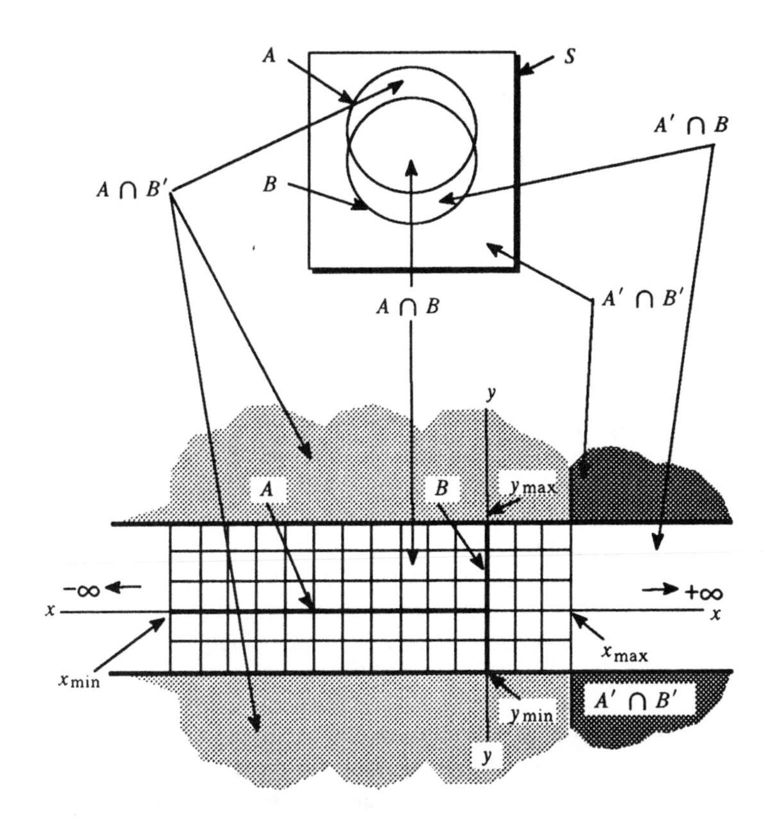

Figure 9–7. Mapping in two dimensions.

$\{\varrho_i : X(\varrho_i) \leq x_i\}$* is an event for any real number, x_i, and (2) that the probability $P\{\varrho_i : X(\varrho_i) = \pm\ \infty\}$ is equal to zero. The function describing the probability distribution of the random variable is called the cumulative distribution function and may be written [2]

$$F_X(x_i) = P\{\varrho_i : X(\varrho_i) \leq x_i\}. \tag{9–11}$$

This equation states that the c.d.f. is a function equal to the probability that the variable, X (representing the elements, ϱ_i, of

*In expressions such as this, the braces define the *set* and the colon is read *such that*.

the sample space), is equal to or less than the value x_i. For present purposes, this equation must meet the following conditions:

(1) It is a real function of a real number.

(2) It is right–continuous; that is, the value of the function $F_X(x)$ at any point $x \leq x_i$ is *equal to or less than* the value given in Equation 9–11.

(3) It is single–valued, monotonic, nondecreasing.

(4) $\lim_{x \to -\infty} F_X(x) = 0; \qquad \lim_{x \to \infty} F_X(x) = 1.$

The random variable, X, may be continuous, discrete, or mixed. When X is continuous, the c.d.f. is continuous. When X is discrete, the c.d.f. is not continuous and, when plotted, appears as a set of steps. These relationships show that $X(\varrho_i)$ may take on values from $-\infty$ to $+\infty$; the c.d.f. correspondingly takes on values from 0 to 1 for $X(\varrho_i) = -\infty$ to $X(\varrho_i) = +\infty$.

If an estimation of a continuous c.d.f. is plotted as in Figure 9–8, the curve looks like an uneven staircase having flat treads and discontinuities analogous to vertical risers. As the number of observations increases and the granularity of readings becomes finer, the treads and risers become smaller. A continuous c.d.f. is a smooth curve as illustrated in Figure 9–9.

It should be noted that the plot of a discrete c.d.f. would also look like Figure 9–8.

Probability Density Function

The derivative of the c.d.f. is defined as the probability density function. It may be written

$$f_X(x) = dF_X(x)/dx. \qquad (9\text{–}12)$$

The function is illustrated in Figure 9–10.

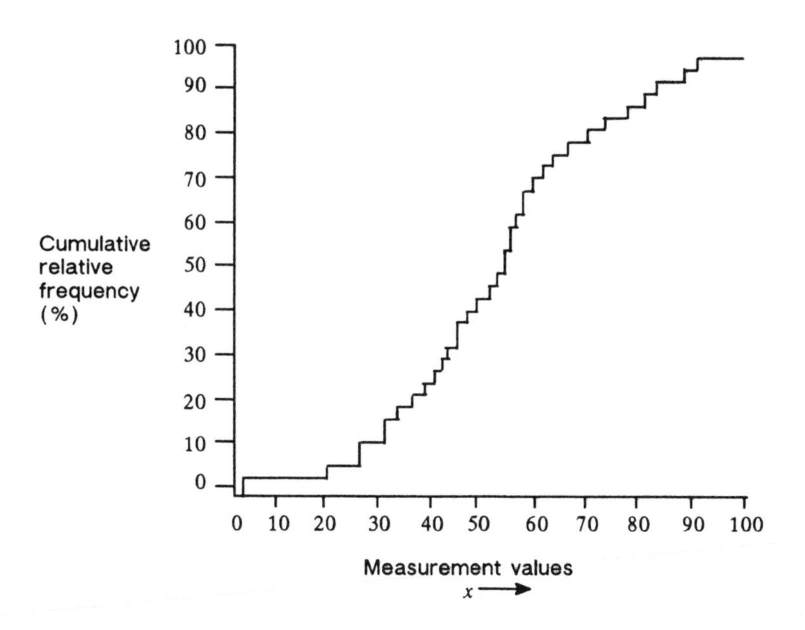

Figure 9–8. Approximation to a continuous cumulative distribution function.

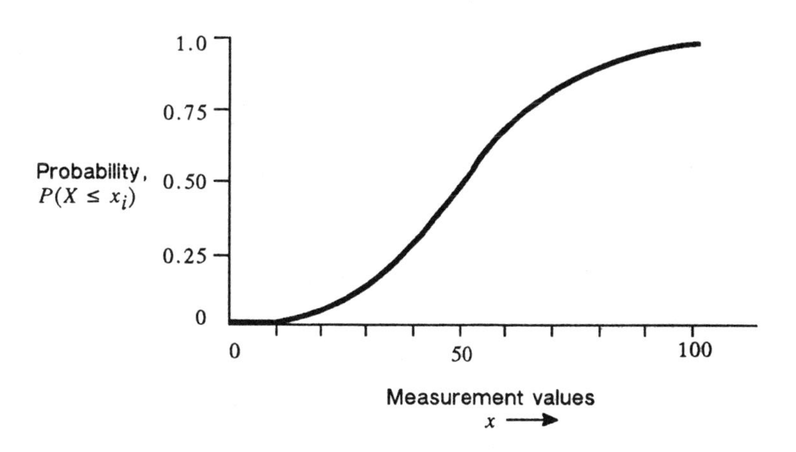

Figure 9–9. A continuous cumulative distribution function.

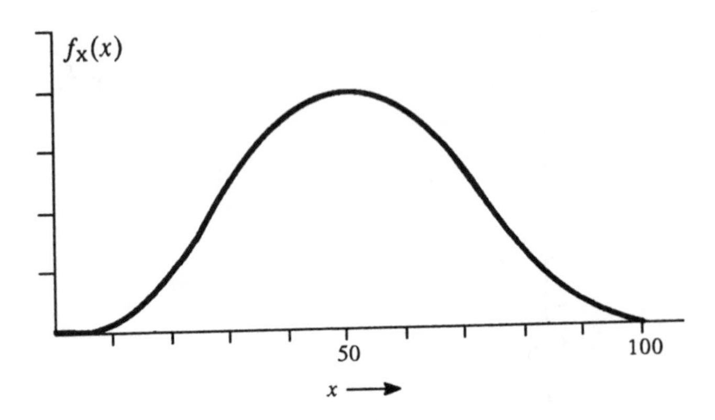

Figure 9–10. Probability density function.

9–3 THE PRINCIPAL PARAMETERS

Expected value, variance, and *standard deviation* are terms that define the important and useful characteristics of random variables. The definitions of other parameters, sometimes useful in statistical studies but seldom used in probability theory, are given later. When available data permit the use of approximations or estimates of the random variable, estimates of expected value, variance, and standard deviation may also be used for statistical analysis.

Expected Value

The expected, or mean, value of a random variable, X, may be estimated from repeated trials of an appropriate experiment by

$$\overline{X} \approx \frac{\sum_{i=1}^{n} x_i}{n} \tag{9–13}$$

where $\sum_{i=1}^{n} x_i$ is the sum of the values x_i, and n is the total number of values of x. Here, x_i is the numerical value that the random variable, X, takes for the ith trial.

If the random variable is discrete, the expected value may be found by

$$E[X] = \overline{X} = \sum_{x_i = 1}^{n} x_i \, P\{X = x_i\}, \tag{9–14}$$

where x_i represents the discrete values assumed by the variable, X. If the random variable is continuous, the expected value is found by

$$E[X] = \overline{X} = \int_{-\infty}^{+\infty} x \, f_X(x) \, dx. \tag{9–15}$$

The term *expectation* has been extended to include the expectation of any function of X, provided X has a probability function. The expectation of $g(X)$ is

$$E[g(X)] = \int_{-\infty}^{+\infty} g(x) \, f_X(x) \, dx,$$

where $f_X(x)$ is the p.d.f. Of particular interest is $g(X) = X^2$, the mean squared value, or

$$E[X^2] = \int_{-\infty}^{+\infty} x^2 \, f_X(x) \, dx = \overline{X^2}. \tag{9–16}$$

Variance

The expected value of a random variable gives no information regarding the variation or range of values that may be assumed by a random variable. The most useful measure of this parameter is the *variance*, defined as the expectation of the square of the deviations of observations from their mean, or expected, value. The expression for the variance, which may be derived from the function of the random variable, is

$$\sigma_X^2 \approx \int_{-\infty}^{+\infty} (x - \overline{X})^2 \, f_X(x) \, dx.$$

By substituting Equations 9–15 and 9–16 and noting that $\int_{-\infty}^{+\infty} f_X(x)\,dx = 1$ (since the probability of the entire sample must be in unity), the above expression may be written

$$\sigma_X^2 = \int_{-\infty}^{+\infty} x^2 f_X(x)\,dx - 2\overline{X} \int_{-\infty}^{+\infty} x f_X(x)\,dx + \overline{X}^2 \int_{-\infty}^{+\infty} f_X(x)\,dx$$

$$= \overline{X^2} - 2\overline{X}^2 + \overline{X}^2$$

$$= \overline{X^2} - \overline{X}^2. \quad (9\text{–}17)$$

The variance may be estimated, from repeated trials of an experiment, by

$$\sigma_X^2 \approx \left(\frac{\sum_{i=1}^{n} x_i^2}{n} \right) - \left(\frac{\sum_{i=1}^{n} x_i}{n} \right)^2. \quad (9\text{–}18)$$

Since expectation is a sum or integral, it obeys the same laws as sums or integrals. The expectation of a constant is that constant. The expectation of a constant times a random variable is the constant times the expectation of the random variable. The expectation of a sum is the sum of the expectations. The mean or any other statistical average is a constant and not a random variable.

Standard Deviation

The square root of the variance is often a convenient parameter to use as a measure of variation or dispersion. It is called the *standard deviation*. For the approximation given in Equation 9–18, it is

$$\sigma_X \approx \sqrt{\left(\frac{\sum_{i=1}^{n} x_i^2}{n} \right) - \left(\frac{\sum_{i=1}^{n} x_i}{n} \right)^2}. \quad (9\text{–}19)$$

The exact expression for the standard deviation is found from Equation 9–17,

$$\sigma_X = \sqrt{\overline{X^2} - \overline{X}^2}. \qquad (9\text{–}20)$$

Example 9–2:

The approximation to the continuous c.d.f. of Figure 9–8 is a plot of the available data concerning the sample space. From the data, determine the expected value, the variance, and the standard deviation. The first three columns in Table 9–3 represent the data from which the figure was constructed; the last two are computed values that are summed.

The multiplier, n, in the last two columns reflects the fact that all x_i points are not different; n is the number of readings of each value (column 2).

From Table 9–3, the expected value may be computed by Equation 9–13 as

$$\overline{X} \approx \frac{\sum_{i=1}^{n} x_i}{n} \approx \frac{5264}{1000} \approx 52.6.$$

The variance may be computed from Equation 9–18 as

$$\sigma_X^2 \approx \left(\frac{\sum_{i=1}^{n} x_i^2}{n}\right) - \left(\frac{\sum_{i=1}^{n} x_i}{n}\right)^2$$

$$\approx \frac{312,628}{100} - \left(\frac{5264}{100}\right)^2 \approx 355.$$

The standard deviation, from Equation 9–19, is

$$\sigma_X \approx \sqrt{355} \approx 18.8.$$

Table 9–3. Continuous Cumulative Distribution Function

Value (Abscissa)	Data n	Data Cum. n	nx_i	nx_i^2
3	1	1	3	9
21	4	5	84	1764
27	3	8	81	2187
29	2	10	58	1682
31	2	12	62	1922
33	6	18	198	6534
36	2	20	72	2592
37	2	22	74	2738
39	2	24	78	3042
42	6	30	252	10,584
43	2	32	86	3698
45	3	35	135	6075
46	7	42	322	14,812
47	3	45	141	6627
49	3	48	147	7203
50	2	50	100	5000
52	2	52	104	5408
53	3	55	159	8427
54	2	57	108	5832
56	8	65	448	25,088
58	3	68	174	10,092
60	4	72	240	14,400
61	4	76	244	14,884
64	2	78	128	8192
67	2	80	134	8978
69	3	83	207	14,283
74	3	86	222	16,428
77	2	88	154	11,858
79	2	90	158	12,482
80	2	92	160	12,800
87	3	95	261	22,707
88	2	97	176	15,488
98	3	100	294	28,812
			5264	312,628

9–4 SUMS OF RANDOM VARIABLES

In statistical analysis and in applications of probability theory, it is possible to make use of certain relationships between several sample spaces or between a sample space and subsets of that sample space. One example of many such useful relationships is the summing of random variables.

If two independent random variables are known, a new random variable may be derived by adding together repetitively one member from each of the two original random variables. The mean value of the random variable is the sum of the mean values of the original two; that is,

$$(\overline{X+Y}) = \overline{X} + \overline{Y}. \qquad (9\text{–}21)$$

The variance of the derived random variable is the sum of the original variances. This may be written

$$\sigma^2_{(x+y)} = \sigma_x{}^2 + \sigma_y{}^2. \qquad (9\text{–}22)$$

These relationships for the random variable derived from the sum of the two independent random variables are valid provided the values of all means and variances are finite. It is also assumed in the derivation of Equations 9–21 and 9–22 that, in addition to the first two random variables being independent, there is equal probability of one member of one random variable combining with any member of the other. The equations may be extended to apply to any number of variables, provided the universes are all independent.

If the two random variables are subtracted, the means subtract but the variances add.

Example 9–3:

In this example, it is assumed that telephone connections may be established between switching machines in two cities, *A* and *C*, by way of a switching machine in city *B*. The trunks between *A* and *B* have a mean loss of 2.7 dB and a standard deviation of 0.7 dB. The trunks between *B* and *C* have a mean loss of 1.6 dB and a standard deviation of 0.3 dB. When connections are established from *A* to *C*, there is in each link (*AB* and *BC*) equal likelihood of connection via any trunk in the group. Determine the mean loss of connections from *A* to *C* and the standard deviation of the distribution of loss between *A* and *C*.

The standard deviation of the distribution of overall losses may be found from Equation 9–22. It is

$$\sigma_{AC} = \sqrt{\sigma_{AB}^2 + \sigma_{BC}^2}$$

$$= \sqrt{0.7^2 + 0.3^2} = 0.76 \text{ dB}.$$

The mean value of the derived random variable (the mean loss from A to C) is found from Equation 9–21 to be

$$\overline{X}_{AC} = \overline{X}_{AB} + \overline{X}_{BC}$$

$$= 2.7 + 1.6 = 4.3 \text{ dB}.$$

Example 9–4:

Assume the distribution of A to C trunk losses determined in Example 9–3, i.e., $\overline{X}_{AC} = 4.3$ dB and $\sigma_{AC} = 0.76$ dB. Assume further that the distribution of talker volume at A is given by $\overline{X}_{\text{vol } A} = -15$ vu, and $\sigma_{\text{vol } A} = 2$ vu. The mean of the distribution of volumes at C may be determined by

$$\overline{X}_{\text{vol } C} = \overline{X}_{\text{vol } A} - \overline{X}_{AC}$$

$$= -15 - 4.3 = -19.3 \text{ vu}.$$

The standard deviation of volumes at C is given by

$$\sigma_{\text{vol } C} = \sqrt{\sigma_{\text{vol } A}^2 + \sigma_{AC}^2} = \sqrt{2^2 + 0.76^2}$$

$$= 2.14 \text{ vu}.$$

This type of computation, involving the difference between mean values, is applicable to the determination of grade of service.

9–5 DISTRIBUTION FUNCTIONS

A number of different distribution functions of random variables are used to represent various phenomena in the field of telecommunications. Each may be expressed mathematically and graphically to illustrate its applicability and general characteristics.

Gaussian or Normal Distribution

A random variable is said to be normally distributed if its density function is a Gaussian curve, i.e., if the function can be written in the form

$$f_X(x) = Ae^{-ax^2}, \quad a > 0.$$

The density functions of many random variables are found to take this form and may be expressed by

$$f_X(x) = \frac{1}{\sigma_X\sqrt{2\pi}} e^{-(x-X)^2/2\sigma_X^2}, \quad -\infty < x < +\infty \qquad (9–23)$$

where e is the base of natural logarithms. If it is assumed that $X = 0$ and $\sigma_X = 1$, Equation 9–23 represents the unit (standard form) normal density function. It may be written

$$f_X(x) = \frac{1}{\sqrt{2\pi}} e^{-x^2/2}. \qquad (9–24)$$

The corresponding unit normal c.d.f. is

$$F_X(x) = \frac{1}{\sqrt{2\pi}} \int_{-\infty}^{x} e^{-u^2/2}\, du \qquad (9–25)$$

where u is the dummy variable of integration. To illustrate these functions, Equations 9–24 and 9–25 are plotted as Figures 9–11 and 9–12.

The density function of the normal distribution is written in rather simple form, as shown in Equation 9–23. The c.d.f., which is its integral, cannot be written in closed form. Its values have been computed by numerical techniques with considerable precision. Values are given in Table 9–4 for the unit normal c.d.f., Equation 9–25. Percentages of the normal distribution that lie within and outside certain symmetric limits of the normal density function are illustrated in Figure 9–13.

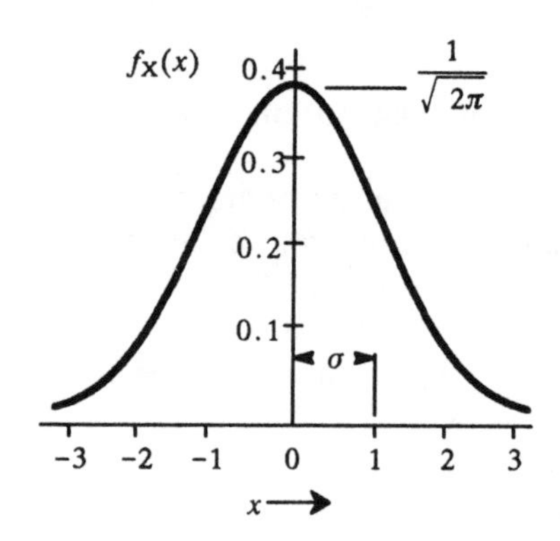

Figure 9–11. Unit normal density function.

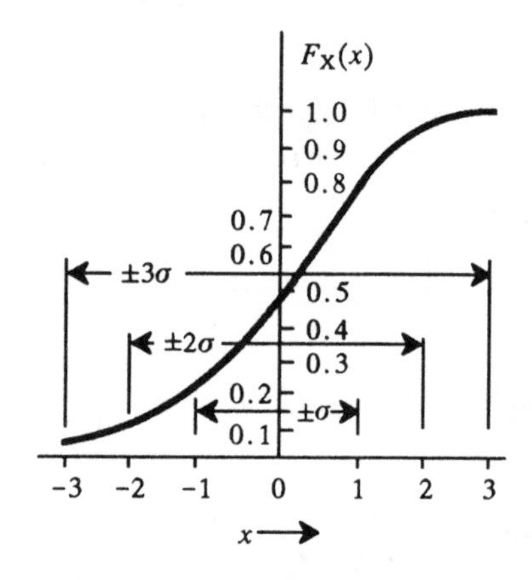

Figure 9–12. Unit normal cumulative distribution function.

It is often useful to plot the c.d.f. from collected data. For the normal distribution this gives an S–shaped curve, called an ogive, such as that illustrated in Figure 9–12. By suitable distortion of the cumulative probability scale, the ogive can be made to appear as a straight line. Commercially available graph paper, called arithmetic probability paper, having just such a distorted scale, has been designed for use with normal distributions. When a set of observations has been plotted on such paper, it is a simple matter to estimate the mean by reading the 50–percent point and the standard deviation by reading the values at the 16–percent and 84–percent points, which are separated by approximately 2σ.

Table 9–4. Normal Probability Distribution Function Values

x	$F(x)$	x	$F(x)$	x	$F(x)$
-4.0	0.00003	-0.9	0.1841	1.1	0.8643
-3.301	.0005	-0.842	.2000	1.2	.8849
-3.090	.0010	-0.8	.2119	1.282	.9000
-3.0	.0013	-0.7	.2420	1.3	.9032
-2.9	.0019	-0.674	.2500	1.4	.9192
-2.881	.0020	-0.6	.2741	1.5	.9332
-2.8	.0026	-0.524	.3000	1.6	.9452
-2.749	.0030	-0.5	.3085	1.645	.9500
-2.7	.0035	-0.4	.3446	1.7	.9554
-2.652	.0040	-0.385	.3500	1.8	.9641
-2.6	.0047	-0.3	.3821	1.9	.9713
-2.576	.0050	-0.253	.4000	1.960	.9750
-2.5	.0062	-0.2	.4207	2.0	.9772
-2.4	.0082	-0.126	.4500	2.1	.9821
-2.326	.0100	-0.1	.4602	2.2	.9861
-2.3	.0107	0	.5000	2.3	.9893
-2.2	.0139	0.1	.5398	2.326	.9900
-2.1	.0179	0.126	.5500	2.4	.9918
-2.0	.0228	0.2	.5793	2.5	.9938
-1.960	.0250	0.253	.6000	2.576	.9950
-1.9	.0287	0.3	.6179	2.6	.9953
-1.8	.0359	0.385	.6500	2.652	.9960
-1.7	.0446	0.4	.6554	2.7	.9965
-1.645	.0500	0.5	.6915	2.749	.9970
-1.6	.0548	0.524	.7000	2.8	.9974
-1.5	.0668	0.6	.7257	2.881	.9980
-1.4	.0808	0.674	.7500	2.9	.9981
-1.3	.0968	0.7	.7580	3.0	.9987
-1.282	.1000	0.8	.7881	3.090	.9990
-1.2	.1151	0.842	.8000	3.301	.9995
-1.1	.1357	0.9	.8159	4.0	.99997
-1.036	.1500	1.0	.8413		
-1.0	.1587	1.036	.8500		

It can be shown that (1) with certain constraints, if n samples are drawn from a sample space, the mean values of the samples constitute a random variable whose density, $f_{\overline{X}}(x)$, is concentrated near its mean and (2) as n increases, $f_{\overline{X}}(x)$ tends to a normal density curve regardless of the shape of the densities of the samples of n. The constraints are that n must be large (usually greater than 10) and that the standard deviation of the random variable must be finite. This is the *central limit theorem*.

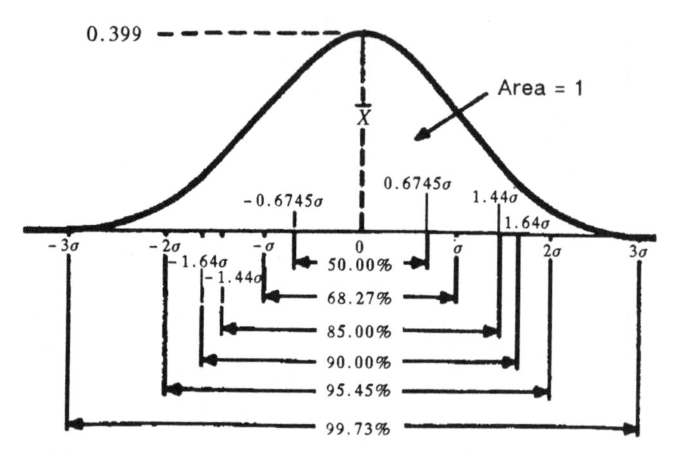

(a) Areas between selected ordinates of the normal curve

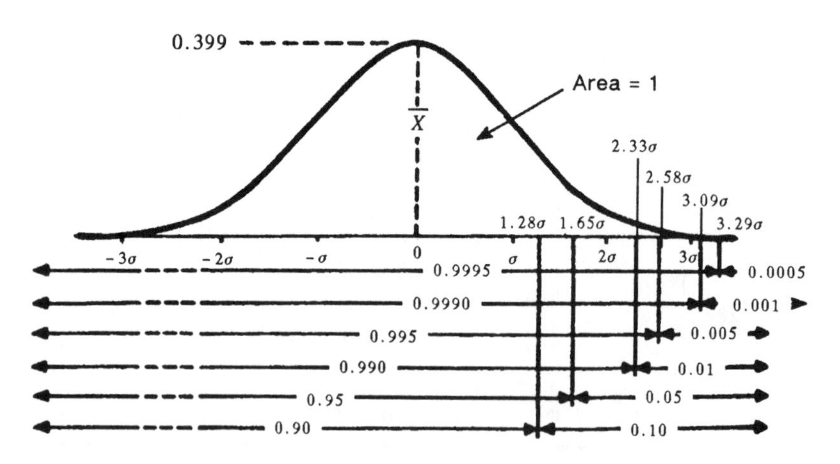

(b) Areas beyond selected ordinates of the normal curve

Figure 9–13. Areas between and beyond selected ordinates of the normal curve.

Poisson Distribution

The Poisson distribution is a discrete probability distribution function that takes the form

$$F_X(x) = e^{-\lambda} \sum_{n=0}^{x} \frac{\lambda^n}{n!}, \quad x = 0, 1, 2\ldots, \qquad (9\text{–}26)$$

$$\lambda > 0.$$

The corresponding probability density function is a sequence of impulses expressed

$$f_X(x) = \frac{e^{-\lambda}\lambda^x}{x!}. \qquad (9\text{–}27)$$

In these equations, λ is a constant. The derivation of the Poisson distribution is based on the assumptions that the number of observations, n, is large (usually greater than 50), that the probability of success, p, is small (less than $0.075n$), and that the product of the two, np, is a constant. Among the properties of the Poisson distribution are the facts that λ is equal to the mean value of x and that the variance, σ^2, is also equal to λ. The Poisson distribution is illustrated in Figure 9–14 for $\lambda = 3$.

This distribution is useful in studying the control of defects in a manufacturing process, the occurrence of accidents or rare disease, and the congestion of traffic, including telephone traffic. It has also been used to represent the statistics of discontinuities in a transmission medium due to certain manufacturing processes and to damage caused by rocks falling on a cable during installation.

Binomial Distribution

A combination of n different objects taken x at a time is called a *selection* of x out of n with no attention given to the order of the arrangement. The number of combinations of such a selection is denoted by $\binom{n}{x}$. It is defined as

$$\binom{n}{x} = \frac{n!}{x!\,(n-x)!}.$$

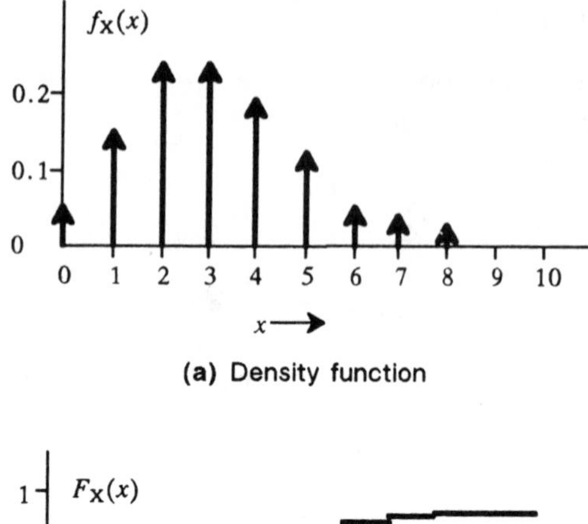

(a) Density function

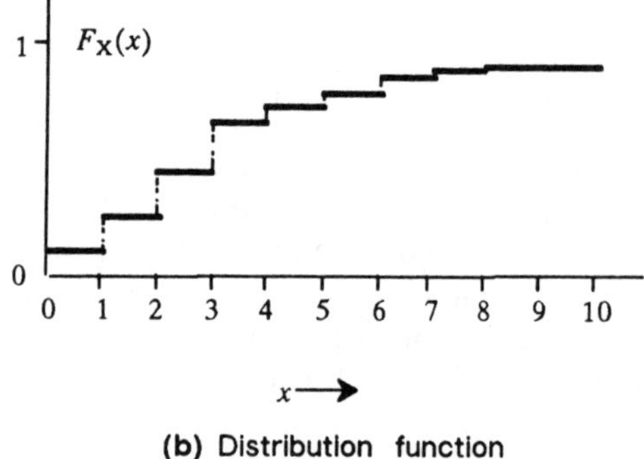

(b) Distribution function

Figure 9–14. Poisson distribution.

If p is the probability of success in any single trial and $q = 1 - p$ is the probability of failure, then the probability of success for x times out of n trials is given by

$$P(x) = \binom{n}{x} p^x q^{n-x} = \frac{n!}{x!\,(n-x)\,!} p^x q^{n-x} . \tag{9–28}$$

This is known as the binomial distribution. Its density function may be written

$$f_X(x) = \binom{n}{x} p^x q^{n-x}. \tag{9–29}$$

This function is a sequence of impulses as illustrated by Figure 9–15, where $n = 9$ and $p = q = 1/2$.

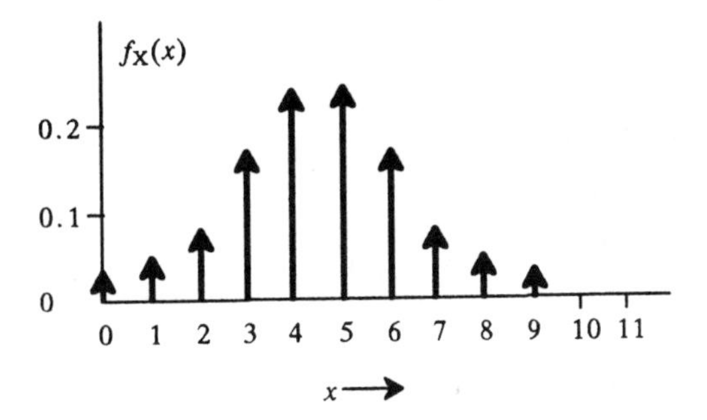

Figure 9–15. Density function for binomial distribution.

The mean value of the binomial distribution is equal to np and the variance is $\sigma^2 = npq$.

Example 9–5:

Consider the 1000 trunks of Example 9–1. Recall that 850 of these trunks meet both noise and loss objectives. Thus, 150 trunks fail to meet the loss or the noise objective or both. In five consecutive connections using these trunks, where equal probability of using any trunk is assumed, (1) what is the probability that all five connections will be satisfactory with respect to both noise and loss and (2) what is the probability that two out of five calls will be unsatisfactory?

(1) $p = \dfrac{850}{1000} = 0.85$

$q = 1 - p = 0.15$

$n = 5$ trials

$x = 5$ successful trials.

Using Equation 9–28,

$$P(x) = \frac{5!}{5!(5-5)!} (0.85^5)(0.15^{5-5}).$$

Since it can be shown that $0! = 1$ and $x^0 = 1$, $P(x) = 0.85^5 = 0.44$.

(2) $p = 0.15$

$q = 0.85$

$n = 5$

$x = 2.$

Again using Equation 9–28,

$$P(x) = \frac{5!}{2!(5-2)!}(0.15^2)(0.85^{5-2}).$$

$$= 0.14.$$

Binomial-Poisson-Normal Relationships

The three distributions described so far are related to one another. If np and nq are both greater than 5, the binomial distribution can be closely approximated by a normal distribution with standardized variable

$$z = \frac{x - np}{\sqrt{npq}}.$$

The unit normal density function, Equation 9–24, may then be written

$$f_Z(z) = \frac{1}{\sqrt{2\pi}} e^{-z^2/2}.$$

If, in the binomial distribution, n is large and the probability, p, of an event is close to zero ($q = 1-p$ is nearly 1), the event is called a *rare event*. In practice, an event can be considered rare if $n \geq 50$ and if $np<5$. In such a case, the binomial distribution is very closely approximated by the Poisson distribution with $\lambda = np$. For this case the Poisson density function, Equation 9–27, may be written

$$f_X(x) = \frac{e^{-np}(np)^x}{x!}.$$

Log-Normal Distribution

Here the random variable is normally distributed when expressed in logarithmic units, for example, decibels. Commercially available graph paper is designed so that a log-normal distribution plots as a straight line.

The log-normal distribution is often encountered in transmission work. In some cases, where the phenomena to be analyzed are multiplicative, the treatment of log-normal distributions is straightforward because in logarithmic form the phenomena are additive and so may be treated as any other random variable in which additive combinations are under consideration. An example is the evaluation of the overall gain or loss of a circuit containing many tandem-connected components, each of which may be represented by a random variable whose distribution is log-normal. A transmission system having a number of transmission line sections and a number of amplifiers in tandem can be so analyzed.

In some cases, the phenomena are individually log-normal but are combined in such a way that the antilogarithms must be considered as the random variables. An example is found in the analysis of signal voltages of combinations of talker signals in multichannel telephone transmission systems. Here, the individual talker distributions are log-normal. The distributions, however, combine by voltage (not log-voltage) to produce a total signal which must be characterized with sufficient accuracy to evaluate the probability of system overload. The analysis, which must be made by graphical or mathematical approximations, has been applied to load-rating theory for transmission systems [3].

Uniform Distribution

This distribution, sometimes called a rectangular distribution from the shape of the density function, is represented by the density function

$$f_X(x) = \frac{1}{x_b - x_a}, \quad x_a \leq x \leq x_b \qquad (9\text{–}30)$$

$$= 0, \text{ elsewhere.}$$

This function and the corresponding distribution function are shown in Figure 9–16.

Example 9–6:

Given a manufacturing process for an amplifier having 6–dB gain with acceptance limits of ± 0.25 dB and given that the random variable (the gain) is uniformly distributed between the two limits, what is the probability that the gain, G, is between 5.9 and 6.1 dB?

From Equation 9–12, it can be shown [1] that

$$F_X(x_2) - F_X(x_1) = \int_{x_1}^{x_2} f_X(x)dx$$

and that

$$P\{5.9 \leq G \leq 6.1\} = \int_{5.9}^{6.1} f_G(x)dx.$$

From Equation 9–30,

$$f_G(x) = \frac{1}{x_b - x_a} = \frac{1}{0.5}.$$

Then,

$$P\{5.9 \leq G \leq 6.1\} = \frac{1}{0.5}\int_{5.9}^{6.1} dx$$

$$= \frac{0.2}{0.5} = 0.4.$$

Thus, about 40 percent of all amplifiers of this type will have gain values between 5.9 and 6.1 dB.

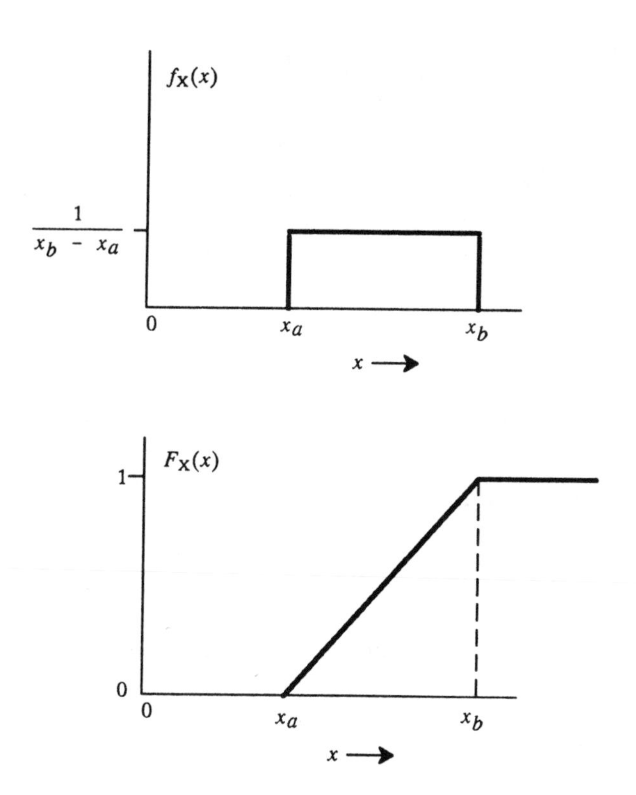

Figure 9–16. Density and cumulative distribution functions of a uniform, or rectangular, distribution.

Rayleigh Distribution

The density function for the Rayleigh distribution may be written

$$f_X(x) = \frac{x}{\sigma^2} e^{-x^2/2\sigma^2}, \quad x \geq 0 \qquad (9\text{–}31)$$

$$= 0, \qquad x < 0 .$$

This density function, illustrated by Figure 9–17, is often used to approximate microwave fading phenomena.

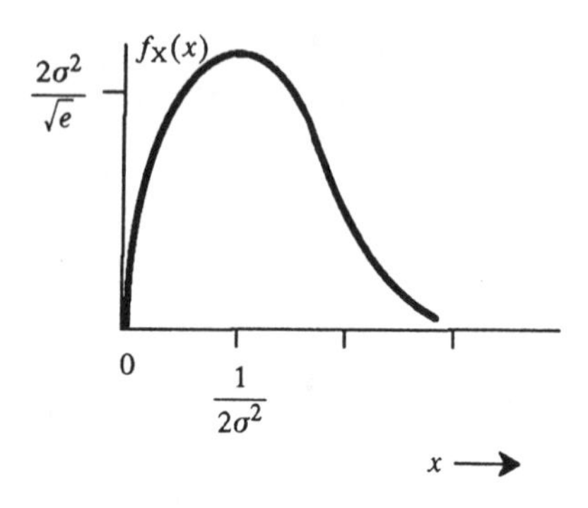

Figure 9–17. Rayleigh density function.

9-6 STATISTICS

The subject of statistics covers the treatment and analysis of data and the relationships between the data and samples taken from the data. Statistics also includes methods of evaluating the confidence in the accuracy of the relationships inferred from data samples.

Central Values and Dispersions

For many purposes, statements of the central value, $\overline{X}$, and the dispersion, σ, provide an adequate summary of a set of observations. Estimates of central values and dispersions, based on experimental outcomes, are frequently used instead of functional relationships, which are often not known.

There are a number of ways of expressing both the central value and the dispersion of a random variable. Since the mean and variance are most easily treated, it is frequently convenient to transform other measures of central values or dispersions to the mean and variance. A specific reason for this convenience is that the relationships of the mean and variance of a subset of samples to the mean and variance of the sample are simple and essentially independent of the nature of the density and distribution functions representing the sample space.

Central Values. A central value may be regarded as an average, where the word *average* is used in its broadest sense.

Following is a list of expressions for the central value of a set of observations that might be used in various circumstances.

(1) The *median* is a central value of the random variable defined such that, in a set of observations, half the observations have values greater than the median and half less than the median. If a number of discrete observations are arranged in order of magnitude, the median is the middle one if there is an odd number of observations. If there is an even number of observations, the median is the arithmetic average of the two middle observations.

(2) The *midrange* is one–half the sum of the largest and smallest of the observations.

(3) The *mode* is the most common value of the variable. It is an estimate of the value of x at the maximum of the density function, Equation 9–12. If the density function has two or more maxima, the distribution is described as bimodal or multimodal, respectively.

(4) The *geometric mean* is the nth root of the magnitude of the product of all n observations.

(5) The *root mean square* (rms) is the square root of the arithmetic mean of the squares of the observations.

(6) The *arithmetic mean* is the measure having the greatest utility in probability theory. It is sometimes called the mean value or the average, where *average* has a narrower connotation than used earlier. Arithmetic mean is an estimate of mathematical expectation. As shown previously in Equation 9–13, the estimate of the mean may be written

$$\overline{X} \approx \frac{\sum_{i=1}^{n} x_i}{n}$$

where $\sum_{i=1}^{n} x_i$ is the sum of the values of observations, x_i, and n is the total number of observations.

Dispersions. A complete description of the dispersion might consist of a tabulation of all deviations. The deviation of any observation, in turn, is the magnitude of the difference between that observation and some stated central value of the observations. Some central value of deviations may be defined as a measure of dispersion. Several commonly used expressions and definitions for dispersions are given in the following:

(1) The *range* is simply the difference between the smallest and largest observations in the sample.

(2) The *mean deviation* is the arithmetic mean of absolute deviations about the mean central value. It is seldom used.

(3) The *standard deviation* is the measure of dispersion that is used as the basis for most of the mathematical treatment of dispersion values in probability theory and in statistical analysis. It is sometimes called the rms deviation. It is given the Greek letter σ as its symbol. The square of the standard deviation is the *variance*.

Histogram

Sometimes it is desirable to display graphically the number of observations that fall in certain small ranges or intervals. The entire range of observations is divided into cells, and the number of observations falling in each cell is listed. The upper bound of each cell is included in the cell. A graphical representation, shown in Figure 9–18, may be prepared by showing values of x as the abscissa and constructing at each cell a rectangle having an area proportional to the number of observations in the cell. The resulting diagram is called a histogram. If the ordinate is expressed in terms of the fraction or percentage of total observations, the histogram is also called a relative frequency diagram. This is illustrated by the right-hand ordinate scale in Figure 9–18. The illustration is a plot of the data of Example 9–2 and of the c.d.f. illustrated by Figure 9–8.

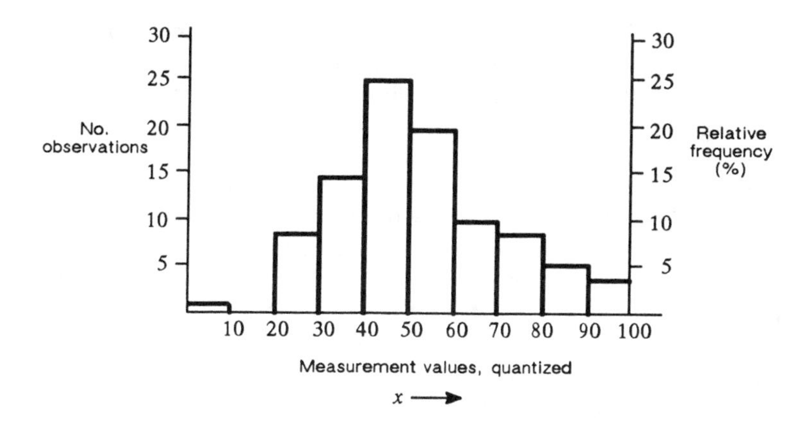

Figure 9–18. Histogram.

Sampling

Sampling involves the measurement of some elements of a universe or sample space. By proper choice of sampling procedure, parameters describing the samples can be used to establish relationships between the parameters of the samples and the parameters of the universe from which they were drawn. Thus, sampling is useful in the estimation of the parameters of the sample space.

Sampling theory is also useful in the determination of the significance of differences between two samples. Tests of significance and decision theory depend on sampling theory [4].

To assure the validity of the results of a sampling procedure, samples must be chosen so that they are representative of the universe. One such process is called random sampling. This may be accomplished physically, for example, by drawing the samples from a bowl in which the universe is represented by properly identified slips of paper or other elements. The bowl is agitated before each drawing of a sample to guarantee random selection. Tables of random numbers are also available and can sometimes be used to advantage.

If a sample is drawn from the universe, recorded, and then placed back into the universe before the next is drawn, the process is called sampling with replacement. If it is not returned, the process is called sampling without replacement. Both processes are used, the choice depending on circumstances. The process of determining the method of sampling is involved in the design of the experiment.

As previously discussed, the sum of random variables may be expressed in terms of probability theory as the addition of two or more functions of the random variable. Summing may also be a statistical process. A sample may be drawn from one sample space and another sample from the same sample space or from a second one. The two values are added and recorded, and the samples are returned and mixed. The process is repeated many times, and from the recorded data the mean and variance may be computed for the summed data by Equations 9–21 and 9–22. Such a process is often necessary when the parameters of the two original sample spaces are not known.

If one value is drawn from each of several similar universes or if several values are drawn from a single universe, independence among the members of the sample is maintained by sampling with replacement. The relationships among the samples can then be used to estimate the parameters of the original universe(s). Such a sample is called a random sample. The size of the sample is designated by the number of members or values, n.

The mean of the sample is the sum of the sample values divided by n. Thus, the mean of the sample means is equal to the mean of the universe. The variance of the sample means is equal to the variance of the universe divided by the sample size.

Estimation. Estimation involves the drawing of inferences about a universe or sample space from measurements of a sample drawn from the sample space. Estimation is sometimes broken down into point estimation and interval estimation. In point estimation, a particular parameter is sought, such as the mean of the unknown universe. In interval estimation, two values are sought between which some fraction (such as 99 percent) of the unknown universe is believed to lie. In some respects, interval estimation is simply two problems in point estimation.

In choosing the methods of estimation, the sample and the set of observations based on it may contain extraneous material that does not belong but that may have a serious effect on the estimate. Estimation may have as one of its objectives the identification and elimination of such invalid data.

In experimentation and production, measurements are made to be used as samples from the potential universe of measurements that might be made. The universe as a whole is nonexistent; its parameters are not and cannot be known, and there is no way to determine what they really are. However, it is expected that if many measurements are made and the results are expressed statistically, the computed values will very nearly represent "true values" for the universe. Since the universe is in fact nonexistent and since the "true values" cannot really be defined, there is no way to define a best estimation. It is necessary to rely on statistical analyses to determine that results are consistent and unbiased.

While there are similarities between the methods of estimation and prediction, there are significant differences too. The scope of this chapter does not permit a more thorough discussion of estimation, but confidence limits should be mentioned.

Confidence Limits. In addition to making estimates of sample space parameters, it is often desirable to express a measure of the limits of confidence in the values. If there is a sample of n observations having a mean value of $\bar{x}_n$ taken from a sample space having a standard deviation of σ, the mean of the sample space may be said to have a value $\bar{x}$ lying between limits of $a\sigma/\sqrt{n}$ and $-a\sigma/\sqrt{n}$ about $\bar{x}$. This interval is called a confidence interval and its two end values, $\bar{x} \pm a\sigma/\sqrt{n}$, are the confidence limits. If by using this estimation procedure to set the interval it is expected that the right answer is obtained 99.7 percent of the time (the area under the normal density curve between $\pm 3\sigma$ points, i.e., $a = 3$), the limits are called 99.7–percent confidence limits. Methods are available for determining limits for various levels of confidence for different types of distributions [4].

References

1. Papoulis, A. *Probability, Random Variables, and Stochastic Processes* (New York: McGraw–Hill Book Company, Inc., 1965).

2. Thomas, J. B. *An Introduction to Statistical Communication Theory* (New York: John Wiley and Sons, Inc., 1969).

3. Dixon, J. T. and B. D. Holbrook. "Load Rating Theory for Multi–Channel Amplifiers," *Bell System Tech. J.*, Vol. 18 (Oct. 1939), pp. 624–644.

4. Spiegel, M. R. "Theory and Problems of Statistics," *Schaum's Outline Series* (New York: McGraw–Hill Book Company, Inc., 1961).

Additional Reading

Feller, W. *An Introduction to Probability Theory and Its Applications*, Vol. I (New York: John Wiley and Sons, Inc., 1957).

Mendenhall, W. and J. E. Reinmuth. *Statistics for Management and Economics*, Fourth Edition (Boston, MA: Duxbury Press, 1982).

Smith, T. E. and A. W. Wortham. *Practical Statistics in Experimental Design* (Columbus, OH: Charles E. Merrill Books, Inc., 1959).

Chapter 10

Information Theory

The impact of information theory on the conception, design, and understanding of communication systems has been very large in the years since the publication in 1948 of Shannon's first paper on the subject, later published in book form [1]. While the subject has its genesis in abstract mathematical thinking, its importance is so great that it cannot be bypassed or overlooked here on the excuse that its thorough understanding requires a full knowledge of underlying mathematical principles that are beyond the scope of this book and the assumed level of academic background of its readers.

The transmission and storage of information—by human speech, letters, newspapers, machine data, television, and countless other means—are among the most commonplace and most important aspects of modern life. The processes have at least three major facets: syntactic, semantic, and pragmatic.

The *syntactic* aspects of information involve the number of possible symbols, words, or other elements of information, together with the constraints imposed by the rules of the language or coding system being used. Syntactics also involves the study of the information-carrying capabilities of communications channels and the design of coding systems for efficient information transmission with high reliability.

In communications engineering, the technical problems of the syntactic aspects of information are of primary concern. While this may appear to restrict the engineering role to one that is relatively superficial, it must be recognized that the semantic and pragmatic aspects of information transmission may be seriously degraded if excessive syntactic errors are introduced. Therefore, the importance of these other aspects of information transmission must be appreciated while the technical problems of transmitting and storing information are being solved.

The *semantic* aspects of information often involve the ultimate recipient of the information. The understanding of a message depends on whether the person receiving it has the deciphering key or understands the language. The problems of semantics generally have little to do with the properties of the communication channel per se.

The *pragmatic* aspects of information involve the value or utility of information. This is even more a function of the ultimate recipient than semantics. The pragmatic content of information depends strongly on time. For example, in a production management system, information on production, sales, inventories, distribution, etc., is made available at regular intervals. If the information is late, its value may be significantly decreased; indeed it may be worthless to the recipient.

Ultimately, the value of any information system depends on all three aspects of storage and transmission of information. The user's willingness to pay for a system is a function of its practical utility. A more complex and expensive system can be justified only by the increased utility of faster response times or greater accuracy.

The purposes here are (1) to present a brief historical sketch of the mathematical background to Shannon's work, (2) to provide some appreciation for the subject in terms of what is meant by *information* and its important relationships to probability theory, (3) to present enough mathematical background to illustrate the importance and power of information theory, and (4) to present the fundamental theorems of information theory and discuss their relationship to transmission system design and operating problems.

10-1 THE HISTORICAL BASIS OF INFORMATION THEORY

The basis upon which most modern communication theory is built has an extensive, implicit background in the work of Fourier. Early in the nineteenth century he demonstrated the great utility of sinusoidal oscillations as building blocks for representing complex phenomena. By his studies on heat flow, Fourier revealed the nature of factors governing response time in physical systems. This led to the modern description of communications systems in terms of available bandwidth, which in turn is related to the impulse response of the system when signals more complicated than sine waves are impressed. The signal itself is regarded as having a spectrum defining the relative importance of different frequencies in its composition and a bandwidth determined by the frequency range. If the bandwidth of the system is less than that of the signal, imperfect transmission occurs.

During the late 1920s, Nyquist and Hartley made significant contributions [2, 3, 4]. Hartley's work quantified the relationship between signalling speed and channel bandwidth. Nyquist's analyses led to conclusions that are now well known throughout the communications industry as Nyquist's criteria for pulse transmission [3]. They apply to the suppression of intersymbol interference in a bandlimited medium. The criteria may be stated as follows:

(1) Theoretically error–free transmission of information may be achieved if the signalling rate of the transmitted signal is properly related to the impulse response* of the channel. These conditions are met if the time of occurrence of any pulse corresponds to the zero amplitude crossings of pulses received during any other time interval. When the proper conditions are met, the maximum signalling rate is $2f_1$ bauds, where f_1 is the cutoff frequency of the channel expressed in terms of a low–pass filter characteristic.

(2) Equally valid error–free transmission of information may be accomplished if the channel characteristic produces zero amplitude crossings of the received pulses at intervals corresponding to those halfway between adjacent signal impulses. In this case, the receiving circuits are adjusted to detect transitions in the signal at intervals corresponding to those times halfway between adjacent signal impulses. (At the cost of doubling the bandwidth, criteria 1 and 2 can be met simultaneously by providing a

---

* The impulse response, defined as the function $h(t)$ found at the output as a result of applying an impulse to the input terminals, is discussed in Chapter 6, Part 3.

certain channel characteristic, namely the so–called raised cosine characteristic. This channel characteristic is often used because it provides margin for departures from ideal in filter design, in the timing circuits needed to perform the detection function, and in protection against external sources of interference.)

(3) The third criterion for error–free transmission is that the area under a received signal pulse should be proportional to the corresponding impressed signal pulse value. The response to each impulse, therefore, has zero area for every signalling interval except its own.

Nyquist and Hartley were concerned with maximum efficiency (highest speed) of transmission of digital signals in a bandlimited system; the rate of transmission must take into account performance limitations due to intersymbol interference and interrelated channel and signal characteristics. Insofar as external sources of interference were concerned, they assumed an ideal, noise–free transmission medium. As a result of his work, Nyquist's name has found a place in the technical vocabulary in such terms as *Nyquist bandwidth, Nyquist rate,* and *Nyquist interval.*

Applying their research efforts to a generalized channel and to considerations of performance in the presence of noise and interference from external sources, Wiener and Shannon made significant contributions to communications theory during the 1940s and 1950s [1, 5, 6]. Shannon, particularly, is credited with initiating the science of information theory.

10–2 THE UNIT OF INFORMATION

To be useful, information must be expressed in some symbolic form that is known and understood by both the originator and the recipient of a message. The symbology may be spoken or written English, French, or German; it may be the dots and dashes of Morse code; it may be the varying waveforms of a television video signal; it may be the *0*s and *1*s of a binary code, etc. Although *information* is popularly associated with the idea of knowledge, in information theory it is associated with the uncertainty in the content of a message and the resolution of that uncertainty upon receipt of the message.

If a message source forms messages as a set of distinct entities, such as Morse code symbols, the source is called a *discrete* source. If the messages form a set whose members can differ minutely, such as the acoustic waves at a telephone set or the light variation picked up by a television camera, the message source is said to be a *continuous* source. In either case, it is possible to express the information in terms of equivalent discrete symbols. If the message produces a continuous signal, the translation from a continuous to a discrete format is accomplished by the use of the sampling theorem and a process of quantization (see Chapter 8, Part 3).

The simplest discrete format is binary; that is, the information is expressed in symbols that can attain one of two equally likely values. The unit used to express the binary format is the *bit* (*bi*nary dig*it*). In binary terms, the number of information symbols generated may be expressed as

$$m = \log_2 n \text{ bits,} \qquad (10\text{–}1)$$

where m, the amount of information, is a function of the logarithm of the number of outcomes, n, that may be attained by the message source. In Equation 10–1, the logarithm is taken to the base 2; this has been found to be the most convenient in solving theoretical communication problems because most practical system applications are binary.* Therefore, the unit most generally used in information theory is the bit.

Although bit was derived from binary digit, the two are really different and care should be exercised in their use. The *bit* is a measure of information, while the *binary digit* is a symbol used to convey that information. To illustrate the difference, consider a channel capable of transmitting 2400 arbitrarily chosen off–or–on pulses per second. Since a pulse can take two states (off or on) there are $n = 2^{2400}$ possible configurations of pulses each second. The information capacity is then

$$m = \log_2(2^{2400}) = 2400 \text{ bits per second.}$$

*Information can, of course, be measured in logarithmic units other than those to the base 2. If base 10 is used, the information is measured in decimal digits, or hartleys; if the base e is used, information is in natural units, or nats.

If, however, the channel is only used to transmit a completely repetitive sequence of off–on pulses, then only one configuration of pulses ever appears and the actual rate of information transmission is

$$m = \log_2(1) = 0 \text{ bits per second}$$

despite the fact that the channel is then transmitting 2400 binary digits per second. To say the channel is transmitting 2400 bits per second under these conditions is to misuse the word *bit*.

10-3 ENTROPY

It should be recognized now that the information contained in a message is a matter of probability. The message is a set of symbols taken from a larger set. If there is no uncertainty about what the message is (what set of symbols is expected by the recipient), the message contains no information. If there is uncertainty and, by successful receipt and decoding of the message, the uncertainty is resolved, an amount of information has been transmitted equal to that defined by Equation 10–1. Something more is needed, however, some measure of the uncertainty in a message before it is decoded. This measure is called *entropy*.

Since a coded message is chosen from among a set of code symbols, there are more choices and, therefore, is more uncertainty in long messages than in short messages. For example, there are just two possible messages consisting of one binary digit (*0* or *1*), four messages consisting of two binary digits (*00, 01, 10,* or *11*), 16 consisting of four binary digits, and so on. The entropy in the message increases for each of these cases as the number of choices increases. If the freedom of choice and the uncertainty decrease, the entropy decreases.

If a message source is not synchronous, that is, not producing information symbols at a constant rate, it is said to have an entropy, H, of so many bits per symbol (letter, word, or message). However, if the source does produce symbols at a constant rate, its entropy, H', is expressed as so many bits per second.

The relationship between information content and entropy may be motivated as follows. Equation 10–1 may be rewritten in the form

$$m = \log_2 \frac{1}{p} = -\log_2 p \; ; \; p = \frac{1}{n}$$

where p is the probability that a particular one of the n possible outcomes will occur. If we now have outcomes whose probabilities are $p_1, p_2, \ldots p_n$, then the expected (or average) value of the information content may be written in the general form

$$H = -p_1 \log_2 p_1 - p_2 \log_2 p_2 + \ldots - p_n \log_2 p_n$$

$$= -\sum_{i=1}^{i=n} p_i \log_2 p_i \text{ bits/symbol} \qquad (10\text{–}2)$$

where H is the entropy of the transmitted message. In this equation, there are n independent symbols, or outcomes, whose probabilities of occurrence are $p_1, p_2, \cdots p_n$.

Equation 10–2 may be used to illustrate the effect on the entropy of a source when probability p_1 is changed. For the simple case in which there are just two choices (X with the probability p_1 and Y with probability $p_2 = 1 - p_1$), the value of H is plotted in Figure 10–1. Examination of the figure makes it clear that for this case the entropy, H, is a maximum of one bit per symbol

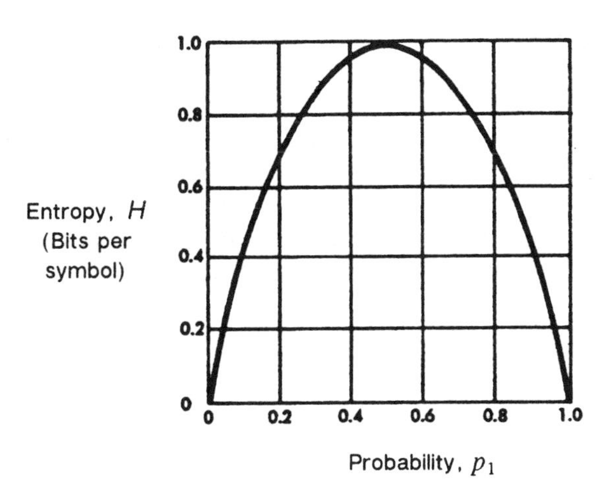

Figure 10–1. Entropy of a simple source.

when $p_1 = p_2 = 1/2$ and that the entropy is zero when $p_1 = 0$ or 1 (no uncertainty in the message).

It may be shown that this situation is typical even when the number of choices is large. The entropy is a maximum when the probabilities of the various choices are about equal and is a small value when one of the choices has a probability near unity.

Example 10–1:

Given an honest coin, if it is tossed once, there are two equally probable outcomes, namely, head or tail. The entropy is computed by Equation 10–2 as

$$H = -[0.5(-1.0) + 0.5(-1.0)]$$

$$= 1.0 \text{ bit.}$$

If the coin has two heads, the probability of a head is unity. Then,

$$H = -(1.0 \log_2 1.0)$$

$$= 0 \text{ bit.}$$

Thus, the tossing of a two–headed coin gives no information. The outcome has no uncertainty; it is always heads.

Example 10–2:

Given an honest die, one roll of such a die can result in any one of six equally probable outcomes. Thus, using Equation 10–2 again, it is seen that the entropy of the source (the die) is

$$H = -\log_2 (1/6)$$

$$= 2.58 \text{ bits.}$$

Now, assume the die is loaded; in this case the outcomes are not equally probable. Assume the following probabilities for the various possible outcomes:

Die Face	Probability
1	0.4
2	0.2
3	0.1
4	0.1
5	0.1
6	0.1

$$H = -(0.4 \log_2 0.4 + 0.2 \log_2 0.2 + 0.4 \log_2 0.1)$$
$$= 2.32 \text{ bits.}$$

An example of a more complex relationship between probability of occurrence and entropy is illustrated by an evaluation of the information content of the written English language. As a first approximation, the occurrence of the 27 symbols representing the 26 letters of the alphabet and a space may be assumed to be equally probable. Such an approximation sets the upper bound at

$$H = \log_2 27 = 4.75 \text{ bits per symbol.}$$

This approximation, however, is inaccurate since the probabilities of occurrence are quite different for different letters. For example, in typical English text the letter E occurs with a probability of about 0.13, while Z occurs with a probability of only about 0.0008. By considering such probabilities and other refinements, the information content of English text is estimated to be about one bit per symbol.

10-4 THE COMMUNICATION SYSTEM

The term *information* has been defined in terms of logarithmic units (bits), and the measure of information has been defined as entropy in bits. The concept of information transfer from a source to a destination has been described as a probabilistic phenomenon involving the probabilities of message generation by the source and the resolution of the uncertainties at the destination. These concepts may now be more specifically related to the communication system.

The general communication system is commonly represented in one simplified form by a sketch such as the block diagram of Figure 10–2. The system is made up of a source of information and a destination for the information. Between the source and destination are a transmitter, a channel to carry the information, and a receiver. The function of the transmitter is to process the message from the source into a form suitable for transmission over the channel, a process frequently referred to as *channel coding*. The receiver reverses this process to restore the signal to its original form so that the message can be delivered to the destination in suitable form. This process is called *decoding*. The channel in such a system interconnects the physically separated transmitter and receiver. It is often assumed to be ideal, introducing no noise or distortion. This is, in fact, never achievable; every channel introduces some noise and distortion. It should also be recognized that there are noise sources that enter the system at places other than the channel; however, it is convenient to assume that all the perturbations on ideal signal transmission are introduced into the channel by a source external to the channel, as shown in Figure 10–2.

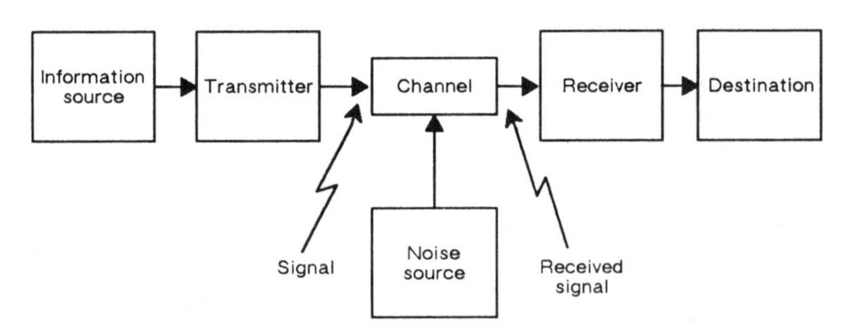

Figure 10–2. Block schematic of a general communication system.

Coding

The coding (and decoding) of messages in the transmitter (and receiver) of the communication system of Figure 10–2 may take any of a large number of forms and may be done for a number of different reasons. At the source, the signal is often encoded for the purpose of increasing the entropy of the source. Morse's dot–dash coding of alphabetical symbols is an example; he assigned short dot–dash symbols to those letters most frequently found in English text and longer symbols to those less frequently encountered. Encoding is found in the transmitter where the purpose may be to increase the efficiency of transmission over the channel, i.e., to increase the rate of transmission of information. Encoding may also provide error detection, error correction, or both. A thorough review of coding principles and techniques is beyond the scope of this chapter; however, practical applications of coding techniques appear in Chapter 8 (Modulation) and in several chapters of Volume 2 in which terminals for specific systems are described.

Noise

Noise contains components whose characteristics generally can be defined in probability terms and thus theoretically in terms of information. The presence of noise in a communication system adds bits of information and increases the uncertainty of the received signal; therefore, one might erroneously conclude that noise is beneficial. However, since the added noise perturbs the original set of choices, it introduces an undesirable uncertainty.

Figure 10–3 illustrates a simple case of how noise may introduce errors in transmission. The transmitted message, shown in Figure 10–3(a) as a series of *0*s and *1*s, is transformed into a series of plus (for *1*) and minus (for *0*) voltages in Figure 10–3(b). This signal is perturbed by an interfering noise depicted in Figure 10–3(c). The signal and noise voltages add as in Figure 10–3(d), and the receiver translates the received composite signal into the received message of Figure 10–3(e), which contains three errors.

The noise introduced in a communication system, such as that illustrated in Figure 10–2, may consist of any of a number of kinds of interference introduced in the transmission path. Its characteristic may be well defined by some probabilistic expression; Gaussian or Poisson distributions are examples. Intersymbol interference, resulting from gain or envelope delay distortion (or both), may be regarded as noise. Signal–dependent distortion of the signal itself, due to nonlinear devices, may also be considered as noise. Other possible types of interference (noise) include frequency offset or frequency shift, sudden amplitude or phase hits due to external influences, echoes (perhaps due to impedance mismatches), or crosstalk due to some other signal being superimposed on the wanted signal by way of an unwanted path. Any or all of these interferences can produce transmission errors.

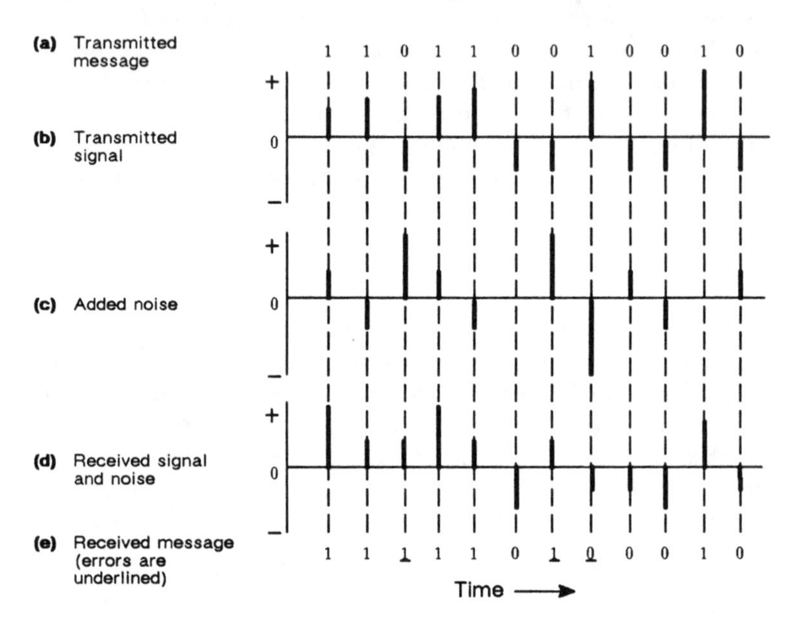

Figure 10–3. Effect of noise on transmission.

10-5 THE FUNDAMENTAL THEOREMS

The value and broad scope of information theory are expressed succinctly by Pierce: "To me the indubitably valuable content of information theory seems clear and simple. It embraces the ideas of the information rate or entropy of an ergodic message source, the information capacity of noiseless and noisy channels, and the efficient encoding of messages produced by the source, so as to approach errorless transmission at a rate approaching the channel capacity. The world of which information theory gives us an understanding of clear and present value is that of electrical communication systems and, especially, that of intelligently designing such systems" [7].

Shannon presents a large number of theorems, all of which pertain to the comments quoted above. There are three, however, that are basic and sufficiently important to discuss here. These are (1) the fundamental theorem for the noiseless channel, (2) the fundamental theorem for the discrete channel with noise, and (3) the theorem for the channel capacity with an average power limitation.

The Noiseless Channel

Let a source have an entropy of H bits per symbol and a channel have a capacity of C bits per second. Then it is possible to encode the output, m', of the source in such a way as to transmit at the average rate of $C/H - \epsilon$ symbols per second over the channel, where ϵ is arbitrarily small. This may be written

$$m' = \frac{C}{H} - \epsilon \text{ symbols per second.} \qquad (10\text{–}3)$$

It is not possible to transmit at an average rate greater than C/H.

It is necessary to distinguish carefully between the m of Equation 10–1, expressed in bits, and the m' of Equation 10–3, expressed in symbols per second. The quantity m as used in Equation 10–1 is a measure of information produced by a source (in the simple binary case, the number of *1*s and *0*s). Then,

$$H' = m'H \text{ bits per second,} \qquad (10\text{–}4)$$

where m' is the average number of symbols produced per second, H is the entropy produced by the source in bits per symbol, and H' is the entropy in bits per second.

For cases of interest, $H' \leq C$ and the number of symbols per second is

$$m' = \frac{H'}{H} \leq \frac{C}{H} \text{ symbols per second.} \qquad (10\text{–}5)$$

Equations 10–3 and 10–5 are equivalent since ϵ in Equation 10–3 is the amount by which C/H exceeds m' in Equation 10–5.

Thus, a source may produce symbols at a rate of m' symbols per second. The entropy may be such that the information produced is only H bits per symbol or H' bits per second. The theorem shows that as long as $H' \leq C$, the source may be coded so that a rate of $C/H - \epsilon$ symbols per second may be transmitted over the channel of capacity C. A rate greater than C/H cannot be achieved by any coding without error in transmission.

The Discrete Channel with Noise

Consider a discrete channel with a capacity, C, and a discrete source with an entropy, H'. If $H' \leq C$, there exists a coding system such that the output of the source can be transmitted over the channel with an arbitrarily small frequency of errors (or an arbitrarily small equivocation). If $H' > C$, it is possible to encode the source so that the equivocation is less than $H' - C + \epsilon$ where ϵ is arbitrarily small. There is no method of encoding that gives an equivocation less than $H' - C$.

It seems strange to find a theorem relating to a "discrete channel with noise" that has no explicit mention of noise in its statement. However, this situation arises from the manner in which Shannon leads up to the theorem. Shannon defines the capacity, C, of the discrete channel with noise as

$$C = \text{Max}\ [H'(x) - H'_y(x)] \text{ bits per second.} \qquad (10\text{–}6)$$

In this equation, C is given as the maximum value of the source entropy $H'(x)$ minus the conditional entropy $H'_y(x)$. The conditional entropy (also known as the equivocation) is a measure of the degree of uncertainty associated with correlating the source signal x with the received signal y. If there is no noise in the channel, then there is no uncertainty in correlating source and received signals, and $H'_y(x) = 0$. As noise increases, the correlation between source and received signal becomes weaker, uncertainty increases, and the conditional entropy increases. Thus, noise is included by implication.

Channel Capacity with an Average Power Limitation

The capacity of a channel of band W perturbed by white noise* of power P_{noise} when the average transmitter power is limited to P_{max} is given by

$$C = W \log_2 \frac{P_{max} + P_{noise}}{P_{noise}} \text{ bits per second.} \qquad (10\text{–}7)$$

Shannon explains, "This means that by sufficiently involved coding systems we can transmit binary digits at the rate $W \log_2 \frac{P_{max} + P_{noise}}{P_{noise}}$ bits per second, with arbitrarily small frequency of errors. It is not possible to transmit at a higher rate by any encoding system without definite positive frequency of errors" [1]. In Equation 10–7, W is the bandwidth in hertz; P_{max} and P_{noise} are signal and noise powers that may be expressed in any consistent set of units (as a ratio, the units cancel out in the equation). As a final restriction, Shannon points out that to approximate this limiting rate of transmission, the transmitted signals must approximate white noise in statistical properties. Coding, used to improve the transmission rate, is accomplished only at the expense of introducing delay and complexity. To achieve or approach the limiting rate may introduce sufficient delay in practice as to make the process impractical.

Other theorems of Shannon give the rate of information transmission for other sets of conditions. For example, the condition of peak power rather than average power limitation is covered. For noise other than white noise, the transmission rate cannot be stated explicitly but can be bounded. The bounds are usually near enough to being equal that most practical problems can be solved satisfactorily.

10-6 CHANNEL SYMMETRY

It may be shown that the maximum rate of transmission of information (the capacity) can be determined for a symmetical

---

* White noise has a flat or constant power spectral density.

channel by straightforward means, but that the computation for an unsymmetical channel becomes complicated [7]. A symmetrical channel is one in which the probability, p, of a *0* from the source being received as a *0* is equal to the probability that a *1* from the source is received as a *1*. Thus, the probability that a transmitted *0* would be received as a *1* and the probability that a *1* would be received as a *0* are both equal to $(1-p)$. Most practical problems involve symmetrical channels.

Example 10-3:

Given the symmetrical channel of Figure 10–4(a) having a transmitter, x, a receiver, y, and additive noise; given channel performance such that $p = 0.9$ [as shown in Figures 10–4(b) and 10–4(c)]; and given the statistics of the transmitter such that the probability of a 1 is 0.6 and of a 0 is 0.4. What is the entropy of the signal received at y?

From Equation 10–2, the entropy of the transmitter is

$$H(x) = -(0.6 \log_2 0.6 + 0.4 \log_2 0.4) = 0.97 \text{ bit per symbol.}$$

The rate of transmission, R, may be shown by an expression similar to Equation 10–6,

$$R = H'(x) - H'_y(x) \text{ bits per second.}$$

The equivocation, $H'_y(x)$, may be found by

$$H'_y(x) = -p \log_2 p - (1-p) \log_2 (1-p)$$

$$= -0.9 \log_2 0.9 - 0.1 \log_2 0.1 = 0.469 \text{ bit per second.}$$

Thus,

$$R = 0.97 - 0.469 = 0.501 \text{ bit per second.}$$

This is the entropy of the signal at y for the situation in Figure 10–4(b).

Next, assume that the transmitter may be encoded differently so that $p_1(x) = p_0(x) = 0.5$, as illustrated in

Figure 10–4(c). What is now the entropy of the signal received at y?

For this condition,

$$H'(x) = -0.5 \log_2 0.5 - 0.5 \log_2 0.5 = 1.0 \text{ bit per second.}$$

The channel is the same as in Figure 10–4(b). Therefore, the new rate is

$$R = 1 - 0.469 = 0.531 \text{ bit per second.}$$

Thus, the entropy of the signal received at y has been increased by increasing the entropy of the transmitter.

For the channel assumed, one having $p = 0.9$, this can be shown to be the maximum rate and thus the channel capacity, C, of Equation 10–6.

It may be shown that if the symmetrical channel has performance such that $p = 0.99$, the maximum rate improves to 0.92 bit per second when the source entropy is unity.

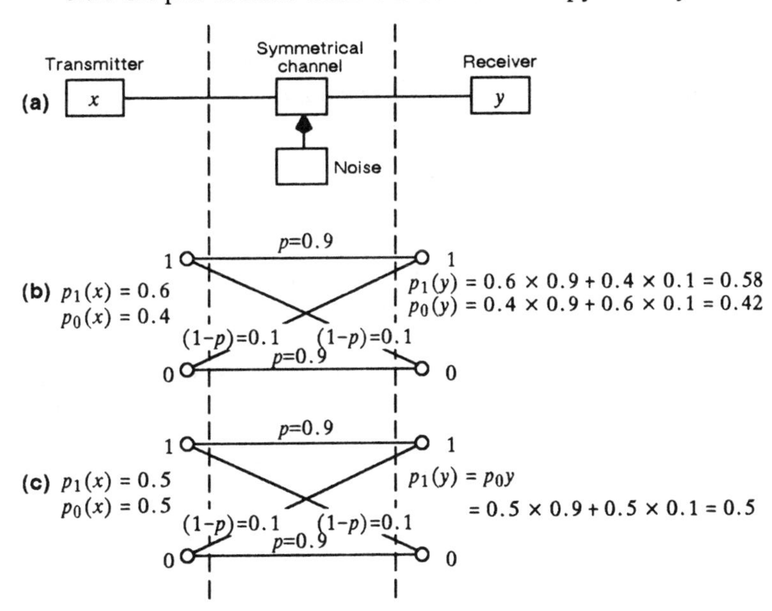

Figure 10–4. Transmission over a discrete symmetrical channel with noise.

References

1. Shannon, C. E. and W. Weaver. *The Mathematical Theory of Communication* (Urbana, IL: The University of Illinois Press, 1963).

2. Nyquist, H. "Certain Factors Affecting Telegraph Speed," *Bell System Tech. J.*, Vol. 3 (Apr. 1924).

3. Nyquist, H. "Certain Topics in Telegraph Transmission Theory," *AIEE Transactions,* Vol. 47 (Apr. 1928).

4. Hartley, R. V. L. "Transmission of Information," *Bell System Tech. J.*, Vol. 7 (July 1928).

5. Wiener, N. *Cybernetics* (Cambridge, MA: Technology Press, 1948).

6. Wiener, N. *Extrapolation, Interpolation, and Smoothing of Stationary Time Series* (New York: John Wiley and Sons, Inc., 1949).

7. Pierce, J. R. *Symbols, Signals and Noise: The Nature and Process of Communication* (New York: Harper and Brothers, 1961).

Additional Reading

Bennett, W. R. and J. R. Davey. *Data Transmission* (New York: McGraw–Hill Book Company, Inc., 1965).

Hyvärinen, L. P. *Information Theory for Systems Engineers* (New York: Springer–Verlag, 1968).

Raisbeck, G. *Information Theory—An Introduction for Scientists and Engineers* (Cambridge, MA: M.I.T. Press, 1963).

Chapter 11

Engineering Economy

11-1 FRAMEWORK OF THE ECONOMY STUDY

The process of planning for the telecommunications network is dynamic and includes both near-term and long-term concerns. Long-range planning, also called strategic or fundamental planning, is concerned with the orderly evolution of the network to long-term corporate goals. The long-range planner recommends architecture and design strategies based on corporate goals and objectives, existing and expected technologies, current and planned service offerings, the regulatory environment, and projected customer demands. The current planner, on the other hand, uses the long-range recommendation as a guideline, along with more solid estimates of costs, customer demand, technology availability and planned service offerings to recommend changes and additions to the network.

These current plans become the basis for the construction program. The construction program typically defines capital and expense expenditures, and material and force requirements over a three-year period. The construction program is reviewed and adjusted during the three years and carefully monitored during the final year.

The planning process has been defined in many ways in various professions and corporate climates. However, all of these processes include the elements that scientific literature refers to as the scientific method and business literature calls the rational problem-solving process. These have been blended into the following eight steps:

(1) *Monitor the System*

This is the data-gathering stage of the planning process. Here the planner (or engineer or analyst) keeps track of what is happening in the network (the growth, the service requirements and maintenance/repair activity). Data consists of facts and assumptions. Facts are those things known to be true that generally occurred in the past or are currently occurring. Assumptions are things that may occur in the future. The planner understands that assumptions will change over time and therefore must be carefully monitored.

(2) *Analyze the Situation*

Once data has been collected, which implies that a problem exists or that there is a potential opportunity for improving the network or offering additional services, the analyst must define the problem or opportunity. The problem statement must be clear and concise and should flow from the data collected in step one. Problem/opportunity identification is important since solving the wrong problem is no solution.

To ensure that the problem is clearly stated, it helps to begin the statement with the words "How to...."; for instance, "How to provide interoffice voice and data facilities between office A and office B." In this way, the statement does not simply describe a symptom of the problem (e.g., traffic measurements showing all trunks busy). The statement should not limit the possible alternatives that could be considered. For instance, the statement "How to provide digital fiber optic facilities between office A and office B" limits the alternatives to optical fiber cable when digital radio may be a better solution.

After the problem has been defined, the planner must set objectives against which alternative solutions will be measured. Objectives should include applicable corporate policy, service requirements, and economics. Where possible, objectives should be quantified and prioritized. In this way, it will be clear whether or not an alternative meets a specific objective.

(3) *Derive Alternatives*

When deriving alternatives, the planner should concentrate on solutions to the problem; objectives should be ignored until the alternatives are compared. There are many methods of deriving alternatives; brainstorming and decision tree techniques are the most common. Regardless of the derivation system, each alternative should be mutually exclusive from the other alternatives. Choosing one alternative precludes the selection of any other alternative. The planner should try to identify all alternatives that will solve the problem.

(4) *Compare Alternatives*

Many alternatives can be eliminated from consideration by inspection. That is, some will not meet high–priority objectives; these can be removed from consideration before detailed economic analysis begins. The recommended alternative is generally the one that best meets the most objectives.

Once a preliminary decision is made, the recommended alternative should be tested for sensitivity to changing assumptions. The most common method is to vary one assumption at a time to determine whether that variation would cause the selection of a different alternative. If the recommended alternative is sensitive to the variation of one or more assumptions, the planner should define a monitoring system, including the amount of change in an assumption that should occur before the problem and alternatives are reexamined. Also, the planner should recommend a contingency alternative that can be implemented if key assumptions change.

Risk analysis may include a Monte Carlo simulation method, which varies all sensitivity and critical assumptions at the same time. This will show the probability distribution of an economic indicator and will show the probability that the project will produce a positive net present value. The Monte Carlo simulation method will not be developed in this text but may be found in several vendor packages, such as EME and CUCRIT.

(5) *Make a Decision*

The planner is responsible for making an economically sound final recommendation in a timely manner.

(6) *Document the Plan*

Documentation is one of the most important parts of the planning process. The documentation should contain enough information to support inquiries by legal and regulatory authorities. Also, documentation should be complete enough for another analyst to duplicate the study.

(7) *Obtain Approval*

Approval should be at the minimum level required for implementation.

(8) *Disseminate the Plan*

The plan should be sent to all interested parties, especially those responsible for implementation.

One final concern in the framework of the study is the study length. The study should be long enough to include all significant cash flows. Thus, all equipment placements and major operating expenditures should be included. The engineering economy study must take into account the interval over which the problem is to be studied and its relation to the life of plant involved. If these intervals are not the same, adjustments must be made. The choice of interval is a matter of judgment; it may be relatively short, such as two or three years, or it may continue, at least theoretically, indefinitely into the future.

11-2 ENGINEERING ECONOMY

Solutions to engineering problems are usually considered complete only after economic analyses have been made of several alternative solutions and the results are compared. This is as true in the field of transmission as it is in any other field. Often a choice must be made on the basis of incomplete information and it becomes necessary to exercise engineering judgment with respect to the impact of intangible aspects of the problem. The company's objectives are to provide economically the best possible service. To meet these objectives, engineering economy studies must be made to demonstrate the value of new systems, new services, and specific proposals for network expansion; service and performance improvements must also be evaluated.

Financial accounting and engineering economy are fields that appear to be quite remote from each other; however, there are numerous points of contact in the paths followed by the two professions. Both are concerned with the use of capital and expense funds. The major difference is that in financial accounting these funds are dealt with in retrospect by examining the results of expenditures, while in engineering economy one of several alternative future courses of action must be selected to use available funds most effectively.

There are two broad categories of expenditure that consume most of the funds available to a company. One is the cost of operating the business and maintaining the plant in service. These expenses are charged in the period in which they are accrued; they are planned, budgeted, controlled, and paid out of current revenues. The second is the capital required to construct new plant to satisfy growing service demands or to replace outdated equipment. Capital expenditures are paid out of funds accumulated as retained earnings from current revenue, the sale of stock and bonds, depreciation, and deferred taxes. Funds are planned, budgeted, controlled, and spent in accordance with procedures generally categorized as the construction program.

Planning and implementing the construction program involves many factors that affect the choice of a course of action. Relative service and performance capability, operating conditions, maintenance complexities, revenues, and costs must all be considered. Costs are given considerable weight because they provide a tangible and quantitative measure of relative worth in terms that most people understand.

Many types of cost studies are made to determine the effects of an action on pricing policy, financial position, or accounting results. Some are made after a course of action has been determined.

Engineering economy studies are intended to show which of several alternatives is economically most attractive in fulfilling service requirements. Therefore, they are important aids in making decisions that cumulatively result in the formulation of the construction program. In the field of transmission engineering, as in many other areas, there are often several possible courses of action that may be feasible. Therefore, familiarity with the principles of engineering economy is necessary in fulfilling transmission engineering functions.

An engineering economy study may be made (1) to determine which of several plans or methods of doing a job will be the most economical over a given time interval, (2) to prepare cost estimates for studies of new and existing service offerings or special service arrangements, (3) to establish priorities for discretionary investment opportunities, and (4) to establish revenue and capital requirements over long periods of time as major projects are programmed and initiated [1]. Objectives such as these are often satisfied by studies in which engineering data are used as input information. The provision of the necessary data for such studies requires an understanding of basic engineering economy principles and often contributes valuable perspective on the total engineering problem.

11-3 TIME VALUE OF MONEY

Engineering economy studies deal with money to be spent or received in various amounts and at various times. The objective of such studies is to evaluate the benefit in relationship to the cost of each plan under consideration. Therefore, it is essential to understand the basic rules that govern the comparison of money spent or received at different times. Simply, a dollar today has a different purchasing power from a dollar a year ago or a dollar next year. One means of expressing dollars at different times in equivalent terms is the time value of money. The concept of the time value of money is implicit in the basic rules of economy studies and in the application of sound principles to the conduct of such studies.

The Earning Power of Money

The money to purchase new equipment ultimately comes from investors (stockholders and bondholders). The incentive for these people to invest is to increase the value of their investment (i.e., the sale price of the stock), and to get a return on the investment in the form of dividends on stock and interest on bonds. From the investor's viewpoint, the minimum amount this return can be to still be attractive is sometimes called the minimum attractive rate of return (MARR). From the company's viewpoint, this value is called the cost of money or the composite cost of capital, i_c. The cost of capital is the composite of the cost of equity (stock) and the cost of debt (bonds), proportionally.

In engineering economy studies, the cost of capital is sometimes called the discount rate, since it is used to compensate for the time value of money. In other words, i_c is called by several equivalent names (discount rate, cost of money, cost of capital, and return rate). The effect of return on the time value of money may be evaluated for a particular time relative to another (usually taken as a reference, t=0) by the expression

$$\frac{D_{t=0}}{D_{t=1}} = \frac{1}{1+i_c} \qquad (11\text{–}1)$$

where D is the value of money at times denoted by the subscripts and i_c is the discount rate for the period. Thus, if one dollar is needed one year from now (t=1) and if the discount rate, i_c, is 0.10, only 91 cents need be invested now (t=0).

The discount rate may be compounded at intervals of typically (although not necessarily) one year. Compounding involves the computation of the value of money on the basis of the cost of money of the original amount plus the accrued cost during the compounding period. Thus, to determine the value of money where the cost of money has been compounded over n intervals:

$$\frac{D_{t=0}}{D_{t=n}} = \frac{1}{(1+i_c)^n}$$

or $\qquad (11\text{–}2)$

$$D_{t=n} = D_{t=0}\,(1+i_c)^n \,.$$

So, if an amount D is to be made available at a future time, $t = n$, a smaller amount of money may be made available now, $t = 0$, when it is invested at a cost of money, i_c.

11-4 EQUIVALENT TIME VALUE EXPRESSIONS

Since the true value of money is different at different times, economic analysts use several time value equivalencies; the choice depends on the nature of the study. Typically, all terms are expressed in terms of equivalent amounts at a time arbitrarily chosen and defined as "the present." In other studies, equivalent costs in terms of some time in the past or future are more convenient. When money in equal amounts at equal intervals must be modeled, the analyst expresses the costs as an annuity equivalent to a lump sum at a specified time. The expressions used are all equivalent and may be converted from one form to another. The following are the most commonly used expressions:

future worth of a present amount,

$$F/P = (1+i_c)^n , \tag{11-3}$$

present worth of a future amount,

$$P/F = \frac{1}{(1+i_c)^n} , \tag{11-4}$$

future worth of an annuity,

$$F/A = \frac{(1+i_c)^n - 1}{i_c} , \tag{11-5}$$

annuity for a future amount,

$$A/F = \frac{i_c}{(1+i_c)^n - 1} , \tag{11-6}$$

present worth of an annuity,

$$P/A = \frac{(1+i_c)^n - 1}{i_c(1+i_c)^n} , \tag{11-7}$$

and annuity from a present amount,

$$A/P = \frac{i_c(1+i_c)^n}{(1+i_c)^n - 1} \quad , \tag{11–8}$$

where

i_c = composite cost of capital

n = number of intervals

F = the future worth of an amount

P = the present worth of an amount

A = an annuity.

Today, these equations can easily be calculated using scientific or financial calculators. Also, most texts have reference tables containing the factors. Figure 11–1 illustrates the application of these factors. Here, the value of $1,000 in year 8, point A, may be traced through various processes to the future and past, and finally back to the same value, $1,000 in year 8.

11–5 THE NATURE OF COSTS

One of the major considerations in planning for the telecommunications network is engineering economics. This primarily involves consideration of a number of aspects of costs. A variety of capital costs are incurred because investors provide funds needed to acquire plant, and operations costs are incurred by the existence of the plant. Changing technology and inflation must also be considered for their effects on costs.

Capital Costs

The costs associated with the acquisition of property are called capital costs; accounting procedures are used to monitor, control, and recover such costs. The property is an asset assigned for

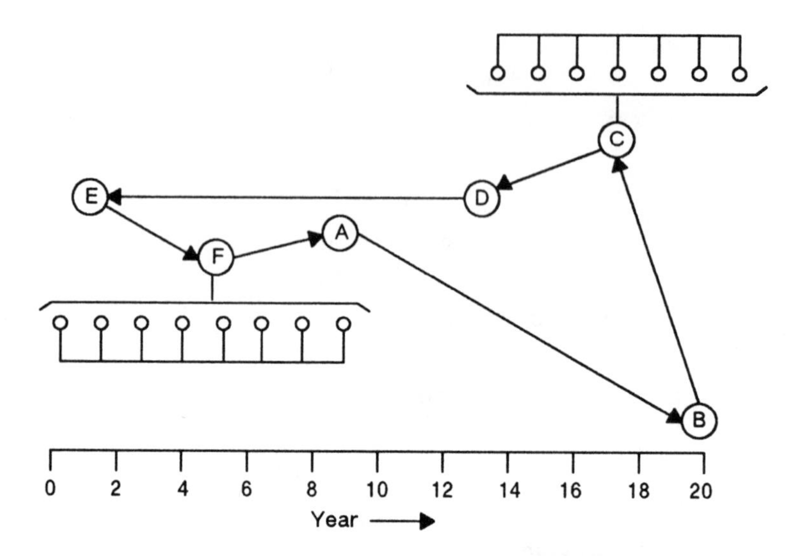

Conversion	Description	Factor	Equivalence	Equation
A → B	Single amount in year 8 to single amount in year 20	*F*/*P* (*i* = 0.1, *n* = 12)	B = 3.1384A = $3138.40	11-3
B → C	Single amount in year 20 to equal amount in years 14 to 20	*A*/*F* (*i* = 0.1, *n* = 7)	C = 0.1054B = $330.79/yr	11-6
C → D	Equal amounts in years 14 to 20 to single amount in year 13	*P*/*A* (*i* = 0.1, *n* = 7)	D = 4.8684C = $1610.42	11-7
D → E	Single amount in year 13 to single amount in year 0	*P*/*F* (*i* = 0.1, *n* = 13)	E = 0.2897D = $466.54	11-4
E → F	Single amount in year 0 to equal amounts in years 1 to 8	*A*/*P* (*i* = 0.1, *n* = 8)	F = 0.1874E = $87.43/yr	11-8
F → A	Equal amounts in years 1 to 8 to single amount in year 8	*F*/*A* (*i* = 0.1, *n* = 8)	A = 11.4359F ≈ $1000	11-5

Figure 11-1. Time-value equivalence.

accounting purposes to a specific plant account. The plant accounts conform to the Uniform System of Accounts (USOA) prescribed by the Federal Communications Commission (FCC) [2]. Capital costs related to engineering economy studies include the concept of the composite cost of money, the first–cost investment, and the sources of capital used to recover the cost of the investment. The process of capitalization of money invested in plant triggers recurring annual costs having three elements: return to the investor, capital repayment, and income taxes.

The Composite Cost of Capital. Money needed to pay the initial cost of plant investment is obtained from a number of sources; the basic external sources of money are *debt capital* and *equity (owner's) capital*. The ratio of debt capital (obtained by borrowing) to the total capital (debt plus equity) carried on the books of the company is called the *debt ratio*. An understanding of the relation of the debt ratio to the composite cost of money is necessary in making engineering economy analyses.

The composite cost of money, or return rate, may be expressed by an equation that relates the composite cost of money, i_c, to the debt ratio, r, the interest paid on debt, i_d, and the return on equity (stock dividends and retained earnings), i_e.

These are expressed:

$$i_c = (r)i_d + (1-r)i_e \ . \qquad (11\text{–}9)$$

Thus, if the debt ratio is 45 percent, the composite cost of money is

$$i_c = 0.45i_d + 0.55i_e \ .$$

The composite cost of money may also be derived using the *after–tax* method:

$$i_c = r\,(1-T)\,i_d + (1-r)\,i_e$$

where T = the tax rate.

The debt ratio, debt interest, and equity return are forward–looking assumptions that may be expected to change with time. Most engineering economy studies, however, assume a stable cost of money. Possible variations may be treated via sensitivity analysis.

First Costs. The first cost of a property is the sum of the costs of materials, transportation, labor and incidentals related to installation, supervision, tools, engineering, and other miscellaneous items. The first cost of a project is the invested capital upon which the rate of return is initially calculated. These costs are accumulated during the construction interval and do not recur during the life of a plant item.

Capital Recovery. Physical plant ultimately reaches the end of its useful life via wear and tear, technological obsolescence, or some other reason. The original cost of the plant, offset by salvage value, is a cost that must ultimately be received from the customer. The investor's interest is protected by the transfer of investment from old to new plant, in installments, as the old plant is used up in service. Capital is recycled by a process called depreciation, through which capital is repaid annually over the life of the plant out of current revenues.

Life of Plant

In conducting an engineering economy study, the analyst should use the actual anticipated life of the plant involved in the study rather than the broad average life used for accounting purposes. Sometimes the life may be established by the conditions of the problem. For instance, a study may look at alternative equipment configurations for providing an interim item of plant that will be replaced after three years. In this case, plans must provide for repayment of all capital expenditures, adjusted for salvage, by the end of the study period. At other times, the life of an item of plant may depend on the life of another item. The life of a radio system may depend on the life of the tower that supports it, or the timing of a capacity–upgrade program that will eventually replace it.

If the problem conditions do not give an indication of the life of the plant, life must be estimated based on past experience and engineering judgment. Even here, study life will rarely coincide with accounting life.

Salvage

Net salvage may be a significant factor in an engineering economy study. Net salvage is the difference between gross salvage (the sales value of the plant) and the cost to remove the plant. If the gross salvage value is greater than the cost of removal, the net salvage can be considered as a benefit to the alternative. If it costs more to remove the plant than can be recovered from salvage, the net salvage is a disbenefit or cost to the alternative.

As technological change speeds up, it is becoming more difficult to estimate both of the variables in net salvage. Many conservative analysts assume gross salvage to be equal to the cost of removal; i.e., net salvage is equal to zero. Occasionally, the best estimate of cost of removal exceeds the best estimate of gross salvage, and net salvage is negative.

Book Depreciation

The USOA requires that common carriers use straight–line depreciation for financial statements. Because it is used for the book records of the firm, it is also called book depreciation. This procedure writes off an equal amount of the capital investment in each year over the life of the plant. That is, in each year, an amount of revenue is shown as having been used to account for the depreciation of that item of plant. Annual depreciation may be calculated by (first cost – net salvage)/service life.

Table 11–1 illustrates the straight–line depreciation of a $1000 investment over a 10–year service life, along with the balance (book value) of the investment.

Book depreciation of a capital investment is related to the term *book value*. Book value is computed as the gross plant investment minus the accumulated depreciation. Sometimes the gross plant investment is called the *book cost* and undepreciated plant is called *net plant*.

Table 11–1. Straight–Line Depreciation and Book–Value Accounting

Year End	Undepreciated Balance Before Year-End Charge	Year-End Charge	Book Value
0			$1000
1	$1000	$ 100	900
2	900	100	800
3	800	100	700
4	700	100	600
5	600	100	500
6	500	100	400
7	400	100	300
8	300	100	200
9	200	100	100
10	100	100	0

Tax Depreciation

As an incentive to companies that make capital investments, U.S. tax code permits the use of accelerated depreciation for tax purposes. Higher depreciation is allowed in earlier years of the life of an asset and lower depreciation in later years. Several different methods of accelerated tax depreciation have been used in the past; the Tax Reform Act of 1986 (TRA 86) requires that all assets purchased after January 1, 1986 use the Accelerated Cost Recovery System (ACRS) as modified by TRA 86. Per TRA 86, all assets must be categorized into six classes, as shown in Table 11–2. The classes are keyed to an asset depreciation range (ADR) of lifetimes. The ten–percent investment tax credit previously allowed in the first year is no longer permitted. Real property is not depreciable.

For each class, a specified depreciation is allowed in each year (see Table 11–3). A comparison of tax and book depreciation is shown in Figure 11–2.

Table 11–2. Asset Classification According to Tax Reform Act of 1986

Class	Description
3–year property	All property with a midpoint life ≤4 under the ADR system, *excluding cars and light trucks*.
5–year property	All property with an ADR midpoint > 4 and < 10. *The new law specifically includes: cars, light trucks, computer–based telephone C.O. equipment (ADR midpoint = 9.5 years),* semiconductor manufacturing equipment, renewable energy and biomass properties, qualified technological equipment, and research (experimentation) equipment.
7–year property	All property with an ADR midpoint ≥10 years and < 16 years, *and* all property that does not have an ADR midpoint and is not assigned to another class. Included: *office furniture, fixtures, and equipment;* railroad tracks; and single purpose agricultural structures.
10–year property	All property with an ADR midpoint ≥16 years and < 20 years.
15–year property	All property with an ADR midpoint ≥20 years and < 25 years. Specifically included: *telephone distribution plants* and comparable equipment used by nontelephone companies for the two–way exchange of voice and data communications (ADR midpoint = 24 years). Excludes cable TV equipment used primarily for one–way communication.
20–year property	All property with an ADR midpoint ≥ 25 years, other than "1250" property.

Table 11-3. Tax Reform Act of 1986—Annual Recovery Table

Recovery Year	3-Year Class (200% d.b.)[1]	5-Year Class (200% d.b.)	7-Year Class (200% d.b.)	10-Year Class (200% d.b.)	15-Year Class (200% d.b.)	20-Year Class (200% d.b.)
1	33.00	20.00	14.28	10.00	5.00	3.75
2	45.00	32.00	24.49	18.00	9.50	7.22
3	15.00[2]	19.20	17.49	14.40	8.55	6.68
4	7.00	11.52[2]	12.49	11.52	7.69	6.18
5		11.52	8.93[2]	9.22	6.93	5.71
6		5.76	8.93	7.37	6.23	5.28
7			8.93	6.55[2]	5.90[2]	4.89
8			4.46	6.55	5.90	4.52
9				6.55	5.90	4.46[2]
10				6.55	5.90	4.46
11				3.29	5.90	4.46
12					5.90	4.46
13					5.90	4.46
14					5.90	4.46
15					5.90	4.46
16					3.00	4.46
17						4.46
18						4.46
19						4.46
20						4.46
21						2.25

Notes: 1. Declining balance.
2. Year of switch to straight-line.

Income Tax

Cash flows associated with any investment must include income tax. U.S. tax code specifies the method for computing income tax; the method changes from time to time. In general, income tax is calculated by taking a percentage of revenue less deductions; deductions include such things as debt interest, operating expenditures, other taxes, and tax depreciation. Since taxes represent a significant part of the cost of any plan, the analyst should verify that current law is modelled in any economic analysis tool used.

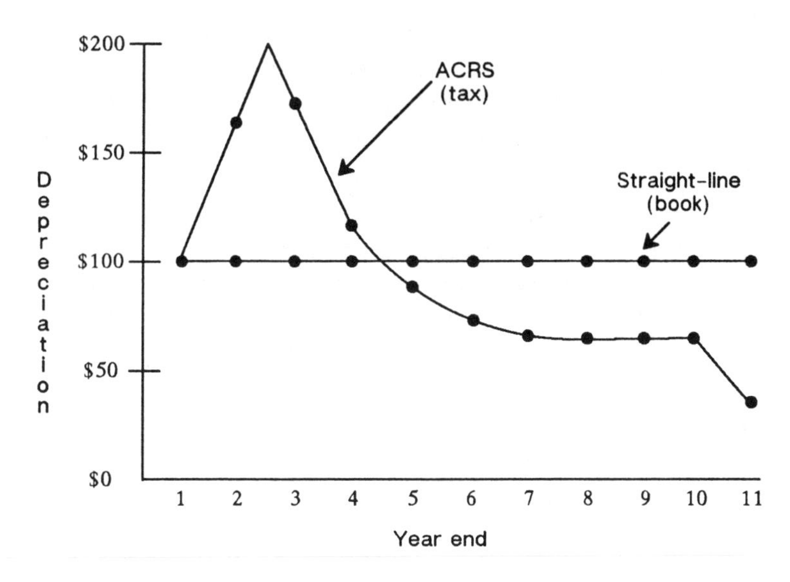

Figure 11-2. Accelerated tax depreciation per TRA 86.

11-6 PLANT OPERATIONS COSTS

Plant operations costs are those associated with the day-to-day practices of operating the company. Funding for operations costs comes from revenues in the year that the cost is incurred.

Included in operations costs are maintenance, administration, taxes other than income taxes, billing costs, marketing and accounting, etc.

(1) *Maintenance Costs*: These are frequently important ingredients of engineering economy analyses. They include labor and material associated with plant upkeep, the related costs of training, testing of facilities, and miscellaneous items such as shop repairs, tool expenses, and building maintenance work.

(2) *Operating Costs*: These include a wide range of costs primarily related to operator services, marketing, engineering, accounting, and administrative work. When these costs are common to alternative plans, they would

normally not be included in the engineering economy study. More and more often, however, one alternative includes significant savings in operations costs. In these cases, the costs are not common and would be included in the study.

(3) *Lease*: Leasing is an important consideration in many studies. It is important to remember that the telecommunications company can be the lessor as well as the lessee. Leasing is considered to be an alternative form of debt financing and requires some specialized modeling techniques.

(4) *Energy*: The cost of energy must be considered where it is not common to alternative plans. Primary power increases with the size of the plant.

(5) *Miscellaneous Taxes*: Sales, occupation and use, and ad valorem taxes must all be considered where appropriate. These taxes are especially important when considering taxable income for income tax purposes. Social security and unemployment taxes are usually treated as loading factors on labor costs.

In general, capital expenditures made in the past cannot be undone or affected by decisions made today. They must be recovered over the life of the plant through depreciation. However, when plant is retired, property tax, maintenance, rent, energy, and many other operating expenditures are no longer incurred.

11-7 DYNAMIC EFFECTS ON ANALYSIS

Since it is necessary to deal with the future, engineering economy studies are bound to be subject to all the uncertainties of prediction. Among these uncertainties are changes in technology, unanticipated demands for new services, and variations in the economic climate, such as inflation.

New Technology

Advances in technology result in improved electronic components, design techniques, and operating efficiencies. Thus, equipment that is less expensive, takes less space, uses less power, and provides more capacity at higher speeds of operation becomes available and must be considered as a replacement for existing equipment. The partial obsolescence and early retirement of older equipment become subjects of engineering economy studies.

Services and Service Features

New services and new service features also have an impact on engineering economy studies and on problems of early retirement due to functional obsolescence. If the new services are to be introduced in an area where present equipment is incapable of providing them, alternatives must consider replacement, overlay strategies, and what happens if the company doesn't offer the service.

Inflation

Inflation affects the cost of doing business in the future. There are several ways to account for the effect of inflation on costs. One straightforward method is to estimate future costs explicitly as of the time of the occurrence and to use these estimates in the study. For example, if an item currently costing $1,000 is needed now and another is needed one year from now, $1,070 should be used in the study for the second item if a seven–percent change in cost due to inflation is anticipated. This is sometimes called the explicit method.

Let X represent the uninflated value of a cash flow occurring at time n. Let Y represent the present worth of the inflated value of this cash flow. Using the explicit method, X and Y are related by the equation:

$$Y = X(1+I)^n \left(\frac{1}{1+i_c}\right)^n \qquad (11\text{–}10)$$

$$= X\left(\frac{1+I}{1+i_c}\right)^n$$

where I = the rate of inflation.

The convenience rate is an arithmetic shortcut to do both inflating and discounting in one step. Instead of using the cost of capital as the discount rate, the convenience rate, i_g, is used for discounting; inflation is ignored since it is built into i_g.

Since the present worth of the inflated cost must be the same calculated by the explicit method or the convenience rate method, the two previous equations can be combined to give a definition of the convenience rate, i_g.

$$X\left(\frac{1}{1+i_g}\right)^n = X\left(\frac{1+I}{1+i_c}\right)^n \qquad (11\text{–}11)$$

or

$$\left(\frac{1}{1+i_g}\right) = \left(\frac{1+I}{1+i_c}\right) .$$

Solving for i_g gives

$$1+i_g = \left(\frac{1+i_c}{1+I}\right) ,$$

$$i_g = \left(\frac{1+i_c}{1+I}\right) - 1 = \left(\frac{1+i_c}{1+I}\right) - \left(\frac{1+I}{1+I}\right) = \frac{i_c - I}{1+I} .$$

Since several different convenience rates would be required in a study involving items that are subject to different rates of inflation, this procedure is sometimes difficult to apply manually. Also, the present worth of taxes and book depreciation must be determined by the composite cost of capital rate and not the convenience rate.

While care must be used in applying the convenience rate, it is a valuable concept when properly used.

11–8 ECONOMIC EVALUATORS

Engineering economy studies generally look at two kinds of economic evaluator: primary and secondary.

Primary Evaluators

Several evaluators can be calculated using discounted net cash flow techniques. They all use the net cash flow (NCF) as a basis.

NCF = revenues − operating expenditures − first cost − taxes + net salvage

at each point in time. The most useful evaluators, called primary evaluators, provide the most complete description of an individual project and the relative value of alternatives. The most widely used primary evaluators are net present value (NPV) and net present worth of expenditures (NPWE).

Net Present Value. Net present value is a long–term after–tax evaluator. It is the present worth of the NCF discounted at the corporate discount rate.

$$NPV = \sum_{n=0}^{N} NCF_n \left(\frac{1}{1+i}\right)^n \tag{11–12}$$

where n = years
N = the project's impact period
i = cost of money.

NPV can also be defined using the cumulative discounted cash flow (CDCF). CDCF is simply the sum of the discounted cash flows (DCFs) at any time point. So NPV is the final CDCF at the end of the project's impact period. (This is sometimes called the CDCF–EOL, for end of life.)

NPV Interpretation. A positive NPV means that a project's long–term rate of return exceeds the corporate cost of capital (discount rate). Similarly, a negative or zero NPV implies that the project's long–term rate of return is less than or equals the corporate cost of capital.

In an incremental analysis, the NPV measures the relative benefit of one alternative over the other, rather than the profitability of either. In fact, if two alternatives with negative NPVs are compared, it is possible to get a positive incremental NPV. This is actually a way of determining which project costs less.

Therefore, the NPV measures the net benefits of the project and it is desirable to *maximize* NPV.

Some of the general advantages of the NPV over other evaluators are that it:

(1) includes the total project impact period

(2) measures the value of a project in dollars, thus avoiding the difficulties of using rates, ratios, or time as surrogate measures

(3) represents an easily calculated, natural by–product of cash flow data

(4) enables the direct assessment of the relative value of two alternatives where other economic assumptions are consistent.

Net Present Worth of Expenditures. Net present worth of expenditures is a long–term before–tax evaluator. It is the present worth of the expenditures required to support the project less the present worth of anticipated revenues.

$$\text{NPWE} = PW(OE + D + DI + NI + T_{RR}) - PW(R) \qquad (11\text{–}13)$$

where PW = present worth
R = anticipated revenues
OE = operating expenses
D = depreciation
DI = debt interest
NI = net income (return to stockholders)
T_{RR} = taxes on revenue requirements.

NPWE Interpretation. Since the NPWE is a before–tax revenue requirement, a plan whose NPV is positive would have the effect of reducing the company's before–tax revenue requirements and the NPWE would be negative. Conversely, a plan with a negative NPV would require additional revenues before taxes to support the project and, therefore, would have a positive NPWE.

The following is the relationship between the NPWE and NPV:

$$NPWE = \frac{NPV}{[(1-t_f) \times (1-t_s) \times (1-t_g)]} \qquad (11\text{–}14)$$

where t_f = federal income tax rate

t_s = state income tax rate

t_g = gross receipts tax rate.

Example: The NPWE for a project with an NPV of \$155 (assuming a federal tax rate of 34%, a state tax rate of 5%, and gross receipts tax rate of 1%) is calculated as follows:

$$NPWE = \frac{\$155}{(1-.34) \times (1-.05) \times (1-.01)} = \$250.$$

The project that enhances the value of the firm by \$155 can also be described as avoiding the need to raise \$250 in revenues. It is in this sense that NPWE represents the present worth of revenue requirements. Regulators often prefer to express project value in terms of revenue requirements.

In fact, NPWE is "unique" to a regulated business. It was developed to give project information from the regulators' (and customers') point of view.

Another way to describe the relationship between NPV and NPWE is that the NPV identifies the project with the largest dollar benefit, whereas the NPWE identifies the project with the lowest dollar cost. If the NPV is positive, the NPWE will be negative and about 1.6 times the absolute value of the NPV due to tax rates of approximately 34 percent. If the NPV is negative, the NPWE will be positive and about 1.6 times the absolute value of the NPV. These evaluators will always identify the same project as the most attractive. Therefore, a basic goal is to minimize NPWE.

Secondary Evaluators

The primary evaluators (NPV and NPWE) are indicators of the long-term attractiveness of the alternatives. Secondary evaluators can provide additional information about risk, sensitivity, and efficiency.

Although many secondary evaluators can be defined, three are of main interest: the discounted payback period (DPP), the long–term economic evaluator (LTEE), and the internal rate of return (IROR).

Discounted Payback Period. The project's discounted payback period is defined as that time (in years) when the cumulative NPV (cumulative present worth of the incremental net cash flows, discounted at the composite cost of capital) is equal to zero and remains equal to or greater than zero for the balance of the project. This evaluator is used to evaluate the "risk" associated with a particular undertaking. The shorter the payback period, the lesser the risk.

Example: Calculating DPP

The DPP for the project in the previous example can be calculated as follows, using a discount rate of 10%:

End of year	1	2	3	4	5
Cash Flow	−120	80.0	80.0	80.0	80.0
DCF	−114.4	69.4	63.1	57.3	52.1
CDCF	−114.4	−45.0	18.1	75.4	127.5

The project balance (cumulative CDCF) becomes positive in year 3, so DPP = 3 years.

Often, the characteristics of the cash flows are such that the cumulative present worth of the cash flows will cross the zero axis (become greater than zero) more than once during a project's study life. These time points represent intermediate payback periods that require subsequent analysis.

Plotting the CDCFs on a project balance diagram (Figure 11–3) can provide additional information about the risk of a project.

The DPP represents the point when all funds (expenses, manufacturing and service costs, taxes paid, capital

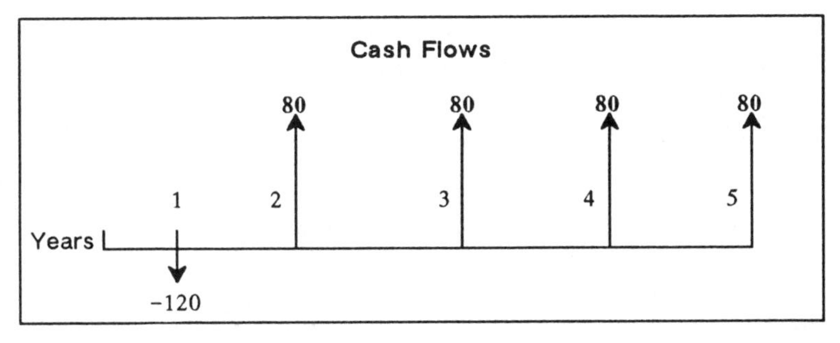

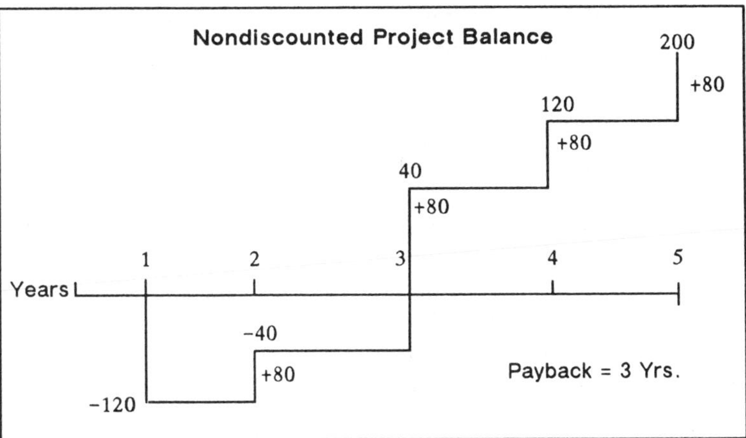

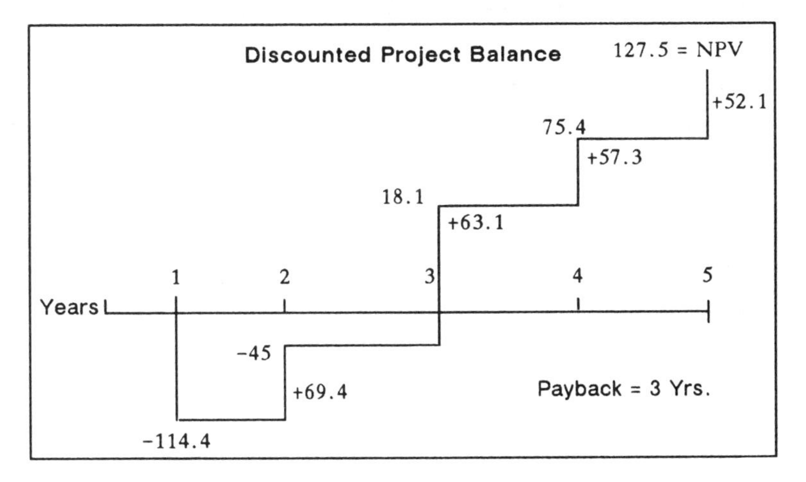

Figure 11-3. Project balance diagrams.

expenditures) have been recovered by the project's benefits, and the objective earnings on capital have been achieved. Many managers have the mistaken impression that the project is not producing any benefits (cash inflows) until the DPP point is reached.

When interpreting the DPP, it is important to remember that this indicator ignores all events that occur after the payback point. A project may break even very early, but achieve significantly lower total benefits to the company than other projects with longer DPPs but greater NPVs. Misapplication of this indicator can lead to a bias toward short–life projects at the expense of good long–term projects.

DPP is only a rough indicator of the economic risk associated with a project. A project with a short payback period may be risky due to the relative uncertainty of cash flows. Sensitivity and risk analysis methods provide better indicators of the risk of a project.

Long–Term Economic Evaluator. The long–term economic evaluator is a measurement of the relative efficiency of a project. LTEE is defined as a benefit–to–cost ratio.

$$\text{LTEE} = \text{efficiency} = \frac{\text{output}}{\text{input}} = \frac{\text{benefit}}{\text{cost}} . \qquad (11\text{–}15)$$

We can extend this definition to include time–value–of–money considerations. So LTEE restated in equivalent present worth values becomes:

$$\text{LTEE} = \frac{(+\,\text{DCF})}{|-\text{DCF}|} .$$

If the LTEE is greater than one, the alternative is generating more in benefits than it is incurring in costs. The higher this ratio, the more efficient the project.

However, to the extent that noncapital negative cash flows contribute to the denominator, the LTEE is not a measure of *capital* efficiency. Rather, it measures the efficiency with which *all* cash flows are used (capital, expense, and taxes).

The LTEE is useful in measuring the efficiency of individual project alternatives for an accept/reject decision. It should not be used as a ranking indicator for selection between project alternatives. The LTEE does not indicate how much benefit was obtained. This shortcoming requires that an incremental analysis be used to select between competing projects. After the incremental differences between projects have been determined, an incremental efficiency can be calculated. It should be noted that LTEE is extremely sensitive to the timing of cash flows.

Example: Calculating LTEE

If NPV is . . .	Then LTEE is . . .	And . . .
> 0	> 1	Benefits > Costs
=0	=1	Benefits = Costs
<0	<1	Benefits < Costs

For the previous example, the LTEE is calculated as follows:

End of Year	**1**	**2**	**3**	**4**	**5**
NCF	−114.4	76.3	76.3	76.3	76.3
DCF	−114.4	69.4	63.1	57.3	52.1

$$\text{LTEE} = \frac{(+\text{DCFs})}{(|-\text{DCFs}|)} = \frac{69.4 + 63.1 + 57.3 + 52.1}{|-114.4|} = 2.1 .$$

The present worth of the cash benefits of the project is estimated to be more than twice the costs incurred to achieve them.

Internal Rate of Return. The internal rate of return is the critical cost of money that a project could tolerate and still break even. It is that interest rate that will cause the final balance of the discounted net cash flows to equal zero.

$$\text{NPV} = 0 = \sum_{n=0}^{N} PW(\text{NCF}) \text{ at IROR for } n \text{ years.} \quad (11\text{–}16)$$

The analyst solves for IROR where NPV=0. This is illustrated by Figure 11–4, which gives NPV versus the cost of capital. The IROR is the point where the curve crosses the X–axis.

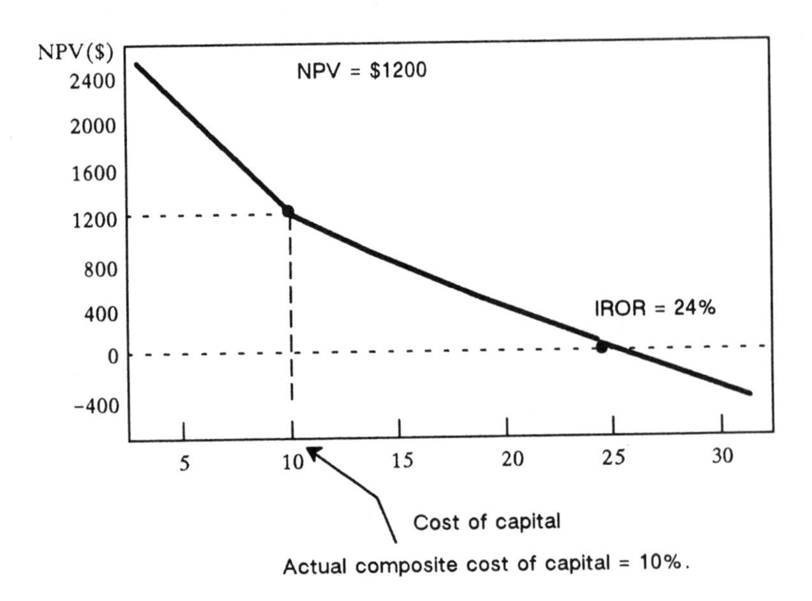

Figure 11–4. NPV versus cost of capital.

This project is attractive since its IROR of 24% is greater than the composite cost of capital of 10%. IROR can be used for this type of "accept or reject" decision. *It is dangerous to use the IROR alone to select between two alternatives.* Managers often believe that if project A has an IROR greater than that of project B, then project A is the more attractive alternative. This assumption is *not valid*, as shown in Figure 11–5. The graph shows project A from the previous charts along with a competing plan, project B.

Figure 11–5 provides the NPV curves constructed by varying the discount rate to determine the point where NPV = 0 for each alternative. At the composite cost of capital of ten percent, NPV project A = $1200 and NPV project B = $1600.

The NPV criterion favors project B, even though the IROR criterion appears to indicate project A is more attractive. This condition could occur even if projects A and B used the same amount of capital investment.

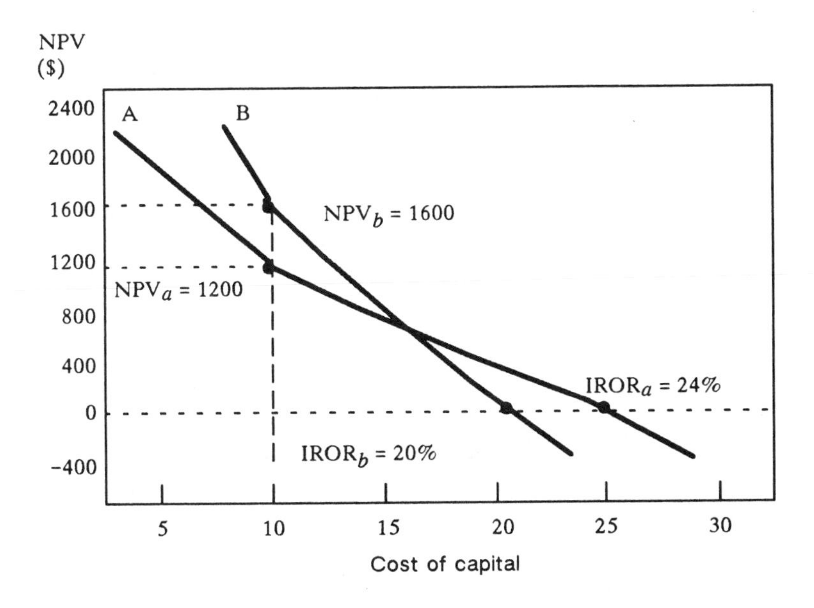

Figure 11-5. IRORs for two projects.

When NPV and IROR give conflicting signals, NPV is used to choose between alternatives. NPV is the fundamental measure of economic merit. Furthermore, the NPVs for competing alternatives can be compared directly.

An incremental comparison is the only way to use IROR safely to compare alternatives. The technique is to calculate IROR based on the differences between the two alternatives' cash flows. This single IROR can be compared to the cost of capital and used to choose between alternatives. It is inappropriate to use IROR to make decisions if plans are compared in reverse order, no mathematical solution exists, or multiple solutions exist. Many financial management texts explain the causes of these conditions in greater detail.

Cautions: (1) IROR does not measure return on investment.

(2) IROR is often misinterpreted as a measure of relative profitability.

(3) IROR gives no more information than does the sign of the NPV (only whether a project is good, not how good).

(4) IROR of two competing alternatives cannot be compared to one another directly. The IROR must be calculated for the incremental comparison.

IROR is used as a sensitivity measure; the only additional information IROR provides can be seen in the project A graph of Figure 11–5. The slope of the NPV line reveals how sensitive project A is to changes in the cost of capital. The steeper the slope, the more sensitive the project to these changes. If the NPV curve crosses the zero line very near the estimated cost of capital, it implies that even a small change in the cost of capital could make the project unattractive. If an IROR is much larger than the estimated cost of capital, the project is not very dependent on the accuracy of the estimated cost of capital.

The project rate of return (PRR) is being considered as a replacement for IROR in some economic evaluation tools.

11–9 SUMMARY

The combination of primary and secondary evaluators can be used to evaluate the economic desirability of alternatives. Table 11–4 provides an easy reference for identifying the evaluators.

Table 11–4. Economic Selection Criteria

Indicator	NPWE	NPV	IROR	DPP	LTEE
When "Plan A" is Winner →	NPWE < 0	NPV > 0	IROR > i_c	DPP ≤ SP *	LTEE > 1
	NPWE > 0	NPV < 0	IROR < i_c	DPP > SP	LTEE < 1
	NPWE = 0	NPV = 0	IROR = i_c	DPP = SP	LTEE = 1
Criteria	Minimize	Maximize	Greater than i_c	Minimize	Maximize; Greater than 1.
"Type"	← Ranking Mutually Exclusive Alternatives →		Accept/ Reject	Risk	Efficiency; Benefit to cost ratio
When Used	← "Unconstrained" Capital →		To eliminate inefficient alternatives; to judge sensitivity to changing cost of capital.	To help choose among mutually exclusive alternatives.	To maximize overall efficiency of limited available capital; not to be used alone; may be inconsistent with other indicators; sensitive to cash flow timing.

* SP = study period.

References

1. Engineering Department, American Telephone and Telegraph Company. *Engineering Economy*, Third Edition (New York: McGraw–Hill Book Company, Inc., 1977).

2. Federal Communications Commission. "Uniform System of Accounts for Class A and Class B Telephone Companies," *Rules and Regulations, Title 47, Code of Federal Regulations, Part 31* (Washington, DC: U.S. Government Printing Office, 1987).

Additional Reading

Grant, E. L., W. G. Ireson, and R. S. Leavenworth. *Principles of Engineering Economy*, Seventh Edition (New York: John Wiley and Sons, Inc., 1982).

Riggs, J. L. *Engineering Economics*, Second Edition (New York: McGraw–Hill Book Company, Inc., 1982).

Smith, G. W. *Engineering Economy Analysis of Capital Expenditures*, Third Edition (Ames, IA: Iowa State University Press, 1979).

Telecommunications Transmission Engineering

Section 3

Signal Characterization

Telecommunications involves the transmission and, in many cases, the switching of many types of signals that differ materially from one another. To facilitate the evaluation of transmission objectives, the nature and magnitude of various impairments, the performance provided by different facilities, and the way all of these interact, it is necessary to describe the various types of signal in terms that permit the expression of mathematical relationships among all these factors. This section provides such characterization for the principal forms of transmitted signals—speech, address and supervisory, data, and video. It also covers the characterization of combinations of signals that are found in a frequency–division multiplexed load on an analog carrier system.

Chapter 12 covers the characteristics of speech signals typically found in an analog* telephone channel, i.e., a loop or trunk. Bandwidth, amplitude, phase, and frequency variations for telephone speech, as well as the characteristics of a multichannel speech signal transmitted on analog carrier systems are described. A brief discussion of radio and television program signals is also given.

Wherever telecommunications signals must be switched, signals must be transmitted for directing and controlling the switching apparatus. These signals, called address and supervisory signals, are of many types. The most important are described in Chapter 13. The proliferation of this variety of signals has resulted from the increasing number of switching system types and features that have been provided. The signal characterization given in this chapter is provided with minimum discussion of the equipment or switching features involved.

---

* Analog is an obsolete technology being replaced by a digital networking architecture.

The material in Chapter 14 represents the characterization of a number of the more important types of data signals found in telecommunications systems. These signals are, in many cases, digital in format; they involve the provision of channels ranging from bandwidths of tens of hertz to several megahertz. Amplitude, frequency, and phase shift keying techniques are employed in multilevel formats ranging from 2 to 64 levels or more. There are some signals that are analog in nature and, as such, may achieve an infinite number of values over a restricted but continuous range.

The transmission of video signals is among the telecommunication services provided by exchange carriers. While the number of video circuits in service is small compared to the number of voice–frequency circuits, the video circuits use a substantial portion of transmission facility capacity because of the large bandwidth most of them require. Characteristics are described in Chapter 15 for telephoto, video, and television signals.

One reason for the extensive and detailed attention given to signal characterization is that signals and transmission systems interact in important ways. It is rare that only one type of signal is to be found in any one transmission system. This is especially true in broadband carrier systems, which carry simultaneously a large variety of signals. Some of the effects of such signal combinations are characterized in Chapter 16, where a qualitative discussion of such combinations is presented.

Chapter 12

Speech Signals

A channel in the message telecommunications service (MTS) network must carry a wide variety of signals; the most common and, therefore, most important is the telephone speech signal. Much research effort has been devoted to an understanding of all of the details of the processes of speech and hearing [1,2,3,4]. The concern here, however, is with the electrical signal analog of the acoustic message. This signal and its characterization are related primarily to the processes carried out in the transmitter (microphone) of the telephone station set and the effects on the signal produced by interactions between it and the channels on which it is carried.

The problems of speech signal characterization are made complex by the large number of variables involved and the resulting difficulties of defining and measuring important parameters explicitly. To overcome these difficulties, signal parameters are defined in terms of their statistical properties, such as average value, standard deviation, and activity factor. These parameters are defined first for a hypothetical single continuous talker of constant volume, V_{0C}. This value, expressed in vu, is next modified to account for breathing intervals and intersyllabic gaps and to define the single constant–volume talker in terms of power in dBm, P_{0C}.

Variables are next introduced to cover the effects of the speaking habits and gender of the talkers, circuit losses, the automatic compensation introduced by talkers to overcome impairments, station set variability, etc. Consideration of these variables introduces the concept of the variable–volume talker, one whose average volume is V_{0C} and whose volume has a standard deviation, σ.

The definition of these parameters is relatively straightforward, but the determination of their values by analytic means is not.

Measurements are usually made in working systems to determine speech power averages and standard deviations. These measurements must be expressed, of course, in terms of a well–established reference point, such as zero transmission level point (0 TLP).

A continuous talker signal is not ordinarily found in a telephone message channel. Activity factors associated with the efficiency of trunk use and talk–listen effects must be evaluated. With these factors accounted for, the statistics of talker signals in multiplexed broadband systems can be evaluated and used for the determination of signal–dependent impairments such as intermodulation noise, crosstalk, and overload.

12-1 THE SINGLE-CHANNEL SPEECH SIGNAL

Whereas tones with single frequencies are easily specified by just a few numbers—one for frequency, one for amplitude, one for phase—in addition to a functional expression such as sine, a telephone speech signal is not so easily specified or defined. It consists of many frequencies varying in amplitude and relative phase. Its average amplitude fluctuates widely, and even its bandwidth may vary with circumstances. Consider first the speech signal generated at a telephone station set and the way in which it is modified by the transmission elements of the channel between the transmitter and receiver.

Energy Distribution of the Speech Signal and Channel Response

The electrical analog of the acoustic speech signal is generated in the station transmitter. Sound waves from the speaker are impressed on the transmitter. Common–battery direct current, supplied from the central office or digital loop carrier over the loop conductors, provides power to the transmitter circuit. The varying pressure of the speech waves causes current variations in the transmitter circuitry and, in effect, modulates the direct current in the loop. While older telephones used transmitters with carbon granules directly modulating the loop current, newer electronic telephones use microphones and active networks.

Human speech contains components extending from roughly 30 to 10,000 Hz. The distribution of the long-term average energy for continuous speech approximates that shown in Figure 12–1. The actual spectral energy density and bandwidth are, of course, highly variable parameters. Nearly 90 percent of the speech energy lies below 1 kHz.

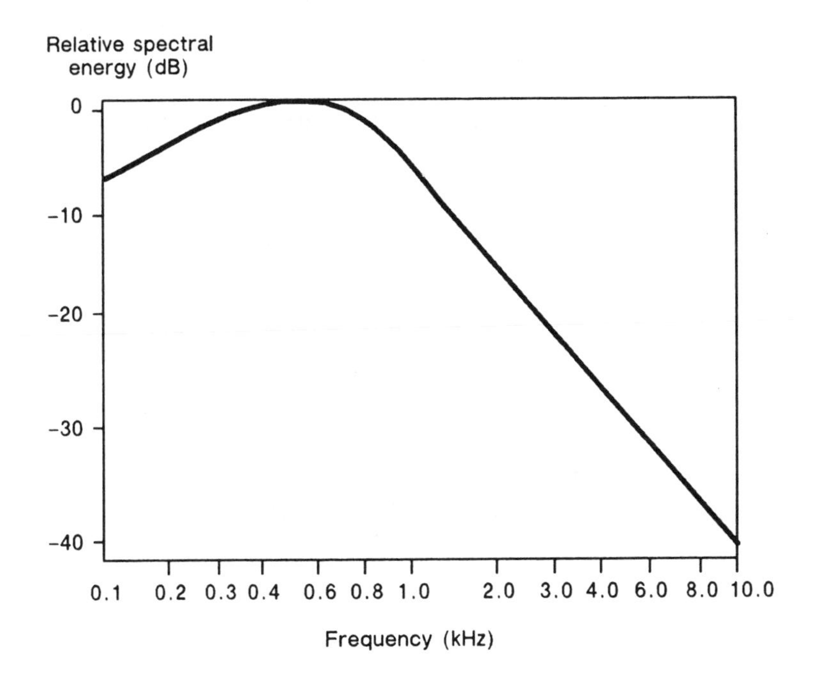

Figure 12–1. Approximate long-term average spectral energy density for continuous speech.

A detailed analysis of speech sounds indicates that most of the power is found in the low–frequency range (vowels) while other sounds carrying less energy (consonants) are important to articulated speech. Figure 12–2 illustrates the importance of the various speech frequencies to the transmission of articulated speech. Both high and low frequencies are necessary where the lower frequencies, although having the most power, provide for naturalness of speech, and the higher frequencies provide articulation.

Listener hearing, the perception by the ears of sound waves, is affected by the frequency and amplitude of sound, or, in terms of sense perception, its pitch and loudness. A tone of some frequency at a very low amplitude will not be perceived by the human ear. As the amplitude is slowly increased, a point will be reached at which the sound will be just audibly perceptible. The amplitude measured at this point marks the "threshold of audibility." As the amplitude is increased, the sound becomes louder until a point is reached at which the sense of feeling becomes physically sensitive, and a slight further increase causes pain. This level of sensibility indicates the "threshold of feeling."

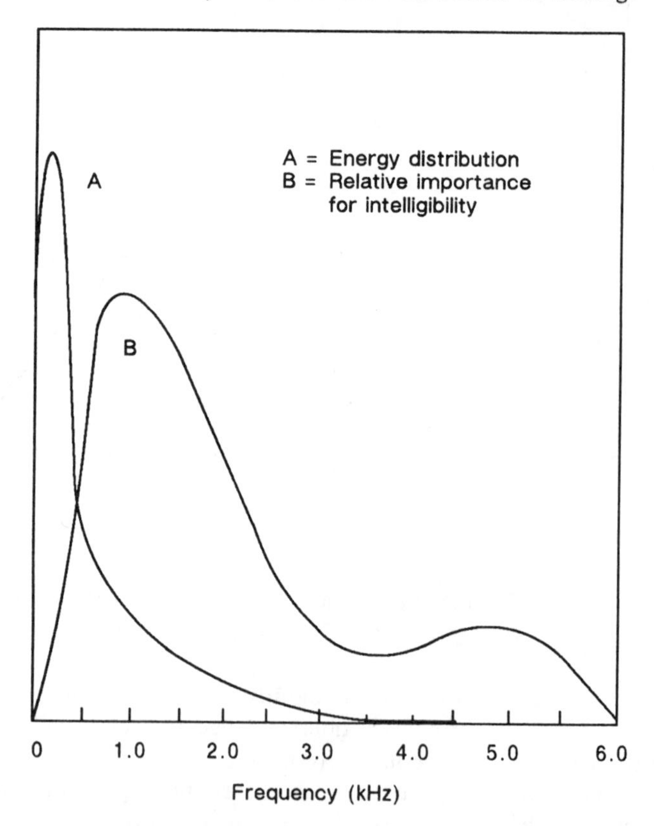

Figure 12–2. Frequency characteristics of speech.

Figure 12–3 illustrates the results for frequencies over a very wide range. The threshold lines intersect at the very low and very high ends of the speech spectrum. Outside of these points of intersection, the sense of feeling becomes noticeable before that of hearing. These outer regions are therefore useless for sound perception and the intersection points mark the frequency limits for the range of hearing. The oval area enclosed by the two curves contains all the simple sounds of variable pitch and loudness that can be perceived by the human ear, and is known as the auditory sensation area. The curves of Figure 12–3 are the empirical results of experiments on a considerable number of individuals, any one of whom might exhibit considerable departures from parts of the smooth curve.

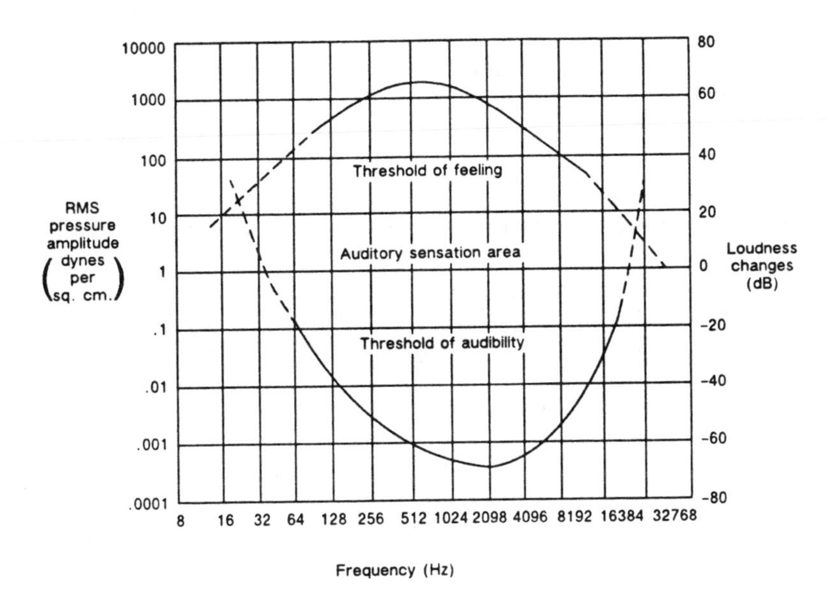

Figure 12–3. Auditory sensation area.

Also in Figure 12–3, the frequency range of the ear is shown to be approximately from 30 to 20,000 Hz, while the effective pressure amplitude is from below 0.001 to above 1000 dynes per square centimeter.

If, at a given frequency, the energy of a simple tone is gradually increased, the ear will not perceive the increase in loudness until a definite value over and above the initial value has been made. These increments of energy that cause changes in loudness perception break up the range for any one frequency into a finite number of steps, although the energy changes are continuous and contain an infinite number of energy values.

A similar effect is observed with variations of frequency along given pressure lines. A definite increment in frequency is necessary before the ear perceives a change in pitch, although the loudness is held constant.

These increments of energy and frequency, necessary for the perception by the ear of differences in volume or pitch, are not constant but increase when higher values of energy and frequency are introduced. These effects can be stated in terms of Weber's law, which defines variables in sense perception as the logarithm of the physical stimulus. This is illustrated in Figure 12–3, where the frequencies on the horizontal axis are spaced off logarithmically; the vertical axis is scaled by decibels, which are 20 times the logarithm to the base 10 of the pressure ratio. Zero point on this scale corresponds to one dyne rms per square centimeter.

For convenience, the reference pressure is generally taken as the pressure at the level of audibility. This value is different for various reference frequencies. Figure 12–3 also illustrates that two tones of different frequencies may sound equally loud, although the pressure values may be appreciably different. This is further illustrated in Figure 12–4, where the pressures of single frequencies equal in loudness are plotted versus the frequencies of the tones. Above 700 Hz, the curves are at the lower pressures.

Another important effect on auditory response is the masking of one tone by another of greater intensity. Tones of given intensity are quite effective in masking tones of higher frequency and less intensity, but hardly effective in masking tones lower in frequency.

Because of its nonlinear characteristic, the ear acts as a modulating device and supplies overtones, difference frequencies, and summation frequencies of any two or more tones in the received sound waves. This permits the ear to interpret the fundamental frequency where pronounced overtones are present and the fundamental frequency is suppressed. This characteristic of hearing has the advantage of supplying lost frequencies in transmission systems having limited bandwidth.

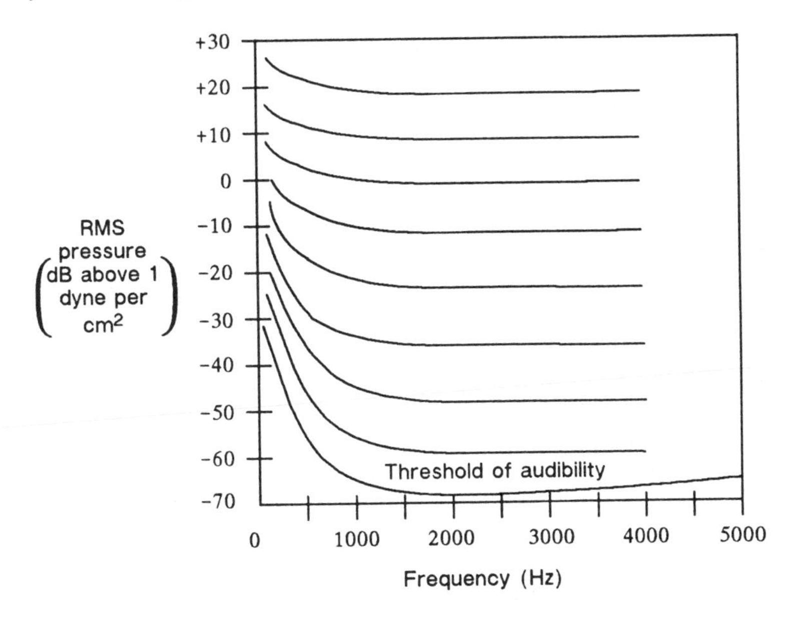

Figure 12-4. Loudness of single-frequency tones.

When considering all the aspects of speech reception by the human ear, the general adequacy of satisfactory speech reception is found to depend on three factors: volume, distortion, and interference. All three affect the intelligibility and the naturalness of received speech. These three factors, interrelated but controlled separately, are considered and their importance weighed when providing economically designed commercial voice transmission systems.

Most of the requirements of satisfactory speech transmission will be met with a system or circuit capable of transmitting a band of frequencies approximately 3000 Hz wide. In practice, a band extending approximately from 0.25 to 3.0 kHz has been found to provide acceptable quality for telephone communications. The transmission response at several points in a simple connection is depicted in Figure 12–5. Generally, the transmission response at several points in a simple connection is defined as being between points that are 10 dB down from the reference frequency, usually taken as 1000 Hz. Figure 12–5 shows that, even for the simple connection depicted, the band is already restricted to approximately 0.25 to 3.0 kHz.

Single Constant–Volume Talker

To develop an understanding of speech signal characterization from the point of view of practical applications of transmission design, layout, and operation, consider first a hypothetical case of a single continuous talker of constant volume. The volume of this talker's telephone speech has, by definition, a value of V_{0c} vu as measured at a zero level point.

However, a continuous talker is not capable of producing truly continuous speech signals. Pauses due to the thought process, to breathing intervals, or to intersyllabic gaps in energy result in an activity factor, τ_c, of 0.65 to 0.75.

The value of power in dBm in a speech wave is defined as the value of volume in vu corrected by the activity factor. Thus, the power for a continuous talker may be written

$$P_{0c} = V_{0c} + 10 \log \tau_c \text{ dBm.} \qquad (12\text{–}1a)$$

For example, if $\tau_c = 0.725$, the power in such a signal is

$$P_{0c} \doteq V_{0c} - 1.4 \text{ dBm.} \qquad (12\text{–}1b)$$

This value agrees with an empirically derived relationship between vu and dBm for speech signals, which is generally accepted as the power corresponding to a zero vu talker.

Sources of Volume Variation

A constant–volume talker is a rarity. Therefore, consideration will be given to some of the important sources of volume variation. First, the telephone speaking habits and gender of the

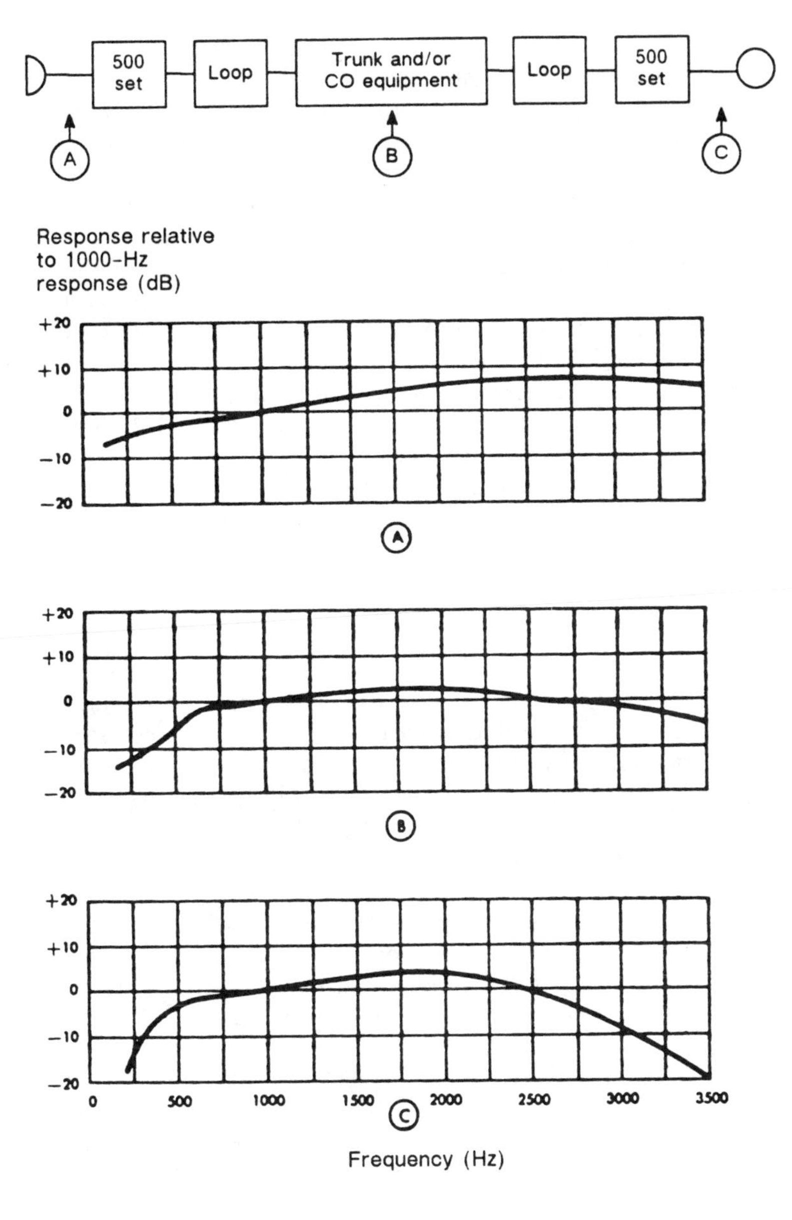

Figure 12-5. Transmission response, normalized at 1000 Hz, at points along a typical connection (trunk and central-office equipment assumed distortionless).

speaker introduce wide variations. He or she may be loud or soft-spoken and may hold the telephone transmitter close to the mouth or at a distance. In addition, telephone station sets have a range of efficiencies with which they transform acoustic waves to electrical waves and vice versa. Further, their efficiencies may be, by design, variables that depend on the value of direct current fed to them from the central office. The length of the loop, the wire gauge used, the presence or absence of irregularities such as bridged taps or bridged stations, and the possible use of loading on the loop all contribute to variations from loop to loop. These variations affect the average losses in the loop and the amount of current fed to the transmitter. In addition, these variations affect differently the attenuation at different frequencies. Variations in average loss and in frequency-dependent attenuation are also found in trunks and central-office wiring and equipment that may be used in a built-up connection. Furthermore, impairments such as improper sidetone, echo, circuit noise, room noise, and crosstalk have subjective effects on speaking habits, as do distance, trunk loss, and type of call.*

Some of the variable losses involved in a simple interlocal telephone connection are illustrated in Figure 12–6. Telephone set efficiencies for conversion of sound pressure to electrical signal conversion and vice versa are such that, with typical losses in the local connection, a speaker producing at the microphone a sound pressure of 89.5 dBRAP (dB above reference acoustic pressure) would be heard at a sound pressure of 81.5 dBRAP. Reference in this case is an acoustical pressure of 0.0002 dyne per square centimeter. The previously mentioned variables are such that received sound signals have a wide range of values with a standard deviation of nearly 8 dB from the average value of 81.5 dBRAP.

In Figure 12–6, the noise impairments shown as introduced in loops, central-office equipment, and the trunk might be picked up at any of these points. Room noise at the speaker's end of the connection enters the circuit through the transmitter and appears at the distant receiver along with the speech signal. Room noise

* It has been observed on traditional analog facilities that volume increases about 1 vu for each 3 dB of trunk loss and about 1 vu for each 1000 miles of distance. Volume on business calls tends to be somewhat higher than on social calls.

at the listener's end affects that person's hearing directly and, in addition, enters the transmitter and appears in the receiver by transmission through the sidetone path. The decreasing powers in the signal and in each noise component, caused by the increasing circuit loss illustrated at the bottom of the figure, are not assigned values in the figure because they are variable on different connections and under differing circumstances. Even though impairments are not discussed here in detail, the noise and loss impairments are shown qualitatively in Figure 12–6 to illustrate their sources. They have an indirect, subjective effect on talker volumes as previously mentioned.

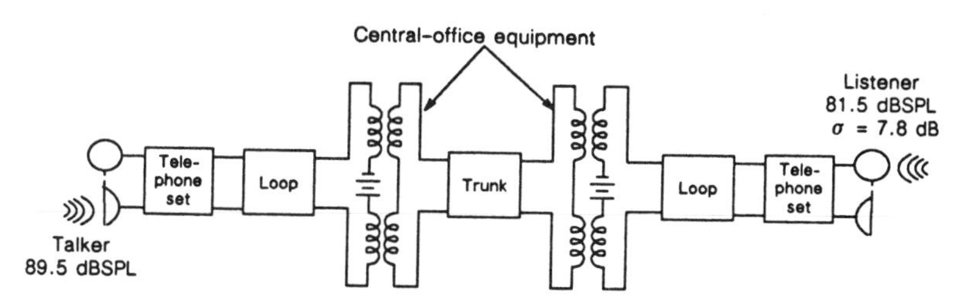

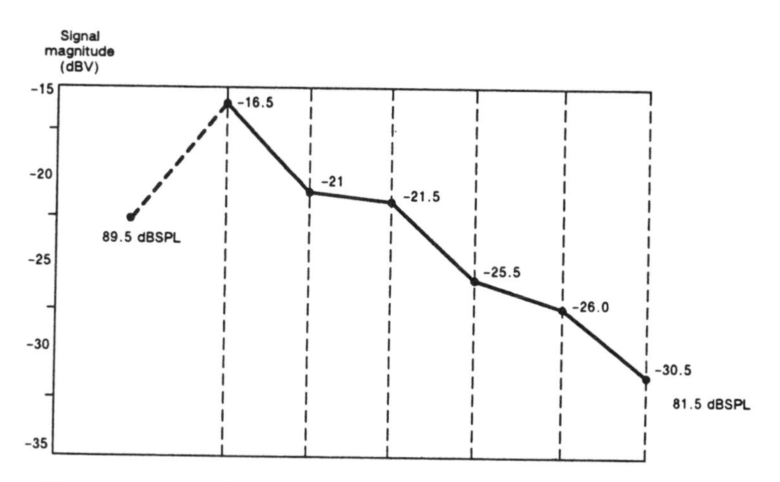

Figure 12–6. Circuit losses and impairments in a typical interlocal telephone connection.

Single Variable–Volume Talker

As it has been pointed out, the single constant–volume talker is a hypothetical case. The aforementioned variables are so numerous and so difficult to evaluate precisely that it is necessary to rely on measured data in order to characterize the single variable–volume talker. The measurements from the 1960 survey of speech volume made on real–time line traffic using vu meters, essentially an evaluation of the variable–volume talker, are compared in Table 12–1 with the 1975–1976 survey results that were translated from equivalent peak level (EPL) measurements at end offices [5,6]. In such surveys, many variables must be considered, and studies are being made to determine which are important in present day plant [7,8].

Table 12–1. Near–End Talker Speech Volumes; 1975–1976 Survey Compared to 1960 Survey

Type of Connection	**Speech Volume (vu*)**			
	1960		1975–1976	
Call destination	Avg.	Std. dev.	Avg.	Std. dev.
Intrabuilding	–24.8	7.3	–22.2	4.6
Interbuilding	–23.1	7.3	–22.5	4.7
Toll	–16.8	6.4	–21.6	4.5

* Measured at the end–office transmitting switch.

It is interesting to note that the 1975–1976 speech volumes for toll calls, which averaged –21.6 vu, are much lower than those found in the 1960 survey, which averaged –16.8 vu. The ranges of volumes in all call destination categories in the 1975–1976 survey were much narrower than in the 1960 survey. Changes introduced in the telephone plant between 1960 and 1975 tended to increase the uniformity of service from the viewpoint of speech volumes. Changes included the replacement of the 300–type telephone set with the 500–type, an increase in direct trunking between end offices, loss plan improvements, and the discontinuance of special "toll–grade" battery on long connections.

The characterization of speech power on message telephone circuits was essential for the design of broadband analog carrier systems and remains important for a wide variety of telecommunications equipment. As an example, the 1960–1976 decrease in talker volume on toll connections made it feasible to uprate one microwave radio system from 1500 to 1800 voice channels. Knowledge of the average power per talker of a group of talkers, all of whose volumes vary with time, is needed to determine signal power loading and crosstalk limitations and establish objectives. System and equipment design must be based on total signal power, determined from the mean value and standard deviation of each of the speech signals to be carried. These signals do not combine as normal distributions, even though each is normal in dB. The average power values must be added by converting from dBm to milliwatts, determining the average value, adding the averages in milliwatts, and reconverting the result to dBm.

Consider a probability density function, normal in dB, having an average value of 0 dBm and a standard deviation of 3 dB. This type of function is plotted in Figure 12–7(a). If the dBm values are converted to milliwatts, the density function of Figure 12–7(b) results. Note that this function is skewed and that its mean value is greater than 1 mW. The difference, δ, between the average value in dBm (0 dBm or 1 mW) and the mean value of the distribution increases as σ increases. The necessary correction to express the power under a log normal probability density function has been derived elsewhere and is equal to $0.115\sigma^2$ [9]; i.e., to obtain the average power in dBm, $0.115\sigma^2$ must be added to the mean value.

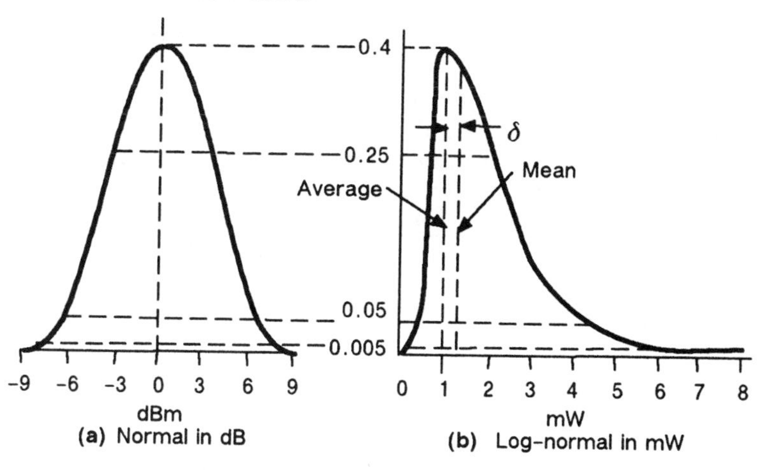

Figure 12–7. Density functions.

Thus, the average power of a variable–volume talker signal having a log–normal density function and a standard deviation of σ may be expressed by

$$P_{0p} = P_{0c} + 0.115\sigma^2 = V_{0c} - 1.4 + 0.115\sigma^2 \text{ dBm.} \quad (12\text{–}2)$$

This equation is derived from Equation 12–1(b) and from the discussion above which relates the average value of power to the mean value of the density curve, normal in dB.

The mean value of volume for toll calls is given on the last line of Table 12–1, but further manipulation is necessary to make the data useful for toll system analysis. The first step is to translate the data from the outgoing switch of the end office, where the measurements were made, to a comparable point in a toll system. Between the point of measurement and the entrance to the analog toll portion of the telecommunications network are, for each connection, a tandem–connecting trunk and certain items of central office equipment. Under the via net loss (VNL) plan, these have a loss of (VNL + 2.5) dB, which includes 2 dB assigned to the trunk and 0.5–dB allowance for the analog switching system equipment. If 0.5–dB is allowed for VNL (a typical value), the average −21.6 vu volume for toll calls shown in Table 12–1 may be translated to toll system values (at the −2 dB TLP) as

$$V_{\text{toll}} = -21.6 - (2.0 + 0.5 + 0.5) = -21.6 - 3.0 = -24.6 \text{ vu.}$$

Typically, the losses of the tandem–connecting trunks have a standard deviation below 1 dB. Thus, when combined with the standard deviation of measured toll volumes at the end office, the standard deviation of volume for the toll system is

$$\sigma_{\text{toll}} = \sqrt{4.5^2 + 1^2} = 4.6 \text{ vu.}$$

For the values of toll call volumes, the average continuous talker power is

$$P_{0p} = -24.6 - 1.4 + 0.115\sigma^2 = -24.6 - 1.4 + 2.4 \approx -23.6 \text{ dBm.}$$

One further correction is needed. In Chapter 3, Part 2, the switch at which the analog tandem trunk is terminated is defined as a −2 dB TLP. Therefore, the average power must be converted to a value at the 0 TLP by adding 2 dB; i.e.,

$$P_{0p} = -23.6 + 2 = -21.6 \text{ dBm0}.$$

This approximates the mean values of near–end speech signal power of toll calls in the 1975–1976 survey.

The above discussion relates to volume and power averaged subjectively over an interval of three to ten seconds. In reading the volume indicator, occasional very high and very low readings are ignored. High peaks, however, do occur, and their magnitude is sometimes of considerable interest. The peak factor for a typical continuous talker is approximately 19 dB. For a talker of lower activity, peak magnitudes are not affected, but the average power is reduced relative to the continuous talker.

12-2 MULTICHANNEL SPEECH

The need for characterizing the speech signals in multichannel systems arises primarily from the need to control overload performance in analog systems. The characteristics of the multiplexed combination of speech signals are determined by extrapolation of the analysis of single signal characteristics.

If there are a number, N_a, of independent continuous talker signals of distributed volumes simultaneously present in a broadband system, with each signal occupying a different frequency band but at the same TLP, the total power represented by the N_a signals is

$$\begin{aligned} P_{av} &= P_{0p} + 10 \log N_a \\ &= V_{0c} - 1.4 + 0.115\sigma^2 + 10 \log N_a \text{ dBm.} \qquad (12\text{–}3) \end{aligned}$$

In a system containing N channels, the maximum number of simultaneous signals that could be present is $N_a = N$; however, such an event is extremely unlikely, especially when N is large. Thus, it is necessary now to examine the factors that enter into an evaluation of the probable number of simultaneous talkers in such a system.

The speech activity factor for a continuous talker, τ_c, was included in Equation 12–1 for the evaluation of P_{0c}. In evaluating P_{av} (Equation 12–3), other forms of activity, such as the increasing use of data transmission, must be taken into account. The assumption is made that during the average telephone conversation, the person talks half the time and listens half the time. Thus, the value of the talk–listen activity factor, τ_s, may be taken as 0.5. Furthermore, during the time a call is being set up, there is low speech activity on the trunk. These effects may be accounted for by a trunk efficiency factor, τ_e. For domestic circuits, τ_e is usually taken as 0.7. For overseas calls, this value may be as high as 0.9.

The two activity factors discussed above are usually combined into a single telephone load activity factor,

$$\tau_L = \tau_s \tau_e = 0.5 \times 0.7 = 0.35.$$

Other activity considerations not specifically evaluated have led to a commonly accepted value of $\tau_L = 0.25$ for domestic telephone systems. A higher value (usually $\tau_L = 0.35$) is used for transoceanic systems. It must be remembered, however, that these are average busy–hour values. The number of speech signals simultaneously present during the busy hour, when such load considerations are important, varies considerably.

For an N–channel system having a load activity factor τ_L, the number of independent continuous talker signals, N_a, is a variable whose mean value is N_L. A system designed to carry just $N\tau_L$ continuous talkers would be overloaded half the time. A system designed to carry N continuous talkers would be impractical because such a signal load would occur only a very small percentage of the time.

It is necessary, therefore, to establish the statistical distribution of channels that would carry continuous talker signal power as a function of time. The variable representing this distribution may be called N_s. The probability that the number of channel carrying continuous talker power N_s may be found from

$$P(N_s) = \frac{N!}{N_s!(N-N_s)!}\tau_L^{N_s}(1-\tau_L)^{N-N_s} .$$

This is a binominal distribution that approaches a normal distribution having a mean value of $N\tau_L$ and a standard deviation

$$\sqrt{N\tau_L(1-\tau_L)}$$

if $N\tau_L \geq 5$.

Generally, for analysis purposes, the number of talkers assumed to generate speech energy simultaneously is the number that may be present one percent of the time. This value, chosen on the basis of experience, shows adequate balance between performance and cost. Thus, N_a is the value of N_s exceeded one percent of the time. From the values of areas under a normal curve (Figure 9–13), this value is

$$N_a \approx N\tau_L + 2.33\sqrt{N\tau_L(1-\tau_L)} \ .$$

Examination of this equation shows that the mean, $N\tau_L$, increases more rapidly than the standard deviation, $\sqrt{N\tau_L(1-\tau_L)}$, as N becomes larger. Thus, for large values of N, N_a approaches $N\tau_L$. Also, it can be seen that the larger the value of τ_L, the smaller N need be for this approximation to be valid. Note that for $\tau_L = 1$, $1 - \tau_L = 0$, and $N_a = N\tau_L$.

Thus, for large values of N, Equation 12–3 can be rewritten

$$P_{av} \approx V_{0C} - 1.4 + 0.115\sigma^2 + 10 \log N + 10 \log \tau_L \text{ dbm.}$$

This approximation can be made an equality, even for systems of small N, by defining a term which takes into account the deviation of N_a from $N\tau_L$.

$$\Delta_{c1} = 10 \log \frac{N_a}{N\tau_L} \ . \qquad (12\text{–}4)$$

When terms are rearranged, this may be written

$$10 \log N_a = \Delta_{c1} + 10 \log N + 10 \log \tau_L$$

and substituted in Equation 12–3 to give

$$P_{av} = V_{0c} - 1.4 + 0.115\sigma^2 + 10\log \tau_L + 10\log N + \Delta_{c1} \quad \text{dBm.} \tag{12-5}$$

For the two values of τ_L given previously (0.25) and (0.35), the relationships among N_a, $N\tau_L$, and Δ_{c1} are shown in Table 12–2 for systems of various sizes. The value of Δ_{c1} is shown to become small as N grows larger. It is often ignored in systems where N exceeds 2000 channels.

Table 12–2. Number of Active Channels and Δ_{c1}

N	$\tau_L = 0.25$			$\tau_L = 0.35$		
	N_a	$N\tau_L$	Δ_{c1}, dB	N_a	$N\tau_L$	Δ_{c1}, dB
6	4.84	1.5	5.1	5.60	2.1	4.3
12	7.37	3.0	3.9	8.78	4.2	3.2
24	11.80	6.0	2.9	14.59	8.4	2.4
36	15.88	9.0	2.5	19.94	12.6	2.0
48	19.84	12.0	2.2	25.19	16.8	1.8
96	34.74	24.0	1.6	45.18	33.6	1.3
300	93.32	75.0	0.9	124.92	105.0	0.7
600	175.55	150.0	0.7	237.89	210.0	0.5
2000	545.91	500.0	0.4	750.34	700.0	0.3

If V_{0c} is evaluated at 0 TLP, the value of P_{av} in Equation 12–3 is in dBm0. In the total speech load of N signals, P_{av} is the average power at 0 TLP exceeded during one percent of the busy hour when all N channels are busy. (A channel is considered busy when a talking connection is established; speech signals need not be present).

From Equations 12–2 and 12–5, the long-term average load per channel may be determined (by substituting the previously derived values $P_{0p} = -21.6$ dBm0 and $\tau_L = 0.25$) for broadband toll systems as

$$P_{av}/\text{chan} = P_{0p} + 10\log \tau_L + \Delta_{c1} \quad \text{dBm0.}$$

For very broadband systems ($N > 2000$), Δ_{c1} approaches zero and the load is

$$P_{av}/\text{chan} = -21.6 - 6 = -27.6 \text{ dBm0}$$

for a telephone signal of variable–volume talkers.*

12-3 LOAD CAPACITY OF SYSTEMS

The load capacity of a multichannel telephone transmission system is the peak power generated by the total number of speech signals the system can carry without producing an undue amount of distortion or noise, or otherwise affecting system performance. The maximum signal amplitude impressed on the system depends on the average talker volume, the distribution of volumes, and the talker activity. Overload may be the result of the signal amplitude exceeding the dynamic range of an amplifier or other active device, of frequency deviations exceeding the bandwidth of an angle–modulated system, or of voltages exceeding the quantizing range of a digital quantizer. A system is often said to be overloaded when the overload point of the system is exceeded by peaks of the transmitted signal more than 0.001 percent of the time. (It is *not* then said to be overloaded 0.001 percent of the time.)

Multichannel Speech and Overload

Overload is defined in a number of ways in Chapter 7, Part 5. These definitions all basically relate to the signal amplitude at which performance is no longer linear enough to satisfy performance objectives. In any of these definitions, it is convenient to use P_s dBm0 to express the average power of a single–frequency signal that causes system overload. The peak instantaneous power of this sinusoid is $(P_s + 3)$ dBm0.

Most systems do not overload on average power but rather when instantaneous peaks exceed some threshold. A

* None of the material in this chapter considers the effects of address, supervisory, or data signals on average channel loading.

multichannel signal with $P_{av} = P_s$ overloads severely because the multichannel signal has a peak factor much larger than the 3–dB peak factor for the single frequency, P_s. The peak factor for multichannel speech is 13 to 18 dB, depending on the number of channels in the system. It has been found that performance is usually satisfactory if the peak power of the multichannel load exceeded 0.001 percent of the time is set equal to or less than the peak power of the sinusoid, $(P_s + 3)$ dBm0. This may be rewritten

$$P_s + 3 = P_{av} + \Delta_{c2} \text{ dBm0}$$

or

$$P_s = P_{av} + \Delta_{c2} - 3 \text{ dBm0} \qquad (12\text{–}6)$$

where Δ_{c2} is the peak signal amplitude exceeded 0.001 percent of the time. The value of Δ_{c2} has been determined and is plotted in Figure 12–8. This figure shows that as the number of active channels, N_a, increases, the peak factor asymptotically approaches 13 dB. This value corresponds closely to that for random noise.

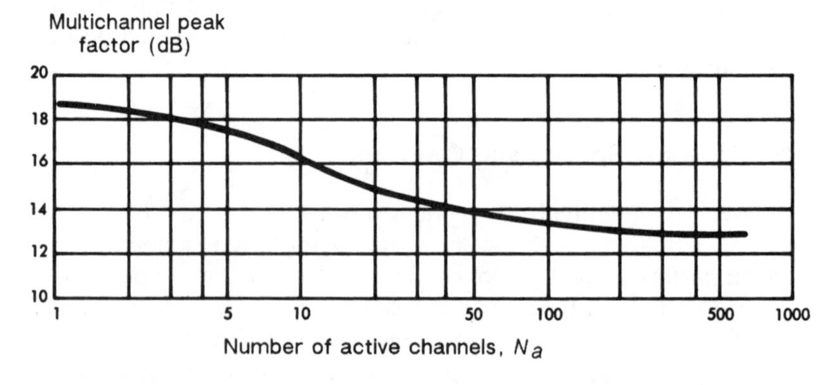

Figure 12–8. Peak factor, Δ_{c2}, exceeded 0.001% of time for speech channels.

If Equation 12–5 is substituted in Equation 12–6,

$$P_s = V_{0c} - 1.4 + 0.115\sigma^2 + 10 \log \tau_L + 10 \log N + \Delta_{c1} + \Delta_{c2} - 3 \quad \text{dBm0} \qquad (12\text{–}7a)$$

or, with $\Delta_c = \Delta_{c1} + \Delta_{c2} - 3$,

$$P_s = V_{0c} - 1.4 + 0.115\sigma^2 + 10 \log \tau_L + 10 \log N + \Delta_c \quad \text{dBm0}. \tag{12-7b}$$

The term Δ_c is known as the multichannel load factor. It is plotted in Figure 12–9 as a function of N and several values of σ for an assumed value of $\tau_L = 0.25$. For other values of σ or τ_L, Δ_c can be found from the empirically derived formula

$$\Delta_c = 10.5 + \frac{40\sigma}{N\tau_L + 5\sqrt{2\sigma}} \text{ dB}. \tag{12-8}$$

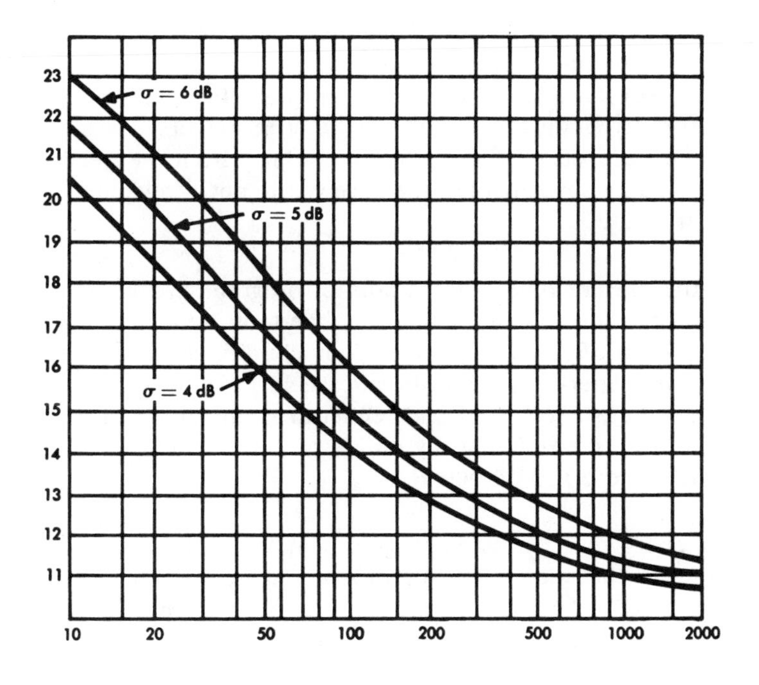

Figure 12–9. The multichannel load factor, Δ_c, for $\tau_L = 0.25$.

Single–frequency signals are used in the analysis of system load capacity, but they are seldom used in load testing. A band of Gaussian noise is frequently used in system testing to simulate a multichannel signal.

12-4 PROGRAM SIGNALS

Program transmission is a service provided by exchange carriers for the transmission of monaural and stereophonic audio programs of radio and television broadcasters between points of program origination and one or more transmitting stations. In addition, "wired music" material is also transmitted for distribution to customers subscribing to such services. Other program services include conference calls and calls connected to public address systems for a large audience. While such signals are audio signals, regular telephone circuits are not generally used for program transmission because of the more stringent objectives generally applied to program service. The more stringent objectives arise from the necessity of transmitting music and from the need for higher fidelity speech when the receiver is not a telephone set receiver.

At the present time, the majority of program circuits used in toll transmission systems employ a band of frequencies from about 200 to 3500 Hz. These specially conditioned voice–grade circuits are generally used for broadcasts in which the program is speech alone, such as newscasts. Other program circuits, less frequently used, cover frequency ranges of 100 to 5000 Hz, 50 to 8000 Hz, and 50 to 15,000 Hz.

The energy distribution in program signals is difficult to specify because of the wide range of program material transmitted—speech, music of different varieties, etc. No generally accepted program spectrum has been established.

The average volume and the dynamic range of program signals are somewhat higher than telephone speech. There are relatively few program channels, however, and contributions to system load effects are generally small enough to be ignored.

References

1. Dudley, H. "The Carrier Nature of Speech," *Bell System Tech. J.*, Vol. 19 (Oct. 1940), pp. 495–515.

2. French, N. R. and J. C. Steinberg. "Factors Governing the Intelligibility of Speech Sounds," *The Journal of the Acoustical Society of America*, Vol. 19 (Jan. 1947).

3. Potter, R. K. and J. C. Steinberg. "Towards the Specification of Speech," *The Journal of the Acoustical Society of America*, Vol. 22 (Nov. 1950).

4. Sullivan, J. L. "A Laboratory System for Measuring Loudness Loss of Telephone Connections," *Bell System Tech. J.*, Vol. 50 (Oct. 1971), pp. 2663–2739.

5. McAdoo, K. L. "Speech Volumes on Bell System Message Circuits—1960 Survey," *Bell System Tech. J.*, Vol. 42 (Sept. 1963), pp. 1999–2012.

6. Ahern, W. C., F. P. Duffy, and J. A. Maher. "Speech Signal Power in the Switched Message Network," *Bell System Tech. J.*, Vol. 57, No. 7 (Sept. 1978).

7. Sen, T. K. "Subjective Effects of Noise and Loss in Telephone Transmission," *IEEE Transactions on Communications Technology*, Vol. COM–19, No. 6 (Dec. 1971).

8. Brady, P. T. "Equivalent Peak Level: A Threshold–Independent Speech Level Measure," *The Journal of the Acoustical Society of America*, Vol. 44 (Sept. 1968), pp. 695–699.

9. Bennett, W. R. "Cross–Modulation Requirements on Multichannel Amplifiers Below Overload," *Bell System Tech. J.*, Vol. 19 (Oct. 1940), pp. 587–610.

Additional Reading

Members of Technical Staff. *Transmission Systems for Communications*, Fifth Edition (Murray Hill, NJ: AT&T Bell Laboratories, Inc., 1982), Chapter 16.

Chapter 13

Signalling

Signalling involves the generation, transmission, reception, and application of information for directing and controlling automatic switching systems and conveying information to and from telecommunications users. With circuit–associated signalling, the signals are carried on the same facility as the voice path. This is in contrast to common–channel signalling (CCS) where the signalling and voice paths are carried on separate facilities. Circuit–associated signalling uses inband or out–of–band signalling systems. Inband signalling shares the channel with voice. Single–frequency (SF), dual–tone multifrequency (DTMF), and multifrequency (MF) signalling are classified as inband. Out–of–band (out–of–slot) signalling does not share the voice portion of a channel, but does use a designated segment of the same channel capacity in the carrier system. Digital carrier systems are considered to use out–of–band signalling. In this system, a portion of the bits in the line bit stream are reserved for voice transmission, while others in the same bit stream are time–shared for signalling. These two modes of signalling provide the following functionally categorized signals:

(1) Address signals

(2) Supervisory signals

(3) Alerting signals

(4) Call progress signals

(5) Control signals

(6) Test signals.

Address signals are used to convey call destination information such as the telephone–station code, central–office code, area code, and desired interexchange carrier (IC), generally in the form of dial pulses or DTMF pulses. These signals may be generated by telephone station sets, switching systems, or other customer terminal equipment (e.g., computer modems).

Supervisory signals are used to convey, to a switching system or to an operator, information regarding the status of a loop or trunk. The five service conditions that supervisory signals convey are as follows:

(1) Idle circuit, which is indicated by the combination of an on–hook signal and the absence of any connection in the switching system between the loop and another loop or trunk.

(2) Busy circuit, which is indicated by an off–hook signal from the call originator and a connection to a service circuit that returns a busy tone.

(3) Seizure, or request for service, which is indicated by an off–hook signal and the absence of any connection to another loop or trunk.

(4) Disconnect, which is indicated by an on–hook signal in the presence of a connection to a trunk or another loop.

(5) Wink–start, which is indicated by an off–hook signal on a trunk from the called office after a connect signal is sent from the calling office.

Historically, the terms on–hook and off–hook were derived from the placement or removal of a receiver from the hook on a candlestick telephone. With present day technology, switch–hook contacts may no longer be used, but the terms off–hook and on–hook are still in use and indicate circuit in use (or request for service) or not in use (or disconnect), respectively. These two conditions produce supervisory signals on the station loop. Therefore, if the station telephone set is conditioned for on–hook, the circuit is idle; if it is off–hook, the circuit is busy. The terms are so descriptive that they are commonly applied to trunks as well as loops. Supervisory signals must be extended over a connection to the point at which billing information can be used by message accounting equipment. Details of how these signals are used are beyond the scope of this chapter and may be found in other documents [1].

Alerting signals are those whose primary function is to alert an operator, a customer, or station terminal equipment to some need, such as an incoming call. Included in this group are such signals as ringing, flashing, recall, rering, and receiver–off–hook (ROH) signals. The most familiar alerting signal device is the telephone bell.

Call progress signals include dial tone, system–generated announcements, audible ring, busy tone and special–identification tones. While many of these signals are normally transmitted at low enough amplitudes or are used infrequently enough that they have little effect on transmission, circuits that convey specific call progress signals require care in their design and maintenance. For example, machine announcement arrangements such as directory assistance, automated coin telephone service, and automatic intercept systems have been engineered so that the customer hears the announcement at about the same amplitude as an operator's voice. This avoids contrast and ensures a good overall grade of service. Also, in order to be compatible with acceptable transmission standards, the design of tone generators for dial tone, audible ringing, busy tone, etc., is controlled by a precise tone plan that specifies the frequencies and amplitudes of all such tones.

Control signals are used for auxiliary functions associated with equipment connections to the point of termination (POT). Typical examples are toll diversion and party identification. These signals are beyond the scope of this chapter and may be found in other documents [1].

Test signals are of many types. Some test signals appear on circuits when the circuit is not in use (idle), while other tests are made during the process of connecting or disconnecting a call. Most switching equipment can apply test signals automatically while repair personnel may apply test signals manually. In either case, design and use of test signals must not promote crosstalk or carrier overload, but provide accurate test results. They are not covered in detail in this chapter, but discussions of several types of test signals are found in later chapters.

The characterization of signals covered in this chapter is important from a transmission standpoint for a number of reasons. There is a great variety of such signals and some are used frequently, in large numbers, and for long periods of time. It is important to know their characteristics if they are likely to affect the transmission performance of other signals sharing the same facility or transmission medium. Furthermore, such signals sometimes have transmission requirements that are more stringent than other "payload" types of signal (e.g., limits on distortion of dial pulses) and, as a result, may be a controlling factor in establishing overall design limits for transmission facilities. In addition, the signalling circuits interconnect with transmission circuits and may contribute to transmission loss and distortion. Finally, on loops, the signalling circuits affect the amount of current delivered to the station–set transmitter, and, if deficient, impair transmission volume and quality.

Incompatibilities between signalling and transmission circuits cause distortion of the address signals. These considerations must be recognized when designing interfaces between telephone companies and customers, or investigating problems. Pulse splitting, a serious form of mutilation that can make a single pulse look like two, is an example. It can occur in four–wire terminating sets as a result of spurious low–frequency oscillation caused by parallel resonance between a transmission capacitor and the inductance of the signalling relay. This type of problem can be corrected with, for example, a diode connected in series with the oscillatory elements to break up the low–frequency oscillations.

13-1 SIGNALLING ON LOOPS

Three aspects of signalling on loops are important from a transmission standpoint. These are supervision, addressing, and customer alerting. All of these aspects of signalling on loops are related to what is known as common–battery operation.

Common-Battery Operation

Most of the equipment associated with an individual telephone switching system is operated from a single large centralized battery. Current supplied from such a battery to the loop connected to the switching system is modulated by speech in the transmitter to form the speech signal electrically on the loop. The same battery is used to implement signalling functions that must be provided from the station equipment toward the switch.

One type of connection of loops to a common–battery supply in an electromechanical switching system is depicted in Figure 13–1. Three loops and station sets are illustrated with the *tip* loop conductors (T) connected to the grounded positive–side bus bar of the battery.* The *ring* conductors (R) are connected to the ungrounded negative side of the battery. The repeat coils (or transformers) and capacitors in each of the battery feed circuits electromagnetically couple the transmission path from the loops into the switches to complete connections to trunks or other loops. Another circuit configuration commonly used as a battery feed circuit is known as the bridged–impedance–type circuit. This circuit, shown in Figure 13–2, couples the loop to the switches by capacitors rather than repeat coils. The line (L) relay shown in Figure 13–2 for an electromechanical switching system can, for illustrative purposes, be replaced by ferrod sensing devices employed in some electronic switching systems.

Supervision on Loops and PBX–CO Trunks

During various stages of a call (request for service, dial tone, dialing, connecting, ringing, talking, one–way or two–way communications, etc.), battery and ground are supplied to the loop or PBX–CO trunk (the term PBX identifies a private branch exchange or any customer switching system connected to the loop at the POT) by a circuit somewhat like those illustrated in Figures 13–1 or 13–2. The battery supply may be a different circuit for each stage of the call and may be different for an incoming or outgoing call. Furthermore, while idle loops always have negative battery on the ring conductor, the battery–ground connections may be reversed during the progress of setting up the call with possible short intervals of no battery, called open switch interval

* Digital switching systems typically use a floating line supply battery where neither side of the battery is grounded [1].

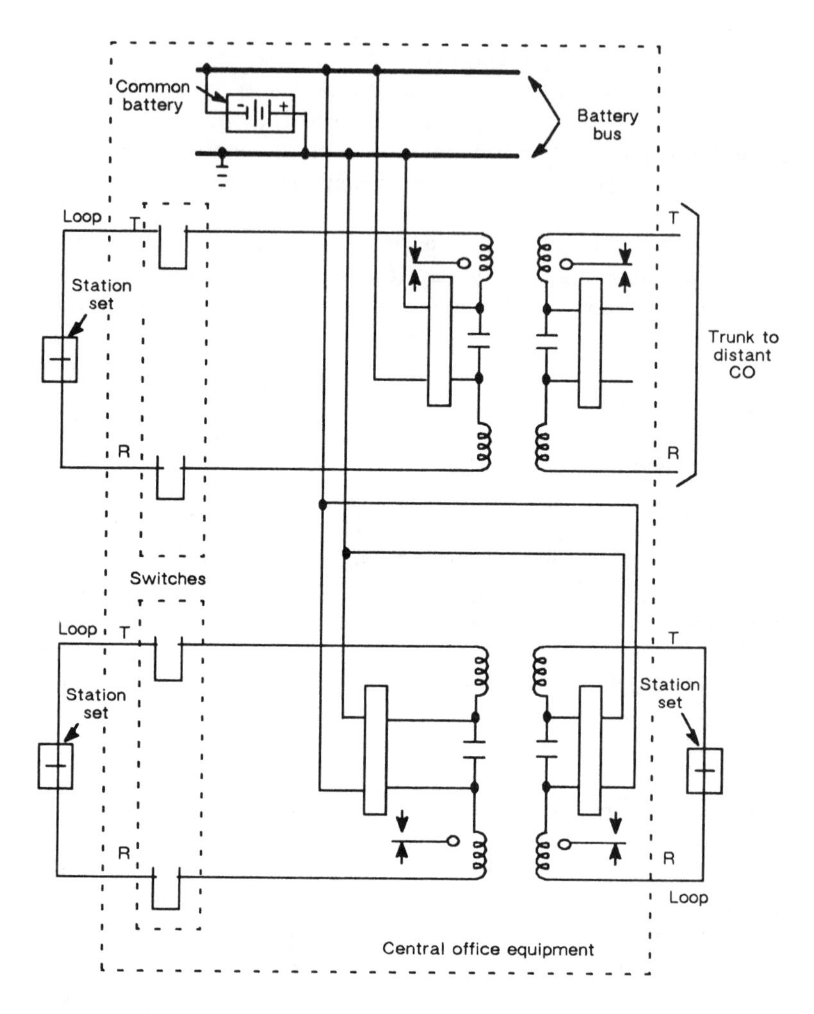

Figure 13–1. Common–battery connections—repeat coil circuits.

(OSI).* Also, in the talking condition, either calling or called party loops may have battery polarity reversed, particularly when served by a step–by–step switching system. Each battery supply

* Generally, OSIs are less than 350 ms, with more than 100 ms between OSIs.

circuit must include a relay or scanning device that can respond to changes in the signalling or supervisory condition on the loop and, in responding, extend the information regarding the changed conditions to monitoring circuits.

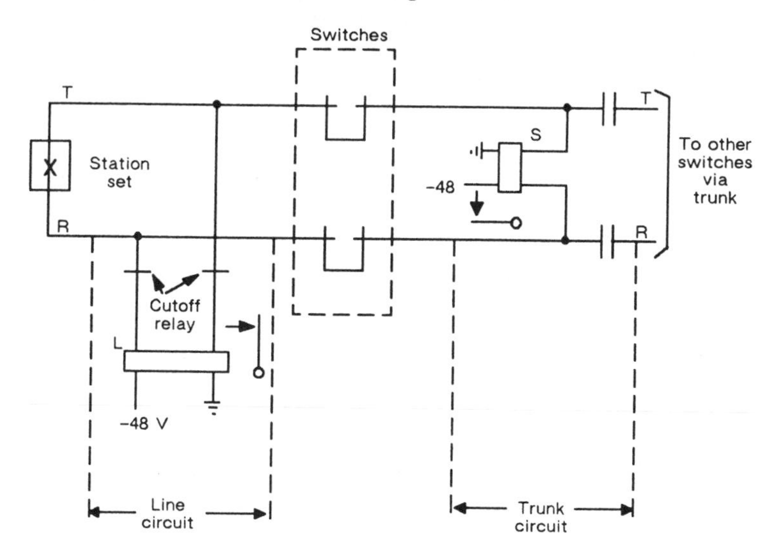

Figure 13–2. Transfer of loop supervision—bridged impedance battery feed.

Figure 13–2 illustrates the process for a normal *loop–start* outgoing call. When the station set is on–hook, battery and ground are connected to the station loop conductors through the windings of the line relay in the loop circuit and the closed cutoff relay contacts. Operation of the L relay, caused by the flow of current through its windings when the station initiates a call by going off–hook, results in switching system operations that disconnect the L relay from the loop (by operating the cutoff relay) and eventually, after processing, connect the loop to a trunk circuit. Thus, during the first part of the call sequence, supervision of the loop is provided by the flow of current through the L relay; during the second part of this sequence, supervision is provided by the S relay in the trunk circuit through whose windings current is supplied to the loop.

It should be stressed that the circuits in Figures 13–1 and 13–2 and the sequence of operation just described are illustrative only.

Although many variations exist in different types of switching systems, the basic function of loop supervision is performed by circuits similar to those described.

The process just described is known as loop–start signalling. Another process used to initiate calls is known as *ground start*. In some cases, for example on two–way PBX–CO trunks, the calling sequence is started by applying ground to the ring side of the loop. In such an arrangement, the line relay or ferrod is wired to accept only this request–for–service signal and responds accordingly. It is used in this application in order to minimize the probability of simultaneous seizure of the trunk from both ends for an incoming and outgoing call.

The simultaneous seizure of the trunk from both ends, a condition called *glare*, would be a serious problem on PBX–CO trunks because there may be an interval of up to four seconds after an incoming call is connected to the trunk before a ringing signal is applied to the line. Additional time may pass until the incoming call is answered and the trunk is made busy at the PBX.

With ground–start operation, the trunk, while in the idle state, has no ground on the tip conductor. Upon seizure by the switching system, ground is applied to the tip conductor, a condition used immediately by the PBX to make the trunk appear busy to outgoing calls at the PBX; removal of the tip ground is recognized as a disconnect by the PBX. When a call is originated at the PBX, ground is placed on the ring conductor. When the switching system connection is established, the normal battery and ground connections are made to the ring and tip. Either state (ground on the ring or loop closure) is recognized immediately by the switching system equipment as a trunk seizure. This equipment later recognizes the opening of the loop as a disconnect signal. However, it is important to note that some station equipment cannot always detect a disconnect before the switching equipment connects a new call.

The parameters that enter into the calculation of loop supervision relationships include the resistance of the station terminal equipment, the resistance of the loop conductors, the resistance of the switching system equipment and wiring, the resistance of the battery supply circuit (nominally 400 ohms in most electromechanical switching systems), the sensitivity of the line relay or other device that must respond to changes in loop status, and the battery voltage itself. These parameters all vary within their respective ranges. The station terminal equipment resistance has manufacturing variations and, in addition, is often designed to be a function of the loop current. The resistance of the loop conductor depends on the distance between the station equipment and switching system, and the gauge of the loop wire employed. In addition, the resistance of the path through the switching system equipment is different according to the circuits used and the type of switching system, and must be accounted for along with manufacturing tolerances. Allowance must also be made for loop leakage currents.

The battery voltage has, in most switching systems, a nominal float value of −50 volts. Provision is sometimes made to increase the supply voltage up to 100 volts when range extension equipment is used to increase loop length. Technological advances in battery feed design, such as in digital switching, provide constant current to station equipment.*

The large number of variables involved in supervisory signal computations makes it necessary to apply a set of rules that can be used universally to determine if signalling or some other function limits loop performance. One such rule for laying out loop plant is that the conductor loop resistance must be equal to or less than 1300 ohms (in most cases, loop resistance may exceed 1300 ohms for signalling). The 1300–ohm limit has been established to assure adequate transmission performance. Some switching systems are capable of operating with a 1500–ohm supervisory range or more. In this case, 1500 ohms is the acceptable range without transmission gain devices. Other rules apply to loading, allowable number of bridged taps, etc. Signalling limits must be determined for each case. However, as digital loop carrier (DLC) becomes the preferred facility for long loops, special treatment of long loops becomes less important.

---

* Some station equipment can adjust its resistance to maintain loop current at a nominal value (26 ma.) for suitable transmission performance.

Address Signalling on Loops

Two modes of generating address signals are used by station terminal equipment. These are dial pulsing and dual–tone multifrequency signalling. They are described in detail because different transmission problems are related to each.

Dial Pulsing. Address signalling occurs when a rotary dial is moved to an off–normal position and then released. Many state–of–the–art telephone sets and terminal equipment employ electronic dial–pulse generators. The address signals consist of pulses that result from the interruption of the loop current by the pulsing contacts of the dial or pulse generator. The number of pulses, determined by the calibrated off–normal position of the dial or the depressed pushbutton, corresponds to the digit dialed. The switching system equipment responds to the dialed digits to establish the desired connection.

Timing relationships are important in this process in a number of ways. Note first that the dial–pulse signals differ from supervisory signals on the loop only with respect to timing. On–hook and off–hook supervisory signals are of long duration while dial–pulsing signals are measured in fractions of a second. The process of transferring address information from the station dial to the switching system equipment depends on these timing relationships and on the design of the dials, the switching equipment, and the loops. The number of ringers bridged on the line can also affect address signalling.

Some basic time relationships are shown in Figure 13–3, where the digits 2 and 3 are assumed to have been dialed sequentially. The first of these time relationships is illustrated by the first pulse in Figure 13–3. The complete pulse cycle is made up of a *break* interval during which the pulse contacts of the dial are open, and a *make* interval during which the pulse contacts are closed. The two intervals are expressed as *percent break*.

$$\text{Percent break} = \frac{\text{break interval} \times 100}{\text{break} + \text{make intervals}} .$$

Switching equipment normally accepts dial pulses that have a percent break from 58 to 64 percent.

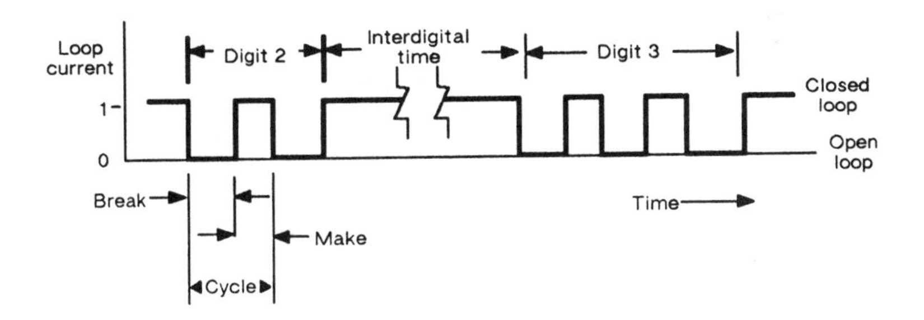

Figure 13-3. Dial pulsing of digits 2 and 3.

The second time relationship of importance is the *pulse repetition rate,* or number of pulse cycles per second that can successfully be transmitted and recognized. Most station terminal equipment is designed to operate at ten pulses per second (pps). While many parameters influence the maximum, the pulse rate of these dials was originally set by the operating speed capabilities of step–by–step switching systems and auxiliary dial repeating equipment.

The two timing relationships given so far, percent break and pulse repetition rate, are governed largely by the operate and release characteristics of switching equipment as they are affected by the loop characteristics. The pulse waveforms of Figure 13–3 are highly idealized. As illustrated in Figure 13–4, impedance characteristics of the loop, station equipment, and ringer circuitry cause distortion of the pulses that must be taken into account when signalling problems are being considered. The dashed–line pulses in Figure 13–4 are again highly idealized; the solid–line pulses show how one form of distortion (caused by charge and discharge of ringer and cable capacitance) causes changes in the percent break of the repeated pulses. Margins for such pulse distortion must be provided in the design of switching system registers.

The third timing relationship in dial pulsing is shown in Figures 13–3 and 13–4 as the *interdigital time.* This is the time from when the loop is closed after a digit has been dialed until the first pulse of the next digit. It includes the time required by the customer to search for the next digit, to pull the rotary dial around to its stop, and to release it (or the electronic timing equivalent in equipment using dial–pulse generators) to start the pulsing of the next digit. The switching system equipment must contain timing circuits to recognize this interval with margins for pulse distortion caused by the loop and other equipment.

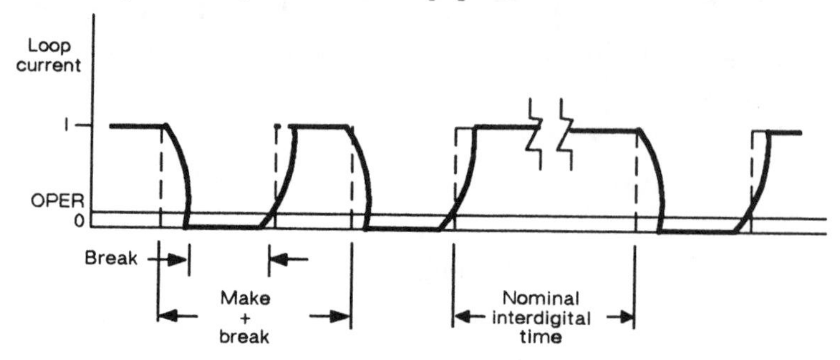

Figure 13–4. Effect of pulse distortion on dial–pulse time relationships.

Dual–Tone Multifrequency Pulsing. A second form of address signalling used on station terminal equipment is implemented by a set of pushbuttons rather than a rotary dial. Other telecommunications equipment, such as computer modems, generate the pulsing tones directly. This form of signalling, called dual–tone multifrequency pulsing, is superior to conventional dial pulsing because it is more accurate, more convenient, and faster (ten digits per second maximum). Operation of any pushbutton results in the generation of two SF tones that may be transmitted either for as long as the button is depressed or for a fixed length of time (i.e., 100 ms). Oscillators within the station equipment pad, activated by pushbutton operation, are powered by the line current supplied over the loop. While the button is depressed, the telephone transmitter circuit is opened, and a resistor is inserted in series with the receiver so that the tones are heard in the receiver at a comfortable sound level.

The layout of the standard 12–button DTMF matrix pad and the frequencies generated by each button are depicted in Figure 13–5. If the number 7 pushbutton is depressed, for example, the 1209–Hz and 852–Hz frequencies are generated. Switching system receivers, different from those used to receive dial–pulse signals, recognize these tones as representing the digit 7. This equipment, called DTMF registers, converters, or receivers, translates the DTMF tones to digital signals to control the switch.

The pushbuttons marked * (star) and # (number) are used for certain specialized signalling functions, such as speed calling or call forwarding.

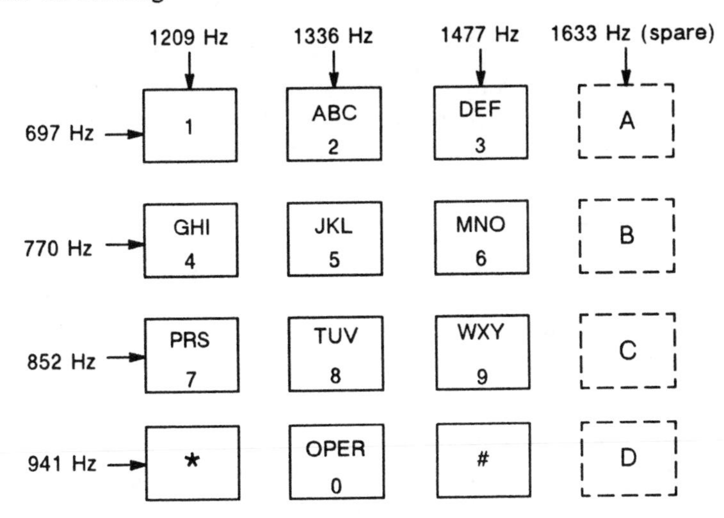

Figure 13-5. Pushbutton layout on a DTMF station equipment pad showing signalling frequencies.

The signals in the low-frequency group, 697 to 941 Hz, are transmitted nominally at -6 dBm; those in the high-frequency group, 1209 Hz to 1477 Hz, are nominally transmitted at -4 dBm. The actual amplitudes depend on the amount of loop current. Electronic switching systems can provide a DTMF transmitter for outpulsing to a PBX. These transmitters generate -7 dBm0 for each of the 12 frequencies used for DTMF pulsing. These relatively high signal amplitudes, and the fact that this type of signalling is not as susceptible to distortion caused by the medium as are dial pulses, make the design of DTMF receivers relatively straightforward. DTMF receivers are typically designed to accept DTMF signals with a power level of -25 to 0 dBm. Although these amplitudes are higher than those of many other signals transmitted in the voiceband, they are considered acceptable because they have a low duty cycle; i.e., they are transmitted only occasionally and are short. Since DTMF signals fall in the voiceband, they may be transmitted through exchange and interexchange carrier networks. Thus, they are capable of sending not only address information, but also calling-party information and control signals, and can be used as a form of data communications.

Alerting Signals on Loops

Two types of alerting signal transmitted toward the station terminal equipment are considered here. They are the ringing signal and the receiver–off–hook signal used to alert the customer that the station receiver has been left off–hook.

Ringing. Conceptually, the alerting signal used to ring the station terminal equipment, whether a bell or other alerting circuitry, is simple. However, details of signal generation, coding for party–line operation, and variables that may affect the ringing process make a conceptually simple process rather complex in practice.

The ringing signal is used mainly on loops, although 20–Hz signalling is used as a ringback and ring–forward signal on some older types of trunk. On loops, it is usually applied at the switching office* as a composite ac and dc signal called superimposed ringing. The ringing signal ac component has a voltage of between 88 and 105 volts and a frequency of 20 Hz. The dc component of the ringing signal may be of either polarity with respect to ground. The dc component, also connected to the loop during the silent interval, provides immediate tripping of the ringing. As a result, the customer does not hear 20 Hz when answering the call, and the called party is connected to the calling party without delay.

In some older switching systems, an ac component of about 420 Hz is superimposed on 40 Hz to provide an indication (audible ringing) to the calling party that the called party has been reached and that ringing has been started. Today, precise audible ringing consists of 440 plus 480 Hz at a level of −19 dBm0 per frequency.

It would, of course, be possible to ring the subscriber's station equipment bell continuously until the call is answered. Early

* One exception, for example, is in DLC systems where the ringing signal is applied to the loop at a terminal remote from the switching system location.

tests, however, indicated that continuous ringing would be undesirable and irritating. The standard switching system ringing cycle has been set as a two–second ringing interval followed by a four–second silent interval. This cycle is sometimes modified to provide coded ringing to alert a particular subscriber on a party line. The ringing cycle used at PBXs and some centrex services is a one–second ringing interval followed by a three–second silent interval. There are other special ringing patterns in some PBXs and centrex customer calling services called distinctive ringing.

It is desirable to set the magnitude of the ringing signals as high as possible in order to maximize the length of the loop over which station equipment operates satisfactorily. However, since the telephone plant is designed generally to operate at low currents and voltages, the maximum ringing–signal voltages are limited to values that do not operate protective devices, cause dielectric failure or overheating of equipment, or present a hazard to personnel.

The station ringer circuit may be connected to the loop in a number of ways, depending on the type of service. On individual lines, the ringer, usually consisting of a capacitor and bell or their electronic equivalent, is connected across the line as illustrated in Figure 13–6(a). With the types of high–impedance ringers presently in use, up to a total of five* ringers can be connected in parallel to the line illustrated in Figure 13–6(a). The number is limited by ringing and dial–pulsing requirements. Ringing ranges vary with the number of ringer circuits connected, the capacitance of the signalling equipment, loop resistance, and the ringing voltage supplied.

For party–line service, other ringer connections are required. The types of service include two–party [Figure 13–6(b)], four–party, and eight–party in some suburban and rural areas. Full selective ringing (only the called party bell rings) can be provided on two–party and four–party lines. Four–party selective ringing requires ringer isolator arrangements poled to ground and ringing current with correctly poled superimposed dc voltage. Semiselective ringing (where only a limited number of parties hear each

---

* *FCC Rules and Regulations, Part 68.*

ring) is provided on some four–party lines and all eight–party lines.

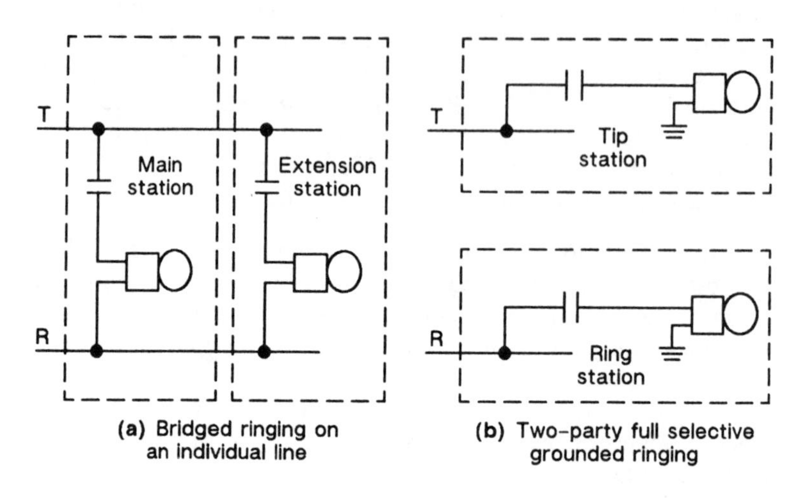

Figure 13–6. Two common types of ringer connection.

As shown in Figure 13–6(b), party–line ringing often involves ringer connections between one side of the line and ground. Due to unbalances caused by different numbers of ringers on each side of the line or very different loop lengths to each, such lines may be relatively noisy and may cause induced 60–Hz noise due to interference currents. In these cases, it may be necessary to use ringer isolators, which balance the lines so that induced currents are not converted to excessively large interferences.

Receiver-Off-Hook Signal. When a station set is left in the off–hook condition, a tone may be applied to the loop to attract the attention of someone at the station to this condition. The ROH tone in an analog switching system is made up of a combination of 1400, 2060, 2450, and 2600 Hz at +5 vu. When applied automatically after a recorded announcement, the ROH signal appears on the loop for up to two minutes, interrupted at a rate of five times a second, 0.1 second on and 0.1 second off. The ROH signal in a digital switching system is made up of the same tones at an equivalent power level of +3.0 to −6.0 dBm per frequency. The ROH signal can also be applied from the local test position as a continuous or interrupted signal.

13-2 SIGNALLING ON TRUNKS

While there are significant differences in detail, most of the same general functions of signalling must be accomplished on interoffice trunks as on loops. These functions include addressing, supervision, alerting, transfer of information, and testing. As may be expected, many of the characteristics of signals used on trunks are similar or identical to those used on loops.

Signals that relate directly to station operation, such as ringing and ROH signals, are not used on interoffice trunks.* On the other hand, the types of switching system that must be controlled and the functional characteristics of the trunks themselves are so diverse that the variety of signals used on trunks is considerably greater than on loops. Two general types of signal are described; they are classified as dc or ac signals. Under each type there are many variations.

The address information required to route a call must be forwarded from the originating switching system either directly to another switching system, through one or more intraLATA (local access and transport area) tandem switching systems, or through an interLATA POT to the terminating switching system.

In general, dc and digital signals are used within the switching systems. Such signals are often unsuitable for transmission over trunks, and it is necessary to transform the signals at one end of a trunk to a form more suitable for transmission and then back to the original form at the other end of the trunk. If the trunk length exceeds the range limits of the dc system or if the trunk cannot pass dc, then ac or derived dc techniques must be used. These conversions require equipment that is described elsewhere.

In the case of modern digital switching systems, trunking between systems normally operates on a straight 24-channel digroup basis. No trunk relay circuits are needed at either end; signalling is contained in the digital bit stream or is handled by an external common-channel signalling system. Thus, signalling is greatly simplified.

---

* In some cases, ring-forward and ringback signals are transmitted over operators' trunks to the local switching system where they are then applied in appropriate form to the loop.

Most signalling systems in older offices, other than loop signalling, are separate from the trunk equipment and functionally are normally located between the trunk equipment and the line facility. One form of signalling interface is used frequently for both dc and ac signalling. The name E and M lead signalling is derived from lead designations historically used on circuit drawings. The E and M lead interface between a signalling path on a transmission facility and a trunk circuit is shown in Figure 13–7. Traditionally, the E and M lead signalling interface consisted of two leads between the switching (trunk) equipment and the signalling equipment. The circuit conditions on the E and M leads are standard in all systems employing this method of connection: the M lead, which carries signals from the switching (trunk) equipment to the signalling equipment, and the E lead, which carries signals from the signalling equipment to the switching (trunk) equipment. As a result, signals from office A to office B leave on the M lead and arrive on the E lead of office B. In the same manner, signals from office B leave on the M lead and arrive on the E lead of office A. (Later systems use paired leads rather than single–wire leads to reduce noise interferences; these applications involve some departures from the simple E and M lead circuit conditions.) Systems that employ E and M leads have the advantage that signals can be transmitted independently in both directions on a trunk.

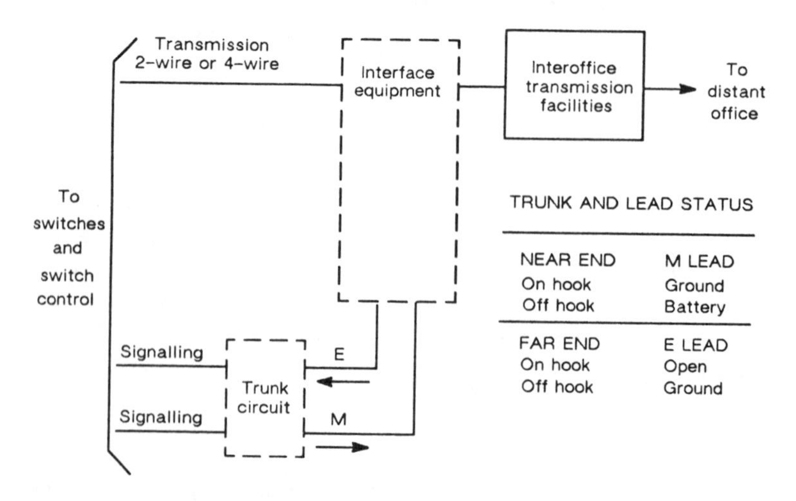

Figure 13–7. E and M leads.

DC Loop Signalling on Trunks

Since the transfer of supervisory signalling information is most economically accomplished by dc signalling, such methods are used whenever feasible. There are two forms of dc signalling. The first is called loop signalling, a name that is derived from the fact that a dc circuit, or loop, is available between the two ends of a trunk. (It is not related to the loop that connects a station with a switching system.) The second form of dc signalling, called derived dc signalling, is discussed subsequently. One or the other of these dc signalling arrangements is used for all trunks that operate with voice–frequency facilities and that are short enough to permit their application.

The dc loop signalling systems operate as a series circuit generally by altering the direct current flow in the trunk conductors. The loop signalling feature is usually combined with other apparatus in a trunk circuit. It provides one signalling state when it is open and a second when it is closed. At one end of a trunk, the current may be interrupted or reversed in polarity. Combinations of open/close and polarity reversal are used for distinguishing those signals intended for one direction of signalling (e.g., DP signals) from those intended for the opposite direction (e.g., answering signals). These changes are detected by relays or ferrods at the other end of the trunk. The signalling systems are known as reverse–battery, battery and ground, and wet–dry.

Signals cannot be transmitted in both directions independently in dc loop systems. Thus, such systems are used on one–way trunks, primarily on local and tandem–connecting trunks.

Reverse–Battery Signalling. Because of its economy and reliability, this is still used as a dc loop signalling method on local trunks. Battery and ground for signalling purposes are furnished through the windings of the A relay at the terminating end of the trunk as illustrated in Figure 13–8. Supervision is provided at the originating end of the trunk, usually by opening (on–hook) or closing (off–hook) contacts in the trunk transmission path under the control of the originating station terminal equipment through relay S1.

At the terminating end of the trunk, supervision is provided by the station terminal equipment and relay S2. Normal battery and ground are connected to the trunk conductors for the on–hook signal and are reversed by operating the T relay to represent the off–hook condition. The CS relay in the originating trunk circuit then operates on the reverse battery to start billing equipment. Address signals may be under the control of the calling station or under the control of the dial–pulsing equipment in the originating switching system.

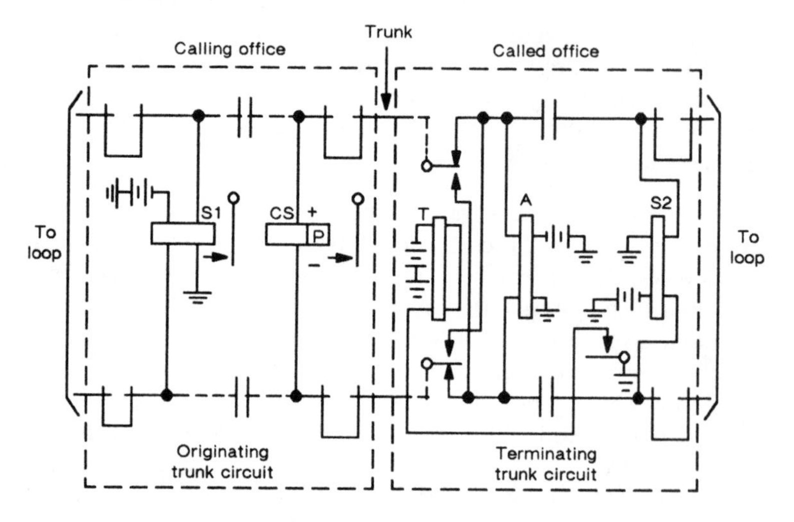

Figure 13–8. Reverse–battery signalling.

Battery and Ground Signalling. This mode of signalling was used to extend the range of loop signalling. It was accomplished by connecting battery and ground at both ends of the loop in a series aiding configuration. This type of connection, illustrated in Figure 13–9, was provided only during the period that addressing information was being transmitted.

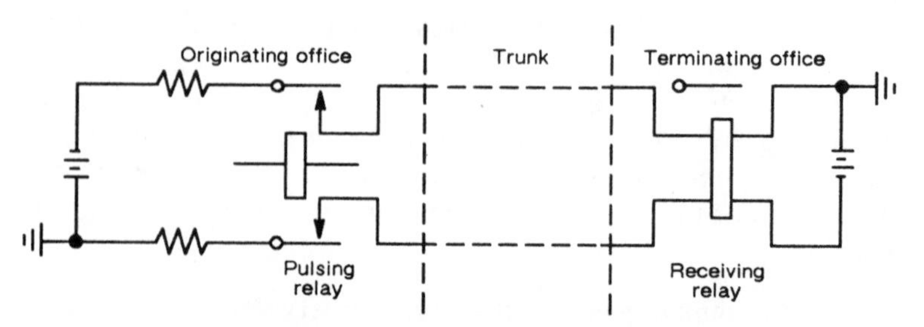

Figure 13–9. Battery and ground pulsing.

Miscellaneous DC Loop Arrangements. A number of other dc loop signalling arrangements were used to provide address or supervisory signalling information on voice–frequency trunks. Most were replaced by the reverse–battery or battery and ground systems previously discussed, and now by signalling in the bit stream. They have so little impact on transmission problems that detailed discussion here is not justified. These include wet–dry signalling and high–low signalling. Both of these signalling methods use dial pulsing or ac signalling for the transmission of address information and may thus be considered primarily as supervisory systems. Wet–dry signalling provides dc loop supervision in the form of the presence or absence of battery and ground. The trunk is "wet" (battery and ground connected) for one set of supervisory states and "dry" (battery and ground disconnected) for the opposite. High–low signalling refers to the impedance bridged across the trunk, with high impedance being used for one set of supervisory conditions and low impedance for the other. This method is still used occasionally to provide supervision at the originating ends of reverse–battery signalling systems.

Revertive Pulsing

Revertive pulsing was developed to satisfy the signalling needs of panel–type switching systems. It was later adapted for use for certain crossbar switching systems because it is capable of operating somewhat faster (up to 22 pps in crossbar and 32 pps in panel) than more conventional dial–pulse systems and because the crossbar systems, designed to replace the panel, had to interconnect with existing panel switching systems. This mode of signalling is disappearing from use. Revertive pulsing had little effect on transmission and its detailed operation will not be covered in this chapter.

Derived DC Signalling on Trunks

Derived dc signalling paths were used extensively at one time for some long local trunks and short–haul toll trunks where a complete dc loop was not available or where extended ranges were desired for dc signalling. In these cases, dc signalling paths are sometimes derived from the transmission path which, of course, had to be a physical facility. Derived systems use E and M lead connections and, as a result, may be used on one–way or two–way trunks. The type of derived path most recently found in use is duplex (DX), which may be used to transmit supervisory and address signals.

Duplex Signalling. DX signalling, illustrated in Figure 13–10, is based on the use of a symmetrical and balanced circuit that is identical at both ends of the trunk. Signalling and transmission are separate functions, but they use the same transmission path. One wire of the trunk conductor pair is used for signalling and the other for ground potential compensation. Its chief advantage is that it can operate on circuits having loop resistance higher than can be tolerated by other systems (i.e., up to 5000 ohms).* However, DX signalling is fading from use as long–wire facilities are replaced by digital facilities and equipment.

AC Signalling on Trunks

Use of ac signalling is common with analog carrier systems. It is theoretically possible to transmit ac signals for address and supervisory information at any frequency in the voiceband, defined as approximately 200 Hz to 3500 Hz. Present inband voice–frequency signalling systems operate in the range of 500 to 2600 Hz. Two are commonly used at present: the 2600–Hz SF signalling system and the multifrequency pulsing system. Each system has an application individual to its function (i.e., inband signalling–supervision, MF–called number).

2600–Hz Signalling. This system is commonly referred to as SF signalling. The system may be used to transmit address and supervisory signals in both directions on trunks, particularly between electromechanical switching systems.

One of the design considerations in voice–frequency signalling is the prevention of mutual interference between transmission and signalling systems. Voice–frequency signals are audible;

* DX signalling became a popular mode of signalling in the long–haul special service arena due to its extended range over voice–frequency cable facilities.

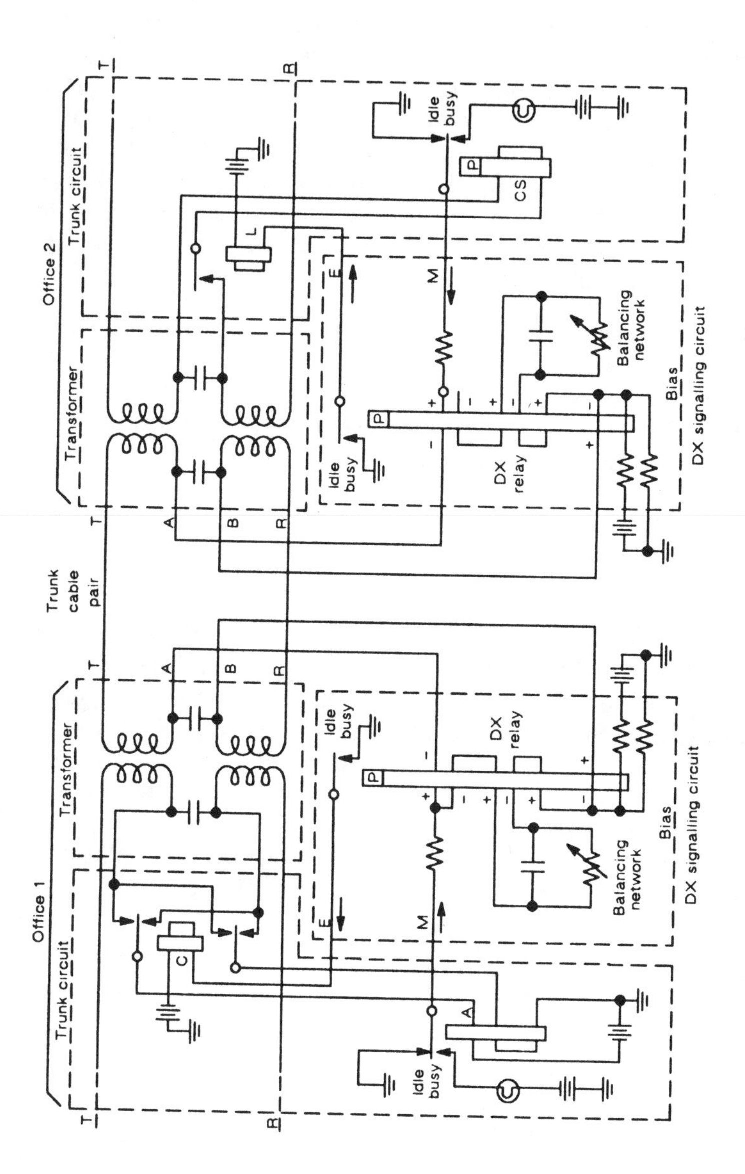

Figure 13–10. Duplex signalling system.

consequently, signalling must not take place during conversation. In most applications of this type of system, the presence of a 2600–Hz signal corresponds to an on–hook condition and the absence of 2600 Hz corresponds to an off–hook. Thus, there is no tone normally present on the line during conversation. Signal–receiving equipment, however, must remain connected during conversation in anticipation of incoming signals and may operate falsely because of speech signal components that resemble the tones used for signalling. Several methods are used to protect against this:

(1) Frequency and level were chosen as not likely to occur in normal speech.

(2) Time delay is used in the signalling circuits so that normal speech currents are ignored.

(3) Speech signal energy, when detected at frequencies other than 2600 Hz, is used to inhibit operation of the circuits in the signalling receiver.

This system, illustrated in Figure 13–11, may be used to signal independently in both directions on four–wire facilities. When a 2600–Hz SF signal is being transmitted to reflect the on–hook, steady–state load supervisory condition, it is applied to the voiceband transmission medium at an amplitude of −20 dBm0 (e.g., −36 dBm at a −16 TLP). When being used to transmit address information, however, the 2600–Hz signal is increased in amplitude by 12 dB to −8 dBm0. Such a high signal amplitude is permissible because of the short duration of the dial pulses and because of the low probability of large numbers of such high–amplitude signals being simultaneously present in a transmission system.

There are a number of SF signalling characteristics that interact importantly with analog transmission systems. Two conditions cause problems: (1) a majority of trunks using a carrier system may be equipped with SF signalling, and (2) most of these trunks may originate at the same central office. In the latter case, there may be high coherence in the relative phases of the many 2600–Hz signals. As a result, the way in which these signals combine may cause overload or excessive peaks of intermodulation noise at certain frequencies and at unpredictable times. In such situations, action must be taken to break up phase coherence among 2600–Hz signals in different channels.

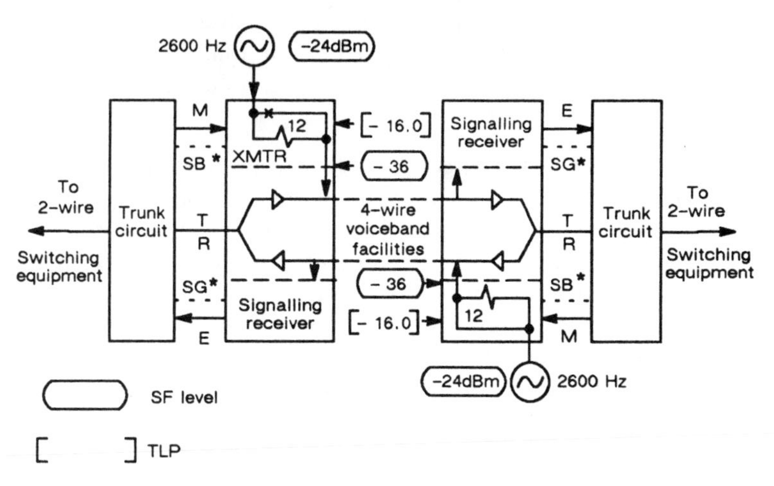

Figure 13–11. Single–frequency signalling.

Furthermore, where large numbers of trunks in a carrier system employ SF signalling and terminate in the same office, serious disruption of the switching system operations can occur as a result of carrier system failure. If most of the trunks are in the idle condition, the carrier system failure causes the sudden interruption of all of the 2600–Hz signals. This is interpreted by the switching system as simultaneous calls for service from many trunks. As a result, the switching system is momentarily overloaded until it can dispose of the disabled trunks. All modern carrier systems employ trunk conditioning circuits, called carrier group alarm (CGA), which causes all affected trunks to appear busy so they will not be seized during the outage. After carrier restoration, the conditioning circuits then remove the busy condition and restore the trunks to service.

A two–frequency adaptation of SF signalling was used for selective signalling in multistation four–wire private–line networks, such as might be used as an order–wire facility for carrier systems or for private–customer communications networks. For example, dial pulses were used to signal selectively any one of a maximum of 81 stations in one type of system. The dc dial pulses were converted at the customer's premises or at the serving office to a frequency shift format using 2400 Hz and 2600 Hz. Later systems, using DTMF signalling, replaced the dial–pulsing arrangements of early–vintage selective signalling systems.

Multifrequency Pulsing. The MF pulsing system consists of transmitting and receiving equipment for communicating information over telephone trunks. Signalling is accomplished by the transmission of two, and only two, of five frequencies in the voiceband. Each combination of two frequencies represents a pulse; each pulse represents a numerical digit (addressing data). An additional frequency is combined with one of the five frequencies to provide control functions. The MF pulses are sent over the regular talking channels and, since they are in the voice range, are transmitted as readily as speech. MF receivers detect the pulses and transfer the digital information to control equipment in order to establish connections through the switching systems and the telecommunications network. MF pulsing is commonly associated with the transmission of called–number information; however, it is also used to transmit calling number information in centralized automatic message accounting–automatic number identification (CAMA–ANI) operations and Feature Group D connections to ICs. In this case, the calling number is MF–outpulsed forward from the originating switching system to the CAMA switching system, whether the called number was originally sent by DTMF or dial pulsing. MF signalling is used for equal access where the called number is sent after the calling number.

The MF system transmits numerical information and control signals; therefore, another signalling system (e.g., SF, loop, bit stream, or DX) must be provided for supervision. Table 13–1 shows how the six MF frequencies are spaced 200 Hz apart and provide 15 possible two–frequency combinations. Ten combinations are used for the digits 0 to 9 inclusive, and one combination each is used for control signals indicating the beginning (KP) and the end (ST) of pulsing. The remaining three combinations are used for other control signals.

Table 13–1. Multifrequency Codes

Frequencies (Hz)	Digit and Control	SIGNALS		
		Expanded Inband	CCITT System 5	Equal Access
700 + 900	1		1	1
700 + 1100	2	Coin collect	2	2
700 + 1300	4		4	4
700 + 1500	7		7	7
700 + 1700		Ringback	Code 11	ST3P(ST"')
900 + 1100	3		3	3
900 + 1300	5		5	5
900 + 1500	8	Operator released	8	8
900 + 1700			Code 12	STP (ST')
1100 + 1300	6		6	6
1100 + 1500	9		9	9
1100 + 1700	KP	Coin return	KP1	KP
1300 + 1500	0	Operator attached	0	0
1300 + 1700			KP2	ST2P (ST")
1500 + 1700	ST	Coin collect Operator released	ST	ST

The principal advantages of MF pulsing are speed, accuracy, and range. MF transmitters operate more rapidly than DP transmitters. Therefore, MF saves holding time and reduces the number of transmitters and receivers in the switching system.

MF Transmitter. MF pulses represent the digits needed by a switch to build up a trunk connection. MF transmitters in electromechanical switching systems generally outpulse at a rate of 8.3 pulses per second. The nominal power output of MF transmitters is −6 or −7 dBm0 per frequency in older and newer equipment, respectively. The level of each of the two frequencies of a digit should be within 1 dB.

MF Receiver. The MF receiver is connected to a trunk as required. It does not respond to MF digit signals until it receives the KP signal. The MF receiver can then receive and pass on the number codes and the ST signal to its associated sender or other equipment. The receiver impedance matches that of the host switching system (i.e., 600 or 900 ohms in series with 2.16 μF). Further, the receiver responds to signal pulses with as much as 6–dB difference in power levels between the two frequency components.

13–3 OUT–OF–BAND/OUT–OF–SLOT SIGNALLING

Any signalling arrangement that uses frequencies out of the voiceband of the trunk over which signalling is taking place may be considered as an out–of–band signalling system. By such a broad definition, dc systems and a CCS system would be considered out–of–band. However, it has been convenient to discuss dc systems as a separate class. In CCS systems, signalling information is transmitted as data packets on a shared channel independent of the associated trunks involved in the connection.

Obsolete carrier systems (i.e., N1, O, and ON) used an SF signal at 3700 Hz associated with each voice channel, but above its nominal passband. Digital signalling arrangements, used in the T–type carrier systems, are considered out–of–band, or more accurately, out–of–slot.

Out–of–Slot Digital Signalling

In the coding of pulse code modulation (PCM) signals for transmission over digital carrier facilities, address and supervisory signals are assigned specified bits in the carrier bit stream. The carrier channel units contain the circuitry necessary to convert the digital signal to a form of dc signal (loop, E and M, etc.) required by the terminating switching equipment. Where digital switching is involved, the 24–channel trunk unit inserts and retrieves the signalling bits directly. In some early cases (e.g., the

D1A and D1B channel banks), this assignment of bits is permanent; as a result, a significant portion of the system's theoretical channel capacity is assigned to signalling (the eighth bit of the time slots assigned to a channel is used for indicating the on–hook state during a signalling sequence in a manner analogous to the transmission of 2600–Hz tone representing the on–hook state in inband signalling). In modern systems, the address and supervisory bits are time–shared in such a way that speech transmission is of higher quality than would otherwise be possible. The signalling bits are used for speech coding when not required for signalling. Digital systems are covered in greater detail in subsequent chapters.

13-4 SPECIAL–SERVICES SIGNALLING

Most of the discussion of signalling and the characterization of alerting, address, and supervisory signals that have been given in this chapter apply equally to signalling in the message telecommunications service (MTS) network and to special–services arrangements. However, there are some significant differences; some are due to the nature of special–services circuits themselves, and some are due to the way the special services are administered.

Service Demands and Plant Complexities

There are many services that compound signalling systems over various types of facility. These circuits must traverse parts of the plant in which a mix of trunk and loop plant occurs. Therefore, circuit layout and assignment of these circuits must consider the most economical and reliable end–to–end arrangements (e.g., SF might be used end–to–end over combinations of analog and digital facilities). A type of service that would merit this kind of consideration would be a foreign–exchange line whose station equipment is located in one central–office serving area but whose home switching system is many miles away. The final loop connection is from the local central office, but the loop must then be extended through cables or carrier systems normally used for interoffice trunks to the distant serving office. Another example is an off–premises extension from a POT at a PBX location that might require a transmission path involving connections through both loop and trunk facilities.

References

1. *Notes on the BOC IntraLATA Networks—1986*, Technical Reference TR-NPL-000275, Bellcore (Iss. 1, Apr. 1986), Section 6.

Additional Reading

Members of Technical Staff. *Engineering and Operations in the Bell System,* Second Edition (Murray Hill, NJ: AT&T Bell Laboratories, Inc., 1983), Chapter 8.

Members of Technical Staff. *Transmission Systems for Communications*, Fifth Edition (Murray Hill, NJ: AT&T Bell Laboratories, Inc., 1982), pp. 201–206.

Chapter 14

Data Signals

The transmission of data signals in telecommunications systems or networks involves the transmission of coded information between computers, between some switching systems, or between humans and computers. In some cases the transmitted information is coded into a digital form that is convenient to the operation of a device, such as a computer, and also convenient to the necessary interpretation by humans at the input and output of computer–like devices. In other, less common, cases the transmitted signal is coded as a direct electrical analog of the information, and digital encoding is not used. Thus, there are two important forms of data signal: *digital* and *analog*. Digital signals are those that can assume only discrete values of the parameter that is varied to convey information; analog signals can assume a continuum of values between given maxima and minima. A common application of digital transmission techniques requires *digital data* signal characterization for transmission over analog systems. However, there is a growing need to integrate the technologies of processing data and transmitting data in bulk and at higher speeds with efficiencies inherent with end–to–end digital transmission.

The increased use of computers in communication networks promoted the development and use of local area networks (LANs), the Digital Data System (DDS), wide area networks (WANs), and packet switching systems (PSSs) which provide fast and efficient data transmission. Of great importance today is the technology and architecture of the integrated services digital network (ISDN), which combines the various data communications and processing techniques into one coherent transmission and switching environment.

Digital data signals are transmitted at signalling rates that range from a few bits per second to millions of bits per second. The most commonly used rates over analog facilities, several thousand bits per second, are those compatible with voiceband circuits. In many private–line applications and in the message telecommunications service (MTS) network, transmission circuits that are normally used for voice communications are alternately used for voice and data. Generally, data signals must be converted to a format compatible with the network. However, ISDN will provide all–digital connectivity over channels that support digitized voice, data, and other digital information.

In many cases of interest here, data signals are those used by computers; they are usually binary signals transmitted serially on a pair of conductors, but they are seldom in a form convenient for transmission. Processing to transform a signal to a suitable format often takes place at two locations—first, at the data station equipment (modem*) to make the signal suitable for transmission over analog telephone loops, and then at carrier terminals to prepare the signal for transmission over a carrier system in a form suitable for modulating and multiplexing with other signals. The processing may involve special coding for error detection and correction. Each of the processes must be reversed so that the signal delivered at the receiving end of the circuit is a faithful replica of the signal accepted originally from the customer's equipment. These signals are similar to those described in Chapter 8.

Since many existing transmission systems were designed to provide analog facilities for the transmission of analog speech signals, the processing of a data signal for transmission over these systems must be such as to make the signal compatible with the transmission system. This compatibility involves loading effects, channel characterization, and intermodulation and signal–to–noise performance. Thus, the nature of the processes must be described here in detail.

Processing of data signals for transmission over digital transmission systems is not covered here because the coding is unique to each digital system and the operation of the digital system is not materially affected by the characteristics of the signals. ISDN

---

* The term modem is derived from data set functional operations of a *mo*dulator and *dem*odulator.

and DDS are covered in detail in Volume 3. On the other hand, the line signal of a digital transmission system (suitably processed) is sometimes transmitted over an analog transmission system. The analytic treatment of digital data signals and digital line signals is identical when they are transmitted over analog systems, and both must be characterized. Hereafter in this chapter, they will be generally referred to simply as digital signals.

14–1 VOICEBAND DATA SIGNAL TRANSMISSION CONSIDERATIONS

A number of considerations related to data signal transmission on analog facilities have had important effects on the design of signal formats. These include restrictions on signal amplitudes, signal–to–noise ratios and error ratios, and the relationships between signal and channel characteristics.

Signal Amplitudes

A number of criteria must be considered in setting the amplitude of a data signal using the MTS network or sharing facilities with it. One such consideration is that the power in the signal should not cause excessive intermodulation or overload in analog carrier systems where service to many customers may be jeopardized. The established requirement is that signals operating in the voiceband are nominally limited to −13 dBm0* (defined as the maximum allowable power averaged over a three–second interval). When the activity factors and other statistics of data transmission are accounted for (e.g., the number of operating half–duplex versus full–duplex channels), this value is equivalent to a long–term average power of −16 dBm0 per 4–kHz channel.

* For data on the MTS network, this value is equivalent to a nominal −12 dBm received from the loop plant at the serving central office. *FCC Rules and Regulations, Part 68* stipulates programmable or fixed–loss data jacks at customer premises with registered equipment to limit signal levels to between −12 and −13 dBm0 at the serving central office. The rules also specify a limit for signals connected to a voice jack at the customer premises. This is set at −9 dBm so that, with the average loop loss at approximately 3 dB, the resultant signal power at the serving office will be −12 dBm0.

The signal amplitude requirement for narrowband data signals, several of which may be multiplexed in a single voice channel, is also −16 dBm0, or a 3-second maximum of −13 dBm0, for the composite signal. The power of each individual signal must be sufficiently lower so that the total power in the channel does not exceed the objective.

For a wideband digital signal, one occupying more than a 4-kHz channel, the amplitude criterion is sometimes expressed differently, namely, that the signal power may not exceed the total power of the displaced channels.

Irrespective of its form or the bandwidth it occupies, one more constraint is imposed on a wideband digital signal. No single-frequency component may exceed an average power of −14 dBm0 [1]. This limit, established to avoid the generation of intelligible crosstalk intermodulation products in analog carrier systems, may sometimes be exceeded on the basis of low probability of occurrence or because of the short duration involved. Where danger of intelligible crosstalk exists, a scrambler or other means of reducing single-frequency components must be used [2].

Signal components at frequencies above the nominal band must be limited in amplitude to low values that cannot interfere with adjacent channels in a carrier system, or interfere through a crosstalk path with a wider-band signal or a cable carrier system that might share the same facility. For example, the limits of unwanted signal components are specified for voiceband signals at frequencies of 3995 Hz and higher [3]. In addition, it is required that the power in the band between 2450 and 2750 Hz not exceed that in the band between 800 and 2450 Hz in order to avoid interference with single-frequency signalling systems.

Error Ratio and Signal-to-Noise Ratio

Unlike speech or video signals, which must be evaluated on a subjective basis because of human responses to various types of signal impairment, digital signals are evaluated objectively. The quality of digital signal transmission is often expressed as a ratio of the number of bits received in error to the number of bits transmitted. This is called *bit error ratio** (BER) and is equated according to the number of errors in a given number of transmitted bits (e.g., one error in 10^{-6} bits or an error ratio of 10^{-6}). Maximum BER can be used as a maintenance tool to determine when a data channel is defective. The error–ratio limit varies according to whether the channel is switched or nonswitched and by the bit rate. It is sometimes convenient, however, to evaluate performance in terms of the signal–to–noise ratio because signal and noise amplitudes are easy to measure. When this is done, the noise characteristics must be specified; usually the Gaussian distribution (see Chapter 17, Part 3) is used because there is a definite and demonstrable relationship between Gaussian noise and error ratio.

While the effects of impairments other than noise (such as gain distortion or delay distortion) may also be expressed in terms of error ratio, transmission studies are often facilitated by converting the impairment into an equivalent signal–to–noise penalty. This is done by evaluating the error ratio for the impairment being studied and determining the reduction in signal–to–noise ratio that would produce the same level of errors in an unimpaired channel. The reduction in signal–to–noise ratio is called *noise impairment*. While noise impairments cannot be added directly to give an overall impairment or error ratio, the technique provides a convenient method of comparing the merits of one mode of transmission with another over a real channel.

There are a variety of methods to determine a figure–of–merit other than signal–to–noise ratio or error ratio. In practice, errors often occur in bursts that produce a high error density for only a short time; the remaining time may be error–free. Such a burst may cause a high apparent error ratio, yet have little effect on the efficiency of transmission. This condition can be translated into a figure–of–merit, called *burst error ratio*, which is the ratio of the number of received error bursts to the number of bits transmitted, where bursts are strings of bits beginning and ending with an erroneous bit and separated by 50 or more error–free bits [4].

* What had been known as a bit error *rate* is now considered internationally as a bit error *ratio* per *CCITT Recommendation G.381*.

Since data messages are usually transmitted in a block format, many data transmission systems reject blocks that contain bit errors and retransmit the entire block of data. In such a case, there is a figure–of–merit that involves determining the number of erroneous data blocks per the number of data blocks transmitted over a period of time. This figure–of–merit is known as the *block error ratio* (BLER) and is of greater practical significance than BER. BLER is the fraction of blocks in error, i.e., the number of blocks received with one or more bit errors divided by the number of blocks transmitted, where a block is a string of bits that is typically, but not necessarily, defined as 1000 bits long [4]. BLER is generally used as a reliability and maintenance tool on higher–speed data circuits.

Another concept of digital channel performance is determined by the fraction of one–second transmission intervals that contain one or more bit errors. This measurement of a channel or system is called errored–seconds. Many of these circuits are evaluated as to the number of *error–free seconds* (EFS) per fractions of a minute or hour. The EFS rate is the ratio of the number of seconds without erroneously received bits to the total number of seconds of transmitted bits [4], usually over a 15–minute period.

Channel Characteristics

The format into which a digital signal is to be processed for transmission on an analog channel must represent a compromise between maximizing the rate of information transmitted (bits per second per hertz of bandwidth) and minimizing the impairments due to extraneous noise or intersymbol interference. The transmission characteristics of the channel bear an important relationship to the design compromises that are made, as does the cost of the terminal equipment.

For a transmitted pulse to retain a rectangular shape, the bandwidth of the transmission channel would have to be very great (theoretically infinite). Bandwidth is expensive and, furthermore, the wide band would admit interference from noise and other perturbations appearing at frequencies outside the band, which contains the major portion of the signal energy. It is desirable, therefore, to curtail the signal spectrum as much as possible without undue impairment of the signal. Nyquist's criteria (1) and (2), defined in Chapter 10, Part 1, give important leads to how the band may be limited and pulses shaped to minimize errors at the receiver. There are several satisfactory ways of shaping the pulses by appropriate design of the channel characteristics [5]. One is the *raised cosine* characteristic; it has the virtues of meeting simultaneously Nyquist's criteria (1) and (2) in response to an applied impulse and does so without undue penalty in added bandwidth. It tends to produce less noise impairment than other channel characteristics and also can be closely approximated by straightforward design techniques. The raised cosine channel characteristic, near optimum for transmission of an impulse, requires some modification to accommodate commonly transmitted rectangular pulses.

Figure 14–1(a) shows an idealized channel characteristic, curve p. Curves r and s are modifications that follow cosine–shaped rolloff characteristics at the high end of the band. They are symmetrical about the frequency f_1 where their values are 0.5 relative to the values at zero frequency. These raised cosine channels are said to have Nyquist shaping. If the characteristic yields zero transmission at frequency $3f_1/2$ (curve r approximately), it has a 50–percent rolloff. If the characteristic yields zero transmission at $2f$ (curve s approximately), it has a 100–percent rolloff. Rolloff is thus defined as excess bandwidth expressed as a percentage of the theoretically minimum requirement.

If the bandwidth is extended beyond f_1 and the rolloff characteristic has Nyquist shaping, a linear phase/frequency characteristic can be closely approximated in practice. When these characteristics (cosine rolloff and linear phase) are provided, the zero amplitude crossings of output pulses resulting from applied impulses still occur at times corresponding to $\pm\, n/(2f_1)$. If the rolloff is extended to $2f_1$, the raised cosine pulses at the output have additional zero crossings at odd multiples of one–half the intervals that occur in the idealized channel transmission.

Rectangular pulses, transmitted through a channel having any of the characteristics in Figure 14–1(a), cannot be readily detected because excessive intersymbol distortion occurs. As discussed in Chapter 6, Part 2, the $(\sin x)/x$ channel response results from the application of an impulse, a signal that has a flat energy distribution. To make a channel respond in the desired fashion to rectangular pulses—that is, so that output pulse waveforms have the desired (sin x)/x format—it is necessary that the spectrum of the applied rectangular pulses be modified to approach the flat spectrum of an impulse. The normal spectrum of the applied rectangular pulse has a (sin x)/x spectrum. To make the spectrum appear flat, the pulse must be multiplied by the inverse function, $x/\sin x$.

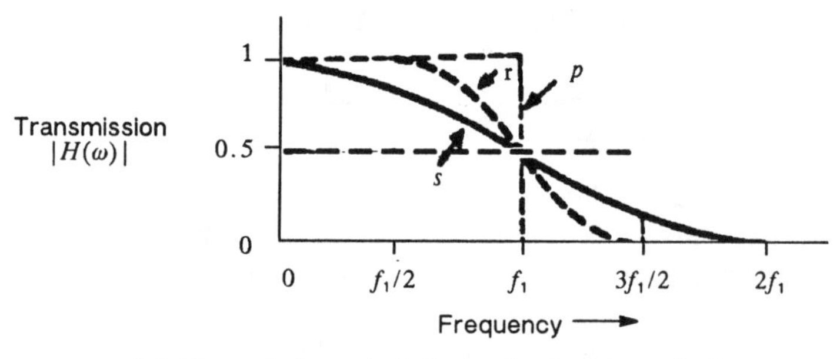

(a) Channel characteristics, raised cosine rolloffs

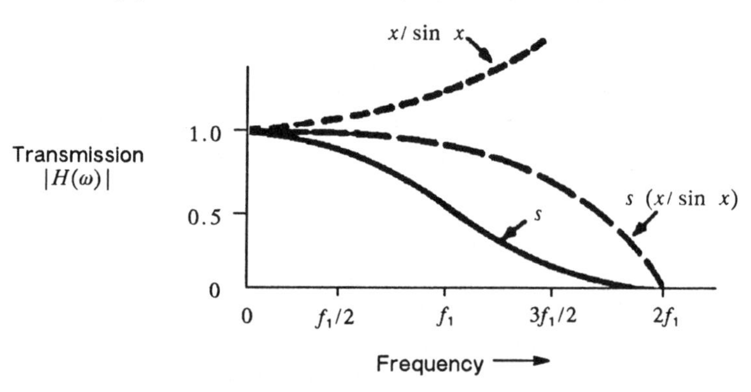

(b) Modified raised cosine characteristic

Figure 14–1. Channel shaping.

The desired modification of the signal can be accomplished by modifying the raised cosine channel by an $x/\sin x$ function. In practice, only the first lobe of the $x/\sin x$ function need be considered. Figure 14–1(b) illustrates this spectrum. Since the (sin x)/x function becomes zero at $2f_1$, $x/\sin x$ theoretically becomes infinity at this frequency. It can be shown, however, that the product of the $x/\sin x$ function and the cosine function representing curve s also becomes zero at $2f_1$.

A stream of rectangular pulses having a 100–percent duty cycle ($\tau = T$) transmitted over a channel having a characteristic, $s(x/\sin x)$, like that shown in Figure 14–1(b), appears at the channel output as shown in Figure 14–2(b), where $T = 1/(2f_1)$. The time delay between the input, Figure 14–2(a), and the output, Figure 14–2(b), is ignored. Note that the $(\sin x)/x$ form of each output pulse is such that the zero crossings correspond to the sampling point of successive pulse intervals.

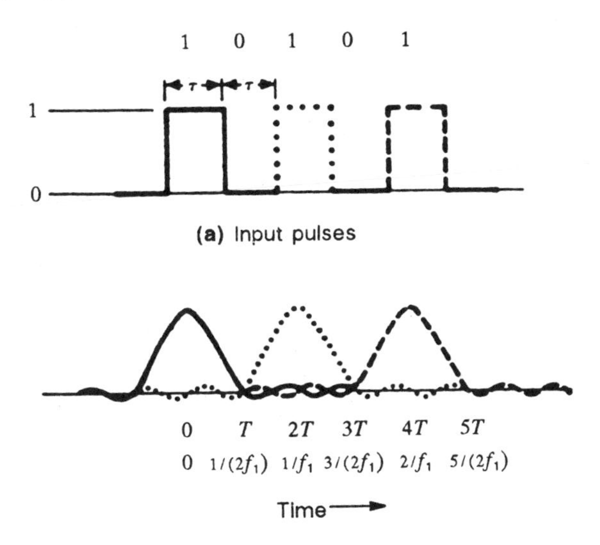

Figure 14–2. Effects of a cosine channel on pulse shaping.

The interval, $1/(2f_1)$, is known as the *Nyquist interval.* In channels having sharp cutoffs, the signalling interval, T, must closely approximate this interval in order to minimize intersymbol interference. With a 100–percent rolloff characteristic, larger departures from $T = 1/(2f_1)$ can be tolerated than in a channel having a sharp cutoff.

The shaping of the channel characteristic may be placed at any point in the channel. If the characteristic of the medium is predictable, its characteristic can be incorporated in the overall channel characteristic. Since there are a number of places where shaping may be used, the detailed effect on characterization of the transmitted signal cannot be generalized.

14-2 DIGITAL SIGNAL CHARACTERISTICS

A large number of digital signal formats have been used in telecommunications. An important stimulus to continued development, in addition to the burgeoning demands of the computer industry, is the desire to make data signal transmission more economical by increasing efficiency, i.e., by increasing the number of transmitted bits per second per hertz of available bandwidth at less cost per unit of information transmitted. Multiplexing techniques, either as a function of a modem or as a specific device, are also used to maximize the efficient use of channels.

Amplitude Shift Keyed Signals

Initially, the transmission of digital information was by amplitude shift keying (ASK) techniques in a binary baseband mode. This mode, basically that used to operate computers, is universally used within computers and telephone switching systems. Because of its simplicity, the ASK binary baseband mode is used for transmission of digital data signals over exchange carrier facilities, but only for relatively short distances. This signal format is neither as efficient for a given bandwidth as other formats nor is it suitable for transmission on facilities that provide no dc continuity or are subject to quadrature distortion, low-frequency cutoff, or significant envelope delay distortion. As a result, where these restrictions are important and signal-to-noise performance is adequate, equipment is installed to process the binary baseband signals into forms more suitable to the environment.

The nature of ASK signals is such that when they are used as baseband signals in the form of simple on-off pulses with an average amplitude of zero, low- and zero-frequency components are important to their characterization and to their recovery by detection circuits in the receivers. Because of their nature, such signals may be regarded as formed by a process of modulation of a direct current. The difficulties associated with transmission of zero– and very–low–frequency components through transmission facilities and networks are among the important reasons for the infrequent use of ASK signal transmission without additional processing. When an ASK mode is used, special provision must be made to eliminate the low– and zero–frequency components of the signal at the transmitter and at the outputs of regenerative repeaters and to restore these components at the repeater inputs and at the receiver. A few very early data modems used this technique for speeds of, for example, 750 or 1600 bits per second.

ASK Signal Waveforms. Digital symbols may be represented by any of a large variety of electrical signal formats. As previously mentioned, the control of computers usually involves the use of binary ASK signals. Logically, the operation of computers relates to the *0* and *1* representation of binary numbers, which in turn corresponds to the two states of a binary signal. Some alternate ways of representing these two states are illustrated by the formats shown in Figure 14–3.

The waveforms of Figure 14–3 have several features in common. First, all of the waveforms represent the same sequence of digits, namely, *0110001101*. Each of the waves depicted represents a synchronous system in which the receiving equipment is timed by a clock circuit so that the incoming signal is sampled at the instants indicated. The sampling is required in most cases in order to determine if the signal amplitude at the sampling instant is above or below one or more of the decision thresholds shown. In Figure 14–3(f), the sampling would take the form of a zero–crossing detector since, in that case, *0*s are represented by transitions in the signal and *1*s by no transitions. Finally, the peak–to–peak amplitudes of the signals are all shown as equal to two units of voltage *V*.

Figure 14–3(a) illustrates the simplest of these signal formats, called unipolar two–level non–return to zero (NRZ), which remains at a particular amplitude for successive symbols of the same type (*0* or *1*). A *1* is represented by the presence of a voltage, and a *0* by the absence of voltage. If the wave is between half and full amplitude at the sampling instant, it represents a *1*; if it is between zero and half amplitude, it represents a *0*. Thus the half–amplitude value is the decision threshold.

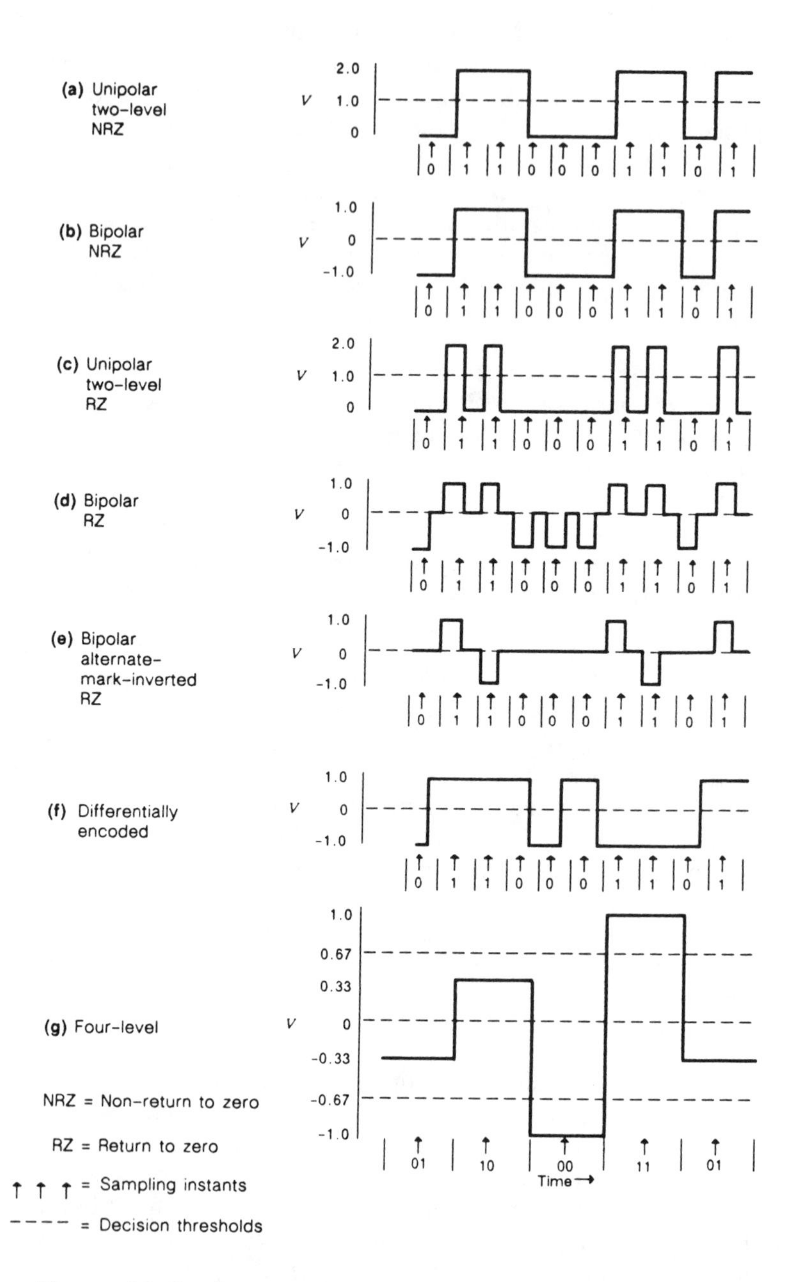

Figure 14–3. Representation of basic ASK waveforms.

Figure 14–3(b) shows a bipolar NRZ signal, which is similar in all respects to Figure 14–3(a) except that opposite polarities of voltage are used to represent *0* and *1* instead of voltage and no–voltage. The decision threshold in this case is zero volts. The bipolar form illustrated here is sometimes adapted to the transmission of asynchronous digital signals.

Figure 14–3(c) illustrates a unipolar two–level return–to–zero (RZ) signal, which differs from the first two in that pulse length for a symbol is less than the time allotted to the symbol interval. The extent to which the pulse length differs from the symbol interval determines the *duty cycle,* defined as $100\tau/T$ percent where τ is the pulse length and T is the symbol interval. As in Figure 14–3(a), voltage is used to represent a *1* and no–voltage to represent a *0* for the RZ signal.

Figure 14–3(d) is the return–to–zero counterpart of the bipolar NRZ signal of Figure 14–3(b). Note that the bipolar RZ signal is really a three–level signal having less than a 100–percent duty cycle.* However, since only two of the values, plus–voltage and minus–voltage, are used to represent the digital information in the signal, it may be regarded as binary. Note that each symbol, whether a *1* or a *0*, is associated with the presence of a pulse. For this reason, the synchronization of the receiving equipment may be accomplished by using the information in the signal, thus making the receiver self–clocking. This feature allows this signal format to be used for the transmission of asynchronous data.

The bipolar signal of Figure 14–3(e) has valuable properties that caused it to be the format chosen as the line signal for the T1, T1C, and T2 carrier lines. The symbol *0* is represented by zero voltage and the symbol *1* is represented by the presence of voltage. However, the polarity of voltage for successive *1* symbols is alternated, providing *alternate–mark–inverted* operation. Two important results are achieved. First, the dc component of the signals is virtually eliminated. This permits transformer coupling of the repeater to the line, facilitates the separation of the signal from dc power, and makes decision threshold circuits more

* The bipolar RZ signals are sometimes called pseudo three–level, or pseudoternary, signals.

practical by effectively eliminating the phenomenon known as baseline wander caused by a varying dc signal component. Second, the concentration of energy in the signal is shifted from the frequency corresponding to the baud rate to one–half the baud rate. This reduces near–end crosstalk coupling, reduces the required bandwidth to about one–half of that needed for a polar signal of the same duty cycle and repetition rate, and makes the design of timing recovery circuits more practical.

In Figure 14–3(f), the information is *differentially encoded* in terms of transitions that occur in the transmitted signal. Successive pulse intervals are compared. If they are identical, a *1* was transmitted in the original signal; if successive intervals show a transition, a *0* was transmitted.

The signals of Figure 14–3(a) through (f) may all be considered binary, whether in the number of values of voltage transmitted or in the significant number of values used to represent binary information. Figure 14–3(g) is not binary; it is illustrative of a class of signals that can be used to transmit data efficiently when the signal–to–noise ratio is high enough to permit signal detection at a number of different decision threshold values, which generally are smaller than those for the binary signals previously discussed. This quaternary signal format has been selected for ISDN access lines.

The signals of Figure 14–3 are used in many ways. In some cases, they are the signals delivered to the station modem by the customer; in other cases they represent the signals transmitted over exchange carrier facilities. The several forms commonly used are characterized more fully in the following paragraphs to illustrate their use.

Wideband Binary ASK Signals.* A limited number of applications of this signal format have been used for data transmission. Data station and carrier terminal facilities have been available to permit the transmission of a polar form of signal, much like that of Figure 14–3(b), at the synchronous rates of 19.2, 50.0, or 230.4 kb/s; for asynchronous service, the signal

* Generally replaced by high–capacity services and optional multiplexing equipment.

elements had corresponding minimum durations of 52.0, 20.0, or 4.0 μs. The three arrangements were developed to permit wideband data transmission in 24–kHz, 48–kHz, and 240–kHz bands found in commonly used frequency–division multiplexing (FDM) facilities. The 50–kb/s arrangement was the one most frequently used; its operation is typical of these arrangements.

As mentioned, the signal is transmitted in a polar form, called *restored polar*, different from the format of Figure 14–3(b) in that the dc component and some of the low–frequency components are filtered out at the transmitting data station and restored at the receiver. As a result of the filtering, the transmitted signal is sharply skewed, as shown in Figure 14–4(b). This mode of transmission obviates the need for high–fidelity transmission at zero and very low frequencies.

(a) Synchronous or asynchronous polar signal

(b) Same signal in restored polar format

Figure 14–4. Restored polar signal.

The power spectral densities for synchronous and asynchronous polar signals and restored polar signals are shown in Figure 14–5. For the synchronous signal, the spectra are those of a signal having a rate of 50 kb/s (T = 20 μs) and random bits having an equal probability of being *1* or *0* (p = 0.5). The asynchronous signal spectra may be used to represent a two–valued facsimile signal in which the average rate of black–white transitions is 4000 per second, and pages are 10–percent black and 90–percent white (the probability of a *1*, p = 0.1). The low–frequency power density spectra are very similar for the synchronous and asynchronous signals.

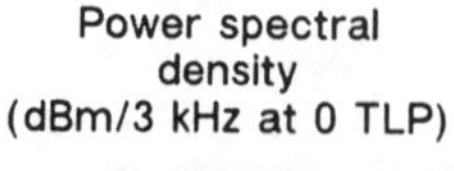

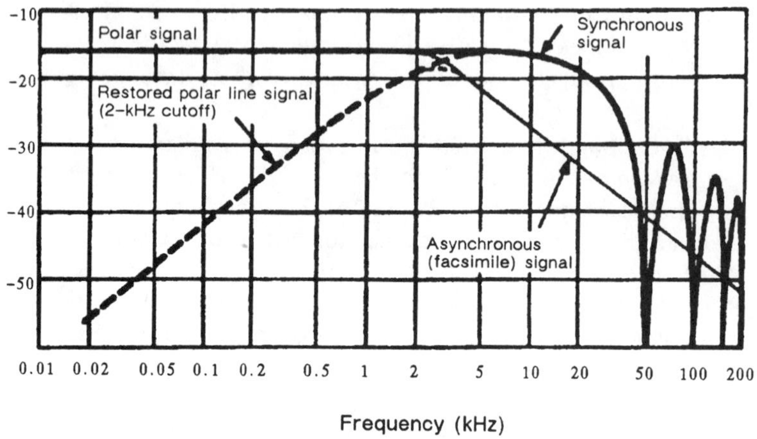

Figure 14–5. Power density spectra of polar and restored polar signals.

Thus, Figures 14–4(b) and 14–5 illustrate, in the time and frequency domains, respectively, the characteristics of the processed 50–kb/s signal as it is transmitted on loops. If a carrier system is used, further processing is necessary. Of interest is the processing for transmission in the 48–kHz group band of the broadband carrier facilities. Transmission is by amplitude modulation with vestigial sideband (VSB).

The frequency allocation, channel characteristic, and resulting signal spectrum for transmission in broadband multiplex are shown in Figure 14–6. The channel transmission characteristic and frequency allocation are shown in Figure 14–6(a). In the 60– to 108–kHz spectrum, a speech channel for coordination of operations may be provided in addition to the data channel for the VSB data signal. The baseband signal spectrum, shown in Figure 14–6(b), is the same as that of Figure 14–5 but modified by the group frequency channel characteristic; the modification is most notable at high frequencies where channel characteristics limit the baseband top frequency to about 37 kHz. The modulation process results in a VSB signal with the carrier suppressed.

However, the fact that signal components at zero and low baseband frequencies have been removed at the station data set permits the reinsertion of a low-amplitude carrier component. This component is recovered at the receiving terminal equipment to control the phase and frequency of the carrier used in demodulation [6].

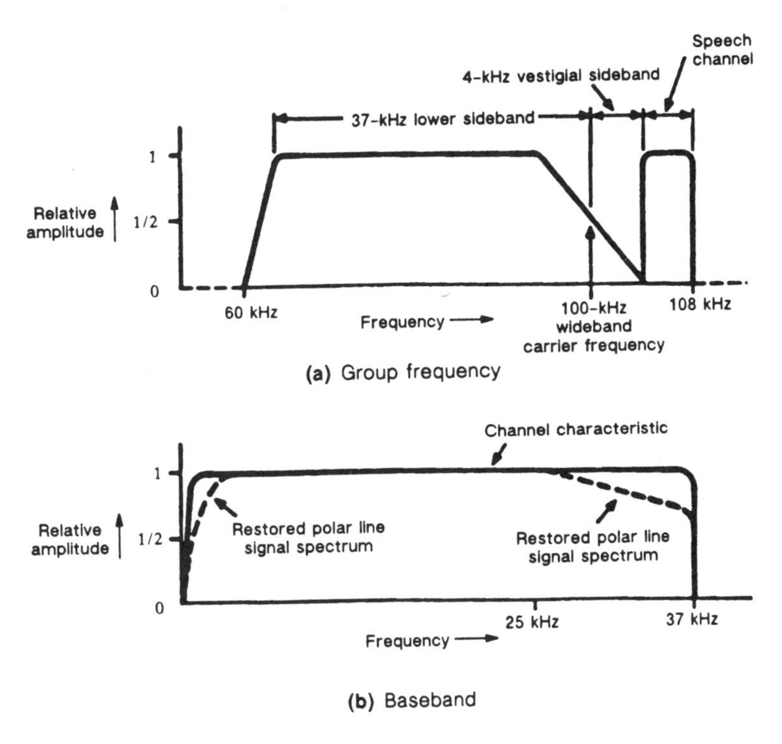

Figure 14-6. Data signals in broadband carrier at 50 kb/s.

Bipolar Line Signals. Bipolar signals find their greatest use as line signals in T-type carrier systems. The line signal in the T1 and T1C carrier systems is bipolar, like that shown in Figure 14-3(e), in all respects. The line signal in the T2 carrier system is similar, but with one important exception; in T2, the line signal is prevented from containing more than five successive *0*s by a method that modifies the bipolar signal format. Termed bipolar with six-zero substitution (B6ZS), this is accomplished by logic circuits in the transmitting terminal, which examines the signal before it is applied to the line. If the signal contains six consecutive *0*s and if the last *1* was a +, a *0*+−*0*−+ signal is substituted for the six *0*s; if the last *1* was a −, a *0*−+*0*+− signal is substituted for the six *0*s. The resulting violation of the bipolar rule (alternate *1*s must be of alternate polarity) is a means for recognizing the need for six *0*s which must be reinserted in the pulse stream at the receiving terminal. The substitution is made in order to guarantee a minimum density of *1*s in the line signal.

This code substitution eases the design and increases the accuracy of repeater timing circuits. The price paid is the additional logic circuits that must be used to accomplish the substitution and the additional complication of ignoring the substituted codes when bipolar violations are used as a measure of system performance.

The timing problem in the T1 carrier system was originally solved by limiting the maximum number of successive *0*s in the line signal but in a manner different from that used in the T2 system. In the T1 line signal, the number of consecutive *0*s that can be transmitted is limited to 15. For example, if encoded speech signals are being transmitted, this limitation is imposed by preventing any eight–bit word containing all *0*s from being transmitted to the line. If such a word is generated, it is modified in the terminal equipment by inserting a *1* in the seventh digit of the coded word. This is the least significant digit of the code representing the amplitude sample. The eighth digit is time–shared between speech and signalling. The code substitution permits the true bipolar feature to be maintained in the line signal. Its cost is a slight increase in channel coding noise. A later approach, termed bipolar with eight–zero substitution (B8ZS), replaces long strings of zeroes with special code words in a manner similar to that for T2. It thus allows 64–kb/s clear–channel capability (64CCC); therefore, the limitation on long strings of zeroes is eliminated. Other methods, such as zero–byte time–slot interchange, may also be used.

The power spectral densities of the T1 and T2 signals are, of course, functions of the statistical makeup of transmitted signals and of the signalling rate employed. The power spectral densities of the two signals are conveniently represented in terms of the probability, p, of a *1* in the signal sequence. The bipolar signal used in T1 is represented in Figure 14–7 for a range of p from 0.4 to 0.6. The T2 signal with code substitution for six successive *0*s is shown in Figure 14–8 for the same range of values of p, 0.4 to 0.6. In both figures the abscissas are normalized to unity, $fT = 1$, where f is in hertz and T is the signalling rate.

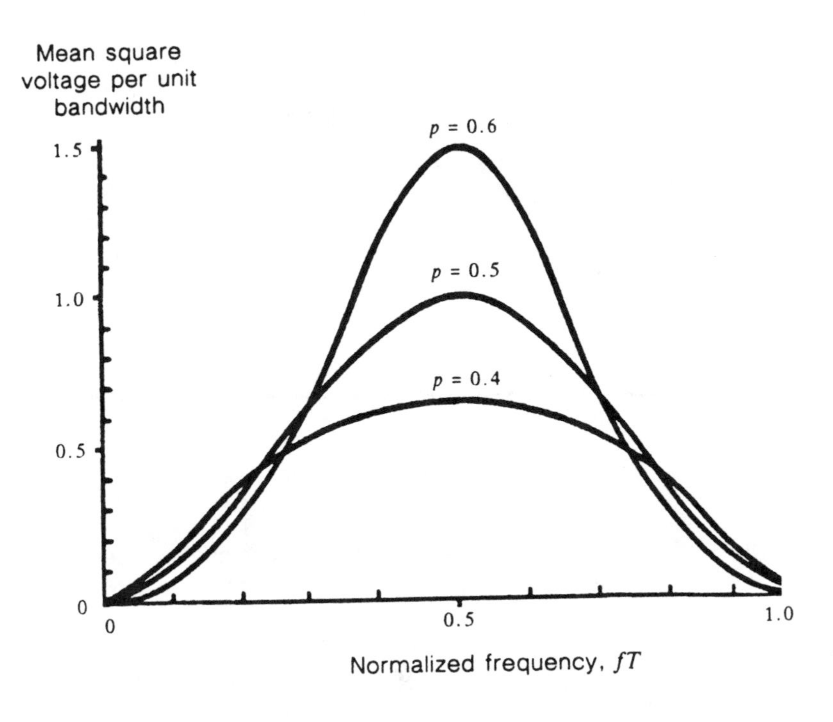

Figure 14–7. Power spectra for T1 bipolar signals.

In the T1 system, the signal is transmitted at 1.544×10^6 bits per second (b/s) with a 50–percent duty cycle. The minimum pulse width is then $\tau = 1/(2 \times 1.544 \times 10^6)$, or about 0.324 μs. T1C uses 3.152×10^6 b/s with a 50–percent duty cycle. In the T2 system, the signal is transmitted at 6.312×10^6 b/s, also with a 50–percent duty cycle. For T2, the minimum pulse width is $\tau = 1/(2 \times 6.312 \times 10^6)$, or about 0.079 μs.

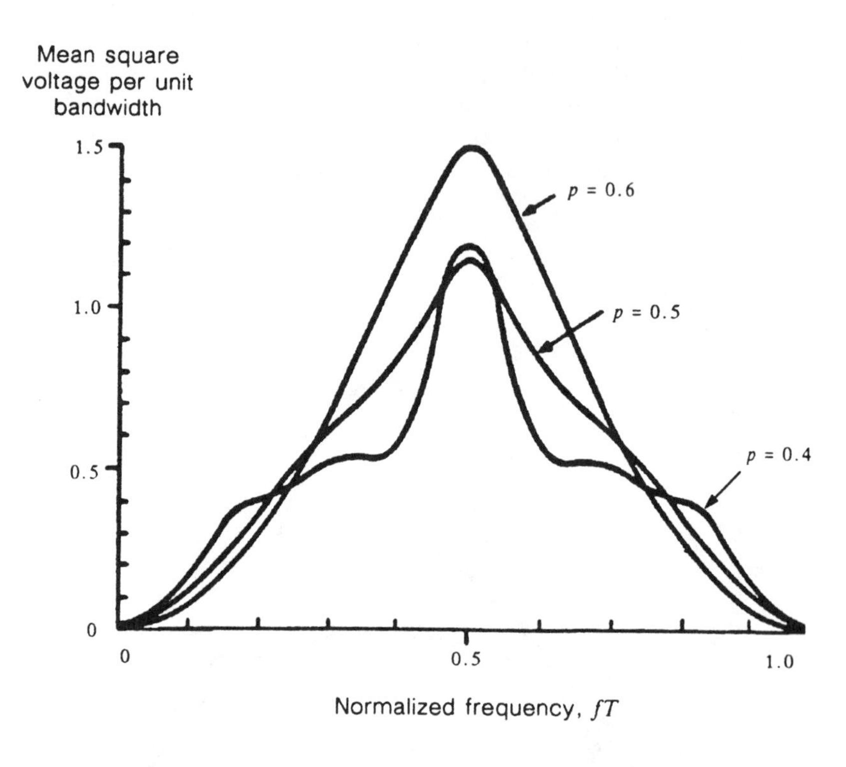

Figure 14-8. Power spectra for T2 bipolar signals coded for restricted number of sequential zeros.

The bipolar nature of two signals and the coding sequence employed result in small discrete frequency components and nulls in the spectrum at zero frequency and at integral multiples of frequencies corresponding to the signalling rate. For each system, the losses above the frequency corresponding to the signalling rate are high; as a result, the spectra of Figures 14-7 and 14-8 may be ignored above the signalling frequency. In T1 carrier, the top transmitted frequency then may be regarded as $f = 1/(2\,\tau) = 1/(2 \times 0.324) = 1.544$ MHz; in T2 carrier, the top transmitted frequency is $f = 1/(2\,\tau) = 1/(2 \times 0.079) =$ 6.312 MHz.

Signal amplitudes vary widely through the system in any of these carrier lines. One reference point often used is the output of a line (regenerative) repeater. The nominal value in T1 is 3 volts peak; in T2, 0.9 volts peak [7].

Multilevel ASK Signals. The ASK signals described thus far have been two–level or, at most, three–level (T1/T1C/T2 line signals). Where channels and transmission facilities exhibit a high signal–to–noise ratio and have well–controlled transmission characteristics (gain/frequency and phase/frequency), the efficiency of transmitting information can be significantly improved by coding the digital signal into a multilevel ASK format such as the four–level signal illustrated in Figure 14–3(g). Such multilevel signals are used in voiceband applications and as the ISDN basic rate interface (BRI). In addition, a system is widely used to encode a 1.544–Mb/s DS1 signal, such as that used in T1 carrier, into a multilevel signal with a bandwidth of 440 kHz for transmission over long–distance microwave systems. This DS1 signal can be added to existing analog microwave routes using special equipment below the normal frequencies allocated to message channels [8]. The designation of frequencies below those used for voice are classified as data under voice (DUV). DUV is covered in greater detail in Volume 2.

The DUV signal, transmitted at the 772–kilobaud* rate, is the digital line signal used for transmission over microwave radio systems. The rate at which information is transmitted for this signal cannot be computed by the expression previously given for voiceband rates because the method of coding is different. The mode involves what is known as class IV partial response coding [9,10], for which the transmission rate is $x \log_2 (n + 1)/2$ bits per second, where x again represents the baud rate and n represents the number of values.

In addition to DUV, there are dedicated digital microwave systems, using solely digital modulation techniques, that permit the transmission of high–level multiplex signals for digital and voice traffic. Multiple DS1 signals can be transmitted over

* The baud is a basic unit of signalling speed. The speed, in bauds, is the number of signalling elements per second. The signalling element is that part of the signal occupying the shortest time interval of the signalling code. It is considered to be of unit duration in the building up of variable signal length combinations.

2–GHz digital radio. The DS3 signal, at 44.736 Mb/s, and the DS4 signal, at 274.176 Mb/s, can also use microwave transmission facilities (generally the 11–GHz and 18–GHz radio systems respectively). Digital radio, proven more economical than analog radio, is finding progressively greater use in short– and medium–haul facility applications.

The basic ISDN user channel (B channel), operating at 64 kb/s, supports voice, data and other information. When two B channels are combined with a common–signalling channel (D channel) in a serial stream of data at 144 kb/s (plus maintenance–channel bits), the structure formed is 2B+D for the BRI using existing nonloaded cable pairs [11]. BRI can support two remote users (simple telephones, multiplexers, etc.) through an associated switch at distances to over 3000 feet using two paired cable facilities. However, two–way BRI transmission using echo cancellers and the four–level type signalling illustrated in Figure 14–3(g) can extend the ISDN two–wire operational range to approximately 12,000 feet.

Signal Spectra. Binary baseband signals that are coded into the multilevel format in a modem or in terminal equipment associated with a carrier system are transmitted in a partial response format and have an energy distribution as shown in Figure 14–9. Energy above frequency f_1 , the Nyquist frequency, is removed by an appropriate cutoff filter, which leaves only a vestige of the second lobe as illustrated in Figure 14–10. A timing signal, P_t , is added to facilitate recovery of the synchronizing, or timing, information at the receiver.

When multilevel baseband signal transmission is appropriate, the signal of Figure 14–9 may be transmitted directly. Note that there is no energy at zero frequency and that low–frequency components are of low amplitude. Only slight signal impairment is suffered when such a signal is transmitted over baseband facilities, which cannot pass direct current and low–frequency components. This mode of transmission (baseband) is used for transmitting the 772–kilobaud signal on microwave radio systems. It is frequency–modulated in the FM transmitter along with the speech channels carried by the system.

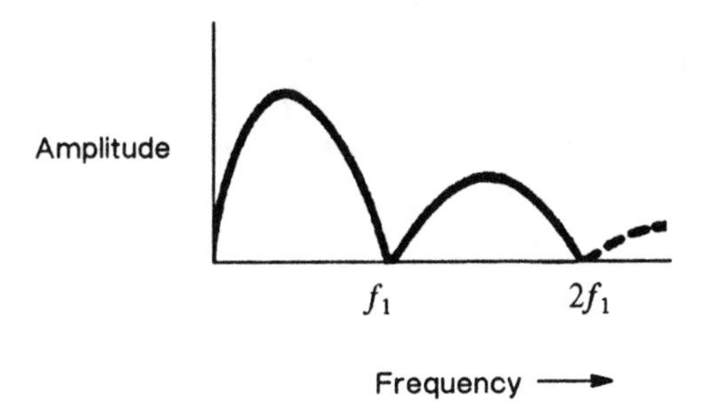

Figure 14–9. Spectral energy distribution, multilevel partial response signal.

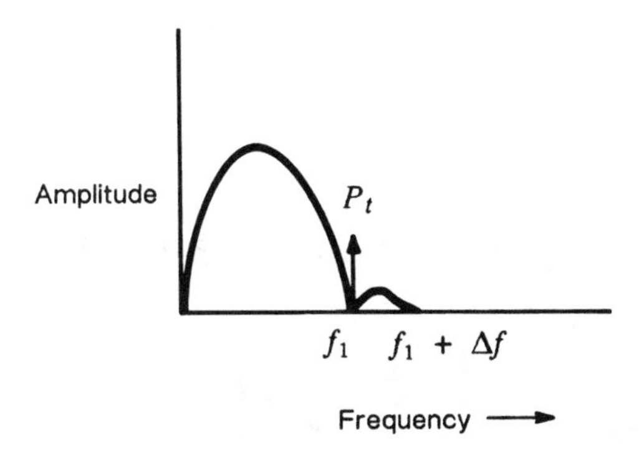

Figure 14–10. Partial response signal—high-frequency lobes removed—timing signal added at f_1.

Most multilevel ASK signals, other than the 772–kilobaud signal, are transmitted by VSB techniques. These may be upper or lower sideband; a lower sideband version with a vestigial upper sideband is illustrated in Figure 14–11.

Phase Shift Keyed Signals

Signal transmission by phase shift keying (PSK) techniques is accomplished in the voiceband by the use of modems that code incoming binary signals into multilevel phase shifts of appropriate carrier signals. In PSK transmission, sideband frequencies are generated by the modulating signal so that the bandwidth required is equal to twice the highest frequency component in the modulating signal. The distribution of energy in the modulated carrier wave follows a pattern like that illustrated in Figure 14–12.

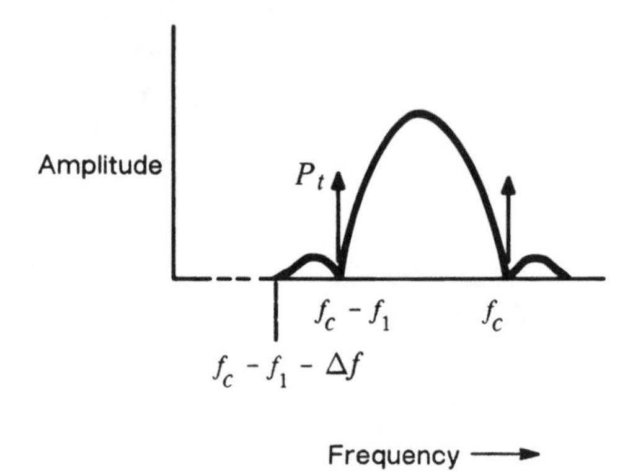

Figure 14–11. Partial response signal at carrier frequency—vestigial upper sideband.

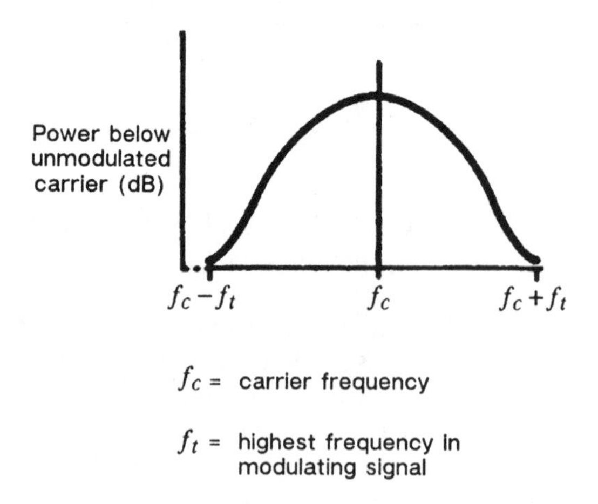

Figure 14–12. Power spectral density of a PSK data signal.

Phase shift keying has an advantage over other modes in certain situations. In a bandlimited channel, PSK signals are relatively immune to amplitude changes. The signal is transmitted at essentially constant power. The mode lends itself well to the recovery of a clock signal at the receiver and offers speed advantages by multilevel coding techniques. This mode has the disadvantage of being sensitive to phase distortions, phase jitter, and impulse noise. These impairments affect the zero crossings of the signal.

In a two-phase modem, an incoming binary *1* (or *0*) would shift the transmitted analog carrier 180 degrees, while a change to *0*s (or *1*s) would be needed to shift the carrier again. Figure 14–13 illustrates the phase shifts that occur at the transition points in the carrier frequency.

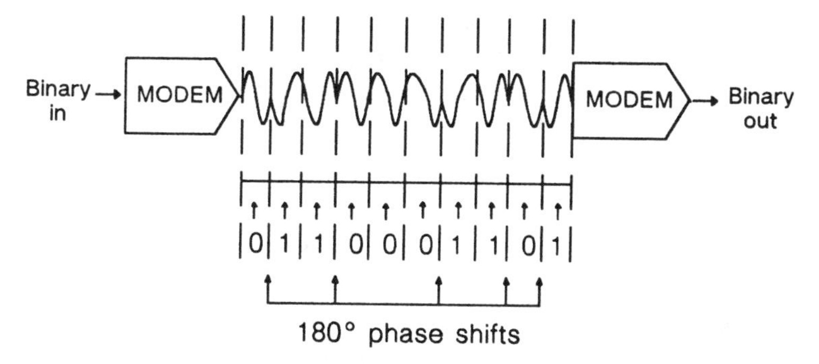

Figure 14–13. PSK line signal illustrating phase shift transition points in the carrier frequency.

There are four-phase voiceband systems that provide for the transmission of data at rates up to 2400 b/s. The baseband signal is divided into pairs called dibits that phase-modulate a carrier by +45 degrees or more relative to its nominal zero-phase condition depending on the previously received bits being recognized as a different dibit. The four recognized dibits and the CCITT recommendations [12] for phase modulation are listed in Table 14–1. Two alternate methods of coding shown may be used on private-line facilities.

Systems using eight-phase modulation techniques use modems that recognize incoming bits in groups of three, called tribits. Each tribit value change is encoded as a phase change in the 1800-Hz transmitted carrier of +45 degrees or more. Table 14–2 lists the phase relationships for different dibit codes. This arrangement permits data transmission at a 4800-b/s rate.

Table 14–1. Recommended Phase Shift for Dibit Coding

Dibit	Phase Change	
	Alternative A	Alternative B
00	0°	+45°
01	+90°	+135°
11	+180°	+225°
10	+270°	+315°

Table 14–2. Phase Relationships for Tribit Coding

Tribit Values			Phase Change
0	0	1	0°
0	0	0	45°
0	1	0	90°
0	1	1	135°
1	1	1	180°
1	1	0	225°
1	0	0	270°
1	0	1	315°

Another system uses 9600–b/s modems that encode quadbits (groups of four consecutive bits of scrambled data) into carrier phase changes of 45 degrees or more and changes in the relative amplitude of the transmitted signal element. The carrier frequency is 1700 Hz. These modems generally have a fallback rate of 7200 or 4800 b/s that are used in the event that impaired transmission is encountered at the higher speed. There are modems that offer multiplexing options that assemble 2400, 4800, and 7200 b/s data subchannels into quadbit assignments for transmission at 9600 b/s [12]. Table 14–3 lists the configurations recommended by the CCITT [13] for various subchannel rates at different transmit data speeds. Subchannel rates of 2400 or 4800 b/s can be accommodated at the 4800–b/s data rate. Subchannel data rates of 2400, 4800, or 7200 b/s can be combined at the 7200–b/s rate while the 2400–, 4800–, 7200–, or 9600–b/s rate can be multiplexed at the aggregate rate of 9600 b/s [13]. Other coding arrangements are used for speeds of 14.4 or even 19.2 kb/s where circuit quality permits.

Table 14–3. Multiplex Configurations for Different Subchannel Rates at Various Bit Rates

Aggregate Data Rate	Multiplex Configuration	Subchannel Data Rate	Multiplex Channel	Modular Bits			
				Q1	Q2	Q3	Q4
9600 b/s	1	9600	A	X	X	X	X
	2	7200	A	X	X	X	
		2400	B				X
	3	4800	A	X		X	
		4800	B		X		X
	4	4800	A	X		X	
		2400	B		X		
		2400	C				X
	5	2400	A	X			
		2400	B		X		
		2400	C			X	
		2400	D				X
7200 b/s	6	7200	A		X	X	X
	7	4800	A		X	X	
		2400	B				X
	8	2400	A		X		
		2400	B			X	
		2400	C				X
4800 b/s	9	4800	A		X	X	
	10	2400	A		X		
		2400	B			X	

Point–to–point data lines can usually accommodate all of the phase–modulated transmission speeds. However, it appears that 2400 and 4800 b/s are the most popular bit rates for transmission over the MTS network and private lines [14].

Frequency Shift Keyed Signals

Modems have been developed to exploit the use of frequency shift keying (FSK) techniques. Figure 14–14 illustrates how, in one such system, the carrier frequency is modulated by teletypewriter signals so that a frequency of 1270 Hz would represent a *1* (mark), and 1070 Hz would represent a *0* (space).*

* Mark and space designations have been handed down from telegraph usage and continued with teletypewriter digital coding.

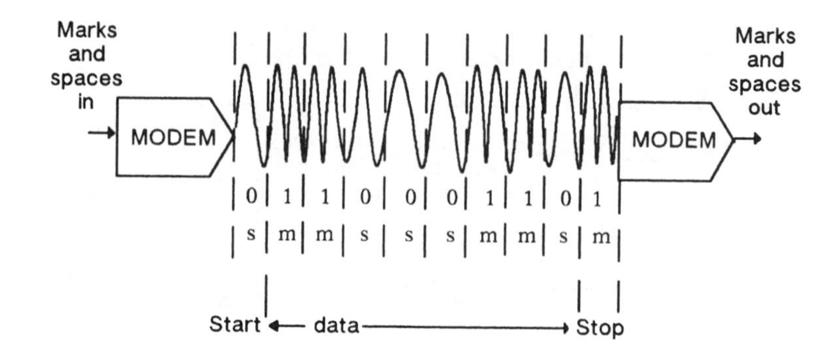

Figure 14–14. Constant amplitude FSK line signal using one frequency to represent a space and a higher frequency to represent a mark.

This modem provides asynchronous data transmission and operates at bit rates up to 300 b/s. The modem (103–type data set) has a channel in another frequency band (2025–2225 Hz) that permits simultaneous two–way data transmission over the MTS network and point–to–point two–wire facilities. Thus, the calling modem would transmit on one channel in the low–frequency band and receive on the high–frequency band while the called modem would send on the high–frequency band and receive on the low. In effect, this system provides equivalent four–wire transmission and allows full duplex operation. Table 14–4 lists the mark and space frequencies for the 103–type modem and the *CCITT V.21 Recommendations* [13].

Table 14–4. Mark (*1*) and Space (*0*) Frequencies for 0–to–300–b/s Modems

	103–type	*CCITT Rec. V.21*
Low–frequency channel 1:		
Mark	1270 Hz	1650 Hz
Space	1070 Hz	1850 Hz
High–frequency channel 2:		
Mark	2225 Hz	980 Hz
Space	2025 Hz	1180 Hz

Modems operating up to 600 baud use FSK techniques employing 1300 Hz for a mark (*1*) and 1700 Hz for a space (*0*).

These frequencies are used for the forward channel while another frequency, with a modulation rate up to 75 baud, provides a channel for coordination and error control.

Other modems, operating up to 1200 baud, provide asynchronous transmission where 1300 Hz and 2100 Hz are used to represent the "mark" and "space" signals, respectively. This system operates at about 1200 b/s over the MTS network and up to 1800 b/s on conditioned private lines. More sophisticated terminal equipment uses FSK techniques to achieve 2400 b/s.

These arrangements display most of the advantages of the PSK mode of transmission and have the additional advantage of being relatively simple to implement. However, FSK signals have the usual feature of achieving high immunity to noise at the cost of wider required bandwidths. As a result, the FSK systems are commonly used in a voiceband or less at low signalling rates relative to those achievable with multivalued PSK and ASK.

Many private–line and switched network telegraph and teletypewriter services have routinely been provided by FSK techniques. In addition, some moderate–speed voiceband data is also transmitted FSK. Rates of 300 b/s and 1200 b/s have become popular with personal computers communicating with other personal computers or information services.

Telegraph Signals. Telegraph signals are commonly transmitted at various rates, including 75 and 150 bauds, by FSK techniques in low–speed channels that are combined into the voiceband by FDM techniques. Seventeen separate channels can be provided with frequencies as given in the top portion of Table 14–5 to transmit 75–baud telegraph signals. For 150–baud signals, the channel assignments are shown at the bottom of the table.

The binary (*1* and *0*) input signals are translated into FSK signals at the terminal equipment. A *1*, or mark, signal corresponds to a frequency in the passband of the channel 35 Hz (for 75–baud signals) or 70 Hz (for 150–baud signals) above the channel center frequency. The *0*, or space, signal is represented by a frequency 35 (or 70) Hz below the channel center frequency. The center frequency itself is not transmitted.

Table 14–5. Voice–Frequency Carrier Data Channel Assignments

SINGLE BANDWIDTH			
Channel Number	**Space Frequency**	**Center Frequency**	**Mark Frequency**
1	390	425	460
2	560	595	630
3	730	765	800
4	900	935	970
5	1070	1105	1140
6	1240	1275	1310
7	1410	1445	1480
8	1580	1615	1650
9	1750	1785	1820
10	1920	1955	1990
11	2090	2125	2169
12	2260	2295	2330
13	2430	2465	2500
14	2600	2635	2670
15	2770	2805	2840
16	2940	2975	3010
17	3110	3145	3180
DOUBLE BANDWIDTH			
21	610	680	750
22	950	1020	1090
23	1290	1360	1430
24	1630	1700	1770
25	1970	2040	2110
26	2310	2380	2450
27	2650	2720	2790
28	2990	3060	3130

Dual–Tone Multifrequency Signalling. DTMF signalling [15] as a means of data transmission is, in effect, also a form of ASK signalling. Described briefly in Chapter 13, Part 1, DTMF signalling was introduced initially as a means for transmitting address signals from telephone station equipment to the central–office switching system. These signals may be impressed on the telephone line after the address signal is sent to permit customers to place calling–card calls from DTMF stations without the assistance of an operator. DTMF signals may also be impressed on the line while a connection is established in order to communicate with terminal equipment associated with banking and other services. Since the DTMF signals fall within the voiceband, they offer the possibility of being used to transmit data to a variety of services. Special receivers are provided for data transmitted by DTMF signalling.

14–3 ANALOG DATA SIGNALS

A few voiceband analog data signals are transmitted by FM techniques. Three of these are found in the exchange carrier plant in sufficient quantity to warrant individual description. They are medium–speed data, facsimile, and low–speed analog data. In each case, signal power is limited to −16 dBm0, long–term average. Many other signals are transmitted in quantities too small to warrant individual characterization. Among these are several types of telemetry and telewriter signals.

Medium–Speed Voiceband Data

One type of analog data signal transmission involves the translation of a 0 to +7 volt continuous signal from the customer to an FM signal that is transmitted over telephone facilities in the voiceband. A zero–volt input signal is transmitted as a 1500–Hz signal on the line, and a +7 volt signal is transmitted as a 2450–Hz signal on the line; intermediate input voltages and line frequencies are linearly related. The baseband signal may contain components from zero frequency to about 1000 Hz.

In the direction of transmission opposite to that used for data transmission, a 60 $\pm$ 1 Hz signal is sent for synchronization. This signal is translated to 600 Hz for transmission over the line. The powers contained in the data and synchronizing signals are equal.

Facsimile. Facsimile (FAX) is a scheme by which pictures, typewritten or printed material, and handwriting can be sent over voiceband or digital facilities. Other business organizations such as banks, utilities, railroads, and stockbrokers use facsimile services widely.

FAX machines are scanning devices that read an image off paper, modulate or code it, and transmit the resultant signal to another FAX machine that reproduces a facsimile of the original document. The FAX machine is both an input (transmitting) and output (receiving) device.* Four groups of FAX have been standardized. Groups I and II, based on outdated technology, are relatively slow and inconvenient to use. Group III, first standardized in 1980, and Group IV, standardized in 1984, represent the latest FAX technology. The later machines use data–compression techniques to avoid wasting transmission time when sending blank portions of the scanned image.

At a transmitting station, material to be sent by facsimile is generally placed on a tray and roller–fed automatically through a scanning device. A beam of light, approximating 1/100 inch in width, scans the material (picture, printed sheet, etc.) with resultant reflections of light or darkness beamed to a photoelectric cell. The photocell converts these variations of lightness or darkness into analog electrical currents. The modulated electric current is converted in a transmitter to an amplitude–modulated or frequency–modulated audio–frequency for transmission over a voiceband facility. In newer machines, the analog electrical currents are encoded into digital signals to be transmitted over analog or digital exchange facilities.

At the receiving end, the signals are demodulated into electric currents that react with either plain or treated paper** to form a black and white image reproducing the material sent from the transmitting station.

The CCITT has recommended group classification of facsimile equipment [16]. Its standards have been widely accepted by the industry.

The Group I amplitude–modulated signal has the greatest amount of power output for black (adjustable between −7 and 0 dBm) and the least for white (15 dB below the black level). The

* Facsimile interface devices permit computer or laser printer communication with FAX machines.

** Recording is accomplished with ink transfer, dielectric, thermal, or electrosensitive paper.

carrier frequency ranges between 1300 and 2100 Hz. These frequencies have the least amount of distortion over dedicated facilities. Group I machines can use frequency modulation for switched connections, the center frequency at 1700 Hz, with the frequency corresponding to black at 1900 Hz and white at 1300 Hz. Other frequencies are used for FAX machine control and response [16].

Group II FAX machines, with double the scanning rate of Group I equipment, use VSB amplitude modulation techniques except those transmitting over leased lines and the MTS network. The carrier frequency is 2100 Hz. The white signal is represented by maximum carrier and the black by minimum carrier.

Group III FAX, with greater image resolution, operates at either 7.2 kb/s or 9.6 kb/s. Group IV FAX, using asymmetrical modems, can deliver a throughput in excess of 14 kb/s over nonloaded metallic facilities. However, over an ISDN 64–kb/s channel, Group IV provides double the image quality and is four to eight times faster than Group III. The Group III and IV machines are, in effect, digital terminals capable of working over either digital or analog channels.

Low–Speed Medical Data

Medical data such as electrocardiograms and electroencephalograms can be transmitted over exchange carrier facilities by FM techniques. Input signals are accepted with components from zero to about 100 Hz and with amplitudes of −2 and +2 volts. Such signals frequency–modulate a carrier at 1988 Hz; the carrier frequency varies linearly with the input voltage at frequencies between −262 and +262 Hz relative to the carrier. A signal at 387 Hz is sometimes transmitted in the opposite direction to permit signalling from the receiver to the transmitter.

The signal generated in medical electronic equipment may be coupled to the telephone line either electronically or acoustically. The latter method has the virtue of allowing for portable equipment such as that generally in use for remote pacemaker testing. Thus, the signal can be coupled to the line at any telephone connected to the MTS network.

References

1. Mahoney, J. J., Jr. "Transmission Plan for General Purpose Wideband Services," *IEEE Transactions on Communications Technology*, Vol. COM–14, No. 5 (Oct. 1966), pp. 641–648.

2. Fracassi, R. D. and F. E. Froehlich. "A Wideband Data Station," *IEEE Transactions on Communications Technology*, Vol. COM–14, No. 5 (Oct. 1966), pp. 648–654.

3. Bell System Technical Reference PUB 41005, *Data Communications Using the Switched Telecommunications Network*, American Telephone and Telegraph Company (May 1971).

4. Members of Technical Staff. *Transmission Systems for Communications*, Fifth Edition (Murray Hill, NJ: AT&T Bell Laboratories, Inc., 1982), Chapter 30.

5. Bennett, W. R. and J. R. Davey. *Data Transmission* (New York: McGraw–Hill Book Company, Inc., 1965).

6. Ronne, J. S. "Transmission Facilities for General Purpose Wideband Services on Analog Carrier Systems," *IEEE Transactions on Communications Technology*, Vol. COM–14, No. 5 (Oct. 1966), pp. 655–659.

7. *ANSI T1.102–1987*, "American National Standard for Telecommunications—Digital Hierarchy—Electrical Interfaces" (New York: American National Standards Institute, 1987).

8. Seastrand, K. L. and L. L. Sheets. "Digital Transmission Over Analog Microwave Radio Systems," *Conference Record*, IEEE International Conference on Communications (Philadelphia, PA: June 1972).

9. Gerrish, A. M. and R. D. Howson. "Multilevel Partial–Response Signalling," *IEEE Conference on Communications Digest* (1967).

10. Kretzmer, E. R. "Generalization of a Technique for Binary Data Communication," *IEEE Transactions on*

Communications Technology, Vol. COM–14, No. 1 (Feb. 1966), pp. 67–68.

11. Browne, T. E. "Network of the Future," *Proceedings of the IEEE* (Sept. 1986), pp. 1–8.

12. Bell System Technical Reference PUB 41213, *Data Set 209A Interface Specifications*, American Telephone and Telegraph Company (May 1974).

13. CCITT, Volume VIII, *Series V Recommendations—Data Transmission Over the Telephone Network* (Geneva: International Telegraph and Telephone Consultative Committee, 1976).

14. "All About Modems," *Datapro Reports on Data Communications, 1986* (Delran, NJ: Datapro Research Corporation, 1986).

15. *Notes on the BOC IntraLATA Networks—1986*, Technical Reference TR–NPL–000275, Bellcore (Iss. 1, Apr. 1986), pp. 6–107 to 6–114.

16. CCITT, Volume VII, *Series T.1 Recommendations* (Geneva: International Telegraph and Telephone Consultative Committee, 1984).

Chapter 15

Video Signals

A wide variety of video services are provided over transmission facilities. These vary from narrowband telephotograph service, operating in the voiceband, to multimegahertz–bandwidth television services provided for the broadcast industry and for educational, industrial, and private distribution systems.

As in the transmission of other types of information, the transmitting and receiving equipment at both ends of a video transmission path must include transducers capable of translating one form of energy to another. In this case, different values of luminance (light intensity), together with color information in the case of color television, are converted to electrical signals at the transmitter; at the receiver, the transducer must translate the electrical signal back to light signals so that the transmitted image can be viewed on a television set or recorded on film, paper, or magnetic tape.

Many natural characteristics of the human recipients of video information have influenced the design of transmitting and receiving equipment as well as the design of the video signals. Among these are the persistence of vision, the preferred viewing distance of visual images, the resolution capability of the eye, the human tolerance to departures from accurate color rendition, and the effects of ambient conditions such as lighting on viewing preferences [1]. These human factors have influenced the rate at which picture images are transmitted, the format of color television signals, the resolution and, therefore, the bandwidth of signals, and many other aspects of video signal transmission.

15–1 TELEVISION SIGNALS

Many part–time and full–time television circuits in North America are provided by exchange and interexchange carriers

(ICs) for the transmission of network broadcast signals* [2]. As a convenience, the black and white (monochrome) signal is used here as the basis of television signal description, even though most television signals now transmitted are in color. The chrominance information in a color signal is regarded as being superimposed on the monochrome signal and is described in this manner.

Standard Monochrome Baseband Signals

Local video signal transmission needs are nearly always provided by exchange carriers using baseband facilities, optical fiber, or microwave radio. Intercity needs are usually provided by microwave relay facilities or satellite relay. These are fed by facilities that interconnect the broadcaster's equipment and the terminals of the microwave radio, satellite**, or optical fiber. The baseband signal received at the intercity facility terminal frequency–modulates the microwave or satellite radio carrier. Baseband signals are also converted to a pulse format, either analog or digital, for transmission over fiber optic transmission systems. Only the baseband signal is characterized in detail here, although some attention is subsequently given to the signal format used in commercial television broadcasting.

The conversion of light signals to electrical signals and the reconversion to light signals at the receiver involves a scanning operation that differs in details for different systems. However, all systems must provide, in the scanning operation, for the synchronization of the receiver with the transmitter (a coding process); they must also provide for the conversion from luminance variations to electrical signal variations (a modulation process).

---

* The signals described here have been standardized in North America. Other standards have been established elsewhere; e.g., 625–line, 50–frame–per–second signals are used in Europe. As a result of satellite transmission, intercontinental facilities have been adapted for transmission of such signals. Conversion of the signal to the American standard is generally the responsibility of the broadcaster.

** Intercity coaxial cable systems were once used for television signal transmission, but they are rarely used for this purpose today.

Scanning and Synchronization. Figure 15–1 illustrates the scanning pattern used for broadcast television signals. The scanning mechanism causes the exploring element and the reproducing spot to move in synchronism across the image field and the receiving field (picture tube) in nearly horizontal lines from left to right. The scanning lines are started at the top; successive scans are made at successively lower parts of the field until the bottom is reached. The spots are then returned to the top to begin a second field. Succeeding scans in alternate fields are interlaced. The interlaced scans of two successive fields make up a frame.

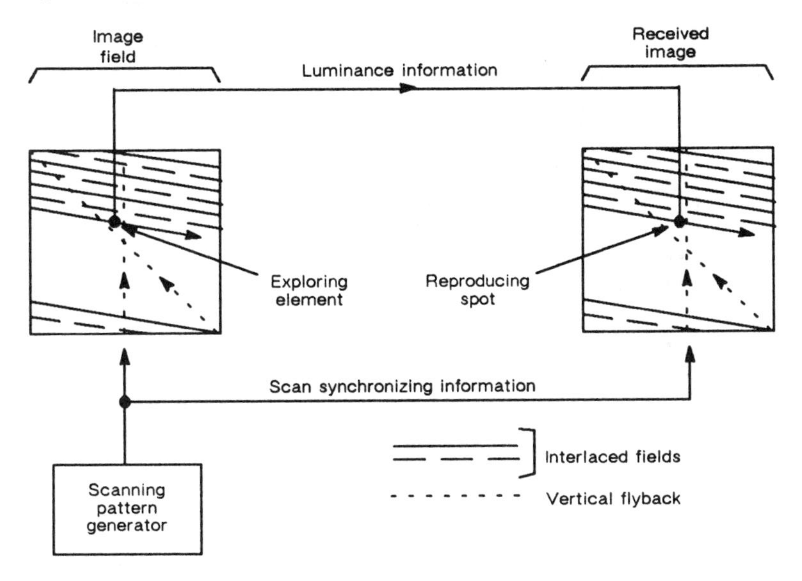

Figure 15–1. Broadcast television scanning process.

The exploring element at the transmitter causes the scanning pattern generator to scan the image field. The scanning pattern must cover every part of the image in a systematic and specified manner. Information regarding the location of the exploring spot and the direction in which it is being moved are coded at the transmitter and sent to the receiver so that the reproducing spot can be located in the received image field at the position corresponding to that of the exploring element in the transmitter. This process of synchronizing the receiver to the transmitter is also illustrated functionally in Figure 15–1.

While Figure 15–1 shows the transmission of scanning/synchronizing and luminance information over separate channels, the information is in reality combined into one composite signal for transmission or broadcast. The two kinds of information, separated in polarity and time, are illustrated in Figures 15–2, 15–3, and 15–4.

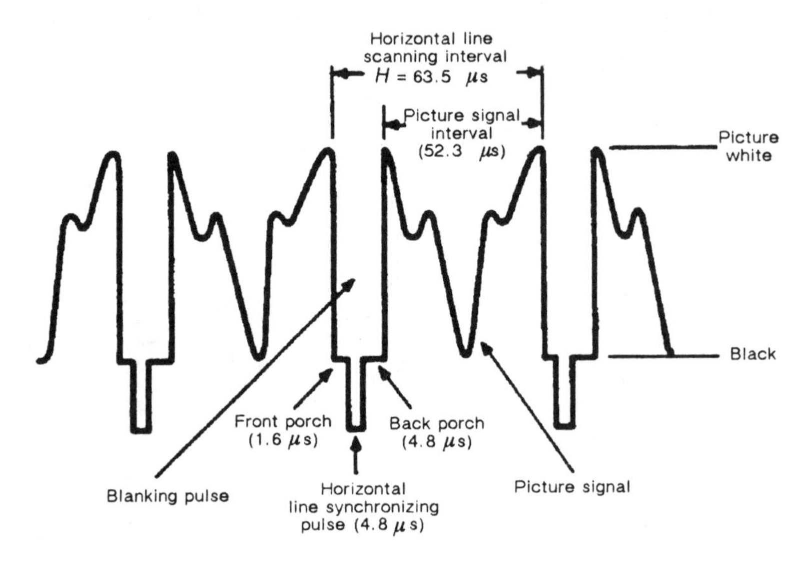

Figure 15–2. Monochrome television horizontal line scanning and synchronization.

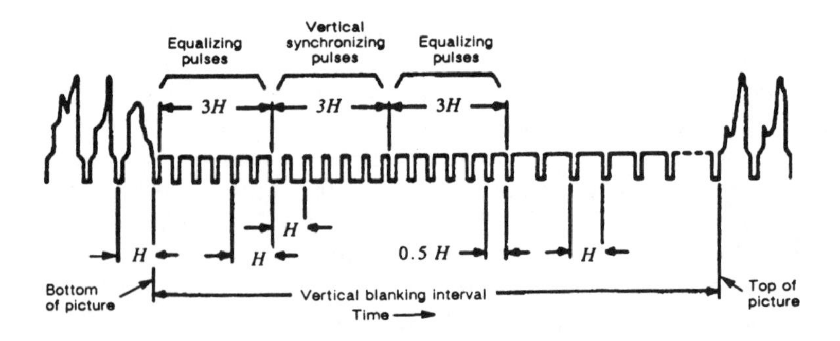

Figure 15–3. Monochrome television vertical synchronization.

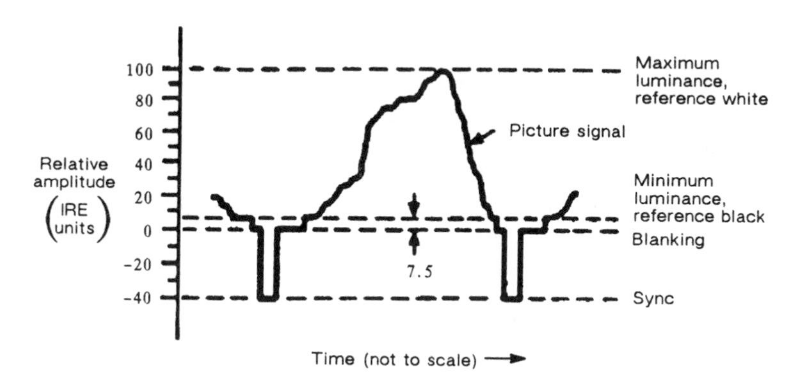

Figure 15-4. Monochrome television signal, relative amplitudes.

Consider first the horizontal line scanning function illustrated in Figure 15-2. The total line scanning interval, *H*, is 63.5 μs, equivalent to about 15,750 line scans per second. Of this interval, 11.2 μs are assigned to a blanking pulse. The blanking pulse interval is divided into a 1.6-μs front-porch interval, a 4.8-μs back-porch interval, and a 4.8-μs horizontal synchronizing pulse interval. The porches isolate the synchronizing pulse from transients or overshoots of the picture signal. The synchronizing pulse, recognized at the receiver by its polarity and duration, triggers circuits that drive the reproducing spot at the receiver to the left side of the image field (flyback). During the blanking pulse interval, the receiving tube is normally blanked out; the luminance of its reproducing spot is driven into the ultrablack region so that the synchronizing pulse and the flyback of the spot are not visible. After each blanking pulse interval, the scanning circuits drive the exploring and reproducing spots in synchronism across the image from left to right.

When the last line of a field scan is completed, the scanning pattern enters a vertical blanking interval as shown in Figure 15-3. This interval is approximately 1200 μs long, equivalent to the duration of about 20 horizontal scan intervals. During a part of this interval, a number of vertical synchronizing pulses, each about 25 μs long, drive the spot to the top of the screen (vertical flyback) to begin a new field scan. A series of horizontal pulses (called equalizing pulses), transmitted at twice the normal line scan rate, precede and follow the vertical synchronizing pulses.

Following the second burst of these equalizing pulses is a series of horizontal synchronizing pulses transmitted at the normal scan rate as a part of the vertical blanking interval. The equalizing and normal horizontal synchronizing pulses transmitted during the vertical blanking interval are provided to condition the synchronization circuits of the transmitter and receiver so that the two are indeed in synchronism, and to guarantee that the interlace pattern is properly implemented in successive field scans.

The timing of the vertical blanking intervals is such that a field is produced every 1/60 second. The interlacing of the next field with the first makes up a frame, one of which is produced each 1/30 second. This 30–per–second frame rate, combined with the 15,750 horizontal line rate, results in a picture nominally formed of 525 lines per frame.

Luminance Signal. As shown in Figure 15–1, an exploring element measures the intensity of the light at a given spot on the image to be transmitted. The light intensity is converted to an electrical signal whose amplitude is a specified function of the measured intensity. This electrical analog of the light intensity is transmitted to the receiver where the inverse process, the conversion of the electrical signal to appropriate light intensity values, takes place. The reproducing spot at the receiver then illuminates the receiving mechanism to the proper intensity.

If the image being scanned at the transmitter is motionless, the light intensity at a given point and its electrical counterpart depend only on the position of the exploring element. Thus, the signal can be regarded as a function of two variables that describe the two–dimensional image field. If the image at any spot involves time variations of light intensity, perhaps because of motion, the luminance is a function of time also, and the corresponding electrical signal is a function of three independent variables.

The electrical signal amplitude is defined by a scale standardized by the IRE (now the IEEE). The scale*, used as a convenience in examining television waveforms on an oscilloscope, is

* This scale can be derived from the table in *FCC Part 73.681*, p. 184 [2].

illustrated in Figure 15–4. Thus, it can be noted in the figure that maximum luminance (reference white) has a relative amplitude of 100 while minimum luminance (reference black) has a relative amplitude of 7.5.

Bandwidth and Resolution. During the development of television, subjective viewing tests were used to determine that a bandwidth of about 4.2 MHz results in a satisfactory received television image. This conclusion involved the combined evaluation of many parameters such as acceptable vertical and horizontal resolution, frame and field rates, equipment costs, etc. Bandwidth in excess of 4.2 MHz does provide somewhat better performance, but the improvement was considered uneconomical.

Low–frequency transmission requirements are set primarily by the low rate of 60 fields per second. Vertical blanking pulses appear in the complex signal waveform at that rate, and because of the complexity of the signal, there are also sideband components around that frequency. As a result, good transmission response must be provided to nearly zero frequency in order to maintain good phase response at and near 60 Hz.

As mentioned, the required bandwidth was determined by subjective tests. These were, in turn, conducted on the basis of previous judgments regarding the desirable vertical and horizontal resolution that was to be provided. These parameters, bandwidth and resolution, are importantly and intimately related.

Vertical Resolution. As previously described, the scanning process produces a standard pattern of 525 horizontal lines per frame. The picture width is 4/3 its height. This ratio is defined as the aspect ratio. Horizontal lines are lost during the vertical blanking period, reducing the effective (visible) number of lines to about 93 percent of the total. Further loss of resolution, inherent in the scanning process, is due to the finite width of the scanning line and to the shape of the scanning spot. The relative position of horizontal image lines and scanning lines affects reproduction. In the extreme, if the image has alternate black and white lines of the same width as the scanning lines and are coincident with them, a faithful reproduction results; however, if the same scanned lines were centered on the boundary between scanning lines, they would produce a flat gray picture. On the average, this effect decreases vertical resolution to about 70 percent. The net effect, then, is that the number of vertical elements that can be resolved is

$$N_v = 525 \times 0.93 \times 0.7 \approx 342.$$

Horizontal Resolution. Horizontal resolution is determined by the highest frequency component that can be resolved along a line. Assume that a simple sinusoid generates a series of black and white dots along the line. If spot size is not limiting, the finest detail that can be resolved is determined by the highest frequency that can be transmitted. If the horizontal resolution is to be about equal to the vertical resolution, the number of picture elements per line scan should be $N_h = 342 \times 4/3 = 456$ (the multiplier of 4/3 is used to account for the aspect ratio).

A sinusoid that would generate 456 alternate black and white dots would go through 228 cycles along a line. As shown in Figure 15–2, the duration of the visible portion of a line scan is about 52.3 μs. Thus, to satisfy the criterion that horizontal resolution should be about equal to vertical resolution, the top transmitted frequency should be $f_t = 228/52.3 \times 10^{-6} \approx 4.3$ MHz, a value close to that mentioned earlier, 4.2 MHz, obtained from subjective tests.

Spectrum. The line–scanning rate of a monochrome television signal determines to a great extent the distribution of energy in the signal spectrum. Thus, strong signal components are found in the signal at 15,750 Hz; since the signal waveform is complex, many harmonics of this fundamental are also produced. The spectral distribution is illustrated in Figure 15–5. Each component varies with time by approximately $\pm$ 3 dB according to picture content and motion; average values are illustrated. No voltage is shown at zero frequency because the amplitude of that component is under design control and is related to the design of the transmission circuits used. However, it is customary to clamp the signal so that the dc component is relatively constant in order to avoid excessive baseline wander [3]. Note that the envelope of the distribution decreases at a rate of 6 dB per octave; i.e., for each doubling of the frequency, the line–scan component is about 6 dB lower (one–half the voltage).

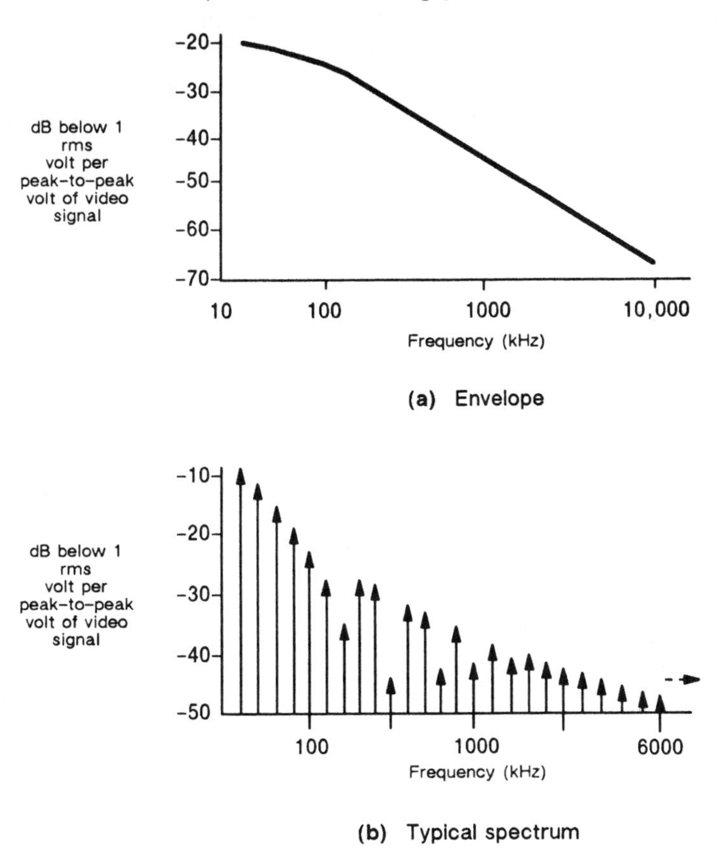

Figure 15–5. Monochrome television signal voltage spectrum.

The 60–per–second field frequency generated by the vertical blanking pulses also influences signal energy distribution. These blanking pulses produce upper and lower sidebands of 60 Hz and 60–Hz harmonics about each multiple of the line–scanning frequency.

Signal Amplitudes. In television signal transmission, amplitudes are limited by signal–to–noise and system overload considerations just as in any other form of signal transmission. These limitations are expressed in terms of average, peak, or single–frequency power at specific transmission level points (TLPs) in the telecommunications system if the signal parameter can be translated to the voiceband. However, TLP relationships cannot usually be directly applied if the signal is wideband (as in television), i.e., if it occupies more than a telephone channel bandwidth.

Generally, television–signal amplitude measurements are more conveniently expressed in voltage than in power. It is important, therefore, to recognize the voltage–impedance–power relationships that exist in television circuits and to recognize the complications inherent in properly translating the voltage expressions and their points of application to the TLPs used in telecommunications system operation. Exchange carriers offer television signals to ICs at a point of termination (POT) using a standard level of one volt peak–to–peak and an impedance of either 75 or 124 ohms [4].

Television signal transmission in North America is generally controlled from television operating centers (TOCs), where signals are received at baseband from the broadcasters. It is from this type center that signals may be switched to other baseband circuits for local distribution or pickup and to terminal locations for transmission over microwave radio or fiber optic channels used in the makeup of a television network. The relation of the TOC to a network is illustrated in Figure 15–6.

In order that the baseband and microwave radio transmission system designs may properly take into account the amplitude–bandwidth–spectral energy distribution relationships, television signal amplitude is maintained at one volt peak–to–peak into 124 ohms at the TOC. While it is never referred to as such, this point is somewhat analogous to the 0–dB TLP found in telecommunication networks. It is sometimes called the 0–dBV point.

Baseband Color Signals

The National Television System Committee (NTSC) was formed by the television industry during the early 1950s [5] to develop a set of color television standards that would be compatible with existing monochrome standards. The signal format that

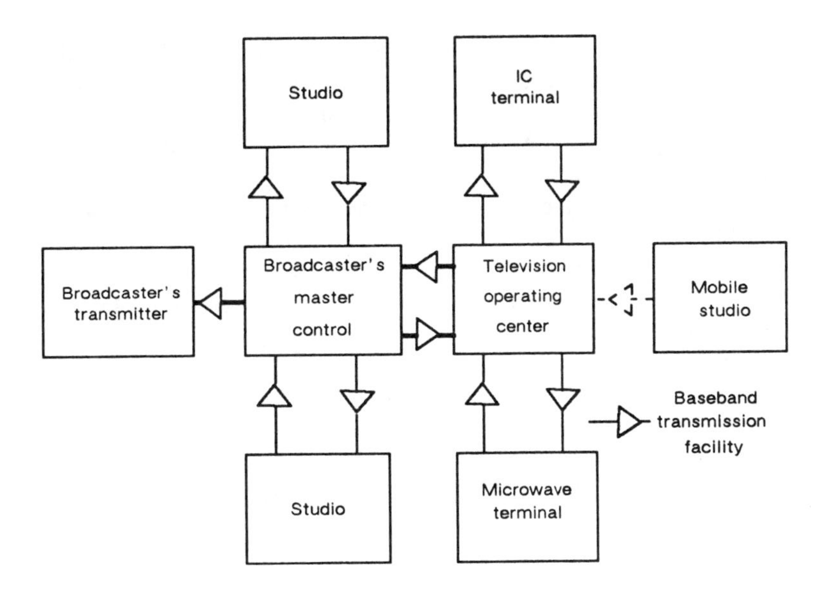

Figure 15–6. Typical intracity television network layout.

finally evolved is one that superimposes information regarding the image color content on the monochrome luminance signal. It has been adopted for use by the broadcast industry in North America and Japan. Uncompressed, the NTSC video signal requires a bandwidth of 4.2 MHz. This same signal transmitted digitally requires about 90 Mb/s for nine–bit coding at a sampling rate equal to three times the color subcarrier frequency. However, it is recognized that the wider the bandwidth, the higher the quality of the received image.

While luminance is transmitted as previously described for a monochrome picture signal, chrominance (hue and saturation) must be coded for transmission and is accomplished by modulating a carrier signal at 3.579545 MHz in the video band. The amplitudes and relative phases of this carrier and its sideband components are carefully controlled and together carry the necessary color information.

Scanning and Synchronization. The scanning for color involves the detection, through the same horizontal scanning procedures used in monochrome picture detection, of the brightness of each explored spot as to the amount of color present. Color is perceived by the amount of luminance of each of three color components. Red, blue, and green light phosphors in the television receiver combine to reproduce the proper picture composition with accurate color representation. The color carrier at the receiver is synchronized to that of the transmitter by superimposing the color carrier frequency at a reference phase on the back porch of each horizontal synchronizing pulse. The color burst signal contains about nine full cycles of carrier frequency (a minimum of eight) as shown in Figure 15–7. The phase relationship of the color carrier and its sidebands to the color burst determines the hue of the color. A scanned line is also illustrated in Figure 15–7 to show how the color information modifies the normal monochrome luminance signal.

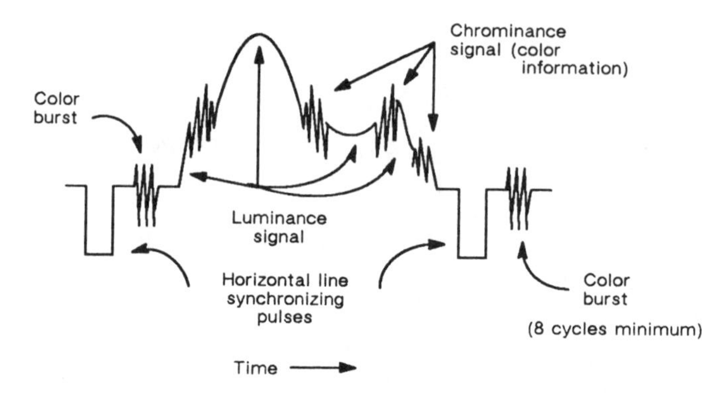

Figure 15–7. Color television composite signal waveform.

Spectrum. Concentrations of energy are found at line scan frequency multiples above and below the color carrier. The color signal carrier, 3.579545 MHz, was chosen as an odd multiple of one–half the monochrome line rate*, so that the color signal components fall in the spectral spaces between components of the luminance signal. The luminance color difference signals are transmitted as double–sideband (DSB) modulations of the color subcarrier called the chrominance signal. The power in chrominance signal components is 10 to 15 dB lower than in the

* For color signals, the line rate is 15.734264 kHz.

corresponding components of the luminance signal. The spectrum of a color signal is illustrated in Figure 15–8.

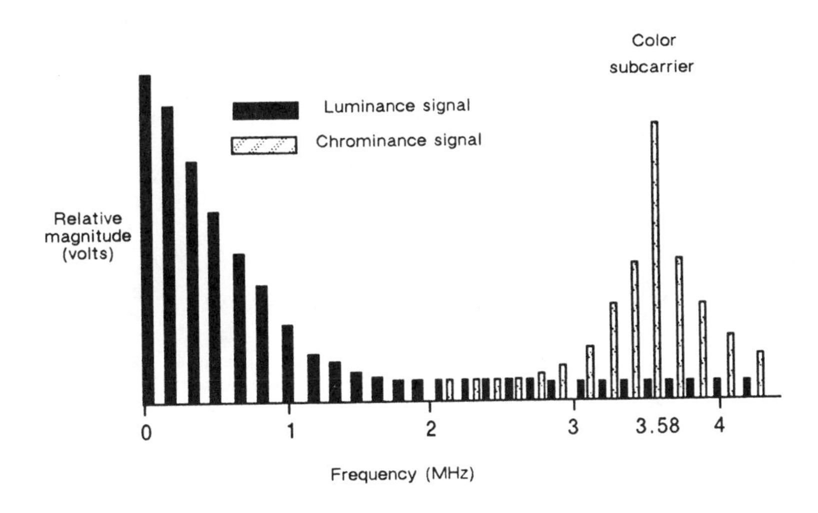

Figure 15–8. NTSC color TV spectrum illustrating interleaving of monochrome and color line-scan components.

Broadcast Signals

Baseband monochrome or color television signals are processed for broadcasting in accordance with standards specified by the Federal Communications Commission (FCC) [2]. The baseband audio signal, which frequency-modulates a carrier at 4.5 MHz, is added to the baseband video signal; the composite video/audio signal is then used to amplitude-modulate a carrier of assigned frequency in the radio-frequency (RF) spectrum. The channel assignments in the RF spectrum, each 6 MHz wide, are shown in Table 15–1. The assigned carrier frequency for each channel is located 1.25 MHz above the bottom frequency assigned to the channel.

Table 15–1. FCC Radio Spectrum Broadcast Television Channel Assignments

Channel No.	Frequency Band (MHz)	Channel No.	Frequency Band (MHz)
2	54–60	36	602–608
3	60–66	37 ......*......	608–614
4	66–72	38	614–620
5	76–82	39	620–626
6	82–88	40	626–632
7	174–180	41	632–638
8	180–186	42	638–644
9	186–192	43	644–650
10	192–198	44	650–656
11	198–204	45	656–662
12	204–210	46	662–668
13	210–216	47	668–674
14	470–476	48	674–680
15	476–482	49	680–686
16	482–488	50	686–692
17	488–494	51	692–698
18	494–500	52	698–704
19	500–506	53	704–710
20	506–512	54	710–716
21	512–518	55	716–722
22	518–524	56	722–728
23	524–530	57	728–734
24	530–536	58	734–740
25	536–542	59	740–746
26	542–548	60	746–752
27	548–554	61	752–758
28	554–560	62	758–764
29	560–566	63	764–770
30	566–572	64	770–776
31	572–578	65	776–782
32	578–584	66	782–788
33	584–590	67	788–794
34	590–596	68	794–800
35	596–602	69	800–806

* Reserved for radio astronomy receivers.

The transmitted signal is a vestigial–sideband (VSB) amplitude–modulated signal with transmitted carrier as illustrated in Figure 15–9. As shown, there is vestigial rolloff shaping at the transmitter. Thus, vestigial shaping must be provided in each television receiver. This mode of transmission tends to optimize overall signal–to–noise performance in the channel.

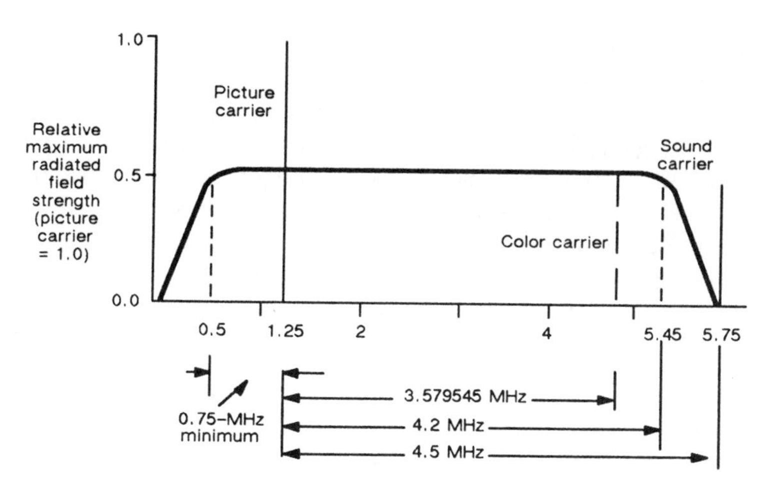

Figure 15-9. Idealized television channel amplitude characteristic—transmitter.

At present, it is general practice for one or two audio channels to be combined with the video signal. This is called diplexing. It is accomplished by transmitting the audio with a frequency-modulated subcarrier just above the video bandwidth. The first 15-kHz audio channel would use a 5.8-MHz subcarrier and the second, when generally provided for stereophonic sound, a 6.4 MHz. Diplexing two audio channels requires a wider bandwidth of about 6.6 MHz [4].

Closed-Circuit Signals

Exchange carriers transmit a variety of closed-circuit television signals such as industrial television (ITV) and educational television (ETV), as well as broadcast signals transmitted over cable television systems (CATV). Because these are closed-circuit arrangements, FCC standards for broadcast TV need not be met; however, the signal format is designed, in most cases, to meet FCC standards so that standard television sets can be used for viewing.

Industrial Television. ITV systems are always operated on a closed-circuit basis; that is, ITV signals are not transmitted over normal broadcast facilities. Because of this, the scanning and synchronizing mechanisms for ITV transmitting and receiving equipment are often less sophisticated than those required for broadcast television, and less operating margin may be provided. Interlaced scanning is sometimes not used because transmission objectives are less stringent than in other types of services. However, the bandwidth provided is usually about equal to that used for broadcast quality service. Amplitude control is provided at a TOC or equivalent when such signals are transmitted over intercity telecommunications facilities.

The overall effect is that for transmission analysis, an industrial or business signal may safely be assumed to be equivalent to a broadcast–quality television signal. Audio signals are transmitted only as required.

Business–quality television transmitted over private networks or video sent over terrestrial facilities may be compressed. Video can be compressed onto a DS1 channel (1.544 Mb/s) and still maintain satisfactory image quality. Some manufacturers offer equipment that uses one–half of a DS1 channel, or 768 kb/s. Compression generally uses a principle known as interframe coding, where only the difference from one frame to the next is transmitted. If the only changes are moving lips and raised eyebrows, satisfactory detail can be transmitted with reduced bandwidth.

At some point below business–quality television, some video conferencing services use "slow scan" technology. Those video systems that transmit at 56 kb/s have somewhat poor quality and have limited applications. However, advances in technology promise acceptable transmission at 64 kb/s.

At the opposite end of the video quality scale from traditional systems is high–definition video. Over the years since the NTSC standard had been developed and approved, video technology has surpassed the standard by displaying, for example, 1125 horizontal scanning lines versus the NTSC 525 lines. Color components are coded separately and transmitted apart from luminance and chrominance information. Over analog facilities, high–definition video requires 20 MHz of bandwidth. The non-compressed signal requires 1.2 Gb/s digital capability. However, commercial applications may use less bandwidth by compressing the video signal without compromising picture quality. The current effort in high–definition television (HDTV) is the development of national and international standards, coupled with research to reduce the bandwidth, to stimulate usage.

Educational Television. While ETV network arrangements may be different from broadcast TV network arrangements, the receiving equipment is a standard television receiver; therefore, the signal format is identical to the standard signal previously described. Sometimes ETV signals are transmitted over coaxial cable systems, which provide multiple channels.

Cable Television. In CATV systems, broadcast signals are received at a common point and distributed by cable to subscribers. In such systems, the signal format is again generally constrained to the broadcast format used by standard television sets. Four exceptions can be found. First, the distribution system need not have the same total band as is assigned in the radio spectrum, although government regulators may prescribe a minimum of channels (usually a greater number of stations than those normally serving the subscriber's area). Second, to ease the design of amplifiers in the distribution system, the sound signal is usually transmitted at a lower amplitude relative to the video signal than in normal broadcast practice. Third, either cable–ready television receivers must be provided by the subscriber or cable converters (tuners) provided by the CATV company in order to receive all the CATV channels. The fourth exception is that the customer may subscribe to additional coded channels in order to receive expanded premium channel services. The coding of the premium channel services is accomplished by the originator and the decoding provided by the CATV company at the subscriber location.

15–2 TELEPHOTOGRAPH SIGNALS

In spite of the fact that telephotograph* is the oldest of the video–type services provided by telephone companies, dating

* The main conceptual difference between telephoto and facsimile is that facsimile systems transmit black and white images while telephoto is capable of transmitting various gray values.

back to the early 1930s, and that new development work has been done in recent years, thousands of miles of telephotograph (also known as wirephoto) circuits are still in operation. These are primarily used to satisfy the needs of the news photo services [6] and users of weather photos taken by satellite.

Several different equipment types are currently in use, but the signal format is sufficiently similar that one general description should suffice. The systems of interest operate on private–line voiceband circuits that are equalized to meet the transmission requirements for picture transmission.

Scanning

In telephotograph systems, the light beam used to scan the image is held in a constant position. The picture being scanned is wrapped around and fastened to a cylinder that is rotated and advanced axially by a synchronous motor.

At the transmitter, the scanning light beam, modulated by the various shades in the picture, is reflected from the image surface into a photosensitive device similar to a photoelectric cell. At the receiving location, the demodulated signal is converted into an electric current that drives a light–emitting device forming a narrow light beam. The light beam is focused on film or photographic paper in proportion to the magnitude of the electric current. The picture is reproduced because the film or paper is light–sensitive.

Synchronism is achieved by absolute control of the motor speed at the transmitter and at the many receivers that can be operated simultaneously. The accuracy of synchronization is maintained to within a few parts per million by sending a sync pulse with each revolution of the sending drum.

Bandwidth

The scanning and modulation processes just described are accomplished by the intensity modulation of a light beam that is itself varied by a sine–wave function at a rate of 1800 Hz, or

2000 Hz in some systems and 2400 Hz in others. The resulting modulated electrical signal at the output of the photoelectric device is then transmitted as a DSB signal (2000–Hz carrier) or VSB signal (2400–Hz carrier). DSB signals are transmitted at a somewhat lower rate than that for the VSB mode. The band must be gain–and–delay–equalized between 1000 Hz and 2800 Hz. This useful band, 1800 Hz wide, is capable of producing the resolution demanded by the scanning rates given.

References

1. Glasford, G. M. *Fundamentals of Television Engineering* (New York: McGraw–Hill Book Company, Inc., 1955).

2. Federal Communications Commission. *Rules and Regulations, Title 47, Code of Federal Regulations, Part 73* (Washington, DC: U.S. Government Printing Office, 1987), Sections 73.681, 73.682, and 73.699.

3. Doba, S., Jr. and J. W. Rieke. "Clampers in Video Transmission," *AIEE Transactions*, Vol. 69, Part I (1950), pp. 477–487.

4. *Television Special Access and Local Channel Services—Transmission Parameter Limits and Interface Combinations*, Technical Reference TR–NPL–000338, Bellcore (Iss. 1, Dec. 1986).

5. Fink, D. G., Editor. "Color Television Standards," *Selected Papers and Records of the National Television System Committee* (New York: McGraw–Hill Book Company, Inc., 1955).

6. Hamsher, D. H., Editor. *Communication System Engineering Handbook* (New York: McGraw–Hill Book Company, Inc., 1967), pp. 2–67 to 2–84.

Chapter 16

Mixed Signal Loading

The signals described in Chapters 12 through 15 are found in various telecommunications facilities, but seldom is any one type of signal the only one ever found at any given location or facility. A variety of signal types are usually transmitted on the same facility; in other instances, the type of signal transmitted on the facility changes with time. Broadband carrier systems, for example, simultaneously carry speech, narrowband and wideband data, facsimile, program, and address and supervisory signals. A single trunk, on the other hand, may carry speech, data, facsimile, and address and supervisory signals at different times.

Important combinations or mixtures of signals must be characterized primarily so that the composite signal may be properly related to overload phenomena in carrier systems. In some cases, overload effects are relatively minor, causing only partial deterioration of performance. However, the effects are accentuated with increased signal amplitudes so that transmission may be seriously impaired and, ultimately, the entire system may fail. Any study of mixed signal loading must then be concerned with the characterization of signals known to be transmitted simultaneously in today's environment. In addition, the characterization must be in terms that permit continuous reevaluation of signals to account for the effects of new technology, the introduction of new terminal equipment or new services, or the transmission of signals originating in customer–provided equipment and interexchange carriers (ICs).

Since a very large number of combinations of signal loads may occur in broadband systems, it is difficult to characterize mixed signal loading effects explicitly. Intermodulation phenomena, whose effects are accentuated as signal amplitudes increase, produce noise, crosstalk, or other distortions such as signal compression. These are due to the nonlinear input/output characteristics of active circuits and devices, such as the line amplifiers in transmission systems. All transmitted signal components interact to form new unwanted signals. In some cases, these combine to form a noiselike impairment. In other cases, components of the interference signal fall directly upon and in phase with the components of a particular wanted signal so that its internal magnitude relationships become distorted (compressed). Sometimes, signal components combine with pilot or control frequencies to fall into other channels as intelligible crosstalk.

16–1 MIXED SIGNALS AND OVERLOAD

As previously mentioned, the effects of intermodulation are sometimes relatively minor, but as amplitudes increase, transmission may be seriously impaired. To avoid this, the signal load on a system must be controlled and limited to well–defined maximum values. When distortion or noise results from excess amplitudes and performance is seriously impaired, a transmission system is said to be *overloaded*. If the effects are so serious that communication is severely degraded, the phenomenon is called *hard overload*. Generally, the effects of overload must be evaluated statistically. Signal and system characteristics interact in ways that strongly depend on how long and by how much a signal exceeds its nominal value, how the system responds to the signal, and how quickly the system recovers from the overload condition.

Signal load criteria have been established to guard against overload. In general, the simplest statement of these criteria for signals transmitted by the exchange carriers is that the long–term average power in any 4–kHz band shall not exceed −16 dBm0. The statistical properties of individual types of signals are applied to specific situations to allow higher amplitudes for short time intervals. The most important of these statistical properties are the variations of signal amplitude with time and the activity factors that may properly be applied to various modes of transmission for each signal mode—simplex, half duplex, or duplex.

The same broad criterion is applied to signals requiring more than a 4–kHz frequency allocation, but it is expressed somewhat differently. In this case, the total power in the signal may be no greater than the long–term average power resulting from a signal of −16 dBm0 in each of the displaced 4–kHz channels. Again, the random nature of the broadband signal amplitude variations sometimes permits excess amplitudes for short periods of time.

Since many of the more serious problems relating to system loading are experienced in wideband systems capable of transmitting signals in 600–channel blocks (a mastergroup) or more, the following considerations of signal loading are mastergroup–oriented. For comparison purposes, the mastergroup speech signal load is first analyzed, and mixed signal loads are then compared to this analysis. For systems wider or narrower than one mastergroup, the same approach may be used by extrapolation and with appropriate care in the treatment of variables that are functions of system capacity. One of the attractive features of digital systems is that they are essentially free of overload effects. Under excessive signal conditions, crosstalk may occur between circuits in a specific 24–channel bank via its pulse amplitude modulation (PAM) bus, but this effect does not spread to larger groups of channels.

16–2 MASTERGROUP SPEECH SIGNAL LOAD

Consider a 600–channel mastergroup loaded with speech signals only. To determine overload relationships, it is necessary to know the average power and the peak power in the composite signal. The average power at 0 TLP (that power exceeded no more than one percent of the time during the busy hour) may be computed by using Equation 12–5.

$$P_{av} = V_{0c} - 1.4 + 0.115\sigma^2 + 10 \log \tau_L + 10 \log N + \Delta_{c1} \quad \text{dBm0}.$$

From Table 12–1, the average value of V_{0c} for toll calls has been found to be −21.6 vu. Since this value must be converted to its equivalent value at 0 TLP, allowances of −3 dB for tandem–connecting trunk loss (VNL + 2.5 dB) and +2 dB for conversion from the outgoing toll switch (−2 dB TLP) are added. Thus, V_{0c} equals −22.6 vu and, from Table 12–1, has a standard deviation of 4.5 vu. A standard deviation of 1 dB is also allowed for tandem–connecting trunk loss variations. Thus,

$$\sigma = \sqrt{4.5^2 + 1^2} = 4.6 \text{ vu, and } 0.115\sigma^2 = 2.4 \text{ vu.}$$

The activity factor, τ_L, is taken as 25 percent for typical speech loads; thus 10 log τ_L = −6 dB. The number of channels, N, is 600; 10 log 600 = 27.8 dB. The value of Δ_{c1}, which accounts for the maximum number of active channels, is found in Table 12–2 to be +0.7 dB for a 600–channel mastergroup. If these values are summed, the average power in a mastergroup of 600 telephone channels carrying only speech is found to be

$$P_{av} = -22.6 - 1.4 + 2.4 - 6 + 27.8 + 0.7 = +0.9 \text{ dBm0.}$$

The peak power can be determined from the average power by adding a correction factor, Δ_{c2}, which represents the rms signal amplitude exceeded 0.001 percent of the time. Thus,

$$P_{max} = P_{av} + \Delta_{c2} \text{ dBm0.} \qquad (16\text{–}1)$$

For N = 600 channels and τ_L = 25 percent, the number of active channels exceeded no more than 0.001 percent of the time is found from Table 12–2 to be N_a = 175.55. From Figure 12–8, the corresponding value of Δ_{c2} is about 13 dB. The 0.001 percent peak, then, is found from Equation 16–1 as

$$P_{max} = P_{av} + \Delta_{c2} \approx +0.9 + 13 \approx +13.9 \text{ dBm0.}$$

Note in Figure 12–8 that for large numbers of active channels ($N_a \geq 100$), Δ_{c2} approaches a constant value of about 13 dB. This relationship permits the use of Gaussian noise (which has the same peak factor) as an excellent simulation for a busy–hour multichannel telephone load for large systems.

The peak power capacity required of a transmission system ($P_s = P_{max} - 3$ dB) is usually expressed in terms of the peak power of a single–frequency sinusoid applied at 0 TLP (see Equations 12–6 and 12–7).

16–3 MASTERGROUP MIXED SIGNAL LOAD

In the present plant, no mastergroup can be expected to carry only speech signals. Thus, it is necessary to consider the effects of mixing various other types of signal with speech.

Speech and Idle–Channel Signals

Where signalling is performed on a trunk–by–trunk basis rather than by using common–channel signalling, the usual signalling system on trunks employing broadband carrier is the 2600–Hz single–frequency system described in Chapter 13, Part 2. When a trunk that is so equipped is idle, 2600 Hz is transmitted continuously as a supervisory signal at an amplitude of −20 dBm0. Thus, theoretically, a mastergroup may carry 600 such idle–channel signals (translated to carrier frequencies). In this case, the total average power in the mastergroup would be equal to −20 + 10 log 600 = 7.8 dBm0, higher than the previously determined average power in a mastergroup carrying only speech signals. However, activity factors affect the mastergroup signal in a complicated manner when the combined speech and idle–channel signal load is considered.

Average Mastergroup Power. It was shown previously that the busy–hour speech (only) load is +0.9 dBm0 if the speech activity factor, τ_L, is 0.25. That value of τ_L is, in turn, based on a trunk efficiency factor, τ_e, of 0.7. Thus, 180 trunks [600 (1 − 0.7)] may carry idle–channel supervisory signals. The power in these signals totals −20 + 10 log 180 = +2.6 dBm0. When this power is combined with the speech power by power addition (Figure 3–6), the total busy–hour load may be found as 2.6 "+" 0.9 ≈ +3.9 dBm0.

As the ratio of speech and idle–channel signals in a mastergroup varies from the busy hour to the nonbusy hour, the total power in a mastergroup varies somewhat from a maximum of +7.8 dBm0 to a minimum of +3.9 dBm0.

To illustrate the way in which the 600–channel mastergroup loading varies with time, consider the load when the busy–hour effect has been reduced so that the equivalent speech load is that of a 300–channel system. The speech load for N = 300 channels may be computed by the same method as that previously used; its value is found to be +1.5 dBm0. It is assumed that the other 300 channels carry idle–channel signals; the power in these signals is −20 + 10 log 300 = +4.8 dBm0. The remaining 300 channels are subject to the trunk efficiency factor, τ_e = 0.7. Thus, an additional 300 (1 − 0.7) = 90 channels carry idle–channel signals.

The power in these additional signals is −20 + 10 log 90 = −0.5 dBm0. The total power is thus 1.5 "+" 4.8 "+" (−0.5) = 5.8 dBm0.

Average Channel Power. In Chapter 12, Part 2, it is shown that the long–time average load per channel in a broadband system is about −27.6 dBm0 when only speech signals are considered. The mastergroup power for speech was found to be about +0.9 dBm0 during the busy hour. When single–frequency signal loading is included with the speech load, the average mastergroup busy–hour load is 3 dB higher, +3.9 dBm0. Therefore, it may be concluded that the long–time channel load is also 3 dB higher, or about −24.6 dBm0. This value is the long–term average channel power based on the average speech volume of −21.6 vu for toll calls measured in the 1975–1976 survey, which also indicated a tendency for volumes to increase slightly, but not significantly*, on longer toll calls [1]. Thus, the speech load on long–haul systems could increase by about 1 dB. Even with the 1–dB allowance, there appears to be a large margin between the speech load with present station sets and the design objective of −16 dBm0.

Maximum Mastergroup Power. In a mastergroup made up only of speech signals, the value of power that is exceeded 0.001 percent of the time may be found by Equation 16–1,

$$P_{\max} = P_{av} + \Delta_{c2} \quad \text{dBm0}.$$

It is now desirable to determine this maximum power value for a mastergroup having a speech and idle–channel signal load. It has been shown that the average value of a composite busy–hour mastergroup load is +3.9 dBm0. A value of Δ_{c2} for the composite signal remains to be determined. As shown in Figure 12–8, Δ_{c2} is equal to about 13 dB for speech signal loads in excess of about 75 active channels. It can be shown [2] that the instantaneous value of the sum of n sine waves of equal amplitude, different frequencies, and random phase relationships also exceeds the root mean square (rms) value of the n signals by about 13 dB

* The results of the 1975–1976 survey testify as to the uniformity of speech signal power in the local and interexchange telecommunication networks.

0.001 percent of the time for values of n in excess of 100. Thus, in a combined signal, one part consisting of 300 channels containing speech signals and one part consisting of 300 single–frequency supervisory signals, each having approximately the same peak factor, it can be safely assumed that the total also has a peak factor of 13 dB. Thus, the maximum power in a mastergroup carrying 300 speech signals and 300 idle–channel signals is

$$P_{max} = 3.9 + 13 = +16.9 \text{ dBm0}. \qquad (16\text{–}2)$$

This result depends on the assumption that the 2600–Hz signals are randomly related to one another in phase. Supposing, for example, that the phases of the 300 single–frequency signals assumed in developing Equation 16–2 were coherent so that their peak amplitudes coincided 0.001 percent of the time (an unlikely event). The peak voltage of the 300 signals then would be 300 times the peak voltage of one, and the peak power would be 300^2 = 90,000 times the power of one such signal. Thus, the peak power would be

$$-20 + 10 \log 90,000 = -20 + 49.5 = +29.5 \text{ dBm0}.$$

This is a peak value 12.6 dB higher than that of Equation 16–2, a value that would surely be expected to cause overload.

While phase coherence would not be expected under past field operating conditions and design practices, it did occur in some large–scale analog toll switching installations. However, a multiplicity of 2600–Hz oscillators and the random physical association of signalling equipment and transmission multiplexing equipment generally causes a dispersion of phase relationships and prevents any significant effects due to coherence. The layout of office equipment is planned and arranged to provide fixed wiring patterns between a single supervisory signal generator and many carrier channels. These wiring patterns, with controlled phase reversals, are designed to guarantee partial cancellation of composite signal peaks.

Speech and Address Signals

Address signals are transmitted at amplitudes higher than the −16 dBm0 long–term average power objective for a channel, but the statistics of these signals do not cause the average power objective to be exceeded. High–amplitude speech signal bursts of short duration also occur; these are limited to a maximum of +3 to +10 dBm0, depending on the system, by channel terminal equipment. The limiting has negligible effect on the individual speech signals; the distortion is masked by other distortions such as that occurring in station telephone sets.

Under normal operating conditions, these high–amplitude address and speech signals do not seriously affect carrier system operation. Their amplitudes are low compared to the total signal power in a mastergroup (+3.9 dBm0), and the frequency of occurrence is so low that such signals do not usually cause trouble. However, abnormal operating conditions or system designs that change the statistical relationships can lead to serious overload troubles related to the transmission of these high–amplitude signals. The misadjustment of channel equipment, operating errors (for example, the improper application of a test tone to a circuit), or the improper maintenance of a carrier system (which might result in some frequencies being transmitted at a higher amplitude than the designed values) are all trouble conditions to guard against. A system design that requires the use of shaped TLP characteristics, such as the preemphasis used in microwave radio systems and the signal shaping used in coaxial systems, results in peak factors that are equivalent to those in systems of fewer channels than are provided in their design (see Chapter 12, Part 3). Thus, peak factors are higher, the effective signal band is smaller, and even a single channel carrying an inordinately high signal can cause system overload.

Speech and Data Signals

Speech channels in carrier systems are frequently used for the transmission of data signals. These signals are routinely transmitted over trunks that are parts of the message telecommunications service (MTS) network or switched private–line networks and over point–to–point private–line circuits. Most of these circuits involve connection with customer–provided and IC equipment at a point of termination. The control of signal amplitudes tends to vary somewhat depending on the source of the signal.

As discussed in Chapter 14, Part 1, the maximum amplitudes of voiceband digital data signals are specified not to exceed −13 dBm0 when averaged over a three–second interval. Allowing for channel activity factors and a mix of duplex and half–duplex operation, the resulting long–term average should not exceed −16 dBm0. Modern systems are designed to operate satisfactorily over a range of data and speech combinations that meet the −16 dBm0 objective. However, a heavy concentration of private–line duplex data channels applied to a given system could cause this average to be exceeded and should be avoided by dispersing these channels over several systems.

Low–speed data signals, as transmitted over carrier systems, may be regarded simply as single–frequency signals insofar as their overloading effect is concerned. Medium– to high–speed data signals (1200 baud and up) are generally transmitted as a scrambled spectrum resembling speech. Therefore, a peak factor of 13 dB for multichannel systems (mastergroup or higher) may be safely assumed, since the peak factors for all contributors—speech, data, and supervisory signals—are equal. Where wideband data signals are involved (e.g., 50–kb/s service in the 48–kHz bandwidth), the data signal is normally scrambled so as to avoid the generation of high–level tones.

16–4 SYSTEM–SIGNAL INTERACTIONS

The characterization of signals in this chapter has been presented to relate the average and peak powers of the signals to carrier system overload. In some cases, system type, design, operation, or maintenance interacts with the transmitted signal to accentuate or mitigate overload effects.

System Misalignment

Multirepeatered broadband analog systems are designed so that the gain of the repeater compensates for the loss of the preceding section of transmission line. Because the compensation is not perfect, the signal amplitudes depart from their nominal values. These departures are called misalignment; they may be positive at some frequencies, causing the signals to be higher than nominal, and negative at other frequencies, causing the corresponding signal components to be lower than nominal. The effects on signal characteristics can be negligible for small misalignment, or they may be quite significant. The analysis of such effects is similar to the shaped TLP concept of Chapter 12, Part 1.

Pilot Signals

The discussion of phase coherence among single–frequency supervisory signals and the importance of guaranteeing random phase relationships among such signals apply also to single–frequency pilot signals. Special wiring designs are sometimes used to introduce phase reversals in order to produce partial cancellation of signal peaks.

Compandors

The advantage in individual channel signal–to–noise performance gained by using syllabic compandors results from the fact that the range of signal amplitudes is substantially reduced for transmission over the medium. The reduction, called the compression ratio, is usually 2 to 1; e.g., a signal amplitude range of 50 dB is reduced to 25 dB, and its standard deviation is also reduced by a factor of 2 to 1, from 6 to 3 dB. The average signal amplitude is a system design parameter. Thus, in a given carrier system using compandors, the value corresponding to V_0 for a noncompandored channel must be selected by the designer to optimize performance. The optimization must take into account the fact that the use of compandors results in a higher average power per channel.

Sometimes, when the multiplexed signal of a compandored carrier system is applied as a portion of the signal to another system of higher capacity that does not normally use compandors, precautions must be taken to avoid overload in the higher capacity system. The requirements of the high–capacity system may be met by reducing its channel capacity, by lowering the amplitude of the total applied compandored signal load, or by using filters to block the carrier components. While some of the compandor advantage may be lost, it is usually not important because the high–capacity systems are less noisy (by design) than the systems for which compandors are provided.

Microwave Radio Systems

The Federal Communications Commission specifies that the frequency deviation of a frequency–modulated microwave carrier be confined to the allocated band [3]. Modern microwave systems, while designed to carry a long–term average per–channel signal power load of −16 dBm0, are tested by noise–loading techniques with a load equivalent to −15 dBm0 per 4–kHz channel. This approach provides some margin against peak excursions of composite signals, which produce the extremes of the microwave frequency deviations. At the same time, signal–to–noise objectives are met for the system when operated at load value equivalent to −16 dBm0 per channel.

References

1. Ahern, W. C., F. P. Duffy, and J. A. Maher. "Speech Signal Power in the Switched Message Network," *Bell System Tech. J.*, Vol. 57, No. 7 (Sept. 1978).

2. Bennett, W. R. "Distribution of the Sum of Randomly Phased Components," *Quarterly Journal of Applied Mathematics*, Vol. 5 (Jan. 1948), pp. 385–393.

3. Federal Communications Commission. *Rules and Regulations, Title 47, Code of Federal Regulations, Part 2* (Washington, DC: U.S. Government Printing Office, 1986), Section 2.202.

Telecommunications Transmission Engineering

Section 4

Signal Impairments and Their Control

This section contains descriptions and definitions of impairments suffered by telecommunication signals as they are transmitted through various channels and media. These descriptions, qualitative for the most part, are related to the sources of impairment, the manner in which the impairments are measured, and the units in which the measurements are expressed.

Signal transmission is subject to impairment by a variety of transmission characteristics of each channel the signal transverses. Channel characteristics include interferences induced from external sources, interferences that are signal–dependent and caused by nonlinear channel input–output characteristics, distortions of the channel transmission characteristics, and indirect effects such as timing and synchronization irregularities. It is possible to control the various impairments to which a signal may be subjected once their effects are understood.

Particular transmission irregularities affect various types of signal differently. For example, a channel that has phase jitter, intermodulation distortion, or impulse noise, which is scarcely noticeable to the human ear, may cause sizable impairment of voiceband data signals. A channel having an impedance discontinuity that may cause bothersome echoes in terms of speech transmission may have negligible effects on the information content of a data signal. A digital signal may undergo sizable amounts of timing jitter with no perceptible effect to an end user.

Sometimes channel irregularities may be dealt with in the design process. For instance, if a particular type of signal is intolerant of frequency shift, the effect can be eliminated by using a method of transmission that permits recovery of the frequency and phase of the transmitted carrier.

Occasionally, substandard performance may be tolerable under emergency conditions. This situation would be typified by an increase in noise during restoration of a major facility that has failed. The ability to furnish some service, even if below normal quality, is preferable to furnishing no service at all.

Some channel transmission characteristics are evaluated subjectively and others objectively. Sometimes the same impairment may be evaluated both ways depending on the type of signal that will be using the channel. For example, random noise must be evaluated in terms of its annoying effect in speech or video signal transmission and for the number of errors it causes in data signal transmission. In either case, the ultimate expression must be in terms that can be stated quantitatively so that meaningful values may later be established for objectives and requirements.

Descriptions of signal impairments and the ways that they can be controlled are discussed in the following chapters:

Chapter 17 – Noise and Crosstalk
Chapter 18 – Amplitude vs. Frequency Response
Chapter 19 – Timing and Synchronization Errors
Chapter 20 – Echo in Voiceband Channels
Chapter 21 – Phase Distortion
Chapter 22 – Maintenance and Reliability.

Chapter 17

Noise and Crosstalk

The transmission of telecommunications signals is affected by the practical limitations of channels and by various types of interference. Interference may be induced from an external source such as power line noise picked up by a voice–frequency circuit, or it may be generated from within, such as quantizing noise generated in a pulse code modulation (PCM) coder–decoder.

The effects of interference depend on the type of signal transversing a channel. For example, bursts of impulse noise are usually of little consequence in the transmission and reception of speech because of the relative insensitivity of the human ear to this type of impairment. However, impulse noise can seriously impair data signal transmission. Some other types of interference, such as thermal noise, affect the transmission of all types of signals. Unwanted signals and interferences, their sources, means of controlling the sources, coupling paths, channel sensitivity, the nature of the impairments, and the method of measurement are all basic to an understanding of impairments to the transmission of signals in the telecommunications network and how to control them.

17-1 COUPLING

Almost all transmission circuits are exposed to external influences and forces because of their proximity to other circuits. Some examples are: a loop or trunk physically close to other circuits in cables or on pole lines, multiplexed message channels sharing the same carrier line, a circuit passing through an electromechanical central office exposed to sizable switching transients, and many circuits having power transmission lines paralleling part of their routes. The exposure to electromagnetic fields created by the currents in these nearby circuits results in many possible interference coupling paths from a *disturbing* circuit to the circuit of interest, the *disturbed* circuit.

Currents and Circuit Relationships

Electromagnetic and capacitive coupling results in longitudinal currents in the disturbed circuit. Longitudinal currents are in–phase currents that flow in the two conductors of a pair of wires. The extent of interference caused by coupling depends on the symmetry, or balance, of the disturbed circuit. A voltage difference between each wire of a pair of wires, from longitudinal currents, results when the impedances to the longitudinal currents are not equal on each wire. A measurement of this voltage difference is called metallic voltage.

Figure 17–1 depicts an induced in–phase voltage source E generating longitudinal currents in a typical exchange circuit. These longitudinal currents are carried to and from the central office, typically through common–battery circuits, which include impedance Z_{c1} and Z_{c2}, as illustrated. Z_{c1} and Z_{c2} impedances may represent supervisory relays, transformers, common–battery supply, etc. The longitudinal current path is complete through distributed exposures to ground on the transmission line and station apparatus represented as impedances Z_{s1} and Z_{s2}.

Figure 17–1(a) is showing a balanced condition where Z_{c1} equals Z_{c2} and Z_{s1} equals Z_{s2}. Longitudinal currents are equal on both sides of the transmission path. A metallic voltage evaluation, represented by E_M, shows equal currents on both paths. At any given point along the circuit, there is zero metallic voltage across the conductors.

Figure 17–1(b), however, is showing that the impedance Z_{s1} is much less than Z_{s2} (note larger arrows), thereby allowing the longitudinal current to be greater on this leg than on the opposite leg. The metallic voltage evaluation indicates the voltage drop difference because of this impedance imbalance.

Coupling Paths and Their Control

The coupling path from a disturbing to a disturbed circuit may result from electromagnetic, capacitive, electrical, or

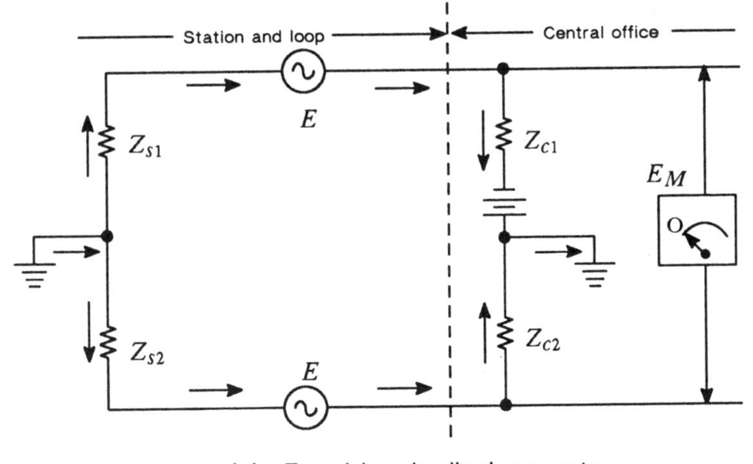

(a) Equal longitudinal currents

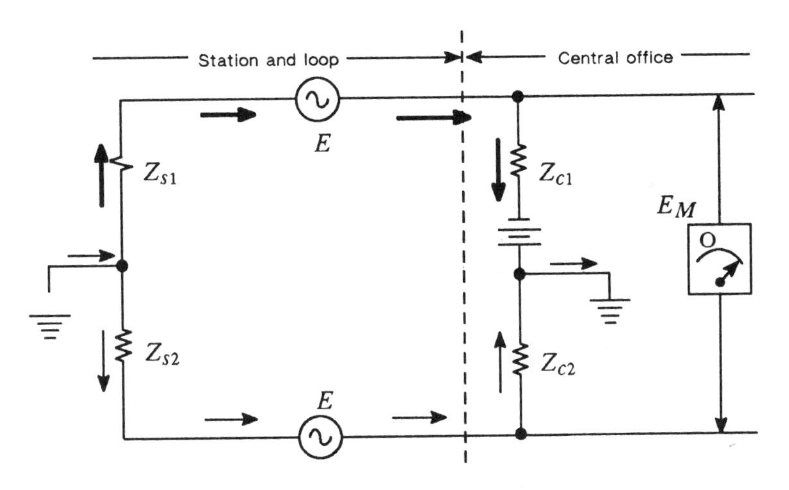

(b) Unequal longitudinal currents

Figure 17-1. Longitudinal currents and metallic voltages.

intermodulation phenomena. These coupling paths are typical of interference problems found in telecommunications systems. Coupling path losses must be controlled so that transmission impairments may be held to tolerable values.

Electromagnetic Coupling Paths. A coupling path is involved when an electromagnetic field resulting from an alternating current carried in one conductor causes a current to be induced in another conductor [1,2]. Figure 17–2 illustrates a simple and idealized case of magnetic coupling developed from Figure 17–1, where A is one conductor of a disturbing circuit equidistant from the two conductors of disturbed circuit B. The magnetic field produced by current I_C in conductor A induces voltages E_1 and E_2 in circuit B. The resulting longitudinal currents in the two conductors of the disturbed circuit, I_1 and I_2, are exactly equal and in phase. However, the longitudinal currents in the disturbed circuit are of opposite polarity to those in the disturbing circuit. Electromagnetic coupling may be called transformer coupling in that the disturbing circuit (A) is the primary of a one–turn transformer and the disturbed circuit (B) is a one–turn secondary.

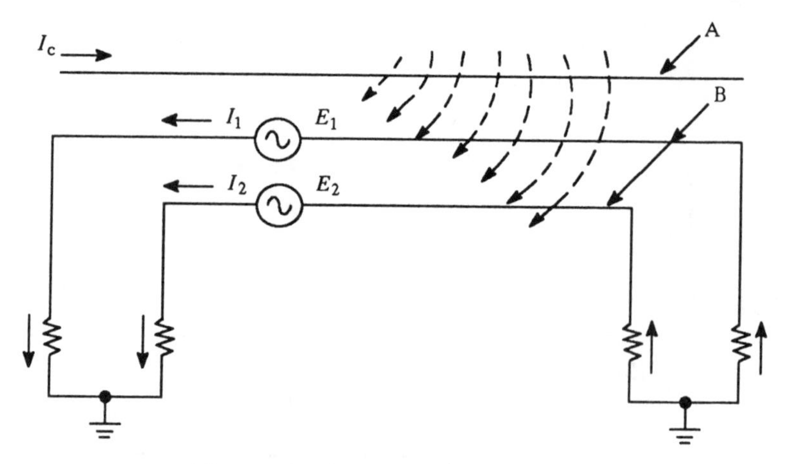

Figure 17–2. Magnetic coupling.

Departures from the idealized conditions assumed in Figure 17–2 may cause the induced currents in the two conductors of the disturbed circuit to be different in magnitude or phase; the result would be a net metallic current and voltage in the disturbed circuit. This may occur if the couplings from conductor A to the conductors of circuit B are unequal, or if the impedances to ground in the disturbed circuit are unequal.

Among the measures employed to control the magnetic coupling between circuits is shielding. Metallic shielding is part of the sheaths of multipair cables. Also, specially shielded wires are used for intracity baseband television signal transmission. More commonly, shielding is used on circuit packs having areas of high component density, especially where magnetic components (inductors and transformers) are used. Where transmission lines are shielded against magnetic coupling, the shields are grounded at both ends. The growing use of digital carrier for long loops brings with it a reduction in the exposure of voice–frequency cable pairs due simply to their reduced length.

Capacitive Coupling Paths. Longitudinal currents in a disturbed circuit may be the result of capacitive coupling between adjacent parallel conductors. This form of coupling is among the most important and most prevalent in communications systems. Capacitive coupling loss is maximum at low frequency and tends to decrease at a rate of 6 dB per octave of frequency. If the capacitance from a disturbing conductor to each of the two wires of a disturbed circuit is different, the longitudinal current flowing in each of the disturbed conductors is different and a metallic voltage results.

Many of the guidelines that govern the disturbing paths in magnetic coupling apply to capacitive coupling as well. Electrostatic shielding of conductors, separation between disturbing and disturbed circuits, and orientation of one circuit with respect to another all affect the result of capacitive coupling paths.

Electrical Coupling Paths. Electrical coupling paths are those produced by circuitry that is common to two or more otherwise independent circuits. These paths include common–battery supplies, their supply leads and filters, and multiplex terminal filters whose passbands overlap.

The manner in which the common–battery impedance can become a source of interference is illustrated in Figure 17–3. Two stations are shown with independent connections to trunks or other stations. The two local stations, STA 1 and STA 2, receive current from the common–battery supply having an internal impedance Z_b. The connections from the stations are over loops to the central–office coupling transformers and then to the battery supply via supervisory circuitry. A signal generated by STA 1 results in signal energy across the central–office transformer. Some

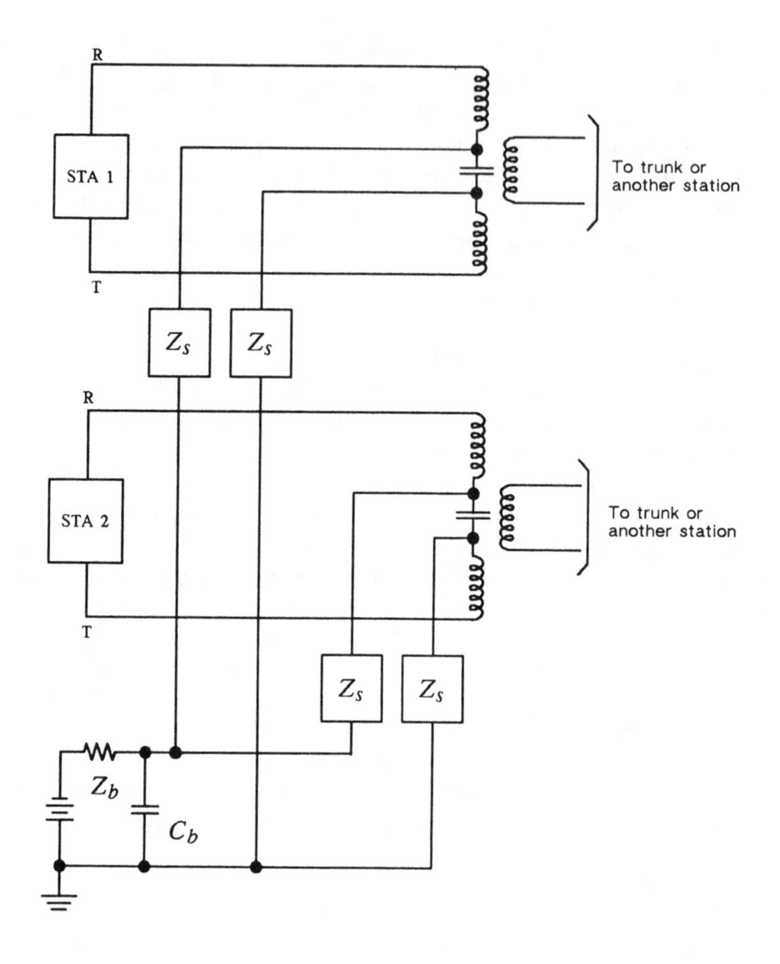

Figure 17–3. Electrical coupling through the common–battery impedance.

of this signal is impressed across the transformer midpoint capacitor, through the supervisory circuits (Z_s), and thus across the battery supply. The value of the battery impedance (Z_b) will determine the degree of interfering signal voltage that is impressed on the supply leads to the central–office equipment of circuit 2. A common method of keeping this battery impedance very low (less than 0.1 ohm) is by shunting the battery impedance with a large capacitor (shown as C_b in Figure 17–3).

This example illustrates how two circuits may be coupled by a common impedance, such as Z_b, and how the coupling may be controlled in design by controlling the common impedance to a much lower value than the coupled transmission circuit impedances. In a real case, the internal battery impedance tends to be very low, typically less than 0.01 ohm, and coupling becomes a problem only where long power leads are common to a number of otherwise independent circuits.

Figure 17–4 illustrates how analog carrier terminal filters may provide an electric coupling path. Two 4–kHz baseband inputs are shown at the left. The input signals modulate carriers at frequencies f_{c1} and f_{c2} (where $f_{c2} = f_{c1} + 4$ kHz) and produce double–sideband signals in bands $\pm$ 4 kHz about the carriers. These double–sideband signals then pass through bandpass filters that pass their lower sidebands and suppress their upper sidebands. At the filter outputs the two signals are combined. Note that the suppressed upper–sideband signal from input 1 falls directly into the band occupied by the lower–sideband signal of input 2. This form of coupling can sometimes be avoided by selecting certain carrier frequencies so that the overlap does not occur; however, bandwidth is wasted. Control is usually attained by designing the filters so that adequate suppression is obtained and interference is at an acceptably low amplitude, 50 to 80 dB below the wanted signal currents.

Intermodulation Coupling. The coupling that results from intermodulation among signals in a frequency–division multiplexing (FDM) system cannot be described in the same physical sense as the electromagnetic and electrical coupling paths just discussed. Intermodulation is inextricably involved in the mathematics of the nonlinear input/output characteristics of the devices; therefore, a detailed discussion of this type of coupling is deferred to a later part of this chapter.

17-2 CONTROLLING INDUCED NOISE AND CROSSTALK

The sources of many types of noise and crosstalk are outside the disturbed channel. Such interferences appear in the disturbed channel by coupling mechanisms and paths discussed earlier in this chapter. Here, some specific induced interferences and methods of control are described.

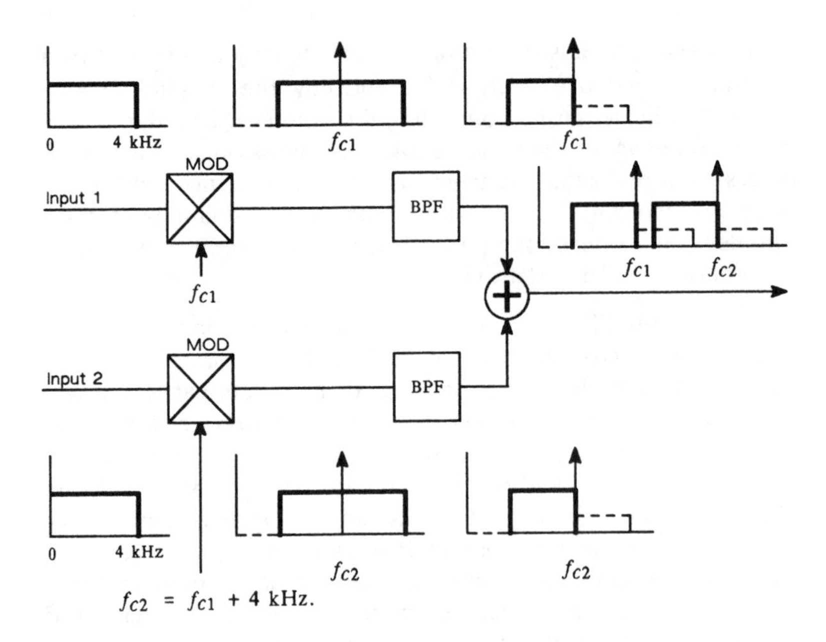

Figure 17-4. Electrical coupling in multiplex terminal equipment.

Power System Noise

Problems involving inductive coupling arise where facilities for the power industry and the communications industry share the same underground or pole line environment. Communication channels using metallic conductors are particularly susceptible to interference from the high–strength magnetic and electric fields generated by power systems.

Nature of Impairments. The characteristics of the 60–Hz wave present in most power distribution systems are of high energy and sizable harmonic content (especially odd harmonics), with significant differences between the currents carried by different conductors of the power line (i.e., unbalanced currents). The 60–Hz fundamental component is at such a low frequency that speech transmission is seldom impaired because most voice–frequency circuits and all carrier circuits have high attenuation at 60 Hz. However, high–amplitude odd harmonics of 60 Hz can cause an unpleasant hum in voice–frequency speech transmission systems. Excessive 60–Hz levels may damage terminal equipment or cause signalling problems.

The 60–Hz, with its harmonic components, can also cause bar–pattern interference in television or other video channels. Video signals require a flat attenuation/frequency response to essentially zero frequency; thus, extraneous signals at low frequencies can cause picture impairment.

Inductive Coordination. This term is applied to the cooperative efforts of the power industry (represented by the Edison Electric Institute), other utilities, and the telephone industry to solve problems that arise where facilities for different types of service share the same environment. In considering problems of interference in communication circuits due to coupling from power circuits, three conditions are considered. These are influence, coupling, and susceptibility. *Influence* refers to those characteristics of power circuits and associated apparatus that determine the character and intensity of the fields that they produce. *Coupling* covers the electric and magnetic interrelations between power and communication circuits. *Susceptibility* refers to those characteristics of communication circuits and associated apparatus that determine the extent of any adverse effects from nearby power circuits. These three conditions form the basis of inductive coordination.

Influence. An unbalance of the currents carried on the individual conductors of a multiphase power distribution circuit is a source of influence on communication circuits. The influence results from the unequal magnetic and capacitive fields that, because of their inequality, result in a metallic voltage. A reduction of power–line influence may be accomplished by transposing the power–line conductors and by balancing power load currents. The influence reduction results from a more complete cancellation of fields generated by the power transmission line. Load balancing, in addition to reducing influence, makes power distribution more efficient. Since an unbalanced power load results in return power to the power substation via ground, a method of analyzing the balance profile of a power–line distribution system is by measuring earth currents along the power line.

Coupling. The distance between power and communication circuits and their mutual orientation are among the factors that must be considered in order to control coupling. On shared facilities, the separation between the potentially interfering source and each communication conductor must be as large as practicable, and the distance between the communication paired conductors must be as small as possible. When power and communication circuits cross one another, a 90–degree crossing minimizes coupling between the lines.

Susceptibility. The susceptibility of communication circuits can be reduced in a number of ways, such as by:

(1) shielding
(2) twisting
(3) transposing
(4) suppressing longitudinal currents
(5) increasing circuit impedance to ground
(6) improving impedance balance to ground.

By *shielding*, and grounding the shield at each end of a run, the induced currents from a power line are of the same polarity and phase as the induced currents in the communication conductors. However, this current in the shield generates a field 180 degrees out of phase with the field generated by the power line, thereby lessening the total power–line field sensed by the communication conductor.

Disturbing metallic voltages on a communication circuit are reduced by *twisting* the conductors of each pair together. Twisting tends to equalize the distance from each conductor of a pair to each power conductor so that induced currents are more nearly equal and metallic voltages on the communication circuit are minimized. Where twisting of pairs cannot be used, the disturbed conductors may be *transposed* at regular intervals to improve balance.

In the design of a communication circuit, careful consideration is given in the choice of supervisory, battery supply, and transmission equipment to *suppress longitudinal currents* and *maximize the longitudinal impedance to ground* without an undesirable effect on the transmission path.

Perhaps the most effective means of minimizing the result of power–line influence on a metallic communications circuit is in the designing, installation, and maintenance of a circuit whose two sides are electrically alike and symmetrical with respect to ground. The *longitudinal impedance balance to ground* of a circuit or component may be evaluated by measurement. This measurement determines the similarity of the longitudinal impedance on each side of the transmission path and is usually expressed in terms of the induced or applied longitudinal voltage and the resulting metallic voltage. This ratio is then converted to dB as follows:

$$\text{Balance} = 20 \log \frac{\text{longitudinal voltage}}{\text{metallic voltage}} .$$

Most noise–measuring sets make both longitudinal and metallic voltage readings directly in dBrn; thus the formula for longitudinal impedance balance becomes

$$\text{Balance (dB)} = N_g - N_m$$

where N_g is the longitudinal noise (voltage) to ground and N_m is the metallic noise (voltage) across the transmission path.

Balance values for equipment and components are commonly greater than 70 dB. For a complete circuit having a transmission facility plus a variety of intermediate and terminating equipment, balance objectives are generally greater than 60 dB.

Some coordination problems are structural and, as such, may result in danger to personnel or communications equipment from high energy coupled from the power source into the communications circuits. While most systems use 60–Hz ac, some high–voltage dc power systems are also in use. Stray direct currents from the latter may cause corrosion if structural problems and appropriate grounding arrangements are overlooked. In addition to corrosion, any voice–frequency harmonic currents from ac–dc converters may travel on the dc line and cause noise as well.

Impulse Noise

Impulse noise consists of spikes of energy of short duration that have approximately flat spectra in the band of interest. The flat spectrum is shaped by channel response characteristics so that, typically, the average spectrum of a large number of observed impulses approximates the frequency response of the channel on which the measurements are made [3]. Some of the more important sources of impulse noise are:

(1) corona discharges in transmission lines

(2) lightning

(3) electrical transients associated with the termination of a connection

(4) dial pulsing and operation of electromechanical switches

(5) microwave radio fading and the associated protection switching

(6) timing and synchronization instabilities in time–division multiplexing (TDM) systems

(7) miscellaneous other operating and maintenance activity.

In speech transmission, impulse noise causes little impairment. Above a certain threshold value, acoustic disturbance caused by impulse noise might be painful; however, circuits are designed to limit amplitudes well below this threshold, and such disturbance is seldom experienced. At amplitudes below the limiting values, the human ear is quite tolerant of impulse noise.

In video transmission, impulse noise takes the form of short–duration interferences such as small light or dark flickers in the picture (sometimes called pigeons) or, in extreme cases, a brief loss of synchronization that causes the picture to tear horizontally or to roll vertically.

The most serious effect of impulse noise is in data transmission. The interfering impulses are short compared with the time between them. As a result, the receiving circuits resolve them as independent events. Depending on the impulse amplitude, polarity, duration, and time of occurrence, individual signal elements may be obliterated, resulting in errors in the received data.

Single–Frequency Interference

Single–frequency signals or wideband signals having discrete single–frequency components can be annoying when coupled into a voiceband channel. If they are of sufficient amplitude and fall between 200 and 3500 Hz, they may produce interfering tones to the human ear. Single–frequency interferences can also disturb single–frequency signalling systems, produce bar patterns in video receivers, and cause errors in a data transmission system.

Crosstalk

Crosstalk was initially used to designate the presence in a voiceband channel of unwanted speech signals from another voiceband channel. The term has been extended in its application to designate interference in one communication channel caused by signals present in other channels. Major issues of current and future interest are, for example, the impairment of the operating margin of one digital signal by crosstalk from many others of the same type or from another of a different type, and degradation of preexisting analog services by crosstalk from newly introduced digital signals. Crosstalk coupling of one signal type to another is often significant in establishing transmission level points (TLPs) for communication systems.

Crosstalk Coupling and the TLP. As discussed in Chapter 3, Part 2, many of the complexities of system design and operation are made manageable by the concept of TLPs. The importance of any coupling path strongly depends on the relative magnitudes of the wanted signal and the interference. It is really a question of the signal–to–interference ratio, which is difficult to define in practice; the TLP approach is found to be a useful way of dealing with signal–to–interference problems.

With this approach, the interference is defined in terms of its value at a specific TLP, and the coupling is expressed as equal level coupling loss (ELCL). The ELCL is defined as the ratio of signal power at a known TLP in the disturbing circuit to the induced power measured at an equal TLP in the disturbed circuit. The ELCL concept is illustrated in Figure 17–5, where two repeatered voice–frequency circuits are depicted as transmitting from left to right. A coupling path having an 80–dB loss is shown from the output of the repeater in the disturbing circuit to the input of the repeater in the disturbed circuit. Thus, the coupling path loss must be adjusted by the gain of the repeater in the disturbed circuit (30 dB) to give an ELCL value of 50 dB. Note that the ELCL is independent of signal amplitude and of the particular TLP used in its determination.

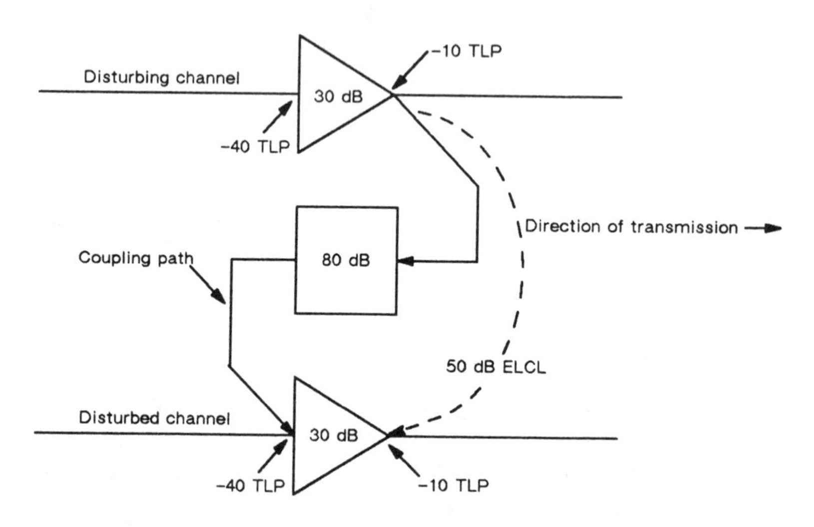

Figure 17-5. Coupling path loss and equal level coupling loss.

Near-End, Far-End, and Interaction Crosstalk. Near-end crosstalk (NEXT), far-end crosstalk (FEXT), and interaction crosstalk (IXT) are subclasses of coupling modes that exist between communication channels or circuits. Although any of these forms of crosstalk coupling may cause an impairment, NEXT and FEXT tend to be predominant. With NEXT coupling, the interference energy in the disturbed circuit is transmitted in the direction opposite to that of the signal energy in the disturbing circuit. With FEXT coupling, the signal and interference travel in the same direction. IXT occurs when energy is coupled to a tertiary path, propagates along that path, and then is coupled to the disturbed circuit. The two stages of coupling may be of the near-end or far-end type or a combination of both.

Speech Crosstalk. When an unwanted speech signal is coupled into another speech channel, the interfering signal may be intelligible or unintelligible, but with syllabic characteristics so that a listener thinks it is intelligible. Such interferences are particularly objectionable because of the real or perceived loss of privacy. Even when this disturbance is clearly not intelligible, it tends to be highly annoying because of the syllabic characteristic. Stringent objectives are applied to minimize these interferences.

When many unwanted speech signals appear in a disturbed channel simultaneously, each at such a low amplitude that neither intelligence nor syllabic variations are conveyed, the net effect may resemble random noise. In rare circumstances, the metallic coupling of large numbers of speech signals through common-battery circuits in a central office might produce such an impairment. However, coupling losses through battery feed circuits are kept at high values by design and maintenance.

Digital-to-Analog-Carrier Crosstalk. Careful coordination of loop assignments is required to avoid digital-line-to-analog-carrier crosstalk, which would otherwise result in noise or tone interference in the analog system. An example of this problem is when an analog subscriber carrier, either single- or multichannel, operating at frequencies in the 10–120-kHz region, is exposed to digital services (like DDS or ISDN) that have most of their signal power in the same region. The analog systems would then be heavily exposed to carrier crosstalk (digital-to-analog).

Video Crosstalk. When a picture signal is coupled to another video channel, the interfering signal may be superimposed on the disturbed picture. This form of interference is rare, however. The coupling path usually has a loss/frequency characteristic that distorts the interfering signal. A more common effect of such coupling is known as the *windshield wiper effect*. The synchronizing pulses of the disturbing signal create a bar pattern across the disturbed picture. Since the two signals are usually unsynchronized, the bar pattern moves across the picture with a windshield wiper effect.

Digital Signal Crosstalk. As in most cases of interference to digital signals, crosstalk may produce errors. Below some threshold essentially no errors result, but a disturbing signal amplitude just slightly higher than the threshold value causes a sharp increase in error rate. In the design of digital transmission systems, the crosstalk due to the presence of the line signals of many systems in one cable is often the limiting factor in the spacing of regenerators or the upgrading of a span line to higher capacity.

17-3 SYSTEM-GENERATED NOISE AND CROSSTALK

Many sources of noise and crosstalk exist within a channel or transmission system. Such interferences are influenced by circuit design and, within the constraints of a particular design, by signal amplitude since the ultimate interfering effect is a matter of the signal-to-noise ratio. If interferences are independent of the signal amplitude, the signal-to-noise ratio is improved by raising the amplitude of the transmitted signal. If interferences are signal-dependent, the signal-to-noise ratio is generally improved by reducing the transmitted signal amplitude, because this type of interference generally changes more rapidly than the signal amplitude. Thus, performance optimization in transmission systems involves the selection of optimum signal amplitudes because both types of noise are usually present.

Random Noise

Random noise is an impairment that appears in all circuits as a result of physical phenomena that occur within the affected circuit or channel. Important noise sources in transmission channels are briefly described in this chapter; the resulting impairments to various telecommunications signals, and methods that are employed to measure and evaluate these types of interferences, are discussed in Reference 4.

The terms *white noise* and *Gaussian noise* are often used to describe random noise. The term white noise has become well established to mean a uniform distribution of noise power versus frequency, i.e., a constant power spectral density in the band of interest. The Gaussian noise distribution, however, is a limiting form for the distribution function of the sum of a large number of independent qualities that individually may have a variety of different distributions. A number of random noise phenomena produce noise having this Gaussian amplitude distribution function.

By definition, Gaussian noise has some probability of exceeding any given magnitude, no matter how large. In practice, however, consideration can be limited to the magnitude attained 0.01 percent of the time. Thus, it is convenient to define the peak factor for random noise (having a Gaussian distribution) 11.8 dB above the rms value (usually rounded to 12 dB for convenience). In cases where the value attained 0.001 percent of the time must be used, the peak factor is 13 dB.

Because there is no phase correlation between noise components from independent sources, the total power may be computed simply by summing the power of the multiple sources.

Thermal Noise. According to the kinetic theory of heat, electrons in a conductor are in a continual random motion, which leads to an electrical voltage whose average value is zero but that has ac components of random amplitude and duration. This phenomenon produces an interference signal called thermal noise [5,6].

The thermal noise spectrum is flat from zero to the highest microwave frequencies used and can be termed white noise. The available noise power is directly proportional to bandwidth and absolute temperature. While thermal noise has a flat power spectrum and a Gaussian amplitude distribution, it should not be concluded that white and Gaussian are synonymous; they are not.

Shot Noise. This type of random noise is found in most active devices. It is similar to thermal noise in that it has a Gaussian distribution and a flat power spectrum. However, it differs from thermal noise in the following respects:

(1) The magnitude of thermal noise is proportional to absolute temperature, whereas shot noise is not directly affected by temperature.

(2) The magnitude of shot noise is proportional to the square root of the direct current through the device. Thus, the shot noise magnitude may be a function of signal amplitude if the signal has a dc component.

(3) The magnitude of shot noise in solid–state devices is proportional to $1/f$, and generally becomes unimportant at high audio frequencies.

For fixed conditions in a particular design, it is often convenient to combine shot noise with thermal noise into a single equivalent source.

Intermodulation Noise and Crosstalk

Intermodulation, caused by nonlinear input/output characteristics of analog and digital system devices, may result in many different types of interference, all of which are signal–dependent. The process is very complex and has many variables that need not be fully evaluated here [4]. However, it suffices to say that the result of the nonlinear characteristics of a device may be expressed as a power series having an infinite number of terms. Often, terms higher than third order are small enough to be ignored. Within a channel suffering this phenomenon (known as intermodulation distortion) there is little disturbance to a voice signal on the same channel. However, modulated data signals can be affected.

When intermodulation occurs within an analog carrier system, the distortion products may overlap into the frequency spectrum of other channels of the same system and result in crosstalk. It is important to note that this occurrence happens not only with poor design or defective equipment but, more often, with an excessive signal level.

Random Intermodulation Noise. If all the signals involved in the intermodulation phenomenon are speech signals, the result in an analog carrier system is an interference very similar to random noise. If a signal is carried in a speech channel, each fundamental signal may be considered as a band of energy 4 kHz wide. As a result the frequency band of the intermodulation product is 8 kHz wide for the second–order products and 12 kHz wide for the third–order products. Thus, more than one channel can be disturbed by these interferences.

If a broadband signal has a large number of speech fundamentals, the number of disturbing products that can be produced is very large. For example, in a system of 10,000 channels, a disturbed channel may receive well over one million third–order products. The probabilistic combination of the large number of contributors, together with the basic characteristics of each fundamental speech signal, generates an interference that is Gaussian in its amplitude distribution and has a flat power spectrum over the band of a disturbed channel.

Other detailed characteristics of speech signals must be evaluated in determining the effects of random intermodulation noise. In addition to the speech signal characteristics, system characteristics must also be considered. For example, in analog cable systems, modulation products of different types accumulate from repeater to repeater. In microwave radio systems, the intermodulation phenomenon is as important as in cable systems but results from different basic causes. Intermodulation noise in amplitude modulation (AM) systems is a function of signal amplitude, but in frequency modulation (FM) systems it is a function of the frequency deviation. In AM systems, the noise results directly from the nonlinear input/output characteristic of amplifiers. In FM systems, it results from gain and phase deviations generated by phase modulation (PM)–to–AM conversion and its inverse, and the effects of envelope delay distortion in the transmission medium. The end result, provided the number of channels in the system is large, is essentially the same—a nearly flat spectrum of noise having a Gaussian distribution.

Intermodulation Crosstalk. The transmission of FDM signals over analog transmission systems produces interchannel coupling, which may yield intelligible crosstalk. The nature of the coupling mechanism may be demonstrated by computation of the interference effect of higher–order terms in the power series expression when considering two signals on a disturbing channel.

Digital Signal Noise Impairments

The various impairments and coupling modes previously discussed apply to digital systems and signals as well as to analog. The nature of the impairments may differ somewhat, but the basic phenomena of interference generation and coupling are similar. However, most interferences to the digital signal of a digital transmission system are essentially nullified by the process of regeneration discussed briefly in Chapter 14, Part 2. In the regeneration process, each pulse of the line signal arrives at a regenerative repeater with various impairments produced in one repeater section only. The function of the repeater is to restore the pulse to its original form, timing, and amplitude, and thus eliminate the impairments. When this is accomplished with few errors, system performance is good; however, adequate margin must be provided to keep error rates low.

One noise impairment is unique to the transmission of analog signals over digital systems. The noise, called *quantizing noise*, is introduced during the process of digitally encoding an analog signal. It results from the assignment of a finite number of quantum steps to limit the total number of codes needed for the range of signal amplitudes to be transmitted. When a sample of a signal is sensed, it is assigned the value nearest its quantum step value. The transmitted value may be in error up to one–half a quantum step. The accumulation of these quantizing inaccuracies results in quantizing noise. Quantizing noise can be reduced to an arbitrarily small value by reducing the spacing between quantizing steps, which increases the total number of code steps. However, this increases the required bit rate or decreases the capacity of a system of fixed bit rate. Thus, it is economical to allow some quantizing noise while maintaining a prescribed signal–to–noise ratio.

When multiple digital systems are connected in tandem, where an analog signal is coded and decoded at each end of each system, each coding–decoding step produces quantizing noise that increases with the number of tandem terminals. This buildup of quantizing noise limits the allowable noise per terminal or, in another sense, limits the number of terminals that may appear in a built–up connection.

Another aspect of quantizing noise to consider in the design of terminal equipment is the size of quantum steps relative to the range of amplitudes to be encoded. If uniform steps are used, the percent quantizing error is greater for small signals than for large signals, thus degrading the relative signal–to–noise ratio for small signals. It is desirable to use quantum steps of decreasing size as the analog signal amplitude decreases so that the percent error remains relatively constant over the expected range of amplitudes. This may be accomplished either by using a complementary nonlinear encoder–decoder arrangement or by using a linear encoder–decoder preceded by a compressor and followed by a complementary expandor. Most practical systems in use employ the former method, which is, in effect, the application of an instantaneous compandor.

The effects of quantizing noise can be aggravated by the introduction of digital loss—the process of converting digital code words representing speech samples into new code words that give, in effect, a lower signal level. This process is found in some digital switches that insert a prescribed loss, depending on the type of connection, using a relatively inexpensive digital technique. The signal-to-noise ratio is inevitably reduced in such a case, typically by an amount comparable to the amount of digital loss and similar to the amount that would arise in an additional step of decoding and recoding. The idle-channel noise is increased compared to that which would occur with an analog pad because an analog pad would naturally attenuate the noise coming from the decoder. The intermodulation distortion and effective level of phase jitter are also noticeably increased.

Since quantizing noise is present only when a signal is applied, it must be measured in the presence of a signal. The technique used is called a C-notched noise measurement in which a holding tone is transmitted at the far end of a circuit. At the near end this holding tone is filtered out in the noise-measuring instrument [7].

Several other forms of distortion arise in the terminal equipment as a result of coding processes. These include harmonic distortion, which may be caused by overload or by poor compandor tracking, and foldover distortion, which may occur if the high-frequency channel cutoff is set too high [8].

17-4 NOISE AND CROSSTALK MEASUREMENTS

There are two general purposes for measuring interferences coupled from sources outside the channel of interest. The first purpose is to determine the magnitude and character of the interference in the disturbed circuit, irrespective of the source or of the coupling mechanism. This type of measurement might be made to evaluate power hum, common-battery supply noise, impulse noise, or crosstalk. Until the magnitude and character of the problem are evaluated, the mode of coupling and means for reducing the interference are of secondary importance. The second purpose is to determine the coupling loss between a disturbing and a disturbed circuit. This measurement establishes the fact that a suspected source of interference involves a particular combination of disturbing and disturbed circuits and determines the increase in coupling loss needed to cure the problem.

Parameters and Units—Noise Measurements

In evaluating interferences, electrical power is most commonly measured by determining the interfering voltage across a known impedance. Power measurements are usually expressed in decibels relative to one milliwatt (dBm) or in decibels relative to reference noise, weighted or unweighted (dBrn or dBrnc). Further, such expressions are often referred to 0 TLP and are expressed as dBm0, dBrn0, or dBrnc0. Most wave analyzers designed to measure single–frequency noise, such as a power frequency harmonic, are calibrated in dBrn. Some interferences that cover a broad spectrum are measured in the voiceband in dBrnc and translated into dBrnc0. The measurement may be made in dBm if the interference is being evaluated for wideband signal impairment or, if signal–to–noise is to be evaluated, by measuring both the signal and noise in dBm. If the interference is impulse noise, the measurement must account for both amplitude and frequency of occurrence. The measurement is often expressed in counts per minute that exceed a set dBrnc threshold, assuming a specific bandwidth. The threshold depends on the type of signal for which the interference is being evaluated and, of course, the TLP at which the measurement is made. (See Chapter 3, Part 3.)

Speech Crosstalk Measurements

Because of subjective effects, special consideration is given to crosstalk between speech circuits. The parameters to be evaluated include the number of exposures to crosstalk, the volumes of interfering speech signals, the coupling loss for each coupling path, the gains and losses of each of the involved circuits (which led to the concept of ELCL), and the hearing acuity of the listener in the presence of noise. The efficiencies of the station transmitter and receiver are implied in the evaluation of crosstalk volume and listener acuity.

Coupling Loss Measurements. The measurement of crosstalk coupling is usually expressed as a loss, in dB, from the disturbing circuit to the disturbed circuit. However, a unit occasionally used in crosstalk computations is the dBx; it is equal to 90 minus the measured coupling loss. The use of this unit is sometimes convenient because, as a coupling becomes tighter (lower loss), the number of dBx increases rather than decreases. As implied, since a test signal must be applied to the disturbing circuit at a known frequency and amplitude and then measured in the disturbed circuit, it is necessary that the disturbing and disturbed circuits both be uniquely identifiable. Unless the coupling is intermodulation, the frequency is the same in both; if the coupling is intermodulation, the frequency may be shifted. Thus, measurements of the received test signal amplitude at the shifted frequency may be necessary, and suitable signal generation and detection equipment is required. Coupling loss measurements are often made across the spectrum of interest because the coupling loss is often a function of frequency.

The concepts of near–end crosstalk, far–end crosstalk, and interaction crosstalk couplings are applied most often to crosstalk problems arising from parallel transmission lines (e.g., pairs in the same cable with metallic conductors). In voice–frequency circuits, where the predominant coupling is usually FEXT and capacitive, the coupling loss tends to decrease at a rate of 6 dB per octave of frequency (20 dB per decade). It has been found that where smooth coupling of this type exists, a single–frequency measurement of coupling loss may be made at 1 kHz; from this measurement a good approximation to the effective coupling loss over the voiceband may be determined by subtracting 2 dB from the measured value [9]. Coupling loss of the FEXT type is also a function of the transmission line length. The loss decreases directly with the length of exposure, whereas NEXT coupling loss tends to be independent of path length and decreases with frequency at a rate of 4.5 dB per octave (15 dB per decade).

The crosstalk coupling between speech circuits in FDM carrier terminals and that resulting from intermodulation in analog repeaters tends to be constant across a speech channel. As a result, a single–frequency measurement is usually sufficient to determine the coupling loss.

Voiceband Data Measurements

The parameter measurements employed to evaluate interference to voiceband data transmission are impulse noise, white noise, single–frequency noise, quantizing noise, intermodulation distortion, phase hits, phase jitter, dropouts, gain hits, and, as an end–to–end measurement, errors to the information content of a data test signal. A direct measurement of error rate involves the transmission of a digital signal of known information content. The receiving equipment used in such an evaluation has stored within it the expected signal. It then compares the received signal, symbol by symbol, with the stored signal to provide the operator with a knowledge of the errors incurred during the test period.

Digital System Measurements

In contrast to a data system that converts an analog signal into a digital signal, the performance of a system that is all–digital is evaluated by measuring an error rate at a point of interest. The errors that are evaluated, however, are not necessarily errors to the information content of the signal, but rather to the "format" or expected excursion of a pulse (positive or negative), pulse pattern (parity), or bipolar violations, if appropriate, at the point of measurement.

Random Noise Measurements

The most common measurement of random noise is measuring the noise power in a given band. Such a measurement must be made at a known TLP (for voice circuits), must cover a band of interest, and must include appropriate weighting factors or characteristics if applicable. The results are expressed in units appropriate to the measurement; dBrnc for voice circuits and dBm for other types of circuit.

Two transmission–system–related measuring techniques are of interest here—the measurement of noise on compandored voiceband circuits and the measurement of analog system performance by noise loading.

Noise in Compandored Circuits. Random noise that appears in analog voiceband channels equipped with syllabic compandors can be regarded from two points of view: one when the circuit is used for speech transmission, and the other when it is used for voiceband data. In the case of speech transmission, the impairment, with or without a compandor, is greatest during silent periods when the noise can be heard best. When speech energy is present, the noise is subjectively far less interfering. Thus, in a compandored system, requirements for noise measured at the expandor input are much less stringent than for an equivalent noncompandored system. However, at the expandor output, where much less noise is measured in the absence of signal, 5 dB must be added to the measured noise to account for the subjective effect of noise during quiet intervals.

For voice–grade data transmission, the noise must be evaluated with a signal present. Compandor action in the presence of a signal results in an increase in noise at the expandor output. To accomplish such a measurement, a single–frequency holding tone is transmitted over the channel at 1004 Hz at an amplitude of −13 dBm0, to simulate a data signal. At the input to the measuring device, a 1004–Hz notch filter is used to suppress the holding tone. The noise, measured in dBrnc and translated to the 0 TLP as dBrnc0, is called C–notched noise.

Noise Loading. The performance evaluation of broadband analog cable and radio systems is difficult, from an analytical point of view, because of the large number of parameters to be dealt with and, from a measurement point of view, because of the lack of control over a true voice system load. Activity, type of signal transmitted, the percent of system equipped for service, and other important parameters are hard to determine or control. In such cases, a technique called noise loading is used to evaluate system performance.

A band of flat Gaussian noise, limited to the spectrum normally occupied by transmitted signals, is applied to the system at a point where the normal multiplexed signal would be applied. The magnitude of the applied noise is adjusted to simulate the loading effect of a normal signal.

In order to measure intermodulation noise, quiet channels carrying no signal are ordinarily used. To simulate quiet channels in a noise–loading measurement, one or more band–elimination filters are used to suppress the noise signal over small portions of the band at the output of the noise generator. At the system output, bandpass filters allow the noise that falls in the measurement band, the quiet channels, to be passed on to the noise–measuring equipment. This noise is due to intermodulation in the transmission system plus the inherent noise of the facility.

The noise–loading technique has three major uses in system test and evaluation. The first is simply to check performance against predicted or specified values. The second is to optimize signal–to–noise ratio by determining the drive level at which the signal–to–noise ratio is a maximum. The third is to provide information as an adjunct to trouble identification and isolation. In the latter case, measured results are compared with a predicted measurement to determine if there is an excess of thermal or intermodulation noise.

The source of excessive noise can often be determined from such a comparison. In microwave radio systems, for example, excessive intermodulation noise at high frequency may be caused by waveguide or radio–frequency (RF) cable echoes, a defective RF amplifier, or a defective intermediate–frequency (IF) filter. Excessive modulation noise at low frequency may be caused by nonlinearity in an FM transmitter or receiver, or in a baseband amplifier. Excessive low–frequency thermal noise may have as its source a defective local oscillator.

References

1. Rogers, W. E. *Introduction to Electric Fields* (New York: McGraw–Hill Book Company, Inc., 1954).

2. Binn, K. J. and P. J. Lawrenson. *Analysis and Computation of Electric and Magnetic Field Problems* (New York: The MacMillan Company, 1963).

3. Fennick, J. H. "Amplitude Distributions of Telephone Channel Noise and a Model for Impulse Noise," *Bell System Tech. J.*, Vol. 48 (Dec. 1969), pp. 3243–3263.

4. Members of Technical Staff. *Transmission Systems for Communications*, Fifth Edition (Murray Hill, NJ: AT&T Bell Laboratories, Inc., 1982), Chapters 7, 8, and 9.

5. Johnson, J. B. "Thermal Agitation of Electricity in Conductors," *Physical Review*, Vol. 32 (1928), pp. 97–109.

6. Nyquist, H. "Thermal Agitation of Electric Charge in Conductors," *Physical Review*, Vol. 32 (1928), pp. 110–113.

7. *IEEE Std. 743–1984*, "IEEE Standard Methods and Equipment for Measuring the Transmission Characteristics of Analog Voice-Frequency Circuits" (New York: Institute of Electrical and Electronics Engineers, Inc., 1984), pp. 12–13 and 21.

8. Bell System Technical Reference PUB 41005, *Data Communications Using the Switched Telecommunications Network*, American Telephone and Telegraph Company (May 1971), pp. 10–11.

9. Sen, T. K. "Masking of Crosstalk by Speech and Noise," *Bell System Tech. J.*, Vol. 49 (Apr. 1970), pp. 561–584.

Additional Reading

IEEE Std. 776–1987, "Guide for Inductive Coordination of Electric Supply and Communication Lines" (New York: Institute of Electrical and Electronics Engineers, Inc., 1987).

Lightning, Radio Frequency, and 60–Hz Disturbances at the Bell Operating Company Network Interface, Technical Reference TR-EOP-000001, Bellcore (Iss. 2, June 1987).

Chapter 18

Amplitude vs. Frequency Response

In the transmission of analog signals, the fidelity of signal reception is strongly influenced by the frequency response of the channel. The amplitude–versus–frequency response is the variation with frequency of the gain (amplification) or loss (attenuation) of a channel. If a channel is linear and time invariant, its frequency response may be expressed as a ratio of output signal to input signal. This ratio reflects the gain or loss of the channel. The desired relationships may be specified by a function that is flat with frequency or shaped in accordance with a specific rule. Whether it be a low–pass, bandpass, or high–pass function, the concern here is with departures from the function specified and from linearity and time invariance.

The effects of bandwidth limitations, gain or loss, and gain variations with time are all related to transmission impairments that may affect speech signals, voiceband or wideband data signals, and video signals. In some cases unique methods are used to measure these impairments. The impairments are often related to transmission system design in significant ways.

18-1 VOICEBAND CHANNELS—SPEECH SIGNAL TRANSMISSION

In normal operation, a connection between two stations, or terminals, may involve as little as two loops for a private line with the addition of a switching system for local exchange service. On the other hand, the connection may be substantially more complex. It may contain several trunks between central offices and may be routed through a number of additional switching systems. The facilities may be all voice–frequency, or may use analog or digital carrier systems. The two terminals may be only a few feet apart or may be halfway around the world. Such diversity in the makeup of connections makes it important to define and control the frequency response of each possible part of the connection and makes it difficult to define and control the overall response characteristic of all station–to–station connections.

Channel Bandwidth

The development of the first mature frequency–division multiplexing (FDM) equipment during the 1930s made necessary the determination of the spacing of voice channels in the spectrum. The problems of designing filters and channel combining and separating networks for the FDM equipment made it necessary also to determine what useful band was to be provided within the assigned channel band (i.e., the rolloff characteristics that could be tolerated).

The bandwidth of a voice channel (established by subjective testing) is conveniently described in terms of the 4–kHz spacing of channels in typical FDM carrier systems. The bandwidth of time–division multiplexing (TDM) carrier systems has also been chosen as 4–kHz nominal. This description, however, is inadequate because it does not account for any of the effects that produce an effective bandwidth of less than 4 kHz, nor does it give the criterion used to define band edges. The band–narrowing effects include the loss characteristics of the loaded or nonloaded cable pairs on loops and trunks, the frequency responses of the filters used in carrier terminals, battery supply repeat coils, station apparatus, etc. In addition, the tandem connection of multiple links introduces a cumulative reduction of effective bandwidth.

The useful band of a voice channel is defined as the band that is between the 10–dB points on the loss–versus–frequency characteristic of the channel (i.e., the points at which the loss is 10 dB greater than the loss at 1000 Hz). The bandwidth varies on message telecommunications service (MTS) network connections from somewhat more than 3200 Hz to about 2800 Hz. This reduction in bandwidth results primarily from technical and economic design compromises made in the past. However, as fewer and fewer trunks use cable pairs and short–haul analog carrier systems in favor of TDM digital carrier, the bandwidth of typical connections will be consistently 3200 Hz or more. Specifically, as digital switching and cross-connection systems become more widespread, the trunking used in a dialed connection will have no effect on frequency response.

Circuit Loss and Loss Variations

The losses in individual loops, trunks, and switching systems must be limited for two reasons. First, if the loss in a circuit or built-up connection is high, the received signal is low in volume and the listener either loses some of the transmitted information or is annoyed because he cannot easily understand. Second, if trunk losses are high, the contrast in received speech volume from call to call may be objectionable due to the many combinations of trunks that may be used. Consider, for example, successive calls to the same destination. On the first call, the connection might be made over one intertandem trunk. On the second call, there might be several intertandem trunks in the connection because of alternate routing. If trunk losses were high, the resulting difference in volume between the two calls would be objectionable.

Ideally, all trunks should be operated at zero loss. This would permit the tandem connection of any number of trunks in a built-up connection, thus simplifying the problems of contrast, low volume, and noise. However, with analog facilities, this mode of operation is impractical. Due to the variability of terminating impedances, many circuits would become unstable (sing) or would be on the verge of singing and thus produce an unpleasant hollow effect in the received signal. Furthermore, echoes resulting from impedance mismatches would impair transmission. Thus, circuit design is based on minimizing losses within the constraints of stability and echo control. In the via net loss (VNL) plan, the amount of loss in each trunk of the network was determined by talker echo considerations. The allocation of loss to tandem-connecting trunks and intertandem trunks was based on the economics of supplying gain, the need for stability margins, and the transmission variations for alternate routing of calls (contrast). The approach to the loss and stability is somewhat different, however, when all facilities are digital with the four-wire-to-two-wire analog conversions occurring only at the end offices or stations. When this is the case, echo control becomes a function of the reflections at the end terminations. The required loss is inserted at the receiving end only. The intermediate trunks are, in effect, transparent.

The same reasons for controlling circuit loss apply to minimizing circuit loss variations with dynamic maintenance. The control of cumulative losses to prevent low received volume, the avoidance of excessive contrast between calls, and the need for controlling circuit stability and echo performance make it mandatory that loss variations with time be held to a minimum.

Amplitude vs. Frequency Distortion

The transmission of speech signals is not seriously impaired by the amplitude–versus–frequency distortion normally encountered in the switched network. The characteristics of cable pairs and carrier channels tend to be smooth across the voiceband and not steeply sloped. There is, however, a sharp rolloff at the high–frequency edge of the voiceband on channels using loaded cable pairs. The filters used in carrier terminals are also relatively smooth and introduce, except at the band edges, a gradually increasing loss as the frequency increases or decreases from the 1004–Hz reference frequency. Thus, inband distortion is usually expressed in dB of slope at 404 Hz and 2804 Hz. These frequencies are near the edges of the useful band. The slope is defined as the dB difference in loss at each of these frequencies relative to the 1004–Hz loss.

Except in the occasional instance of defective apparatus, the slope of an analog voice channel is not a matter of field operating or maintenance control. The channel characteristics are established primarily by design, including occasional use of special equalization techniques. Where broadband analog carrier is involved, the response of an "edge" channel (channels 1 or 12 of a 12–channel group) is affected not only by the performance of channel filters, but also by the presence of group–connector filters used in the overall multiplex layout. This effect is controlled by limiting the number of group–connection points, typically to five. Similarly, the performance of a groupband carrier slot used for 50–kb data is affected by supergroup connectors if it is an edge group (1 or 5).

Measurements

Frequency response measurements of analog channels can be made by single-frequency measuring techniques involving a variable frequency oscillator and a transmission measuring set. By this method, measurements can be made to establish the absence of transmission irregularities, the cutoff frequencies (10-dB point), and the slope at 404 and 2804 Hz. Slope measurements (sometimes called three-tone slope) are commonly performed by automated test equipment.

18-2 VOICEBAND CHANNELS—DATA TRANSMISSION

The specification and control of the frequency response of voiceband channels is generally more critical for voiceband data transmission than for speech because many types of data receiving equipment are less tolerant of distortion in the frequency response than is the human ear. Such distortion is a departure from the designed, or expected, amplitude-versus-frequency response of a channel used for data signals. The ideal response is the anticipated response for which data station equipment has been designed. In general, station equipment is designed to perform with a published frequency response "norm" based on periodic end-to-end measurement surveys.

Where station equipment is designed to process data signals for transmission over the switched network, the processing (coding, rate of transition, signal shaping, etc.) must result in signal characteristics that are compatible with transmission characteristics of the switched network end to end. When channels are dedicated to the transmission of data signals (i.e., private lines), the station equipment is usually designed so that the signal processing is coordinated and made compatible with the dedicated channel characteristics. *Conditioning*, the treatment of such channels to improve their amplitude-versus-frequency response characteristics, involves the provision of fixed or adjustable equalizing networks.

The nature of amplitude-versus-frequency distortion and the related effects of phase-versus-frequency distortion often result in a need for more precise equalization than that provided by conditioning. Additional equalization may also be required because distortions vary with time, different types of facility, and the different numbers of trunks involved in successive connections. This additional equalization, provided by dynamic and adaptive equalizers, may be designed in the form of tapped delay lines or their digital equivalent. Each tap is provided with an electronically controlled attenuator that automatically adjusts the delay line to approximate the inverse characteristic of the channel. The adaptive control is usually based on samples of received signal and an algorithm that uses statistical estimates of the sample signal response as control information. This signal–dependent method of control combines amplitude–versus–frequency and phase–versus–frequency distortion correction and, in addition, can provide automatic gain control to compensate for changes in the overall loss of the connection.

Available Bandwidth

As previously discussed, the bandwidth of intertandem trunks is about 3200 Hz (between 10–dB loss points); when tandem–connecting trunks and loops are included in an overall connection, the bandwidth will be somewhat less. Also, there is variation in bandwidth from connection to connection as the number of and type of links vary because of alternate routing. The chosen data signal format and the nature and magnitude of other impairments will control the sensitivity that the data signal will show to this limited and variable bandwidth. If the baud rate is too high relative to the bandwidth of the channel, serious signal distortion occurs because high–frequency components are attenuated.

Loss and Loss Changes

The transmission loss between transmitting and receiving data stations is fixed on dedicated channels, but is variable on switched channels. On a four–wire dedicated channel the nominal loss is 16 dB. On channels that are switched, no such close control of the loss can be specified because the loss depends on the length of the connection, the number and type of links in the connection, and on the losses of the loops at the two ends. Thus, the design of terminal equipment must take into account the variation of loss and the resulting variation of the received signal amplitude. Automatic gain control is usually employed in the receiver to eliminate the problem of loss variation.

Loss (or gain) changes occur in telecommunication circuits for a number of reasons. Slow changes generally tend to be small and occur over a broad frequency range. They are generally caused by temperature changes or by the aging of active devices. In digital systems, sources of gain change are confined to the terminals, and the gain is usually stable to a quarter of a decibel or better. Where analog carrier facilities are involved, these changes are usually compensated for by automatic regulation. Such changes cause little impairment in the transmission of data signals. On the other hand, sudden gain changes occur sporadically as a result of component failure, protection switching of broadband carrier facilities, maintenance activities, or natural phenomena such as microwave radio fading. Even dropouts (i.e., momentary loss of signal) may occur. All such phenomena may cause data errors.

Dropouts

Dropouts are short–duration impairments in which the transmitted signal experiences a sudden drop in power, often to an extent that the signal is undetectable. They have been defined as any reduction in signal power more than 18 dB below normal for a period exceeding 300 ms. Dropouts occur infrequently.

Inband Distortion

There are two sources of distortion that often occur in voice–frequency facilities as a result of trouble or oversight. Figure 18–1 illustrates the deterioration of the insertion loss due to these sources. Curve 2 depicts the distortion when a cable pair is missing a load coil (see Chapter 5, Part 3). This also produces serious amplitude–versus–frequency distortion due to the impedance discontinuity. Curves 3 and 4 illustrate the deterioration of the insertion loss due to the presence of a bridged tap on this same cable pair. Cable layouts are often made with bridged connections at splice points to increase flexibility in circuit assignment. The bridged connection acts as a shunt capacitor to produce an impedance irregularity. There are other causes of impedance irregularities that occur occasionally, such as double loading, split loads, deficient load spacing, etc., that cause similar irregularities as indicated in Figure 18–1 for bridged tap and missing loads.

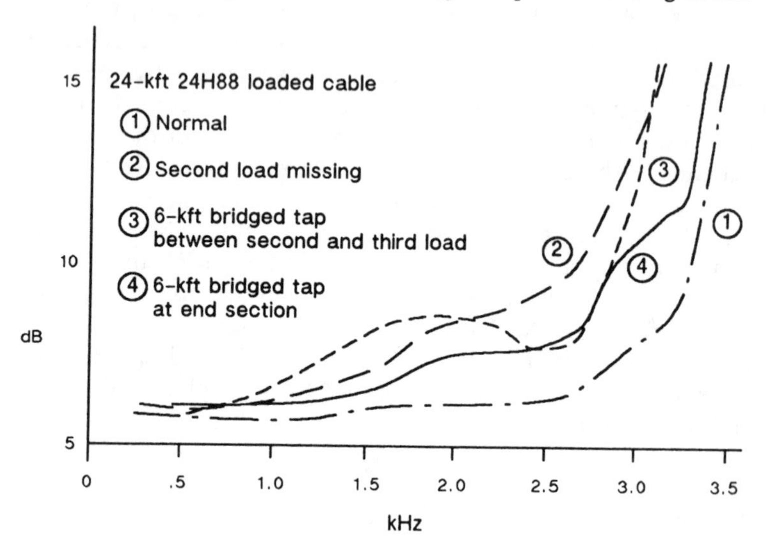

Figure 18–1. Effect of irregularities on insertion loss of a cable pair.

Measurements

A number of techniques are available for the evaluation of the performance of voiceband circuits for data signal transmission. These include the measurement of errors in the transmission of a known message and the use of the peak-to-average ratio (PAR) meter. The PAR meter measures the ratio of an envelope peak to the envelope full-wave average for a closely controlled signal stream transmitted over the channel under test. Distortions in the channel tend to disperse the energy in each signal transition and to reduce the PAR. The measurement of errors in a known message evaluates all transmission parameters and includes the performance of all devices, including the modem. PAR, however, evaluates only attenuation distortion, envelope delay distortion, and message circuit noise of the channel. A good PAR measurement does not insure that the channel will pass a good data signal because many other transmission parameters that PAR does not measure can affect the quality of a data signal. When both methods (errors or PAR) indicate unsatisfactory performance, it becomes necessary to resort to single-frequency measurements to determine the amplitude-versus-frequency distortion in the band.

18-3 WIDEBAND DATA CHANNELS

The amplitude-versus-frequency characteristics of wideband channels tend to have ripple components of higher amplitude than are typical of voiceband channels. Like voiceband channels, wideband channels display increasing loss toward band edges. However, wideband channels generally have less slope across the band. Figure 18-2 illustrates a typical wideband loss-versus-frequency characteristic showing the cumulative ripple and nonuniform loss of about 1000 miles of an analog cable carrier system. The rolloffs at band edges do not appear in this figure because it does not include the effects of bandlimiting filters.

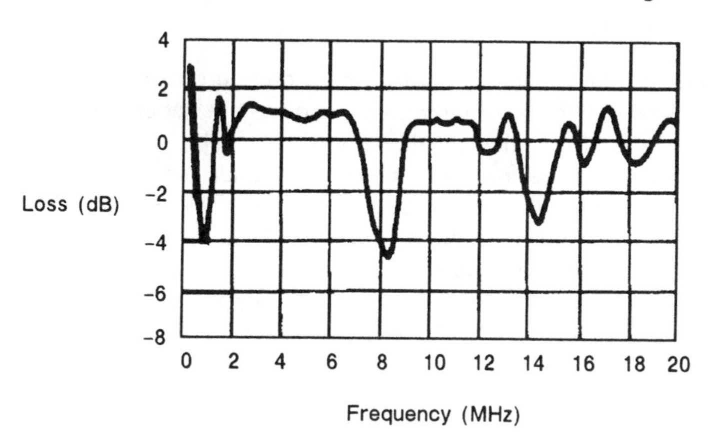

Figure 18-2. Typical frequency response of an equalized analog cable system (approximately 1000 miles long).

Nonuniform losses distort the data signal spectrum and, hence, the waveform, resulting in a tendency toward increased data errors. As with the voiceband channel, the nonuniform amplitude–versus–frequency characteristic of the wideband channel is corrected by fixed or adjustable networks designed to equalize the amplitude–versus–frequency response. It is also sometimes necessary to provide adaptive equalizers for wideband data transmission similar in concept to those used in voiceband transmission.

Available Bandwidths

Equipment has been designed to provide for a number of wideband channel offerings. Standard tariffed services correspond to building blocks in the FDM hierarchy. The blocks used include the 48–kHz group band and the 240–kHz supergroup band, as well as a half–group 24–kHz band. In each case, the signal format is tailored in the terminal equipment (data stations and carrier modems) to use the available bandwidth most efficiently, i.e., to provide the highest rate of information transmission per unit of cost. These analog–based services have been eclipsed by the offering of the Digital Data System and high–capacity digital services.

Multilevel data signals are transmitted on microwave radio systems at a bit rate of 1.5 million per second in a baseband extending from zero frequency to about 500 kHz [1], and at higher rates on pure–digital radio. Three–level signals are transmitted over wire facilities at 1.5, 3.1, and 6.3 million bits per second in digital carrier systems. In each case, the signal format has been designed to coordinate with the available bandwidth. The line signals may be composed of a variety of data signals combined by TDM techniques. Fiber optic systems, however, display no hard bandwidth limitation with the present state of the art.

Measurements

In the wideband systems and channels under discussion, point–by–point single–frequency measurements are often impractical because of their time–consuming nature. The evaluation of frequency response characteristics is therefore often made by examination of a data eye diagram or by an error–ratio measurement. The direct evaluation of the frequency response characteristic may be accomplished by sweeping the band with a sweep–frequency oscillator and then displaying the characteristic on an oscilloscope. A plot of the characteristic may be used to compare with limits established for various portions of the band. These limits are defined so that, when not exceeded, the error–ratio objectives are met, provided other impairments are also held within limits. The terms commonly used to describe qualitatively the principal types of distortion are slope, sag, and peak. *Slope* describes the loss at the high end of the passband relative to that at the low end; *sag* describes the midfrequency bulge in the characteristic; *peak* describes the ripple components in the passband.

18–4 VIDEO CHANNELS

While data signal formats have generally been adapted to available channel bandwidths, video signal requirements have largely dictated the channel transmission characteristics that must be provided to maintain picture fidelity. (An exception is slow–scan or freeze–frame transmission, where the signal format is tailored to the 4–kHz voice channel or the 56–kb data channel.) The significant impairments to a video signal are those associated with bandwidth, cutoff characteristics, loss and loss changes, differential gain, differential phase, and inband amplitude distortion.

Bandwidth

The bandwidth required for video signal transmission is determined by the horizontal and vertical resolution to be provided in the received signal. This bandwidth has been established to be about 4.2 MHz for television signal transmission in North America; it is potentially much wider for higher definition television.

For a picture generated with a specified number of scanning lines in a frame, the first noticeable impairment caused by a reduction in bandwidth is a loss of horizontal resolution, resulting in increasing difficulty in distinguishing between adjacent picture elements along a line. In addition to a loss of horizontal resolution, color information is also lost as the bandwidth is reduced, because the color information is conveyed by a carrier signal at 3.58 MHz.

Cutoff Characteristics

Another aspect of amplitude–versus–frequency response is the nature of the video channel cutoff characteristics. At the low end of the band, it is necessary to provide good transmission to nearly zero frequency. Since frequencies near zero cannot generally be transmitted over ac–coupled facilities, the information in these components must be restored at the receiver by a process called clamping. At the high end of the band, an essentially flat amplitude–versus–frequency response must be provided to at least the color carrier of 3.58 MHz. The channel loss above that frequency must be increased gradually because, if the band is cut off too sharply, a phenomenon called ringing may occur. A signal containing sharp transitions, when applied to such a channel, generates damped oscillations at approximately the cutoff frequency.

Loss and Loss Changes

The absolute gain or loss of a transmission system is set by the constraints of intermodulation, overload, crosstalk, and signal–to–noise ratio. Television signals are impaired only slightly by gain or loss changes, provided these changes do not occur at a regular, low–frequency rate. When this does occur, the result is a flicker effect, a serious impairment of which viewers are quite intolerant.

Differential Gain

Video circuits generally require amplifiers. These amplifiers may have nonlinear input–to–output characteristics that produce a signal–dependent form of nonlinear amplitude distortion called differential gain, an impairment to which color television signals are particularly susceptible. Differential gain is the difference between unity and the ratio of the output amplitudes of a low–amplitude, high–frequency signal (simulating the color carrier) superimposed on a low–frequency signal (simulating the luminance signal) at two different specified amplitudes of the low–frequency signal. The differential gain may be expressed in percent by multiplying by 100, or in dB by taking 20 log of the ratio of the two high–frequency signal amplitudes. The visual effect of differential gain in a color television system is a change in color saturation with changes in brightness (changes in luminance level).

Differential Phase

Differential phase is a second type of nonlinear distortion that can impair color television transmission. It is defined as the difference, in degrees, of the phase shift through a transmission system exhibited by a low–amplitude, high–frequency signal (simulating the color subcarrier) superimposed on a low–frequency signal (simulating the luminance signal) at two different specified amplitudes of the low–frequency signal. The visual effect of differential phase in a color television system is a change in the hue of a color with changes in brightness (changes in luminance level). The high–frequency test signal is usually the 3.58–MHz color carrier; the low–frequency signal is normally the 15.75–kHz line–scan frequency.

Inband Amplitude Distortion

Departures from flat inband amplitude–versus–frequency response cause a number of video signal impairments, two of which are called *streaking* and *smearing*. Both may be caused by transmission distortions in the frequency regions between about 60 and 1000 Hz and between 15 and 200 kHz. Both streaking and smearing cause objects in a picture to appear extended beyond their normal boundaries toward the right side of the received picture. With streaking, object extension appears undiminished. With smearing, the extension, which may be positive or negative in brightness relative to the object, diminishes substantially toward the right edge. The smearing impairment also tends to be more blurred than a streak.

If a channel has excessive gain at high frequencies, sharp signal transitions may experience overshoot. The result is a black (or dark) outline to the right of a white object and a white (or light) outline to the right of a dark object.

Departures from a flat response can be analyzed by Fourier techniques, and the loss characteristics may be expressed in terms of the Fourier components. If the departure from flatness is a simple sinusoid (gain or loss in dB versus frequency), the impairment can be shown to be a pair of echoes (low–amplitude duplicates of the signal displaced in time from the main signal) of the same polarity, one leading and one lagging the signal. Each of the Fourier components of the response characteristic produces a pair of echoes of this type.

The subjective effects of echoes due to distortion in the frequency response characteristic depend on the amount of time displacement of the echo and on the magnitude and shape of the distortion. The effects are also related to the presence of other echoes due to loss–versus–frequency or phase–versus–frequency distortion.

Measurement of Television Impairments

Point–by–point single–frequency measurements to determine the characteristics of a television channel are time–consuming, and therefore impractical, because of the wide channel bandwidth. While sweep techniques are sometimes used, waveform test signals are most commonly employed to evaluate a video channel. Two such test signals, the composite and combination test signals, are shown in Figures 18–3 and 18–4.

Objective measurements are made easier by the transmission of waveform test signals that are transmitted on an in–service basis with the video signal. In–service signals, called vertical interval test signals (VITSs), are sent during the vertical blanking interval so that they are not seen on the receiver. The waveform is displayed on a calibrated oscilloscope or vectorscope for examination and interpretation. Following are brief descriptions of some of the test signals used:

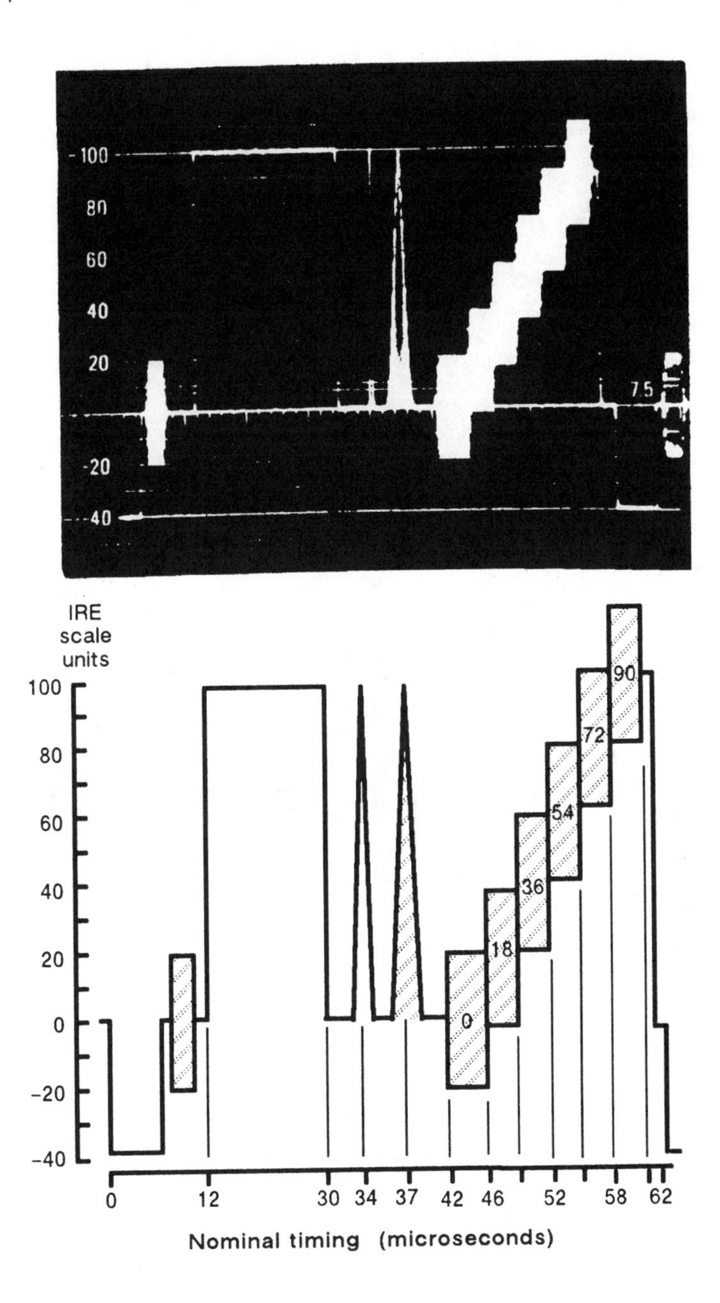

Figure 18-3. The composite test signal.

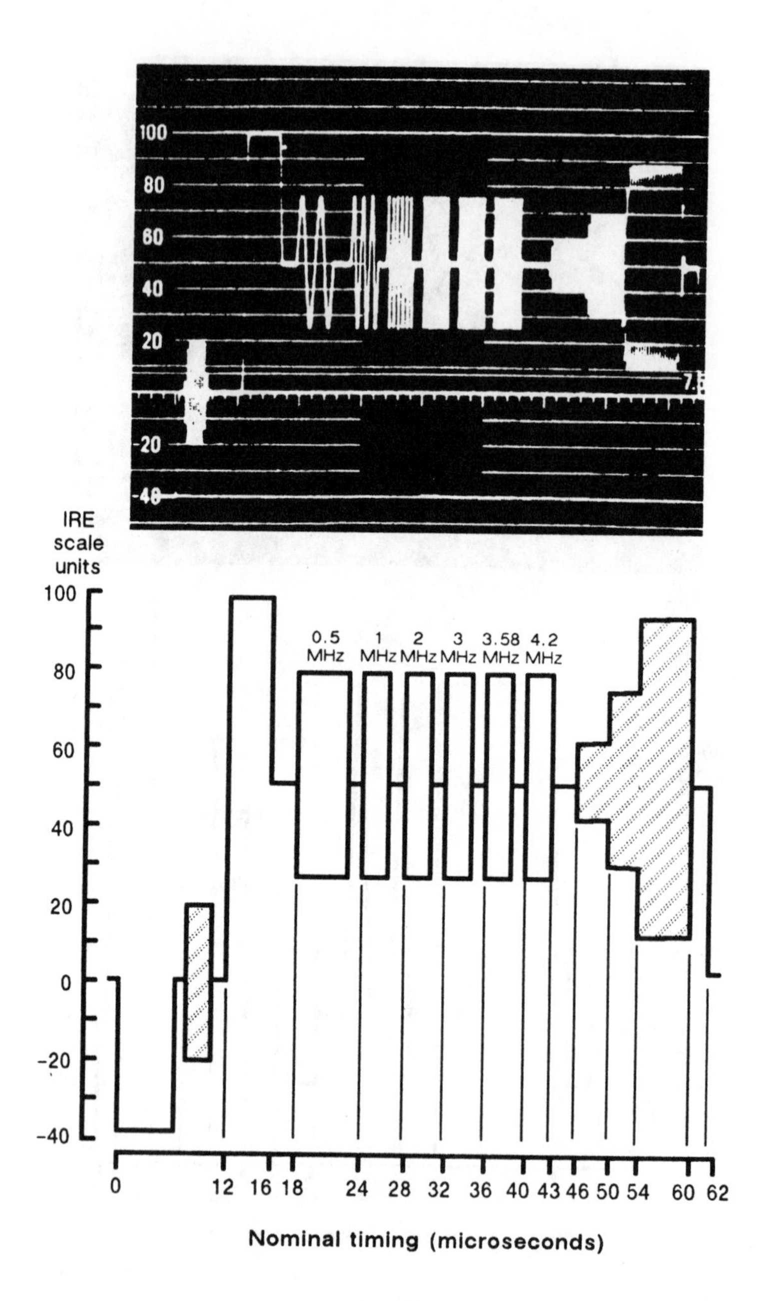

Figure 18-4. The combination test signal.

(1) *Time–domain signals* (first part of a composite signal) include pulse and amplitude step signals. They are used to measure various impairments such as unwanted luminance variations in large–detail sections of a picture, smearing, streaking, ringing, and overshoot.

(2) The *stairstep* (last part of a composite signal) is a signal of increasing amplitude formed of equal "steps." Color subcarrier is added to each step. It is used to measure differential gain and (with a vectorscope) differential phase. Excessive amounts of differential gain cause departures from equality in the step heights.

(3) The *multiburst* (first part of a combination signal) is formed of brief impulses of 0.5, 2.0, 3.0, 3.6, and 4.2 MHz transmitted at equal amplitudes. Their relative amplitudes at the receiving point, measured on an oscilloscope, provide an evaluation of the channel amplitude–versus–frequency response.

The duration, rise time, and transition shapes of the time–domain test signals are often defined in terms of a $\sin^2$ pulse shape. Pulse durations are defined in terms of the time between half–amplitude points, and the time of transition is defined as the Nyquist interval. Video–channel testing has favored the use of time–domain signals because they appear to give a more direct measure of circuit quality. Frequency–domain signals, such as the multiburst signal and the three–level chrominance signal (last part of combination signal), are less commonly used.

References

1. Seastrand, K. L. and L. L. Sheets. "Digital Transmission Over Analog Microwave Radio Systems," *Conference Record*, IEEE International Conference on Communications (Philadelphia, PA: June 1972).

Additional Reading

Davidoff, F. "Status Report on Video Standards; IEEE Video Signal Transmission Subcommittee 2.1.4," *IEEE Transactions on Broadcasting*, Vol. BC–15 (June 1969), pp. 27–32.

Lessman, A. M. "The Subjective Effects of Echoes in 525–Line Monochrome and NTSC Color Television and the Resulting Echo–Time Weighting," *Journal of the SMPTE*, Vol. 81 (Dec. 1972), pp. 907–916.

Schmid, H. "Measurement of Television Picture Impairments Caused by Linear Distortions," *Journal of the SMPTE*, Vol. 77 (Mar. 1968), pp. 215–220.

Schmid, H. "The Sin^2 Pulse and the Sin^2 Step in the NTSC TV System," *IEEE Transactions on Broadcasting*, Vol. BC–18 (Dec. 1972), pp. 81–84.

Sullivan, J. L. "A Laboratory System for Measuring Loudness Loss of Telephone Connections," *Bell System Tech. J.*, Vol. 50 (Oct. 1971).

Chapter 19

Timing and Synchronization Errors

Where signal processing involves time–domain coding or frequency translation, impairments may result from lack of synchronization between the transmitter and the receiver or from deterioration of the timing signal itself.

Some impairments to transmitted digital signals resulting from amplitude–versus–frequency or phase–versus–frequency distortion may lead to difficulties in timing recovery at regenerators or receivers.

In frequency–division multiplexing (FDM) systems, synchronization errors may be caused by incidental periodic, random, or discrete displacement of the carrier, resulting in unwanted modulation of the information signal. The carrier displacement produces frequency offset (or shifting) of the received signal. Other forms of incidental modulation cause carrier signal impairments such as phase hits and jitter.

To limit the impairments caused by synchronization problems, national and regional networks distribute timing signals that synchronize analog and digital systems.

19–1 FREQUENCY OFFSET

In analog transmission systems employing suppressed–carrier FDM, the output signal components may be offset in frequency from their proper values as a result of frequency differences between carriers in the transmitting and receiving terminals. The demodulation process is controlled by carrier supplies at the receiving terminal, which are synchronized with those at the transmitting terminal by some external means. Without this, perfect synchronization cannot be achieved.

In FDM systems that transmit at high frequencies, the control of frequency offset imposes stringent requirements on the accuracy and phase stability of the receiving carrier. If the top frequency of a system is, for example, 100 MHz and if the frequency offset must be held to 1 Hz, the carrier at the receiver must be synchronized to within 1 Hz in 100 MHz or one part in 10^8. Such accuracy requirements made necessary a national synchronization network and also have led to the use of very stable oscillators, sophisticated test equipment, and specialized methods of measurement and control.

By contrast, in some remaining short–haul FDM systems, the carrier is transmitted with the signal. Its frequency and phase are accurately recovered to control demodulation. This practice is followed in the transmission of some wideband digital line signals.

Speech and Program Signal Impairment

Frequency offset reduces the naturalness of speech and program signals. When music is transmitted, the offset is most objectionable to listeners with high aural acuity because many musical instruments produce sounds having high harmonic content. It has been determined by subjective tests that frequency shift should be held to ± 2 Hz to satisfy discerning listeners. For speech, shifts of the order of ± 20 Hz are tolerable.

Voiceband Data Signal Impairment

The way voiceband data signals are impaired and the extent of the impairment are related to the signal format used. In many forms of data transmission, the timing signal used for decoding at the receivers is derived from the signal itself. In such cases, frequency offset is not a serious impairment, especially for the small offsets encountered when channels meet the requirements for speech and program transmission.

Other voiceband data signals, particularly phase shift keying, are prone to error in the face of frequency offset. A frequency offset uses up margin with respect to threshold circuit recognition of the binary information contained in a received signal, thus making the receiver more prone to errors.

Analog System Impairments

The most serious analog system impairment caused by frequency offset is the breakdown of system functions resulting from a large frequency offset that shifts signals outside the passbands of filters. In times of gross synchronizing failure, substantial offsets may be experienced. Signalling may become impossible because the filters used for the single–frequency signalling system are relatively narrow. Large numbers of supervisory signals being shifted out of their passbands simulate a simultaneous call for service from a large number of callers and may cause a massive seizure of switching system equipment and a breakdown of service. Frequency offset may also shift pilot frequencies and cause automatic gain regulators to operate improperly. Such impairments are minimized by redundant synchronizing arrangements.

Digital System Impairment

In most digital transmission systems, the signal formats are designed so that a timing signal is derived from the line signal at regenerators and terminals. The timing recovery circuits are designed to operate within the normal range of expected frequency shift in a stand–alone digital carrier system. Receive timing from the incoming pulse stream is quite satisfactory. However, digital systems rarely stand alone. To interconnect with digital switching and cross–connect systems, a clock synchronization network is used to ensure that the timing of all interconnected terminals is synchronized. It is quite common for the transmit timing in an outlying terminal to be "slaved" to the incoming clock (i.e., to use "looped" timing).

Small differences between the rate of a received signal and that of the master clocking network can be compensated for by buffer stores. However, if the frequency difference persists, a buffer may overflow or underflow. The system is designed to reset the buffer when this occurs, thus causing deletion or repetition of bits from the output signal. The overflow or underflow and the resulting reset cycle of the buffer continues until the frequency offset is corrected. The resulting impairments, called *controlled slips*, cause clicks in speech signals and serious deterioration of digital signal transmission.

19–2 OTHER INCIDENTAL MODULATION

In addition to frequency offset, synchronization and timing signals are subject to other forms of incidental modulation that may cause transmission impairments in the channels they control. These include phase hits and periodic or random jitter. Normally, these forms of incidental modulation have little effect on speech transmission, but they may introduce errors in data transmission by reducing the noise margin.

Phase Hits

Rapid changes in channel phase result in signal impairments called *phase hits*. These hits can be caused by timing signal aberrations. One potential source of phase hits is the switching of transmission facilities or multiplex equipment from working to standby facilities for trouble or maintenance work. If the facility that is switched carries a synchronizing signal or if the switch occurs within the synchronizing equipment, differences in phase between working and spare equipment could cause a hit on the synchronizing signal. As a result, the receiving synchronizing equipment is carefully designed to buffer changes in incoming phase. One widely used clock source has a narrow loop bandwidth of only 32 μHz to prevent rapid changes in phase. However, the switching of transmission facilities may cause a hit directly on the transmitted signal as a result of the difference in phase between the working and standby facility.

Jitter

The generation of an absolutely pure single–frequency signal for use as a carrier is impossible; minute variations in phase always occur. These variations can usually be held to very small values, but from time to time they exceed acceptable limits and cause signal impairments. Continuous and rapid changes in phase, which may be random or periodic, is defined as *jitter*. The principal sources of jitter have been in the power supplies and harmonic generators associated with analog multiplex equipment.

Phase jitter is an unwanted change in phase of a received signal. The problem results from some form of modulation of the wanted signal by another signal. A single–frequency signal that is modulated in this way has sidebands that may be discrete or random and noiselike, depending on the nature of the modulating signal. The amplitude of these sidebands relative to the wanted signal is one measure of the phase jitter suffered by the wanted signal.

Periodic forms of phase jitter sometimes are a result of modulation by power frequency (60 Hz and harmonics) or ringing frequency components (20 Hz and harmonics). Random forms of jitter may result from impulse noise or interfering signals having high–amplitude random components.

Timing Jitter in Digital Systems

An important impairment in digital systems is *timing jitter*. This effect is a random accumulation of variations in the timing of a digital signal. The time variation arises from small timing errors in the clock circuits of each of a string of regenerators and from the action of asynchronous multiplexers that remove "stuffing" bits from a high–speed digital stream in the process of demultiplexing. The spectrum of the jitter tends to be relatively large at low frequencies, falling off at a typical rate of 20 dB per decade as the frequency increases [1]. Jitter effects are controllable by using relatively high "Q" timing circuits in regenerators and, in extreme cases, by applying dejitterizing buffers. The amount of jitter that can be tolerated in practice is relatively large; for example, a peak–to–peak tolerance of 14 full time slots is allowable on 1.5–Mb/s high–capacity channels [2]. Gross amounts of jitter tend to introduce digital errors, and even to introduce distortion into digital voice channels.

19–3 THE SYNCHRONIZING NETWORK

The carrier frequencies used in analog systems and digital timing in North America are each derived from and controlled by a clock of extremely high accuracy and stability. The output of these clocks is transmitted in a variety of ways to all parts of the country. At each location where synchronization is needed, control signals derived from these clocks are used as a master.

The distribution and transmission of clock signals involve many intermediate links and pieces of apparatus. Many of the impairments described in this chapter can occur as a result of impairments suffered by the clock signal in the process of transmission and distribution. Each dependent office has a clock or synchronizing signal source of its own. These local signal sources are controlled as long as a master clock signal is available. Failure of transmission links or apparatus, however, can make a master clock signal unavailable. In this case, the local clock becomes free-running. Its frequency may deviate enough from that of the master to be a source of synchronization impairments. See Figure 19–1.

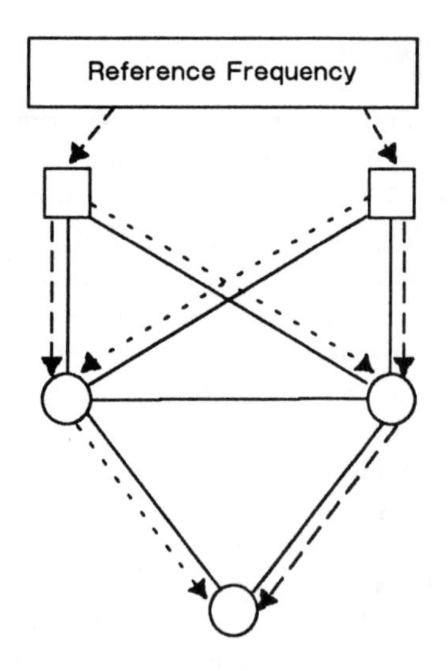

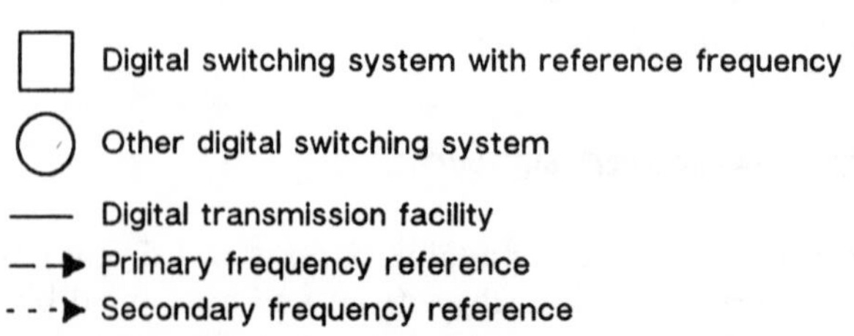

Figure 19–1. Typical synchronization network.

Hierarchical Synchronization

An ubiquitous synchronization network is required to control "slips" as the telecommunications network evolves toward all–digital operation. This requires that a hierarchical synchronization plan be adhered to for a systematic supply of timing signals. One of the dangers of not following a hierarchy system is the possibility of creating a loop within the synchronization network where a clock is being timed by its own output via other clocks in the loop. This would result in timing frequency instability. Clocks must derive synchronization reference from other clocks that are higher in the synchronization hierarchy and never from a clock that is lower in the hierarchy.

Separate transmission facilities need not be allocated for hierarchical synchronization, since existing digital transmission facilities can be used. To enhance reliability, a backup link is furnished between levels of the network, and high–reliability facilities are chosen for both the primary and the alternate links.

Since the overall United States network is made up of subnetworks owned and operated by different carriers, it is now common to have "plesiochronous" operation in which each subnetwork has its own highly precise master clock. Interconnection between subnetworks relies on high–frequency accuracy rather than a hierarchical locking scheme.

References

1. Members of Technical Staff. *Transmission Systems for Communications*, Fifth Edition (Murray Hill, NJ: AT&T Bell Laboratories, Inc., 1982), pp. 326–327 and 728–741.

2. *High–Capacity Digital Special Access Service—Transmission Parameter Limits and Interface Combinations*, Technical Reference TR–NPL–000342, Bellcore (Iss. 1, June 1989).

Additional Reading

Digital Synchronization Network Plan, Technical Advisory TA–NPL–000436, Bellcore (Iss. 1, Nov. 1986).

Chapter 20

Echo in Voiceband Channels

Echo occurs when transmitted signal energy encounters an impedance discontinuity and a significant portion of the signal energy is reflected toward the energy over a return path. Echoes constitute one of the most serious forms of impairment in voiceband channels, whether the channels are used for speech or data transmission. The phenomenon is more difficult to control in switched networks, where terminating impedances may change with every new connection, than in dedicated private circuits, where the impedances are fixed and are more nearly under the control of the circuit designer.

Transmission is impaired by echoes for both talker and listener on an end–to–end connection. A frequently encountered source of echo, one that aptly illustrates talker echo and listener echo, occurs at the junction of four–wire and two–wire circuits. This type of connection and the resulting talker and listener echo paths are shown in Figure 20–1. The transitions between two–wire and four–wire modes of transmission are provided at each end of a four–wire connection by a hybrid circuit, designated HYB in the figure.

20–1 ECHO SOURCES

Consider the circumstances that make the interface between four–wire and two–wire circuits a frequent and difficult–to–control source of echo. The function of the hybrid is to:

(1) provide a transmission path from the receive side of a four–wire facility to a two–wire facility

(2) provide a transmission path from a two–wire facility to the transmit side of a four–wire facility

(3) provide echo control by controlling the coupling path from the receive to the transmit side of the four-wire facility.

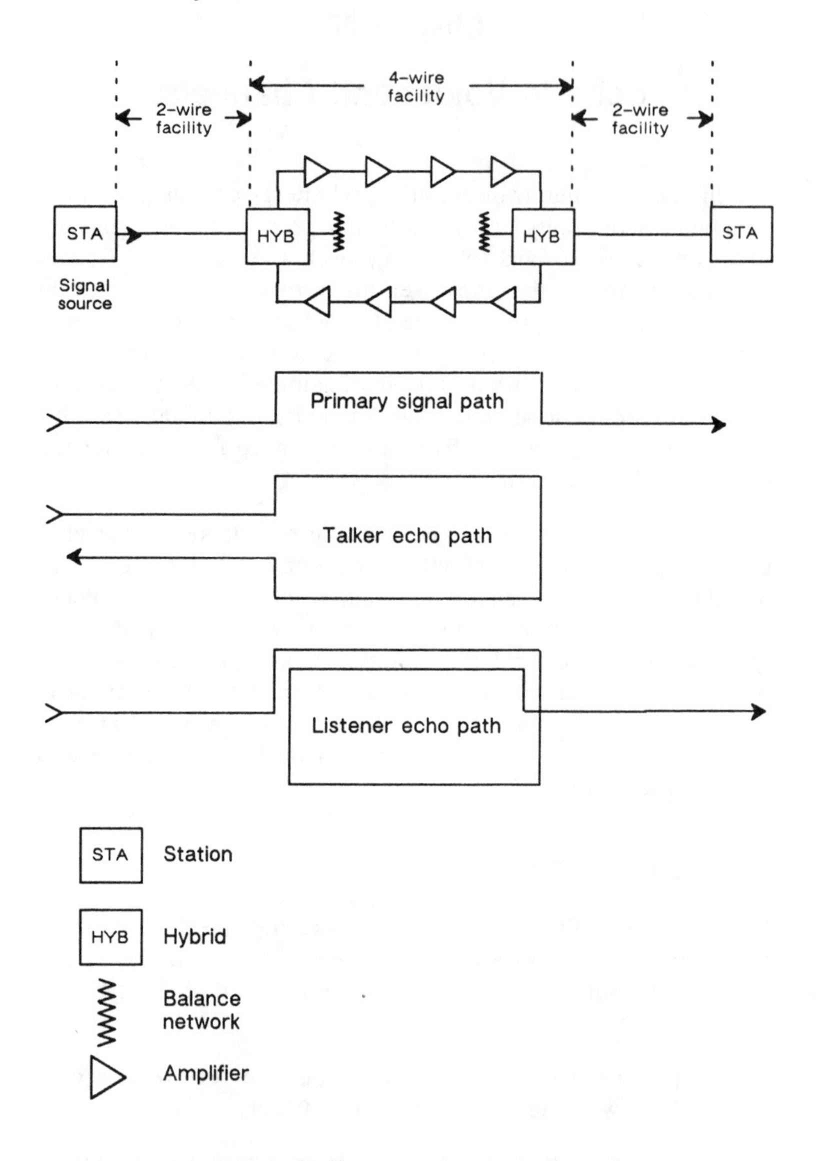

Figure 20-1. Talker and listener paths.

Figure 4–24 and the accompanying text may be used to illustrate the relationship between a passive hybrid and the impedance that must be matched for satisfactory operation. The critical factor in controlling echo in this type of hybrid is the impedance of the balancing network. If a passive hybrid is to be used in a dedicated transmission path, the impedance of the two–wire transmission facility is under design control, or at least has a known value. When this is so, the balancing network may be adjusted to match the two–wire impedance to any desired degree. Thus, in dedicated private–line channels, echo can be anticipated by design and controlled by adjustment.

Now, consider Figure 20–2. Here, office A may be an analog tandem office and office B an end office. The hybrid in office A may be switched to a variety of two–wire trunks. The switch in office B further connects the transmission path to a variety of subscriber loops. This results in a highly variable impedance presented to the two–wire port of the hybrid in office A, depending on gauge, loading, length, etc.

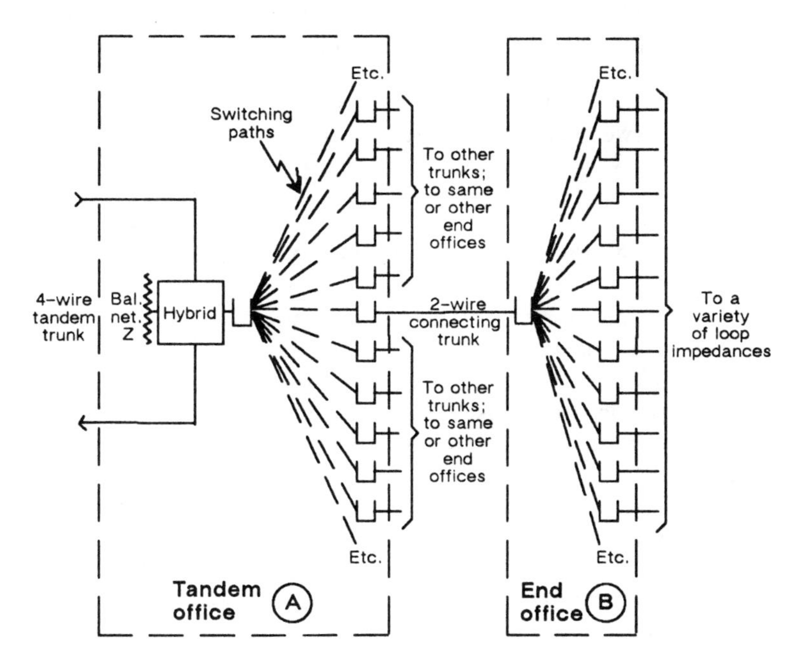

Figure 20–2. Variable impedance paths at tandem office.

Since echo control is a function of the degree of similarity of the two–wire transmission facility and the balance network, and since the two–wire impedance is variable (as shown in Figure 20–2), echo control becomes one of compromise. In fact, in this example, the network is called a *compromise network*; as such, its impedance rarely matches the impedance of the two–wire transmission facility exactly. Thus, some reflection usually does occur.

Where the tandem–to–end–office connecting trunk is two–wire and the hybrid is at the tandem office, loop impedance variations at the end office are somewhat masked at the hybrid by the impedance and loss of the tandem–to–end–office connecting trunk. Where the interface between four–wire and two–wire circuits is at an end office, more serious reflections occur because of the great variability in loop impedances.

While any impedance irregularity produces reflections, such reflections are minor unless the impedance irregularity is in the two–wire path and near a hybrid or a two–wire gain device, so that impedance balancing requirements cannot be met.

Another, normally negligible, source of echo is the crosstalk coupling between the two directions of transmission in a four–wire circuit.

20–2 NATURE OF ECHO IMPAIRMENTS

Transmission impairments caused by echo must be considered in relation to the type of signal involved. In some cases these impairments must be evaluated subjectively, as in speech signal transmission; in other cases the impairments must be evaluated objectively, as in data signal transmission.

Speech Signals

Talker echo usually produces a more serious impairment to speech signal transmission than does listener echo since listener echo is a result of a double reflection and is usually of low amplitude. Therefore, talker echo is stressed in the following.

Talker Echo. If the elapsed time is very short between the production of a speech signal at a station and the reflection of that signal to the speaker's ear, the reflection sounds like sidetone. Unless it is very loud, the speaker may not even be aware of the presence of the reflection. On the other hand, if there is some discernible delay between the initiation of a signal and the reception of an audible reflected signal, the phenomenon is called echo. In extreme cases, the speaker may get the impression that the distant party is trying to interrupt him. This interferes with the speaker's normal process of speech. The overall effect of talker echo depends on how loud it is (which depends on how loudly the speaker talks and how much loss is in the echo path), how long the echo is delayed from the main signal, and the speaker's tolerance to the echo. All of these are interrelated and are best expressed in statistical terms.

It is important to note at this point that the impairment due to echo is related directly to the magnitude of the received echo signal, which can be reduced by the loss in the echo path. Loss in the echo path can be increased by increasing the return loss (RL) through the hybrid and the loss in the transmission path. Any increase in the transmission path loss would have to occur between the talker and the point of reflection and might be accompanied by an unacceptable increase in overall circuit loss. The echo problem must then be solved by a compromise between RLs (balance) and transmission losses. This compromise, based on measured performance and its relationship to subjective evaluations of echo and loss impairments, is achieved in the original via net loss (VNL) transmission design of the switched network and the later fixed–loss plan associated with all–digital networks.

Figure 20–3 summarizes the results of carefully controlled experiments to establish talker tolerance to echo. These experiments found that for any value of delay, there was a minimum echo path loss that was acceptable to the average observer.

The overall echo path loss (EPL) is made up of two important components, return loss and circuit loss from the signal initiation point to the reflection point. RL is the loss (in dB) that occurs between the received and the reflected signal at the point of reflection. Thus, the overall EPL is the circuit loss to the reflection point plus the RL at the reflection point plus the circuit loss from the reflection point back to the initiation point.

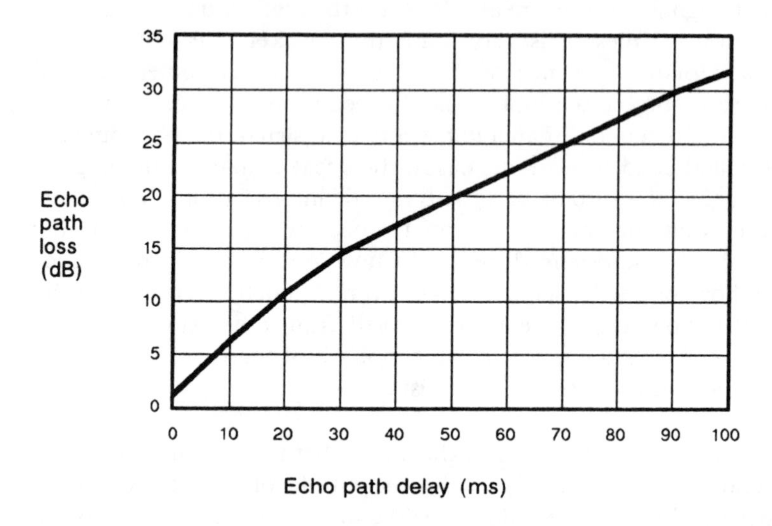

Figure 20–3. Talker echo tolerance with variations in echo path delay and echo path loss.

Via Net Loss Plan. The approximations used in the development of the VNL plan for the analog message network considered:

(1) the average RL at an end office

(2) the estimated loss deviation for each trunk in a built–up connection

(3) the talker echo tolerance of the average observer.

The values that were used were:

(1) 11.0 dB for average return loss with a standard deviation of 3 dB

(2) 2.0 dB standard deviation for round–trip trunk loss

(3) Figure 20–3 for echo tolerance of the average observer with a standard deviation of 2.5 dB.

From these standard deviations for losses and observer tolerance, a value may be derived to represent the standard deviation of the minimum permissible EPL on connections made up of a number of trunks and used by talkers of different echo tolerance. This effort results in the following data:

Number of Trunks	Standard Deviation (dB)
1	4.4
2	4.8
4	5.6
6	6.3

A relationship between loss and talker echo can now be demonstrated. It is convenient to express this relationship in terms of the minimum permissible one–way overall connection loss (OCL) that allows 99 percent of all calls to be completed without echo impairment. The value of the two–way OCL is the average echo tolerance reduced by the average RL at the controlling point of reflection and increased by 2.33 times the standard deviation determined above. The result is divided by two to determine the permissible one–way OCL.

$$\text{OCL} = \frac{\text{Avg. Echo Tolerance} - \text{Avg. RL} + (2.33 \times \min \sigma)}{2}.$$

This equation yields the OCL between end offices. Values of permissible OCL for various numbers of trunks are plotted in Figure 20–4. Linear approximations can be made for these curves that also satisfy echo performance objectives on 99 percent of the connections experiencing the maximum allowable delay and even higher percentages on connections having less than the maximum delay.

Listener Echo and Near–Singing. Listener echo is usually negligible if talker echo is adequately controlled. As shown in Figure 20–1, listener echo is suppressed by a second RL at the talker end of the four–wire circuit and by loss of one additional end–to–end transmit of the four–wire circuit. An exception is found in large multistation private–line circuits. Here, listener echo is often controlling, and special care must be taken in designing this type of circuit.

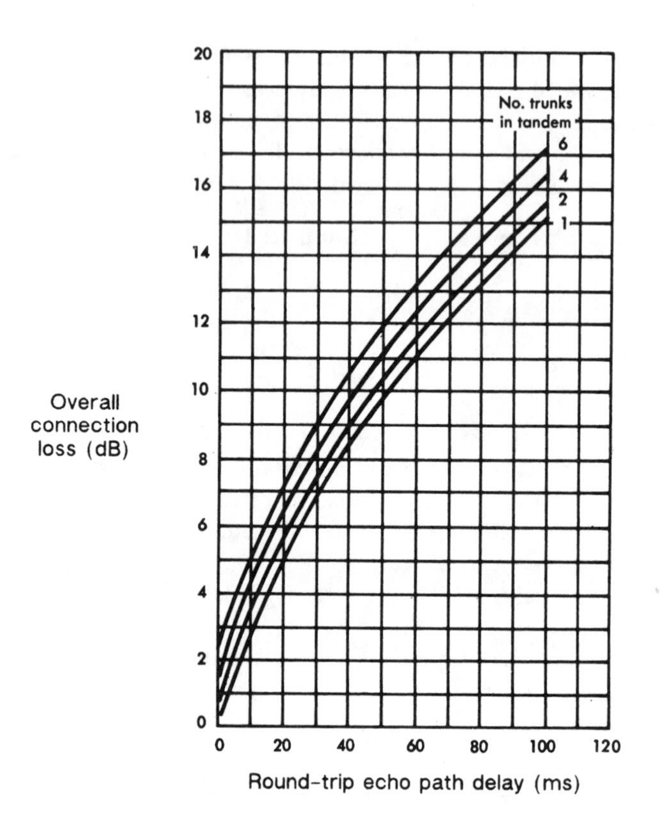

Figure 20-4. One-way loss for satisfactory echo in 99 percent of connections.

There is a close relationship between listener echo and near-singing of a circuit. Both conditions are caused by currents circulating within a transmission path. Interestingly, transmission impairment occurs before singing actually takes place. If the singing margin is too low, the near-singing condition of the circuit causes voice signals to sound hollow, somewhat like talking into a barrel. To avoid this effect, singing margin is usually designed to be 10 dB or more and seldom less than 4 dB.

Data Signals

When data signals are transmitted over voiceband channels, there is greater concern with listener echo than with talker echo since the detector of a data receiver may have difficulty in discerning the appropriate signal with which to synchronize. Listener echo is usually very low in amplitude when talker echo is controlled well enough to satisfy speech signal transmission, but some built-up connections in a switched network may have abnormally poor RLs at both ends of a four-wire intermediate link. As a result, echo performance may be poor and data echoes may result at high enough amplitudes to confuse a data receiver. Listener echo is less of a problem in private lines because they can be designed so that listener echo is of low amplitude.

There are certain instances when talker echo may disturb data set operation. As an example, after a data terminal transmits a circuit assurance inquiry, the internal data set circuitry switches from a send mode to a receive mode to listen for the "handshake" that would be a response from the distant data set. Echo from the end of the inquiry signal may appear as the leading edge of an unwanted receive signal and, since this is not the appropriate handshake that the receiver is expecting, a disconnect is initiated.

When data modems are designed for use on switched network channels, the problems relating to echo must be solved by terminal design. Adequate signal-to-noise margin must be provided in the receiver to cope with listener echo effects. Timing circuits must be provided to avoid talker echo effects in terminal equipment, which may serve as both transmitter and receiver. However, the amount of time delay allowed for in engineering design (called turnaround time) reduces the data transmission efficiency, or throughput rate. When this is critical, full duplex four-wire private-line operation may be required or full duplex operation may be established over the switched network by using two separate connections.

20-3 ECHO MEASUREMENT

The evaluation of echo performance in the telecommunications network is accomplished primarily by RL measurements. RL is usually used as a measure of echo performance resulting from an impedance discontinuity of a two-wire transmission path, and is defined in terms of the ratio of the sum and difference of the complex impedances at the discontinuity or imbalance. However, the complexity of phase relationships in the incident and reflected voltage or current waves makes it impractical to express RL over a band of frequencies except by averaging the performance over the band of interest. Three frequency bands are evaluated using three RL measurements called echo return loss (ERL), singing return loss (SRL)–low, and SRL–high.

Echo Return Loss

A suitably weighted power average is used to express RL over a band of frequencies. For echo evaluation (ERL), this average is applied to the band from 560 to 1965 Hz. An ERL measurement is normally made by applying band–filtered random noise. The weighting is applied by the inclusion of appropriate networks in the test equipment.

Singing Return Loss

As previously discussed, margin must be provided against circuit instability or singing. The measurement that is made to evaluate circuit stability is SRL. This measurement is made at all frequencies at which a circuit might become unstable. Experience has shown that the important bands are those from 260 to 500 Hz and from 2200 to 3400 Hz. Frequencies in the 500–to–2500–Hz range are usually satisfactory from the standpoint of stability if they meet ERL requirements. Instability below 200 Hz and above 3400 Hz is prevented by the increased loss at these frequencies in voiceband circuits.

SRL measurements are usually made by applying shaped random noise to the bands of interest. The 260–to–500–Hz measurement is called SRL–low; the 2200–to–3400–Hz measurement is called SRL–high.

Two–Wire Return Loss

A two–wire RL measurement is a method for evaluating a two–wire facility for impedance irregularities when the facility is terminated in its characteristic impedance. The measurement is made using a test hybrid that uses the same impedance for the balance network. Also, two-wire circuit requirements are usually expressed in ERL and SRL minimums when the circuit termination and balancing network are a 900-ohm resistor in series with a 2.16-μF capacitor (see Chapter 4, Part 5).

Four-Wire Return Loss/Echo Path Loss

When measuring RL at a four-wire test point, the total transmission losses and gains in both directions plus the loss introduced by any echo control devices (hybrid balance networks, echo cancellers, echo suppressors, etc.) are totaled to make up a four-wire RL evaluation. This loss is also called the echo path loss. See Figure 20-5.

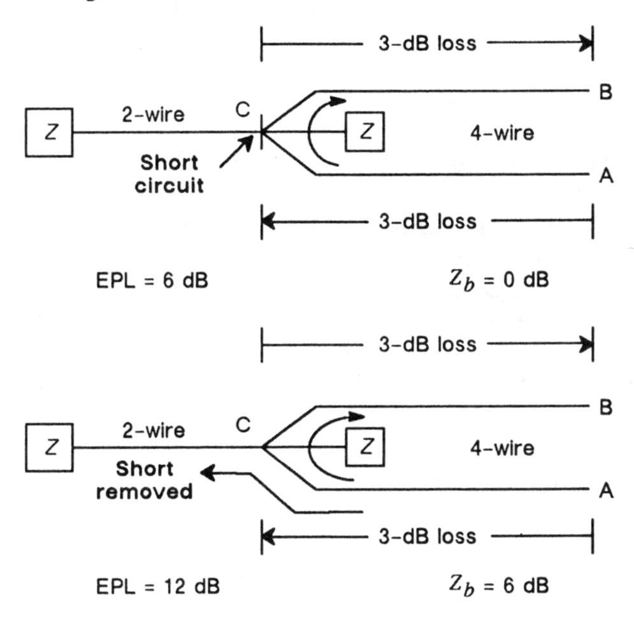

Figure 20-5. Echo path loss and impedance balance.

Impedance Balance

Impedance balance (Z_b) is a term that is used to identify the degree of similarity between the impedance of a hybrid balance network with that of a two–wire "load" or facility. The evaluation of Z_b is made by measuring the ERL and SRL from the four–wire side of the hybrid. The effect of the gains and losses in the echo path that is not related to this evaluation is removed by subtracting the ERL and SRL when total reflection is occurring at the two–wire port of the hybrid by short–circuiting the port.

Equal Level Echo Path Loss

To evaluate the impedance balance of a hybrid from a remote location of the four–wire portion of a transmission facility, without performing the total reflection test as described above, an approximation may be achieved by removing the computed gains and losses of the transmission path from an RL measurement. This technique has been formalized as the equal level echo path loss (ELEPL). It adjusts an EPL measurement by the transmission level points (TLPs) at the test location.

The TLP correction to the measured EPL can be somewhat simplified by adjusting the transmitting test signal to equate to the transmit TLP. Then ELEPL becomes the difference between the level specified as the receive TLP and the measured reflected echo level at the same test point. This relationship between TLP and the echo level will hold at any point in the receive leg of a circuit under test but only when the transmit test level is controlled to be the same as the transmit TLP and the circuit is aligned as designed (see Figure 20–6, Step 3).

A test signal transmitted at a lower level than the specified transmit TLP will produce a lower level return signal measurement by the same degree. Therefore, any difference in transmitted signal level must be considered when calculating ELEPL as follows (see Figure 20–6, Step 4):

$$\text{ELEPL} = \text{EPL} - \text{TLP(transmit)} + \text{TLP(receive)}.$$

A test signal higher than the transmit TLP can cause carrier overload; therefore, higher transmit test levels than the design TLP are not used.

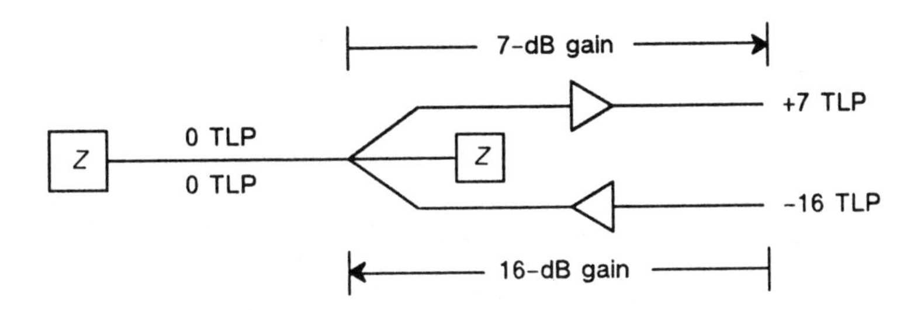

1 –
- Short circuit at 2-wire port
- EPL = −23 dB (gain is negative loss)
- Z_b = 0 dB

2 –
- Short circuit removed from 2-wire port
- EPL = −13 dB
- Z_b = 10 dB   [−13 − (−23) = 10]

3 –
- Test signal at −16 TLP = −16 dBm
- Echo at +7 TLP = −3 dBm
- ELEPL = +7 − (−3) = 10

4 –
- ELEPL = EPL − TLP_T + TLP_R
  = −13 − (−16) + 7
  = −13 + 16 + 7
  = 10

Figure 20–6. Echo path loss, impedance balance, and equal level echo path loss.

Through and Terminal Balance

In the analog switched network, ERL and SRL measurements are made at various switching offices; balance adjustments are made to guarantee the echo performance at each office involved. The measurements and adjustments are usually made with a standard value of impedance as a reference. The measurements are called through balance and terminal balance. In a digital environment, only terminal balance is necessary at the digital–to–analog conversion location.

20–4 ECHO CONTROL DEVICES

Return loss is controlled by impedance adjustments of the balancing network of a passive hybrid and the deployment of echo suppressors and echo cancellers for connections with substantial delay. In private–line applications, the echo consideration may be eliminated with the use of a four–wire configuration with a separate signal path for each direction of transmission.

Echo Suppressors

As shown in Figure 20–3, as circuit delay becomes greater the EPL must also be increased. On circuits having a long delay, increasing the EPL may require undesirable circuit loss. To increase the EPL without increasing the circuit loss, an echo suppressor may be used. An echo suppressor is a signal–activated device that inserts a very high loss in the return path of a four–wire trunk to block the echo. Echo suppressors cause some mutilation of the signal during signal direction transitions and permit some echo to pass during the period when a signal is detected in both directions simultaneously.

Echo Cancellers

An echo canceller enhances RL over an echo suppressor in that it predicts the echo signal and subtracts this prediction from the actual returned signal. Echo is effectively cancelled in both directions during periods of simultaneous signals in opposite directions. The use of cancellers is now preferred wherever echo control is required and is particularly valuable on satellite circuits, whose delay time is too long for the satisfactory operation of echo suppressors. However, an echo canceller can adapt only when a signal is present and there is some discrete time that elapses from a no–signal condition to echo cancellation at the beginning of a signal. This time lag is presently about 250 to 500 ms.

Additional Reading

Cavanaugh, J. R., R. W. Hatch, and J. L. Sullivan. "Models for the Subjective Effects of Loss, Noise, and Talker Echo on Telephone Connections," *Bell System Tech. J.*, Vol. 55 (Nov. 1976), pp. 1319–1371.

CCITT, VIIIth Plenary Assembly, *Red Book*, Vol. III, Fascicle III.1, Rec. G.165 (Geneva: International Telecommunications Union, 1985), pp. 8–19 and 258–279.

Church, T. L., R. F. Koester, and E. H. Mahmoud. *Specifications of 4-kHz Voice and Voiceband Data Network Performance, T1Q1/85–007,* Document Number T1Q1.1/87–052$_r$1, Bellcore (May 1987).

Clement, M. A. *Transmission* (Chicago, IL: Telephony Publishing Corporation, 1969).

Huntley, H. R. "Transmission Design of Intertoll Telephone Trunks," *Bell System Tech. J.*, Vol. 32 (Sept. 1953), pp. 1019–1036.

Chapter 21

Phase Distortion

Some signals, such as data and video signals, are particularly sensitive to departures from linear input/output phase characteristics in the channels over which they are transmitted. Speech signals are not adversely affected by typical amounts of these irregularities because human hearing resolves signal components at different frequencies in a way that has little phase dependence; thus, little attention is given to phase irregularities in speech transmission. However, the transmission of other types of signal requires understanding and coping with phase–related impairments.

21–1 PHASE vs. FREQUENCY CHARACTERIZATION

Chapter 6, Part 2 covers the breakdown of a square wave into its Fourier components and mentions the necessity of maintaining proper phase relations among the signal components. Impairments caused by attenuation distortion are considered in Chapter 18. Here, we consider the impairments that result when departures from the ideal (linear) phase–versus–frequency characteristic of a channel occur.

Linear and Nonlinear Phase

Consider a simple square wave. This wave can be synthesized from an infinite number of odd harmonics of its fundamental frequency. Figure 21–1 (input) depicts a 1000–Hz fundamental and its in–phase 3000–Hz third harmonic with the combined resultant wave; observe that the 1000–Hz fundamental combined with just one of its odd harmonics is already taking on a square shape. Figure 21–1 (output) shows a different resultant wave at the output of 10 miles of 22H88 cable. If the cable exhibited

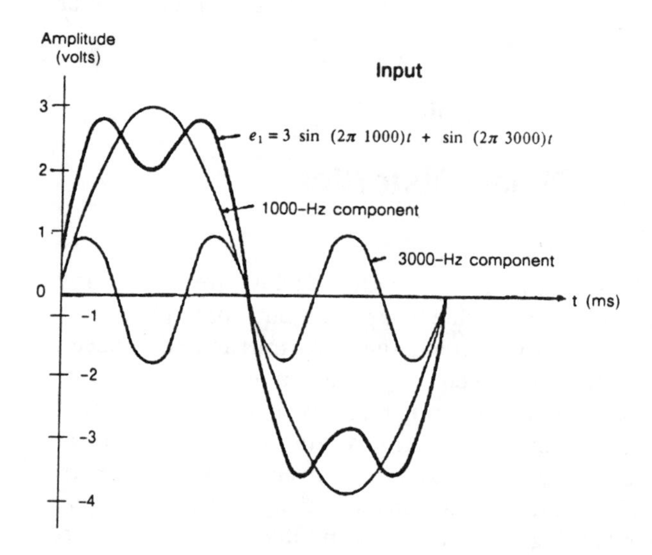

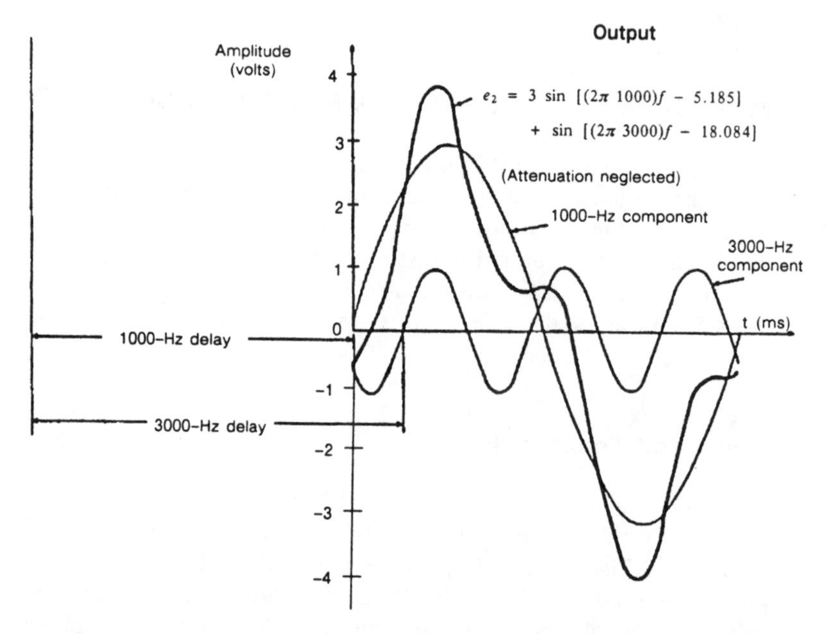

Figure 21-1. Waveforms in and out of 10 miles of 22H88 cable.

linear phase characteristics, the combined waves for the input and output would be identical but displaced in time due to absolute delay (propagation time). However, since the absolute delay for the fundamental and third harmonic differ in this example, a distorted output wave results.

Only the fundamental with its third harmonic has been shown in Figure 21–1 for clarity. Normally a transmission signal has a complex wave having a variety of fundamentals and related harmonics. Phase distortion affects the wave shapes of these signals to varying degrees depending on the amount of nonlinearity.

Phase Delay

Phase delay, group delay, propagation time, and absolute delay are expressions used to define the time delay of a signal or its components and the input and output of a network or transmission line. Since phase delay is a function of frequency, phase delay must be considered with reference to a specific frequency. If the phase–shift characteristic (β) is known, the phase delay (τ_p) at any frequency (ω) can be computed as $\tau_p = \dfrac{\beta \text{ radians}}{\omega \text{ radians/second}}$. If β and ω do not change in direct proportion, the phase delay will change with frequency. The distortion caused by this effect is called *delay distortion*.

Envelope Delay Distortion

Delay distortion is defined in terms of the delay at one frequency relative to the delay at another frequency. Phase distortion is difficult to measure, so to facilitate measurement a more useful parameter called *envelope delay* is used.

To evaluate envelope delay distortion (EDD), a narrowband AM–modulated sine wave is transmitted at various carrier frequencies to evaluate the slope of the phase–shift curve at each frequency selected. The resulting measurements are the envelope delay at each point. EDD is the maximum difference in microseconds of the envelope delay characteristics between any two specified frequencies.

21-2 DELAY DISTORTION IN TRANSMISSION FACILITIES

All of the forms of delay distortion described so far may occur in 4-kHz channels used for voiceband transmission. The sources of signal impairment are largely departures from linear phase-versus-frequency characteristics, the generation of echoes, and signal-dependent distortion. The effects on various types of signal may be quite different depending on the nature and format of the transmitted information.

The three predominant sources of EDD in a voiceband channel are low-frequency rolloff effects in transformers and amplifiers, the high-frequency cutoff of loaded cable, and the low- and high-frequency cutoffs of filters in carrier systems. The exact shape and magnitude of the distortion and the extent of the impairment to transmission (if any) depend on the type of cable and loading, the length of line, and the type and number of carrier channels in tandem. Some examples of the EDD characteristics of typical end-to-end private-line channels are shown in Figure 21-2.

Analog Data Signals

While speech signals are not particularly affected by envelope delay distortion, this is not the case with analog data transmission. Consider a frequency shift keyed (FSK) data signal shifting from one frequency to another to signify the two states of a binary format, and consider that the delay times of the two frequencies are different so that at the data receiver one of the frequencies overlaps into the time slot of the other frequency, thus confusing the state sensing (0 or 1) of the detector. This example of the delay distortion effects on an FSK signal is called intersymbol interference and reduces error margins.

21-3 MEASUREMENT AND CONTROL OF ENVELOPE DELAY DISTORTION

Most transmission parameters can be measured by relatively straightforward methods and by the use of conceptually simple test equipment such as signal generators and detectors. EDD, however, is somewhat more difficult to measure and evaluate.

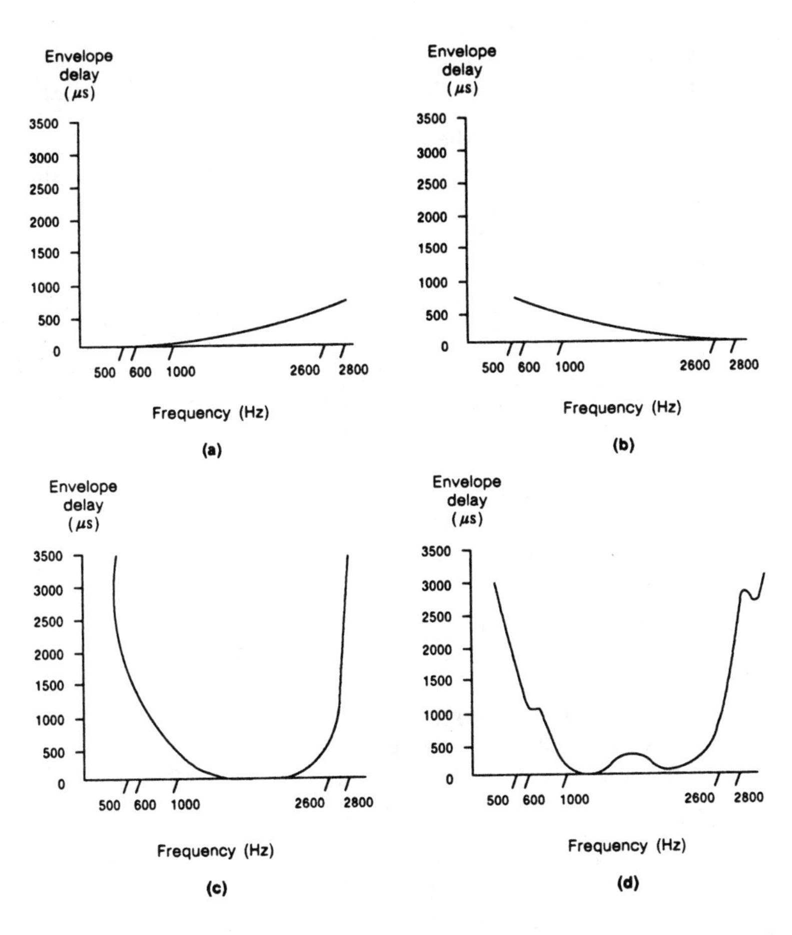

Figure 21-2. Examples of envelope delay on typical channels.

Envelope Delay Distortion Measurements

Envelope delay is measured to a close approximation by transmitting a carrier frequency signal modulated by a low-frequency signal. The envelope delay may be determined by the slope of the phase curve at the carrier frequency used for the measurement. The modulation frequency must be low to make a reasonable approximation. Generally, voiceband–measuring sets have 83–$^{1}/_{3}$ Hz, and occasionally 25 Hz, as the modulation frequency. Sets designed for wideband channel measurements use proportionately higher carrier and modulating frequencies.

The measurements of the type described above may be made on a point–by–point basis by adjusting the test set carrier and reading the envelope delay. The measurement may be automated by sweeping the band with a continuously varying carrier and displaying the envelope delay on a plotter or storing it in computer memory. Equivalent measurements can also be made by using a multicomponent test signal and a fast Fourier transform analyzer.

The Peak–to–Average Ratio System

Time–domain measurements have the advantage of relative simplicity and speed, but have the disadvantage that the effects of individual impairments cannot be separated fully from one another. When poor performance is indicated, the time–domain measurement must be supplemented by frequency–domain measurements in order to identify specific sources of impairment. The common time–domain methods of measurement and evaluation include error–rate measurements by means of pseudo–random data messages and eye–diagram presentations on an oscilloscope, or by the use of the peak–to–average ratio (PAR) system of measurement.

Three parameters are evaluated with a PAR measurement: circuit noise, EDD, and attenuation distortion. The PAR system provides a single–number measure of the combined quality of these parameters with particular sensitivity to EDD. A PAR generator transmits a precise, repetitive pulse at one end of a circuit. The pulses are dispersed by the above impairments on the journey to a PAR receiver, which responds to the pulse envelope peak and its full–wave average. A meter in the receiver indicates the ratio of these two parameters. A reading of 100 denotes an "ideal" channel, while lower or higher values indicate lower quality. A PAR of 50 or more suggests that the channel is suitable for data at 2400 b/s with relatively simple modems.

The PAR approach to voiceband circuit evaluation is valuable because of its simplicity and speed of measurement, but keep in mind that a PAR measurement does not evaluate all transmission parameters that affect data signals and that a poor PAR reading requires further tests to evaluate the transmission parameters listed above.

Control of Envelope Delay Distortion

High–frequency and low–frequency cutoff characteristics are ideally designed to have a linear phase–versus–frequency relationship, most nearly achievable if the rolloffs are gradual. Sharp channel cutoffs are to be avoided, yet are unavoidable in practice. Because digital terminals involve relatively simple filtering compared to analog (SSBAM) equipment, the EDD on digital facilities is relatively low. It is unaffected by the use of intermediate multiplexing or switching.

Another means of controlling the envelope–delay–versus–frequency characteristic is by equalizing. Many types of delay equalizer, both fixed and adjustable, are used for this purpose. If a channel is dedicated to point–to–point operation, specific equalizers may be used to correct that channel for satisfactory performance. If the channel is partly dedicated and partly switched, fixed equalizers may be provided for the dedicated portion of the circuit and for an average characteristic representing the switched portion. Finally, adaptive equalizers, which adjust themselves automatically to gain and delay impairments, may be used within the data modem.

Additional Reading

Bell System Technical Reference PUB 41004, *Transmission Specifications for Voice Grade Private Line Data Channels*, American Telephone and Telegraph Company (Oct. 1973).

Campbell, L. W., Jr. "The PAR Meter: Characteristics of a New Voiceband Rating System," *IEEE Transactions on Communications Technology*, Vol. COM–18 (Apr. 1970), pp. 147–153.

Fennick, J. H. "The PAR Meter: Applications in Telecommunications Systems," *IEEE Transactions on Communications Technology*, Vol. COM–18 (Feb. 1970), pp. 68–73.

IEEE Std. 743–1984, "IEEE Standard Methods and Equipment for Measuring the Transmission Characteristics of Analog Voice–Frequency Circuits" (New York: Institute of Electrical and Electronics Engineers, Inc., 1984).

Lucky, R. W. "Techniques for Adaptive Equalization of Digital Communication Systems," *Bell System Tech. J.*, Vol. 45 (Feb. 1966), pp. 255–286.

Wheeler, H. A. "The Interpretation of Amplitude and Phase Distortion in Terms of Paired Echoes," *Proceedings of the IRE* (June 1939), pp. 359–385.

Chapter 22

Maintenance and Reliability

This chapter relates the basic principles and general application of maintenance and reliability to transmission and service impairments. When maintenance procedures are inadequate, or when unreliable equipment, apparatus, and facilities are used, service will deteriorate. Maintenance and reliability are interrelated; in the extreme, poor maintenance can lead to the ultimate impairment—circuit or system failure.

Maintainability and reliability are carefully planned during the design, development, and installation of equipment and systems. Maintenance is also a continuing concern throughout the service life of each system or item of equipment so that performance standards continue to be met. Awareness of and familiarity with all facets of maintenance systems, maintenance support systems, and test equipment are major elements in the control of transmission performance.

The reliability aspects of transmission systems vary widely in accordance with such factors as accessibility, availability of protection switching and broadband restoration facilities, and the impact of service outages on the kinds of circuits to be routed over the system. A balance is sought among such factors as the degree of reliability improvement obtained, the cost of the improvement, the length of outage deemed acceptable to the customer, and the cost of service restoration when outages do occur.

Economics plays a large role in the design, development, and operation of the maintenance and reliability aspects of equipment and systems. One example may be found in submarine cable systems. The cables and repeaters in these systems are placed in a highly isolated and stable environment, the ocean floor. However, when failure occurs, the recovery of cables and repeaters for repair is a very time-consuming and costly operation. The revenue lost during system outages represents a substantial financial penalty. For these reasons, it is economical in submarine cable systems to spend large sums to provide high system reliability and accurate fault–location equipment.

22–1 MAINTENANCE

Maintenance work is carried out either to correct an existing trouble or to avoid the occurrence of trouble. In the first case, there are various indications that alert maintenance personnel to the need for repairs. The indications may come directly from a customer, an operator, or other observer of a malfunction; or trouble may be indicated by local or remote alarms or measurements that reveal that some parameter fails to meet requirements.

The second case, preventive maintenance, is performed on a routine basis to recognize, limit, and prevent the deterioration of transmission performance and minimize the likelihood of service failure. Preventive maintenance may involve only measurements; if no trouble is indicated by the measurements, further action may be unnecessary.

Many transmission parameters, such as noise, loss or gain, balance, etc., are measured periodically. The results of the measurements are reported to a central point where the data are analyzed and combined. From these analyses, indices are derived and distributed both as a means of comparing performance with other organizational units for which the same indices are derived and as a means of determining trends in performance. By using these indices as guides, it is often possible to see where preventive maintenance routines are inadequate and where their application must be strengthened. In many cases, routine maintenance procedures are prescribed in which a system is temporarily removed from service at specific intervals so that it can be realigned for optimum performance. Later digital systems do not require such readjustment.

Requirements for preventive maintenance and periodic performance measurements have greatly increased since direct distance dialing (DDD) became widespread. With the many facility combinations possible in a connection between two end offices, some performance variance is likely to occur. Design and maintenance plans must be coordinated in order to prevent the normal variations in quality from becoming objectionable to the user. The objective is to maintain transmission performance on all connections between two network points so that the user will notice no difference between the quality provided by successive connections between these same two or similar points. An all–digital network with a fixed–loss plan shows promise in meeting this objective. In the past, most long–distance calls and many local calls were established with the help of an operator. When transmission was unsatisfactory or when there was a service failure, the operator could usually identify the defective circuit and report it to maintenance personnel. Today, the customer often fails to report troubles, particularly those of a marginal nature, and if a report is made, it is difficult to identify the cause of the complaint. In addition, the tremendous plant growth and the need to improve the productivity of maintenance personnel has added emphasis to the use of automated transmission testing.

Three methods have been used to assure that customers seldom encounter unsatisfactory trunks. The first and oldest method is to test the established circuit before connecting the two customers who will use it. This was informally done by operators in the days when switching was a manual operation. In recent years, some switching systems have emulated this process by making a quality check in the circuit establishment process. The earliest of these was a feature of a local switching unit; the test consisted of a continuity test on the trunk before the customer loops were attached. Some four–wire tandem switching systems are equipped with loop–back arrangements that allow a 1000–Hz loop test to be made by the originating office before a through trunk connection is made.

The second method of assuring that defective trunks are not switched into a connection is continuous surveillance. The earliest example of this is the carrier pilot frequencies, which have been standard equipment on analog carrier systems since the open–wire carrier days. With this system, one or more carrier frequency tones are inserted between the carrier channels. The levels of the pilot frequencies are monitored and used to control a variable gain carrier frequency amplifier. When the desired receive carrier level can no longer be maintained by the amplifier, an alarm is triggered to indicate the out–of–limits condition and remove from service all trunks assigned to the system in trouble. Digital carrier systems use a performance–monitoring system that alarms and makes derived trunks busy when bipolar violations or other errors exceed a specified level.

The third widely used method of assuring transmission performance is to test all trunks at specified time intervals (i.e., daily, weekly, monthly). This method is especially adapted to analog facilities and is equally applicable to controlling transmission impairments on voice and data connections. The systems employed make extensive use of computer–controlled testing and test data analysis.

Another network transmission quality control program involves end–office–to–end–office testing on a sample group of connections. The test sample is selected on a random basis from a record of the actual calls placed from an end office. Connection test calls are placed from a subscriber line or simulated appearance in the originating test office and terminated in a far–end appearance connected to a test line that steps through a series of terminations. One part of the sequence provides one to three transmission test tones. Thus, the test simulates conditions on a trunk that might be encountered by a customer making a call to the distant office.

Subscriber loops are the only important segments of the network connections that are not subject to routine transmission testing. Where loop carrier systems are in place, the carrier legs can be tested on a loop basis, but usually are done on an "on–demand" basis or trouble basis. In addition, some switching systems test loops for excessive 60–Hz ac power levels or other foreign voltages before connecting to them.

Most private networks are tested on an "as–required" or trouble basis, although most modern private branch exchanges (PBXs) have computer–controlled testing and analysis programs to monitor the system.

With an analog switched message network, trunk maintenance efforts relied on trunk–by–trunk testing to detect transmission irregularities. This was necessary because of the preponderance of analog transmission equipment common to each trunk. However, as the network evolves into digital transmission and switching, the need for individual trunk testing diminishes. The evaluation of network transmission becomes simplified with surveillance systems that simply monitor digital errors and timing slips. When the integrity of a digital signal is degraded, it affects all channels and trunks in the bit stream; thus, individual trunk testing is unnecessary. However, individual trunk testing is still required when digital–to–analog conversions occur, since impairments at these locations, including any analog extensions, are still common to a trunk.

As the analog message telecommunications service (MTS) network migrates to an all–digital design, maintenance procedures will rely on the surveillance systems that are built into the digital switching systems and digital carrier systems. Operations support systems (OSSs) will be capable of performance–monitoring telecommunications plant so that potential trouble is identified, isolated, and reported before it affects service.

In addition to the switched telecommunications network systems discussed, many specialized surveillance and test systems are available from manufacturers. There are numerous portable microprocessor–based testing units available. Despite this mass of equipment and testing methods available, none can analyze and repair all cases of trouble. They may be able to switch facilities or equipment to bypass trouble or reinitiate a troubled program. However, the effectiveness of these procedures relies on management and personnel who know how and why conditions occur in these systems.

Sources of Deterioration and Failure

Causes of performance deterioration are numerous. As devices age, they often perform less efficiently and cause changes in critical parameters. All devices, active and passive, display some form of deterioration with age. This is most apparent where high mechanical, thermal, and electrical stresses exist.

Where moving parts are involved, electrical performance and reliability deteriorate due to mechanical wear and corrosion. This type of deterioration is most often found in electromechanical switching systems, relay and plug–in unit contacts, etc., and increases in noise or loss adversely affect transmission.

Each year, millions of terminations are changed because people move, equipment is rearranged, and facilities are changed. The resulting plant rearrangements often cause performance deterioration. As examples, undesired bridged taps may be left on cable pairs causing increased transmission loss and attenuation distortion, impedances may change at interface points to produce changes in return loss and echo, and defective workmanship can cause grounds or circuit crosses.

Weather changes may also cause deterioration of transmission performance. Seasonal changes make it necessary in some analog carrier systems to readjust equalizers to compensate for changes in transmission. Expansion and contraction from variations in temperature may loosen joints in wire conductors. Cable shield continuity may be disrupted, thereby eliminating shield currents that are needed to cancel the power induction in the transmission conductors. Impedance unbalances and open shields increase the possibility of increased noise. Moisture from rain or humidity can produce trouble in the loop and trunk plant, particularly in circuits using wire cable with deteriorated insulation so that they are no longer waterproof.

Finally, equipment defects are also a source of impairment or unreliability. These may result from poor control of manufacturing errors, incorrect application (design or installation), poor workmanship in the field, damage in transport or in service, unusual stress, and other causes.

Maintenance Systems and Equipment

Maintenance arrangements are provided for transmission systems to balance the demands of satisfactory cost and service. Sometimes the maintenance equipment is built into the transmission system as a subsystem. In other cases the maintenance or monitoring equipment is centralized and applied to several transmission systems. Maintenance operations may be automatic or manual, locally or remotely controlled, and may involve the use of fixed or portable test equipment.

Integrated Designs. Where continued satisfactory operation of a transmission system depends on the frequent adjustment of system components or where efficient fault–location procedures must be provided to minimize the cost of service failure and repair, maintenance equipment is often built in as a subsystem of a transmission system.

An example of integrated maintenance equipment is found in a typical fiber optic transmission system, where a transmission surveillance center is provided at terminal locations. This equipment provides the capability of measuring remotely the error performance of a large number of fiber optic transmission systems, monitoring protection switch activity, and reporting alarms as they occur. In addition, fault–location equipment may be activated from the surveillance center to assist in the identification and isolation of troubles in remote regenerators. Protection switching functions can also be activated from the surveillance center.

Adjunct Designs. Many designs of maintenance equipment and complete maintenance systems have been provided as adjuncts to transmission systems or to switched networks. These are maintenance facilities that interconnect manually or automatically with transmission systems or with large groups of trunks. Their functions range from simple manually or automatically sequenced measurements of loop–to–ground resistance to complex series of automatic tests of loss, noise, slope, return loss, etc., in both directions of transmission on interoffice trunks. Special test bays are provided to measure automatically or manually the performance of special–services circuits. Manual private–line test boards have largely given way to switched maintenance access arrangements whereby one remote tester can access both ends of the circuit or intermediate points, and make a full series of tests.

The provision of adjunct test facilities has been stimulated by the expanding plant. The large number of circuits that require testing has led to considerable automation; time, cost, and personnel limitations simply do not permit manual testing. The expansion of types of service has created a demand for well–designed test facilities and orderly procedures for their use. For example, the increased use of voiceband channels for data transmission has led to the use of remotely controlled loop–back tests from the end–user's station. This mode of testing has the advantages of minimizing the number of visits that must be made to remote locations to test these circuits and minimizing the amount of portable test equipment that must be carried to customer locations, thus making the maintenance job more economical.

A wide variety of portable test equipment is also required for the maintenance task. The portable equipment must be able to measure loss, slope and attenuation distortion, C–notched noise, impulse noise, envelope delay, intermodulation distortion, peak–to–average ratio, phase jitter, phase and gain hits, echo return loss, singing return loss, noise–to–ground, ac and dc voltage and current, and time–domain wave shapes. Various recording devices to record intermittent trouble phenomena for identification and isolation of problems must also be included in the portable equipment. The portable equipment is used sporadically in the investigation of trouble situations or periodically in preventive maintenance.

Maintenance Support Systems. Maintenance support systems are those that provide communication service for maintenance personnel (order wires), local and remote alarm and telemetry arrangements, and system features that are adapted to the maintenance function. These support arrangements may be integrated into the transmission system or may be provided as adjuncts.

Consider first the communication facilities needed by maintenance personnel. In some instances, the facility may be simply local exchange service to permit direct–dialing flexibility from and to remote locations or to a maintenance center. In other cases, typically at major remote repeater installations, private order wires are installed. These use wire pairs in the same cable as the systems being serviced or may be dispersed–routed to increase reliability. Order–wire systems have become quite sophisticated and may include selective signalling and alternate use for data transmission. These facilities may also be integrated with the transmission system. Examples are found in radio and fiber optic systems that include an order–wire circuit in the basic system design.

Alarms are provided in every system to alert maintenance personnel to real or incipient trouble. This type of maintenance support equipment also varies widely in design and application. An alarm may be as simple as a local alarm actuated when a fuse operates. More typically, alarms are extended from remote, unattended locations to a manned central location. The extension of the alarm usually involves a connection over a data transmission system that collects alarm information from many remote locations and forwards it to the central location. This type of system may provide for the remote control from the central location of maintenance functions at the remote stations.

In systems involving large numbers of circuits or requiring high reliability, automatic protection switching systems are often provided so that a "hot standby" facility is switched into service in the event of failure of a working system. When it is necessary to perform maintenance, measurement, or repair of a working system, service may be temporarily transferred to the spare facility while the maintenance work proceeds. When protection switching facilities are not available, this function is accomplished by a manual patch to spare facilities.

Documentation. The specification of maintenance equipment and the definition of tests and testing intervals on transmission systems are important aspects of maintenance operations. For tests to be meaningful, the test equipment must be properly calibrated and personnel must be trained in its use.

22-2 RELIABILITY

Reliability may be considered with respect to a device, a circuit, a transmission system, or service to the customer. An even broader term, survivability, is used to describe the ability of the telecommunications network to function in the event of enemy attack; however, this subject is not covered here since it is only indirectly related to transmission. The reliability of a device is defined as the probability that the device will continue to function satisfactorily during a specified interval, normally its useful life. Where repair and replacement of failed devices (i.e., maintenance) is feasible, reliability for a system composed of discrete devices is defined as the percentage of time the system is expected to operate satisfactorily over a given time interval.

The opposite of reliability is the probability of failure during a specified time interval. For systems, this measure of unreliability is often expressed as the *outage* time over a given time period. Short–term and long–term outages, and intervals between which outage times are measured, are all important, as in the case of microwave radio systems where short–term outages due to fading differ in their effect from longer outages due to equipment failure. Typically, the objectives for system outage are expressed as minutes per year. Since systems comprise many devices, overall system failure rates and reliabilities are functions of complex combinations of individual device reliabilities. The laws of probability, discussed in Chapter 9, are used to evaluate these combinations. In general, combinations of devices in series are less reliable than the least reliable device; parallel combinations are more reliable than the most reliable device. Increased reliability of parallel combinations is the justification for providing dispersion for critical systems or components of systems.

Sources of Failure

The sources, causes, and mechanisms of service failure may be categorized in many ways. The principal categories can be defined as external and internal. Within each of these there are natural and manmade categories. In the following discussion, these categories will be considered briefly and their effects on the deterioration of signal transmission will be qualitatively evaluated.

External Sources. Among the most common external sources of failure are the effects of weather and other natural phenomena. Lightning, in its direct impact, causes serious damage even to well–protected cables and equipment. Indirectly, it is a source of impulse noise, static in radio transmission, and induced currents in wire circuits that can cause damage and system outage. Ice, snow, and wind can also be destructive; they often bring down aerial wire and cable. They make access difficult where remote equipment and facilities are necessarily exposed. Water does tremendous damage when flooding occurs, but even relatively light rain or humidity can cause deterioration of service where insulation is exposed and weakened by age and the elements. Rain attenuation of some microwave radio signals is a serious source of impairment. Atmospheric layers not broken up by convection or winds are a source of refractive fading in microwave radio systems.

Other natural phenomena, such as sunspot eruptions and the aurora borealis, can create earth currents that temporarily disable coaxial cable systems and can disrupt high–frequency radio transmission. Finally, communication systems are in no way immune to earthquakes, landslides, and fire.

Among manmade sources of failure are the environmental hazards created by nearby power transmission systems. These systems may induce interference currents into communication circuits or, in the event of certain power system faults, may produce damaging currents and expose personnel to high voltages. If communication circuits are exposed to dc power systems, still found in traction company operation, the damaging effects of electrolysis must be considered.

Construction, installation, and maintenance are also frequent causes of failure. Outside plant may be damaged by highway or building construction forces, or service may be disrupted inadvertently during normal outside plant operations. Sometimes, service outages are a result of automobile accidents.

While little damage has occurred to telecommunications plant in North America as a result of enemy action in time of war, this potential source of failure is of great concern to all those responsible for the design, development, and operation of the plant. Little can be done to protect against direct hits of even conventional weapons. The possibility of direct nuclear damage and the effects of electromagnetic pulses due to high–altitude explosions are considered in the design of portions of the present–day plant that carry critical services.

Internal Sources. Within systems, circuits, and devices there are a number of natural or manmade stresses that may be the causes or sources of unreliability. Among these stresses are high voltage (which may produce noise or failure by breakdown), electrostatic discharge (which damages semiconductors), heat (which accelerates the aging process and may cause fire), and mechanical stress (which may cause fatigue failure or breakage due to mechanical shock or long–term vibration in transport or service). Natural aging is, of course, also a source of performance degradation and, ultimately, failure.

Defects due to manufacture, handling, design, or improper installation may cause failure or deterioration. Such defects are sometimes difficult to control because they are so unpredictable.

Another form of internal stress is that of overload. At least two forms of overload can cause transmission impairments. One form, which results when analog signal amplitudes exceed design values, produces serious performance impairments due primarily to intermodulation and, in the extreme, can cause system failure. This form of overload sometimes occurs when test signals are misapplied or when high noise amplitudes are introduced by a feeding system that has failed. The second form of overload, excessively high traffic, has its greatest impact on switching system operation. This form of overload causes blocking of calls and a breakdown of service.

Designs for Reliability

Reliability in a telecommunication network hinges on the design of all elements in the network. The performance quality of apparatus, circuits, systems, and manufacturing designs has a direct impact on reliability.

Apparatus. Certain types of apparatus are commonly used to protect circuits and systems from failure due to high voltage that may result from lightning, contact with power systems, or excessive currents caused by power system faults. Protective apparatus includes carbon–block, gas–tube, metal–oxide–varistor, and solid–state (diode) protectors, which temporarily break down when subjected to excess voltage and carry fault currents safely to ground. These devices are themselves sometimes sources of transmission impairment. When lightning or other faults cause a breakdown, and the device does not completely restore to normal, low resistance to ground or high series resistance in one conductor may result, impairing the circuit by causing excessive attenuation and impedance unbalances. This results in low signal level, noise, distortion, and crosstalk.

Heat coils are used on many cable pairs to protect personnel and equipment from power sources that have voltages too low to operate carbon–block or gas–tube protectors and that produce currents to ground (through office equipment) too low to operate normal fuses. Heat coils are designed to operate when fault currents exceeding specified values continue to flow on a communications conductor for time periods sufficient to cause excessive heating or fire in the equipment. Operation of the device grounds the offending conductor permanently. To remove the ground, the heat coil must be replaced when the fault has been cleared.

Fuses and circuit breakers are devices that also operate to protect personnel and equipment from excessive voltage or current. These devices open the offending circuit.

Circuits. A few examples of the many circuit arrangements furnished to provide reliable operation are described here. Among the most important is the central–office battery arrangement that is used to furnish power to switching and transmission systems associated with each office. The battery supply circuits are designed so that in normal operation the load is supplied from the primary commercial source with a charging current supplied to the battery. The size and capacity of the battery are determined by the load that must be carried in the event of primary power failure and by the amount of time the battery alone must carry the load without service failure.

Power distribution circuits within communication systems must also be designed to guarantee maximum system reliability. The design problems involve the size of conductors, the division of load among the battery feed conductors, and the location and capacity of fuses. The circuits are arranged so that a fault in one part of a system is contained and the whole system is not taken out of service.

Systems. Perhaps the most common feature of system design for reliability involves the "hot spare," i.e., the provision of spare equipment that is powered and ready to operate. Service from a failed line or piece of equipment may be transferred to the spare. The transfer may take place automatically, by action of a switching arrangement designed to recognize failure and to substitute the spare facility, or manually, by patching in the spare equipment. High–capacity systems using coaxial, microwave, and fiber are usually provided with automatic switching. Reliability improvement hinges on the reliability of the protection switch, which may lie idle for long periods of time before it is called into action. New designs of high–speed digital transmission systems are also provided with sophisticated monitoring and switching arrangements. Short–haul carrier systems are provided with flexible patching arrangements. The degree of protection in each case is determined by the reliability of the component parts of the system and the resulting effect on the overall end–to–end reliability of the circuits routed over the system.

Arrangements are also provided for physically replacing damaged plant on an emergency basis. Portable microwave radio repeaters and towers, cable lengths and splicing arrangements, engine–driven generators, spare remote terminals for digital loop carrier, trailer–mounted switching systems, and portable repeater arrangements are all kept in storage, available on demand to furnish emergency service.

"Hardened" systems have also been installed to increase reliability of the network, especially in the event of enemy attack. Cables, structures, and buildings have been built or installed to meet stringent blast–resistance requirements, and equipment is often shock–mounted. Shielding is used on cables, structures, equipment, and buildings to minimize the possibility of service failure from electromagnetic pulses that accompany nuclear blasts.

Grounding of cable shields, apparatus cases, and other outside plant items is controlled to minimize corrosion effects due to electrolysis, especially where dc power systems are used.

Manufacturing Designs. The reliability of apparatus, circuits, and systems is related to the manufacturing processes used. Mechanical, thermal, or electrical stresses can often be avoided by proper design of the manufacturing process. Reliability can be improved by proper test and inspection methods. All of these, however, must be brought into economic balance. Manufacturing costs are usually increased by more stringent reliability requirements. They can be justified only by savings realized in field operation, such as reduced maintenance, less outage time, reduced cost of repairs, etc., all of which are termed "the cost of poor quality."

Network Operating Methods and Procedures

Reliability of service is related finally to network operating methods and procedures. An important element in the layout of facilities for reliable operation is the provision of dispersed routes between switching points. For example, microwave and fiber optic systems allow a variety of transmission paths between two common terminations so that some service will remain in the event of a particular system or route failure.

Alternate–routing features of the local and toll portions of the message network provide a great measure of reliability. If a route is blocked as a result of trouble or excessive amounts of traffic, alternate routes can be found to satisfy most service needs.

Many features of route layouts are selected to maximize reliability. Hardened long–haul transmission systems are laid out so that the backbone route that carries the bulk of the traffic bypasses large cities. These routes are thus less vulnerable to damage by enemy attack. Service into the cities is carried by sideleg systems, which are usually smaller in capacity and less protected against damage.

In certain environments, the provision of appropriate maintenance vehicles is an important element in system reliability. Access to outside plant may be hampered by snow or other vagaries of the weather, long water crossings, or mountainous terrain. Trucks, snowmobiles, barges (for river work), and helicopters all find their places in route maintenance and reliability work, not only for repair activities but also for patrolling the route to detect construction work or other sources of trouble.

Another environmental factor influencing reliability is out–of–sight plant in which cable is buried directly or placed in conduit. In recent years, there has been increased emphasis on the part of the public to improve and beautify the environment, one result of which is increased desirability of out–of–sight plant. While in many cases higher capital costs have resulted, some added benefits in reliability and maintenance cost have been realized. Generally, out–of–sight plant is less susceptible to damage by people, ice, snow, wind, rain, sun, and lightning. Offsetting this advantage, however, is the fact that outages tend to last longer.

Additional Reading

Members of Technical Staff. *Engineering and Operations in the Bell System*, Second Edition (Murray Hill, NJ: AT&T Bell Laboratories, Inc., 1983), pp. 376–385 and 597–602.

Operations Systems Strategic Plan, Special Report SR–NPL–000022, Bellcore (Iss. 3, Dec. 1986).

Telecommunications Transmission Engineering

Section 5

Objectives and Criteria

The design, installation, operation, and maintenance of transmission facilities are based on logically and scientifically established objectives that can be applied throughout the useful life of the facilities. The objectives must also be realistic in that, when met, they lead to customer satisfaction at a reasonable cost. Objectives are dynamic. They must be changed to accommodate changing customer opinion and the introduction of new services. However, the degree of change and adjustment of objectives must be tempered by economic considerations. If objectives are too stringent, excessive costs may be incurred for new system designs and for maintenance of existing systems. If objectives are too lenient, performance may be so poor that customer satisfaction may be low and an excessive number of service complaints may be received. To avoid either extreme, objectives are continually reexamined and reestablished.

There are occasions when relaxation of objectives must be considered for economic or logistical reasons. The importance of the service, an understanding of the basis of the objective, and a firm plan to correct the situation are required before relaxation can be implemented. It is sometimes tempting, for example, to apply more relaxed objectives for a new service on the basis that the new service is temporary or limited in application. This approach risks concentrations of the new service to the extent that damaging effects cannot easily be overcome. There is also the danger that the demand for the service will increase and the problems first introduced as a result of poor judgment will proliferate and require years of effort to correct. At the same time, no new service is assured of success in a competitive environment. Intelligent development of fully mature objectives until the service is established commercially may simply represent good control of business risks.

Many objectives are determined by a process of subjective testing because impairment judgments are based on the sight or hearing mechanisms of the users. Subjective testing is not used much by the exchange carriers; it is carried out primarily under controlled conditions, primarily in a laboratory, and may be supplemented by a controlled field test. A knowledge of subjective testing techniques and methods is desirable in order that results can be properly interpreted and used. This type of testing is covered in Chapter 23.

Customer satisfaction with the transmission performance of a network is conveniently expressed in terms of *grade of service*, a measure of the expected percentage of telephone users who rate the quality of telephone connections as excellent, good, fair, poor, or unsatisfactory when the connections include the effects of transmission impairments such as loss, noise, or echo. The grade–of–service concept is described in Chapter 24, and two applications of the concept are shown.

Transmission objectives are subject to considerable manipulation to make them applicable to various operational situations. After the objectives have been determined, a number of ways of interpreting them must be considered to account for such factors as variability of an impairment and the probability of its occurrence. Also, objectives must be translated into firm requirements for system or circuit performance; then the requirements must be allocated to different parts of the network and to different impairments. These methods of treating objectives are considered in Chapter 25. In addition, this chapter includes a brief description of two standards–setting organizations: the new (1983) Exchange Carriers Standards Association and the older International Telecommunications Union with its two advisory committees, the International Telegraph and Telephone Consultative Committee (CCITT) and the International Radio Consultative Committee (CCIR).

Specific transmission objectives and service requirements for message, operator, digital, and video services are covered in Chapter 26. Several transmission loss plans for message service are described. Many analog and digital parameter requirements affecting loss, noise, and echo grade of service are given.

The economic trends and tradeoffs that must be considered as engineering compromises in the design, application, and operation of transmission facilities are interrelated and involve judicious application of transmission objectives. These relationships are discussed and illustrated by significant examples in Chapter 27. Included in this chapter is a brief overview of a newly proposed economic concept to better estimate economic lives and depreciation rates in the exploding digital technology arena.

Chapter 23

Subjective Testing

Clearly defined transmission objectives must be established for use in setting performance standards and maintenance limits for existing systems and in setting design and development requirements for new systems. The performance standards and maintenance limits must be adjusted to achieve economically feasible performance that yields satisfaction for the majority of customers. Subjective testing of small groups of customers provides the means for estimating customer satisfaction. This chapter briefly describes the general methods, test plans and procedures, and data analysis that are involved in the subjective testing process.

Various methods of subjective testing have been developed to measure customer opinion about the disturbing effects of transmission impairments. Data thus acquired can be used by the application of statistical principles to establish relationships between transmission impairment measurements, the subjective effects of the impairments, and overall customer satisfaction. These relationships can then be applied to the establishment of transmission objectives.

Transmission objectives should be reviewed frequently and brought up to date so they reflect changes resulting from the introduction of new transmission technology, the introduction of new signals or services, and the slowly evolving customer responses to these changes. If customer opinion and objectives are not reviewed regularly, there is the danger that they may be accepted as a matter of habit and not reflect current customer expectations.

Some objectives, whether old or new, can be established or revised in a discrete or quantitative manner because they relate only to machine and equipment performance. Some, such as the error–ratio objectives for digital signal transmission, can be stated discretely because performance thresholds are sharply defined. Other impairments, however, must be judged by subjective testing, and in many cases objectives and performance must be expressed statistically.

In these instances, one of a number of available test methods must be selected, the purposes of the tests must be well–defined, and the test environment must be well–controlled. Also, it must be possible to quantify and express the results in useful terms for application to system design and operation and for use in transmission management. Subjective testing does not provide the answers to many of the questions relating to objectives or grade of service, but it is often the starting point.

23–1 SUBJECTIVE TEST METHODS

Subjective tests of communication system phenomena fall generally into one of three categories: (1) threshold tests to determine threshold values of impairment, (2) pair–comparison tests to compare interfering effects of two different forms of impairment, and (3) category judgment tests to establish subjective reactions to a wide range of impairments (including intelligibility of telephone circuits), a range that spreads from threshold values to unusable values. Circumstances determine selection of the category to be used.

Threshold Testing

In determining an impairment threshold, test procedures are arranged so that each participant, or observer, is given the opportunity to establish a value of impairment magnitude at which the stimulus is "just perceptible" or "just not perceptible." Threshold measurements often are made at the beginning of more extensive tests to establish a base from which other work may proceed. Threshold measurements are valuable also in determining the sensitivity of observers to a particular type of impairment, i.e., to determine if the variation of reactions shows a large or small standard deviation. This information is useful in assessing the importance of other parameters that may affect the result by masking or enhancement.

Threshold measurements may also be used to determine the importance of some newly observed impairment. If, in the normal course of development or operation, the impairment is well below the threshold value, it may be safely ignored. If it is at or above the threshold value, its importance is increased and more extensive testing is indicated. Threshold testing is usually easier to carry out than other types of subjective testing programs. Threshold measurements need only a few test subjects since multilevel opinions are not involved.

Pair–Comparison Testing

Occasionally, a new form of impairment can be evaluated by a test procedure called pair–comparison testing. Pair–comparison testing is designed to establish a magnitude that makes the impairment being tested as disturbing as another type of impairment for which objectives are well–defined. Two typical arrangements for this type of testing, also called isopreference testing, are illustrated in Figure 23–1. In Figure 23–1(a) the signal source is connected to the two receivers, which are adjusted to be as nearly alike as possible with no impairments present. Attenuator (ATT) A is then adjusted to impress on receiver A an impairment such as random noise. The test subject is then asked to adjust attenuator B until the disturbing effect of impairment B (for example, a single–frequency interference) is the same on receiver B as impairment A is on receiver A.

In Figure 23–1(b) the same procedure is followed except that only one receiver is used. The test subject adjusts attenuator B in the same manner as previously described but makes the comparison between impairments by switching back and forth between the two disturbing sources. In the second procedure, the uncertainty of the equivalence of the two receivers and the coupling–decoupling between the ear and telephone receivers are removed. Clearly, this procedure is the preferred way.

The pair–comparison method of testing may be used for either visual or hearing tests. The results are generally considered more valid, or at least more useful, than threshold test results; however, not all the statistical aspects of observer reactions are included. Impairment evaluation is most valid when the impairment is rated relative to some standardized scale that can be applied to all kinds of impairment.

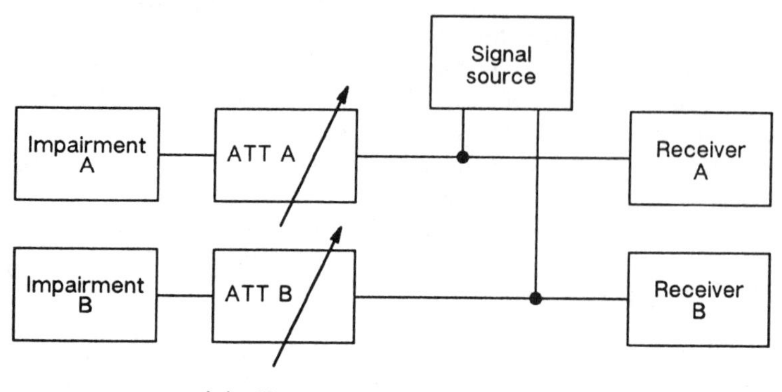

(a) Two-receiver arrangement

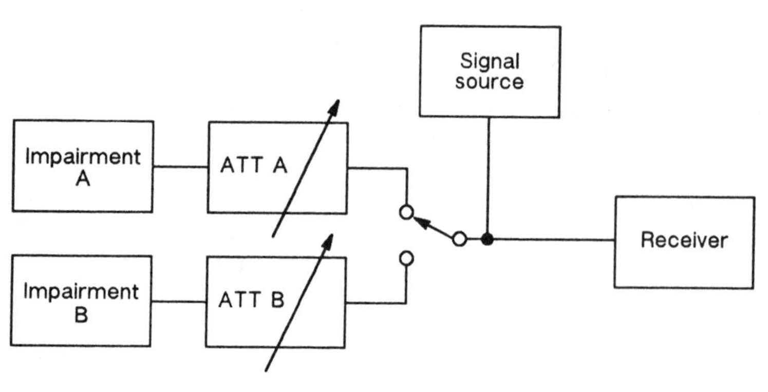

(b) Single-receiver arrangement

Figure 23–1. Experimental arrangements for pair-comparison testing.

Comment-Scale Testing—Category Judgments

The two types of service for which subjective testing has been conducted widely are telephone and television. For each of these, there has evolved a mode of subjective testing that involves a scale that permits quantitative evaluation of judgments of the disturbing effects of impairments. Although they have certain similarities, the two scales are different, and are used somewhat differently.

Telephone Impairment Testing. Bandwidth limitation was one of the first types of telephone transmission impairment for which objectives were established by subjective testing. The test method that evolved, known as articulation testing, measured intelligibility of received speech. The test procedure involved the preparation of stimuli in the form of standard lists of vowel and consonant sounds, syllable sounds, real and nonsense words, and sentences. These stimuli were transmitted by a number of speakers to various listeners who recorded what they heard. Errors in their records were used as the basis of evaluating the effects of variations in the high–frequency or low–frequency cutoffs or in the overall bandwidth of the circuit [1]. Such lists are still used in some related types of testing.

Category judgment evaluations have been applied analytically to the results of articulation tests. At present, other kinds of impairment to the transmission of speech signals over message channels are often evaluated directly by subjective tests in which participants rate their opinions on specific transmission conditions as excellent, good, fair, poor, or unsatisfactory. The tests are conducted so that various impairments are rated under listening conditions selected to be as representative as possible of operating conditions.

In the past, opinion rating tests were carried out on a specific impairment effect such as noise or speech volume. Other impairments were suppressed or enhanced to represent values found in practice. More recently, opinion ratings were carried out with both speech volume (loss) and noise impairments, and were combined with echo impairment tests to provide a composite rating.

Television Impairment Testing. Subjective testing of television impairments has followed the complete cycle of threshold testing, comparative testing, and comment–scale testing. The comment scale shown in Table 23–1 was selected after many efforts to find suitable terms that would adequately describe a wide range of impairments.

Table 23–1. Seven–Grade Comment Scale for Rating Television Impairments

Comment Number	Comment Description
1	Not perceptible.
2	Just perceptible.
3	Definitely perceptible, but only slight impairment to picture.
4	Impairment to picture, but not objectionable.
5	Somewhat objectionable.
6	Definitely objectionable.
7	Extremely objectionable.

23–2 TEST PLAN AND PROCEDURES

The design and administration of a subjective test program involves careful planning and preparation. To have maximum value, such tests must have well–defined goals and must be carefully controlled throughout. The nature of the goals often influences the choice of test method, determines the details and sophistication of test arrangements, establishes the importance of providing a well–designed test environment, and helps to establish the number and qualifications of observers.

Setting Goals

Some carefully considered questions—by definition, the *right questions*—are involved in the determination of the goals of a subjective test program. The determination of the right questions is sometimes an iterative process, because they are not automatically known. However, if the attempt is not made to ask the right questions, the risk is high of getting the right answers to the wrong questions and then to set off in the wrong direction. To illustrate, consider the problem of determining the relative impairing effects of various bit rates and coding algorithms for low-bit-rate digitized voice. If a subjective test program were conducted in which only speech signals were used, the results would give a choice based solely on the preferences of human users. However, since low-bit-rate voice (LBRV) channels may also be used for voiceband data transmission, a system satisfactory to telephone talkers might give unacceptable degradation of high-speed data performance, while another system might be appropriate for either. Speech signal testing could be limited to that which would assure that speech transmission would be acceptable when objectives for data were satisfied.

Who should be pleased and to what degree are two other questions that must be answered. For television signals, transmission objectives that satisfy home viewers may not satisfy the broadcasters or the advertisers. Transmission objectives must be set to satisfy the most critical of these groups. It is often necessary to set objectives at threshold or near-threshold values.

Test Locale

Subjective tests may be conducted in the field or in the laboratory, with the choice normally determined by the test goals and by the relative advantages and disadvantages of each locale.

Field Tests. The principal advantages of conducting tests in the field are that the normal, uncontrolled environment of the operating plant is used and that the test subjects (observers) are the customers who normally use the service. Thus, realistic appraisals of various phenomena can be made under operating conditions. Service impairments such as *slow dial tone* and *all circuits busy* conditions can be evaluated best in the field environment. Subjective evaluations of transmission impairments, however, are seldom made in the field because of the inability to control the environment and the observers.

Laboratory Tests. Most subjective testing of transmission impairments is carried out in the laboratory where test conditions can be controlled and where observers can be selected and trained. Sometimes the entire test program is carried out in the spirit of a laboratory experiment, with all facets of the program (impairment simulation, environment, procedure, etc.) carefully designed and controlled. In other cases, a laboratory test program may be designed to simulate the field environment but in a controlled manner.

If a choice can be made, field tests may be preferred over laboratory tests in that the "real world" is present. However, the choice has to be tempered by the extent of the testing program and the extra expense usually incurred.

Test Conditions

After the initial test plans have been formulated and the procedure has been decided, the next step is to establish all test conditions and facilities. It must be decided whether the tests are to be made under laboratory or field conditions, a source of the impairment (real or simulated) must be provided, the circuits required for conducting the test must be selected or designed, and the necessary test equipment must be procured.

Laboratory Environment. The process of laboratory subjective testing must involve careful control of the test environment, including both the physical environment in which the test is conducted and the environment induced by circuit conditions and arrangements. In both cases, the laboratory environment must realistically simulate the environment in which the impairment actually occurs and must be consistent with the stated goals of the test program.

Physical Environment. It is not possible to specify the nature of control necessary over the physical environment in any given situation; however, the physical environment must be consistent with the goals of the test. If, for some reason, the threshold value of a stimulus must be determined under the most stringent conditions, external distractions must be minimized. For example, if a listening test is required, the environment must be like a soundproof room; if a visual test is required, the environment must be a darkened room.

On the other hand, if a stimulus is to be evaluated by category–judgment–type testing with normal observing conditions, it may be desirable to create a noisy environment by playing recorded street sounds or room noise at appropriate sound levels. If the tests involve television viewing, appropriate ambient lighting may be used.

Circuit Conditions. Circuit conditions that are provided for subjective tests are perhaps even more variable than physical conditions, but they must also be consistent with preestablished goals.

If telephone listening tests are to be made, a number of questions must be answered. First, would the purposes of the tests be served in the presence of impairments other than the one under test (multiple impairment testing)? If so, how loud should they be? For example, if echoes are being evaluated, should there be noise on the test circuits or should they be as quiet as possible? Impairments are interdependent; while subjective test results may apply to one or a combination of several impairments, other interrelated impairments must always be kept in mind. Second, should there be a normal signal present? If so, what kind of signal? If a particular noise impairment is under test, for example, should the observers listen to a simulated conversation while evaluating the noise? Or would it be better to have the observers listen to continuous speech? Or perhaps there should be no speech signal on the circuit at all.

These questions have no general answers and therefore must be considered both separately and collectively. The answers can often be determined from the results of preliminary testing carried out to establish procedures and to determine which parameters affect the results sought.

Source of Impairment. Introduction into the test circuit of the impairment to be evaluated is, of course, a prerequisite to subjective testing. Sometimes this is straightforward, particularly when the impairment is well–defined and easy to simulate. For example, an attenuator can be used to insert loss or a random noise generator can be used to introduce noise into the test circuit.

On the other hand, it is often necessary to record the impairment and then to use the recording as the source during the test. This approach is used in cases where the impairment is intermittent or has some other unusual characteristic that is not easily reproduced except under carefully controlled conditions.

Test Circuits. The principal requirement of test circuits used in subjective testing is that they be capable of delivering signals and impairments to the test area without introducing distortion that might mask the results of the test. It is essential, therefore, that all transmission, distribution, and control circuits be thoroughly tested under all conditions to which they will be subjected in the test program.

Test Equipment. The selection of test equipment is as important as the selection of test circuits. The test equipment must be available before the test program is begun, and effort must be expended to assure that its capabilities and accuracies are appropriate to the task. Consideration must be given to the human engineering of the tests so that the selected test equipment can be used conveniently. Consideration must also be given to the availability of automated measurements and recordings of measurements. In many modern cases like the LBRV example discussed previously, much of the testing can be done by computer simulation in place of physical tests.

Test Procedures

The actual conduct of the test program finally must be worked out in detail. Only general guidelines can be given here because the procedure may be different for each test. In most situations, ten or more observers (sometimes expert and sometimes non–expert) are asked to participate, particularly when comment–scale testing is to be used. For these purposes, an expert observer is defined as a person with good vision (for television) or hearing (for telephone) with experience in judging impairments, who has exhibited consistency in evaluations over a reasonable period of time. Frequently, television testing involves the use of expert observers. For telephone testing, non–expert observers are chosen to be a representative sample of all users.

Usually, a test program is begun by training the observers. The training involves first an exposure to the impairment under test so that each observer knows what to look for or listen for. If test arrangements are such that other impairments are present and cannot be eliminated, the observers may be told to try to ignore them and judge only the impairment under test. The observers then are given the scale of comments to be used in judging the impairment. Often, a few trial runs are used to give the observers an understanding of the test process.

Some pitfalls are to be avoided. One is the introduction of observer bias such as the bias that might result from presenting an ordered sequence, like best–to–worst or worst–to–best test conditions, to all observers in all test sequences. Randomizing the sequence of presentation is desirable. Another pitfall to be avoided is observer fatigue. This condition can sometimes be identified in preliminary testing; experienced observers may become inconsistent after a period of observing, and that period of time may then be used for the duration of final testing.

Before actual testing begins, every step of the procedure should be rehearsed, and every effort must be made to ensure that all observers are exposed to the same or very similar viewing or listening conditions. The circuits, test equipment, television pictures or telephone signals, and impairment sources should be checked and calibrated before each test to eliminate unwanted and unexplained variations. In short, the entire procedure must be conducted with great care and precision to ensure valid results.

23-3 DATA ANALYSIS

For a subjective test program to produce useful results, the accumulated data must be analyzed and presented so that they may be related to performance criteria. For these purposes, the subjective test data and performance criteria are often expressed in terms of mean values and standard deviations.

Methods of analysis have improved as subjective testing procedures have become more scientific and as the importance of test results has become more widely appreciated. The progress made in evaluating, analyzing, and presenting subjective test results of television impairments is a case in point. In the middle and late

1940s, when the television industry was growing rapidly, most of the development and research emphasis was placed on gaining an understanding of the fundamentals of television camera, transmission, and reproduction processes. Later, experiment and analysis were devoted to the evaluation of various impairments. A major current topic is the subjective evaluation of the effects of reduced bit rate and various coding and prediction algorithms for digital television transmission.

The analysis of subjective test results of telephone impairments is still evolving. Over the years, tests have been made on single and combined impairments to cover speech volume, loss, circuit noise, quantizing noise, talker and listener echo, room noise, bandwidth, echo suppressors and cancellers, and other impairments [2,3]. The results of these tests have been used to model various telephone connections to estimate customer satisfaction in grade–of–service terms (see Chapter 24).

References

1. French, N. R. and J. C. Steinberg. "Factors Governing the Intelligibility of Speech Sounds," *The Journal of the Acoustical Society of America*, Vol. 19 (Jan. 1947).

2. Sen, T. K. "Subjective Effects of Noise and Loss in Telephone Transmission," *IEEE Transactions on Communications Technology*, Vol. COM–19, No. 6 (Dec. 1971).

3. Hatch, R. W. and J. L. Sullivan. "Transmission Rating Models for Use in Planning of Telephone Networks," *IEEE National Telecommunications Conference Record* (Dallas, TX: Nov. 1976), Vol. II, pp. 23.2–1 to 23.2–5.

Chapter 24

Voice Grade of Service

Transmission *grade of service* is a calculated measure of the expected percentage of telephone users who would rate the quality of telephone connections as excellent, good, fair, poor, or unsatisfactory when the connections include the effects of given transmission impairments. This chapter describes a statistical method of calculating transmission ratings and the resulting voice grades of service. Equations and curves are included to show the effects of various transmission impairments.

The grade of service is based on the opinion ratings of a sample of telephone users taken during subjective testing of simulated telephone connections. In effect, it combines the distribution of these ratings with the distribution of plant performance parameters to obtain the expected percentage of customer opinions in a given category. The term grade of service is usually applied to overall communication service. However, the grade–of–service concept can be applied in theory to one aspect of communications such as transmission, to one specific impairment such as loss, noise, or echo, or to combinations of these impairments. It is usually expressed in terms such as *a grade of service of "X" percent good or better* or *a grade of service of "Y" percent poor or worse* for the impairment under consideration. Loss, noise, and echo are predominant among the impairments that tend to degrade speech signal transmission performance.

Transmission management of the telecommunications network involves establishing transmission objectives, measuring transmission performance, and estimating customer opinions of the quality of service rendered. The grade–of–service concept is a useful tool for fulfilling these responsibilities. While the concept is usually used directly in establishing objectives, it is also sometimes used in inverse applications to determine what performance must be achieved to meet grade–of–service objectives.

Solutions to many transmission engineering problems can be compared by using the grade-of-service concept. The information obtained by such analyses often provides the basis for making engineering compromises in establishing or allocating objectives, identifying weak spots in performance, evaluating improvements, showing the effects of time on services rendered, or conducting a subjective test program or a field performance survey. Grade of service is also used to evaluate combinations of service parameters and to optimize performance with respect to such combinations.

24-1 TRANSMISSION RATING

Subjective testing involves many experiments, each covering a series of tests with test subjects listening to simulated impaired telephone connections and rating the connections in five categories: unsatisfactory, poor, fair, good, or excellent. In this manner the effects of loudness loss, noise, and echo impairments have been studied, analyzed, and reported [1]. Such studies are continuing and methods have been developed for analyzing the test data. The analysis method has been difficult to develop because of the variability of the test subjects and conditions, and has evolved over the years. One detailed analysis procedure, which has been widely accepted, is reported in Reference 1 for loss, noise, and echo impairments with additional application information contained in Reference 2. The analysis procedure is a complicated modeling process, as many sets of data are included, in order to arrive at a convenient yet accurate model to reflect the test results. In the referenced procedure, the modeling was based on normal probability densities, as this provided flexibility in dealing with deviations found in the test data. The steps involved in the analysis are described briefly in the following:

Step A. For each set of data, a normal density curve was fitted to the opinion score histogram curve so that the mean and standard deviation of the normal curve were the same as those for the raw data. Other normal curves were fitted this way to other sets of data.

Step B. The standard deviations of all the fitted normal curves of Step A for all the data sets were weighted and averaged to obtain a single value for the standard deviation. The weighting was based on an analysis of variations in the standard deviations as a function of the fit mean.

Step C. The single value of the standard deviation found in Step B is used in each test condition as the standard deviation of the corresponding normal curve, and the mean of the normal curve is adjusted so that the mean of this curve and that for the raw data are equal.

Step D. The means of all the normal curves of Step C are fitted on a least–squares–error basis to an appropriate function of the test parameters.

Figure 24–1 exemplifies Step A of the procedure. It is an example of a smooth curve [3] of subjective responses that resulted from one specific set of values of noise and loudness loss in one such experiment. The smooth curve is a normal–distribution curve that was fitted to the data. Note that the area under the normal curve from minus infinity to 2.5 (the boundary between poor and fair) is the probability of a poor–or–worse (PoW) rating, and from 3.5 (the boundary between fair and good) to plus infinity is the probability of a good–or–better (GoB) rating. The PoW designation really means poor plus unsatisfactory ratings and the GoB designation means good plus excellent ratings. Similarly, other normal curves were fitted to histograms of other sets of data for analysis.

Proceeding with the analysis through the other steps of the procedure noted above, the relationship found for the fitted mean, μ, for an experiment labelled MH (Murray Hill) [3] on loudness loss and noise was:

$$\mu = 8.21 - 0.144\sqrt{(L_e - 2.2)^2 + 1} - 0.1215N_F + 0.001297L_eN_F \qquad (24\text{–}1)$$

where L_e is the loudness loss in dB defined in IEEE terms [4,5] and N_F is the total noise, in dBrnc, obtained by power–summing circuit noise, the equivalent circuit noises due to room noise and quantizing noise, and an empirically determined constant equal to 27.4 dBrnc.

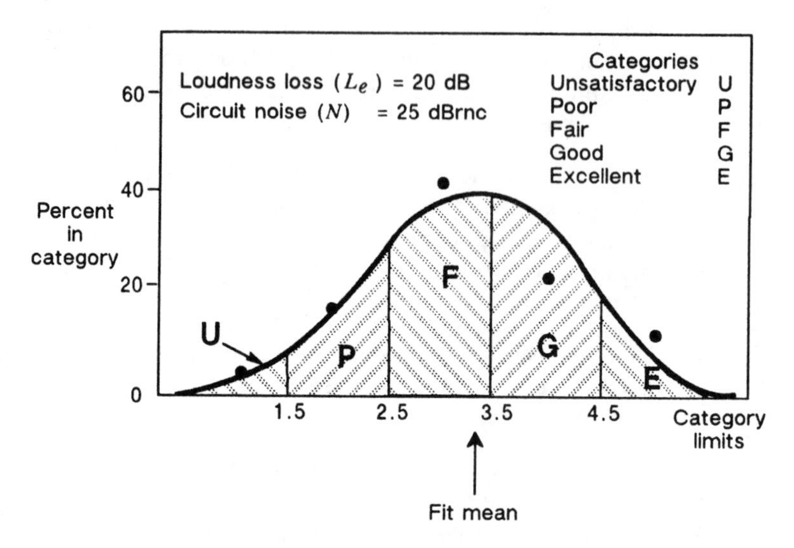

Figure 24-1. Example of smooth curve fit.

The estimated proportion of telephone users who would find the service for a particular experiment of several values of loss and noise to be GoB is given by:

$$\text{GoB} = \frac{1}{\sqrt{2\pi}} \int_{a}^{\infty} \epsilon^{\frac{-t^2}{2}}\, dt \qquad (24\text{–}2)$$

$$a = (3.5 - \mu)/\sigma \qquad (24\text{–}3)$$

where μ = mean
σ = standard deviation = 1.44

In the analysis of subjective test results, however, it was recognized that different tests yielded somewhat different results even when the same impairments were tested. This complicated the combining of results from different tests into a composite model of subjective opinion and led to the concept of a general transmission rating scale, referred to as the *R–scale*, which assigned a single numerical value to any specific impairment.

In analyzing the data, it was found that the mean for any one experiment was nearly a linear function of the mean of any other experiment depending on the range of values used for the loss and noise impairment. This led to defining the R–scale for two anchor points and having a linear function in between. The two values were for R–loss noise ratings of 80 and 40 for transmission conditions of (1) loudness loss of 15 dB and noise of 25 dBrnc and (2) loss of 30 dB and noise of 40 dBrnc, respectively. The first represents excellent transmission typically provided by a short intraLATA (local access and transport area) tandem connection, and the second by a long interLATA analog connection with long loops. The linear transformation of Equation 24–1 into an R–loss noise rating (R_{LN}) was found to be:

$$R_{LN} = 18.8 + 15.7\mu \tag{24–4}$$

and

$$R_{LN} = 147.76 - 2.257\sqrt{(L_e - 2.2)^2 - 1} - 1.907N_F + 0.02037L_eN_F \tag{24–5}$$

where L_e and N_F are defined under Equation 24–1.

Using Equation 24–4, Equations 24–2 and 24–3 can be written:

$$\text{GoB} = \frac{1}{\sqrt{2\pi}} \int_{-\infty}^{A} \epsilon^{\frac{-t^2}{2}} dt \tag{24–6}$$

$$A = (R_{LN} - a)/b \tag{24–7}$$

The values of a and b will vary from experiment to experiment. For the data base "MH," which is applicable to local connections, a = 64.07 and b = 17.57; for the data base "long toll," a = 51.5 and b = 15.71.

Transmission rating (R) relationships [3] have been developed for loss/noise, talker echo, and loss/noise/echo; they are indicated by the equations in Table 24–1. The R relationships have been extended to cover impairments caused by listener echo (Chapter 20, Part 2), loss/noise/listener echo, quantizing noise (Chapter 17, Part 3), bandwidth, room noise, and sidetone. These are indicated by the equations in Table 24–2 [6].

Table 24–1. Transmission Rating *R*–Model for Loudness Loss, Circuit Noise, and Talker Echo

Loss/Noise

$$R_{LN} = 147.76 - 2.257\sqrt{(L_e - 2.2)^2 + 1} - 1.907N_F + 0.02037L_eN_F \;. \quad \text{(1a)}$$

L_e = loudness loss (in dB) of an overall telephone connection according to the IEEE method [4,5], which also defines ROLR* and TOLR**,

N_F = total effective noise (in dBrnc) referred to the input of a receiving system with ROLR = 46 dB,

$= N_C$ "+" N_R "+" N_Q

("+" signifies power addition),

N_C = circuit noise (in dBrnc) referred to ROLR = 46 dB,

N_R = circuit noise equivalent (in dBrnc) of the room noise referred to ROLR = 46 dB (N_R=27.37 dBrnc for the loss/noise tests), and

N_Q = circuit noise equivalent (in dBrnc) of the quantizing noise referred to ROLR = 46 dB.

Talker Echo

$$R_E = 106.4 - 53.46\log_{10}\left[(1 + D)/\sqrt{1 + (D/480)^2}\right] + 2.277E \quad \text{(1b)}$$

where

E = loudness loss (in dB) of the talker echo path, and

D = round-trip talker echo path delay (in milliseconds).

Loss/Noise/Echo

$$R_{LNE} = \frac{R_{LN} + R_E}{2} - \sqrt{\left[\frac{R_{LN} - R_E}{2}\right]^2 + 100} \;. \quad \text{(1c)}$$

* ROLR (receive objective loudness rating) expresses the sensitivity (in loudness terms) of the receiving reference element in converting an electric signal to an acoustic signal.

** TOLR (transmit objective loudness rating) similarly expresses the conversion of an acoustic signal to an electric signal.

Table 24–2. Extension of R Model

Listener Echo

$$R_{LE} = 93(\text{WEPL} + 7)(D_L - 0.4)^{-0.229} \tag{2a}$$

where

WEPL = weighted listener echo path loss (in dB),

$$= -20 \log \frac{1}{3200} \int_{200}^{3400} 10^{-\text{EPL}(f)/20} df, \tag{2b}$$

EPL(f) = echo path loss (in dB) in the listener echo loop as a function of frequency (in Hz), and

D_L = round–trip listener echo path delay (in milliseconds).

Loss/Noise/Listener Echo

$$R_{LNLE} = \frac{R_{LN} + R_{LE}}{2} - \sqrt{\left[\frac{R_{LN} - R_{LE}}{2}\right]^2 + (13)^2} \ . \tag{2c}$$

Circuit Noise Equivalent (in dBrnc) of Quantizing Noise

$$N_Q = 77 - L_e - 2.36Q \tag{2d}$$

where L_e is defined in Table 24–1 and

Q = the speech–to–speech–correlated–noise ratio determined subjectively using a reference device called the modulated noise reference unit (MNRU). Ratio is in terms of speech power and unweighted noise power [7].

Q values for some typical codec pairs (a codec pair produces analog–to–digital–to–analog conversion) can be estimated from

Table 24–2. Extension of R Model (Continued)

Circuit Noise Equivalent (in dBrnc) of Quantizing Noise (Continued)

pulse code modulation (PCM) with $\mu = 255$:

$$Q = 0.78L - 12.9 \tag{2e}$$

nearly instantaneously compandored PCM:

$$Q = 0.74L - 2.8 \tag{2f}$$

adaptive delta modulation:

$$Q = 0.42L + 8.6 \tag{2g}$$

adaptive differential PCM with fixed predictor:

$$Q = 0.98L - 5.3 \tag{2h}$$

adaptive differential PCM with adaptive predictor:

$$Q = 1.04L - 4.6 \tag{2i}$$

where

$L =$ line bit rate in kb/s.

Before using Equation 2d, the total Q for connections with tandem codec pairs is estimated using

$$Q = -15 \log \left[\sum_{i=1}^{n} 10^{-Q_i/15} \right]. \tag{2j}$$

Note: Q values for codec pairs are determined from subjective tests showing that the threshold of detectability for quantizing noise in terms of Q is about 25 dB (higher values were generally not detectable). Equations 2e through 2i predict Q values greater than 25 dB at higher line bit rates. These equations were derived so that the total Q obtained when tandeming codec pairs would reflect subjective test results.

Bandwidth/Attenuation Distortion

$$R_{LNBW} = K_{BW}(R_{LN} - 22.8) + 22.8 \tag{2k}$$

Table 24–2. Extension of R Model (Continued)

Bandwidth/Attenuation Distortion (Continued)

where R_{LN} is from Equation 1a of Table 24–1 and

$$K_{BW} = k_1 k_2 k_3 k_4 \tag{2l}$$

where

$$k_1 = 1 - 0.00148(F_l - 310), \tag{2m}$$

$$k_2 = 1 + 0.000429(F_u - 3200), \tag{2n}$$

$$k_3 = 1 + 0.0372(S_l - 2) + 0.00215(S_l - 2)^2, \tag{2o}$$

$$k_4 = 1 + 0.0119(S_u - 3) - 0.000532(S_u - 3)^2 - 0.00366(S_u - 3)(S_l - 2), \tag{2p}$$

$F_l F_u$ = lower, l, and upper, u, band limits (in Hz) at which the acoustic–to–acoustic response is 10 dB lower than the response at 1000 Hz (for $F_u > 3200$ Hz use 3200 Hz), and

$S_l S_u$ = lower, l, and upper, u, inband response slopes (in dB/octave) below and above 1000 Hz, respectively, that would have the same loudness loss as the actual response shapes.

Circuit Noise Equivalent of Room Noise

$$N_R = N_A - 35 + 0.0078(N_A - 35)^2 + 10\log\left[1 + 10^{\frac{1-\text{SOLR}}{10}}\right] \tag{2q}$$

where

N_A = room noise in dBA and

SOLR = sidetone objective loudness rating (in dB) of the telephone set sidetone path in IEEE terms [4,5].

Table 24–2. Extension of R Model (Continued)

Sidetone

$$R_{LNST} = K_{ST} R_{LN} \tag{2r}$$

where R_{LN} is from Equation 1a of Table 24–1,

$$K_{ST} = 1.021 - 0.002(\text{SOLR} - 10)^2 + 0.001(\text{SR} - 2)^2(\text{SOLR} - 10), \tag{2s}$$

SR = sidetone response in dB/octave below 1000 Hz. (The model assumes that the slope in dB/octave above 1000 Hz is 1.5 times that below 1000 Hz.)

$$R_{EST} = R_E + 2.6(7 - \text{SOLR}) - 1.5(4.5 - \text{SR})^2 + 3.38. \tag{2t}$$

(R_{EST} replaces R_E in Equation 1c of Table 24–1.)

Some of the R relationships have been plotted to indicate the effects of various impairments visually [6]. Figure 24–2 gives the R_{LN} rating for loss and noise values. Figure 24–3 gives the R_{LNE} for loss/noise/echo values. Figure 24–4 gives the circuit noise, N_R, in dBrnc equivalent to room noise, N_A, in dBA. N_R is power–summed with other noises to produce N_F used in the equations. Figure 24–5 gives the estimated percentage of GoB and PoW for the MH and long–toll grade–of–service models versus the R–rating values. The procedure would be to find the R value(s) from the earlier figures (or calculated from the equations in Tables 24–1 and 24–2) and enter Figure 24–5 to find the expected grade–of–service value(s).

24–2 A LOCAL CONNECTION

Grade–of–service calculations are useful in showing the differences in estimating connection quality between different conditions of interest. For example, one might be interested in how an increase in loudness loss or room noise condition might affect the estimated percentage of good–or–better values.

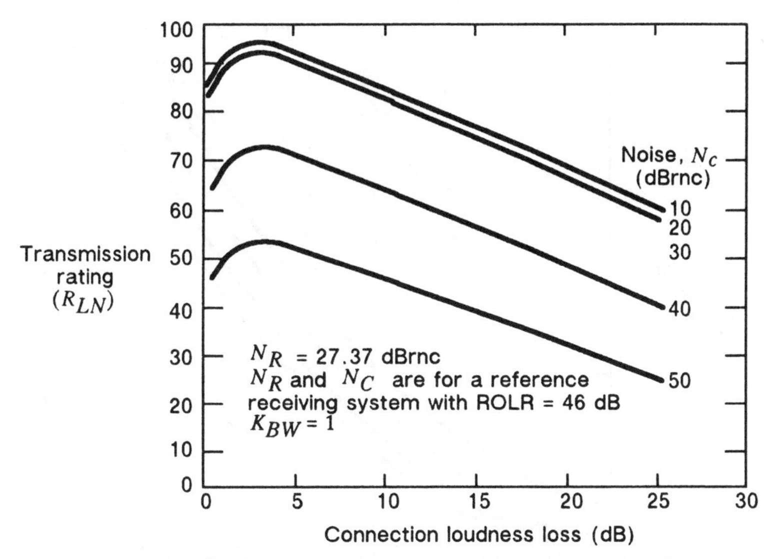

Figure 24–2. Transmission rating for loss and noise.

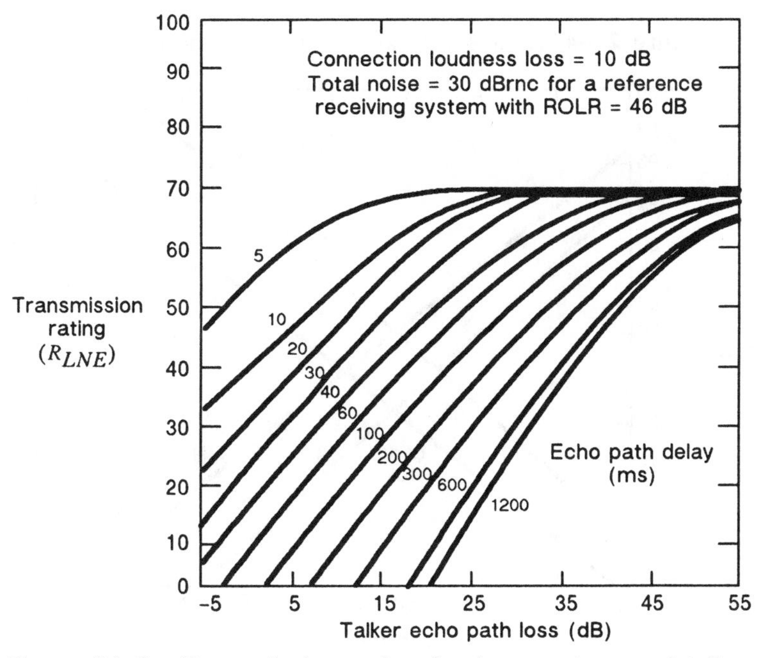

Figure 24–3. Transmission rating for loss, noise, and talker echo.

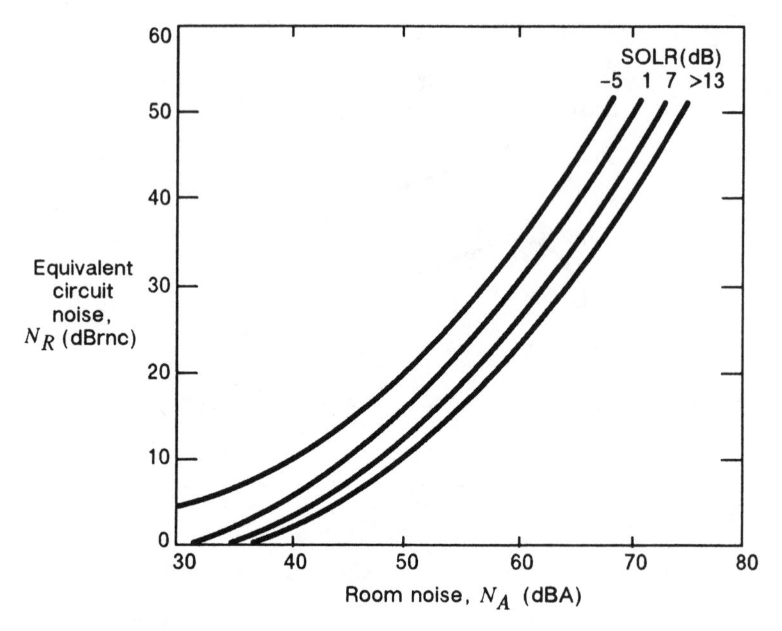

Figure 24–4. Equivalent circuit noise for room noise.

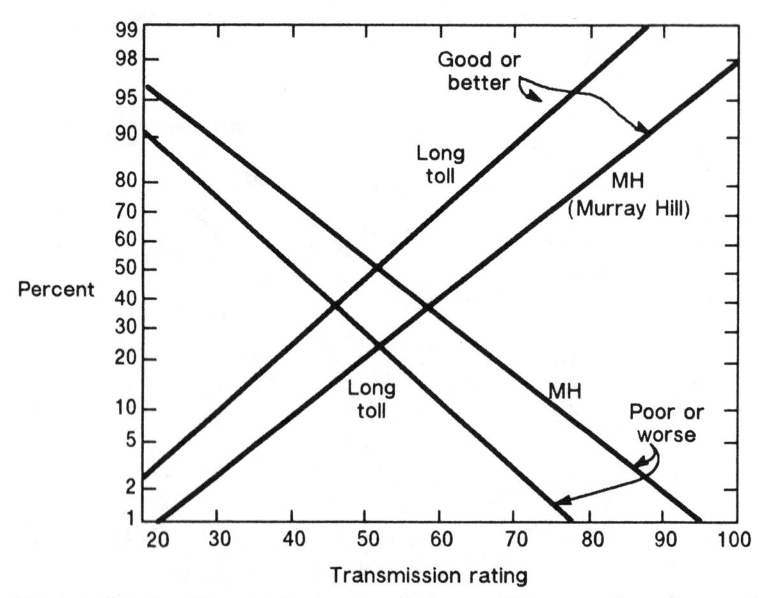

Figure 24–5. Comparison of opinion ratings as functions of transmission rating.

This can be determined by repeating the calculation after the condition is changed and noting the differences in the quality obtained. This can be shown by the following example [6] for a local connection to show the effects of increasing room noise at the receiving end. The connection is assumed to be from an average loop to an average loop, each terminated by a 500–type or equivalent telephone set and connected together through an analog office. The connection is assumed to have the following characteristics in terms of Equations 1a, 2k, 2q and 2r of Tables 24–1 and 24–2:

Loudness loss, L_e	=	3 dB
Circuit noise, N_c	=	25 dBrnc
Room noise, N_a	=	55 dBA and 65 dBA
Sidetone objective loudness rating, SOLR	=	7 dB
Sidetone response, SR	=	2 dB/octave

No quantizing noise, no talker or listener echo.

The calculations for R_{LN} (Table 24–1, Equation 1a) and the grade of service were found to be:

		R_{LN}	% GoB
Room noise	55 dBA	94.0	95.7
	65 dBA	74.3	72.5
	Difference		23.2%

The difference in the GoB quality of service represents a large change and indicates the strong influence of room noise. A rating chart [8] for business office noises suggests that telephone use becomes difficult at room noises above about 66 dB (average sound level in bands at 500, 1000, 2000 Hz re 20 $\mu N/m^2$).

For the older via net loss (VNL) analog voice transmission network, the estimated residential grade–of–service performance is related to mileage (circuit noise and delay) and is usually expressed in loss/noise/talker–echo terms. For the newer fixed–loss digital transmission network, the estimated performance is related to mileage only because of delay (echo). Circuit noise depends on quantizing noise, Q, as given in Table 24–2, Equations 2e to 2i. The performance is usually expressed in loss/noise/talker– and listener–echo terms.

24-3 CUSTOMER/OPERATOR

The R relationship can also be adapted to estimate the quality of service for customer/operator connections. In this case, the overall loudness loss, L_e, of Equation 1a in Table 24–1 would be the sum of (a) the transmit or receive objective loudness rating of the assumed 500–type or equivalent telephone set [9] and associated loop (the ratings change with distance), (b) the 1004–Hz insertion loss of the connecting trunks, (c) the gain or loss of the operator console circuit [10], and (d) the receive or transmit objective loudness rating of the operator's headset [11].

The general influence of operator room noise can be included in the operator grade–of–service calculations for over–the–ear headsets. Room noise is to be converted into equivalent circuit noise by using Figure 24–5 and then power–summed with the other noises as indicated in Table 24–1, Equation 1a. This assumes that the noise conversion of Figure 24–5, based on customers with 500–type sets, is valid also for the over–the–ear operator headset and operator environment.

Some research work using an over–the–ear operator headset on a manikin called KEMAR (Knowles Electronic Manikin for Acoustic Research) and on live listeners has been reported [12]. It was possible with the manikin to measure the combined circuit noise and room noise at the manikin's eardrum for wide ranges of taped operator room noise and bandlimited circuit noise. The results show that high room noise, such as 65 dBA, increases the combined noise at the eardrum by about 10 dB when the circuit noise was 30 dBrnc without room noise. This is a significant impairment. The results also indicated the strong importance of an acoustic seal between earpiece and the ear to reduce leakage of room noise into the ear cavity.

The reference also presents a curve of speech transmission quality as a function of speech–to–noise ratios at the eardrum in terms of mean opinion scores of excellent, good, fair, poor, and unsatisfactory. The curve has an "S" shape. From the curve it was estimated that the speech–to–noise ratios for the good, fair, and poor points were approximately 25, 17, and 11 dB, respectively. These data can be used in basic communications studies.

24-4 GRADE-OF-SERVICE CALCULATIONS

Grade-of-service studies can be made quickly by computer programming. The grade-of-service calculations (GOSCAL) program, which is available to the exchange carriers, proceeds by entering data in a link-by-link basis of a built-up connection. The link characteristics (mean, standard deviation) can be selected from results of field tests or as specified by a designer. The overall results of tandem links are found by Monte Carlo techniques of many repetitions of the end-to-end calculations depending on random selection of the individual links within the specified distribution. The GOSCAL program has been used frequently for network studies.

References

1. Cavanaugh, J. R., R. W. Hatch, and J. L. Sullivan. "Models for the Subjective Effects of Loss, Noise, and Talker Echo on Telephone Connections," *Bell System Tech. J.*, Vol. 55 (Nov. 1976), pp. 1319–1371.

2. Hatch, R. W. and J. L. Sullivan. "Transmission Rating Models for Use in Planning of Telephone Networks," *IEEE National Telecommunications Conference Record* (Dallas, TX: Nov. 1976), Vol. II, pp. 23.2-1 to 23.2-5.

3. Members of Technical Staff. *Transmission Systems for Communications*, Fifth Edition (Murray Hill, NJ: AT&T Bell Laboratories, Inc., 1982), p. 143.

4. *IEEE Std. 661–1979*, "Determining Objective Loudness Ratings of Telephone Connections" (New York: Institute of Electrical and Electronics Engineers, Inc., 1979).

5. *IEEE Std. 269–1983*, "Method of Measuring Performance of Telephone Sets" (New York: Institute of Electrical and Electronics Engineers, Inc., 1983).

6. Cavanaugh, J. R., R. W. Hatch, and J. L. Sullivan. "Transmission Rating Model for Use in Planning of Telephone Networks," *Globecom 1983 Conference Record* (San Diego, CA: Nov. 1983), Vol. 2, pp. 20.2-1 to 20.2-6.

7. *CCITT Temporary Document No. 31*, "Draft Recommendation PMN–Modulated Noise Reference Unit," Study Group XII Meeting (May 1983).

8. Peterson, A. P. G. *Handbook of Noise Measurement*, Ninth Edition (Concord, MA: GenRad, Inc., 1980).

9. Bell System Technical Reference PUB 48006, *Functional Product Class Criteria—Head Telephone Sets (Headsets)*, American Telephone and Telegraph Company (Jan. 1980).

10. *OSSGR, Operator Services Systems Generic Requirements*, Technical Reference TR–TSY–000271, Bellcore (Iss. 2, Dec. 1986 with Rev. 3, Mar. 1988), Vol. 1, Section 7 and Vol. 2, Section 21.

11. *Generic Requirements for Telephone Headsets Used at Operator Consoles*, Technical Reference TR–NPL–000314, Bellcore (Iss. 1, Dec. 1987).

12. Silbiger, H. R. "Human Factors of Telephone Communication in Noisy Environments," *Proceedings of the National Electronics Conference, 1981*, Vol. 35, pp. 170–174.

Chapter 25

Objectives, Requirements, and Standards

Transmission objectives must be processed in a number of ways to make them useful in system design and operation. They can be considered as goals that are established as criteria for the achievement of a quality of service that is economical and ultimately satisfactory to nearly all customers. This chapter reviews the process by which objectives are developed to produce a set of requirements, i.e., performance parameters that *must* be satisfied if the objectives are to be met.

25–1 OBJECTIVES

Ambiguity between the terms *objectives* and *requirements* may be noted. The distinction may be a matter of definition, point of view, or terminology. When an expression such as "X percent good–or–better grade–of–service objective" is used, there can hardly be any doubt that an objective is being discussed. If it is stated that "the transistor must have a single–frequency fundamental–to–third–harmonic ratio of 50 dB for a milliwatt output," there is little doubt that a requirement is being stated. In between, there are uncertainties and shades of meaning. The processing of requirement and objective data, therefore, is often similar and involves what shall here be called *determination, interpretation,* and *allocation*.

These are all broad subjects. It is not the intent here to discuss in detail the many possible applications. Rather, general considerations of the process are reviewed and several examples are given. It must be recognized, also, that the processes are generally reversible. The measurement of performance of devices, circuits, and parts of systems can often be extrapolated to determine or estimate the overall performance of a transmission system.

Determination

Determining transmission objectives most often begins with a subjective test program designed to establish the relationship between an impairment (or combination of impairments) and observer opinions of the effects of various amounts of the impairment. The data representing impairments are then combined with observer opinion ratings so that the combination can be expressed in terms of grade of service. Finally, the grade–of–service relationships are analyzed through engineering economy studies to determine relative costs of furnishing and maintaining various grades of service. On the basis of these studies, a value for a trunk or end–link objective can be selected that represents a reasonable compromise between providing customer satisfaction and the cost of providing the service.

Another way of processing data in order to form an objective is to derive an objective index for a given impairment; an index is a single number, based on a scale of 1 to 100, that can be used as a broad measure of plant performance. Indices are used mainly as tools for transmission management; they are particularly valuable in showing trends of performance for large geographical or administrative units of the plant. For example, one widely known maintenance index uses a scale of:

99 –100	Excellent
96 – 98	Fully satisfactory
90 – 95	Fair to mediocre
Below 90	Unsatisfactory

A newer measurement plan involves measurement of up to five transmission characteristics. The measurements are compared with the maintenance limits (MLs) and immediate action limits (IALs). The percentages of trunks exceeding these limits are calculated and used to determine an index for each transmission characteristic.

The results are summarized into three bands:

(1) Band 0 – indicates the maintenance performance is within the objective level:

Band 0 = 95.5 – 100.

(2) Band L – indicates the performance is lower than objective level:

Band L = 89.5 – 95.4.

(3) Band U – indicates the performance is unsatisfactory:

Band U = 89.4 or lower.

Requirements that must be met in order to satisfy the objectives are usually derived from them. The dependency of the one on the other distinguishes requirements from objectives. An *objective* is a goal; a *requirement* must be met to satisfy the goal.

Interpretation

In whatever form it may be stated, an objective is subject to a great deal of interpretation to make it applicable to a particular set of circumstances. The interpretive treatment of objectives may call for identification of the following: (1) the static characteristics of the impairment (for example, a pure single frequency versus one having variable sideband content), (2) the statistical characteristic of the impairment that might cover the probability of occurrence or some other statistical property of a varying impairment, (3) the simultaneous effects of multiple impairments of the same or different types, (4) the way of expressing the impairment and the objective for various types of signal, and (5) the establishment of a requirement in terms of a *limit*. The term limit implies that corrective action must be taken if the requirement is exceeded.

Interpretation Examples. The interpretation of an objective often depends on some characteristic of the impairment. For example, the transmission of television signals in the presence of speech signals introduced new types of interference to telephone transmission. Some of these interferences may be regarded as single–frequency interferences for which objectives have long been established. However, since they result from line–scan frequency components of the television signal, there is sideband energy at multiples of 30 and 60 Hz on both sides of each multiple of the line–scan frequency. These sidebands produce a subjective effect that makes the single–frequency interference sound distorted. The objective for this distorted single–frequency interference is a matter of interpretation and had to be established by subjective tests designed to compare the interfering effect of the distorted tone with that of a pure single frequency. The distorted tone was found to be less interfering than the pure single frequency, so that 2 dB more interference power could be tolerated. Thus, the single–frequency interference objective may be interpreted for application to television tone interferences so that if the single–frequency objective is x dBrnc0, the television tone objective is $x + 2$ dBrnc0.

Another example of interpretation concerns crosstalk objectives, usually expressed in terms of minimum allowable crosstalk coupling loss derived from crosstalk indices. In some cases, the nature of the coupling path results in nonintelligible crosstalk due to frequency inversion. In the past, the objectives applied to nonintelligible crosstalk were the same as those applied to intelligible crosstalk. The reason for this interpretation was that when nonintelligible crosstalk is heard, the syllabic character of the interference is quite recognizable, and the listener finds it as annoying as if it were intelligible. More recently, this type of crosstalk has been treated as noise, with the objective made about 3 dB more stringent than that for random noise.

Probabilistic Characteristics

A statement of objectives for impairments having probabilistic characteristics must include appropriately qualifying phrases to account for the variability, as in the following discussion.

In digital signal transmission, one of the most common ways to express transmission impairment is in terms of bit error ratio (BER), which is the number of bits in error received in a specified time interval to the total number of bits received in the same interval. When the effect of impulse noise is evaluated, performance must be related to error ratio in such a way that the rate of signal transmission, the amplitude distribution of the interference, the probability of occurrence of the interference, and certain characteristics of the transmitted signal must all be considered. Simplistic statements of error–ratio objectives are often inadequate. The block error ratio (BLER), which is the ratio of the blocks received in error in a specified time period to the total number of blocks received in the same period, is usually more meaningful because error detection codes and the basic coding of the signal often permit retransmission of impaired blocks. A burst of errors might completely ruin a single block of information, causing an apparently excessive average error ratio. The desired result, however, would be the retransmission of that single block of information instead of many blocks. Thus, a BER objective simply expressed as an objective of 10^{-6} must be further interpreted to account for signals and interferences having various parameters. As a result, additional expressions are in use to describe error performance over short time intervals: *error–free seconds* (EFS), which is the number of seconds in which received bits are error free, and *severely errored seconds* (SES), which is the number of one–second intervals having a BER greater than or equal to 10^{-3}.

Multiple Impairments

Another situation that involves interpretation of objectives occurs when multiple impairments are simultaneously present. When they are of different types, a portion of the objective must be allocated to each impairment. If the impairments are of the same type, the objectives are usually established or interpreted in terms of the combined effect.

One illustration of the way multiple impairments may be treated is the handling of multiple sources of intelligible crosstalk. The performance and objectives for intelligible crosstalk are usually expressed in terms of the crosstalk index. The index is derived from mathematical relationships that include the probability of hearing intelligible crosstalk. If the sources are independent, the number of sources is included as a parameter in the derivation. If the crosstalk paths can cause simultaneous exposures to the same source, the coupling is increased by 10 log N or 20 log N (N is the number of paths), as appropriate, and the crosstalk is treated as if it were from a single source.

Sometimes there are multiple random interferences whose combined interfering effect is best evaluated by summing the powers of the individual contributors. In digital transmission systems, for example, the effective circuit noise is determined by adding the equivalent circuit noise powers of thermal noise, quantizing noise, and room noise. The objective must be interpreted as applying to the power sum of the contributors.

25-2 REQUIREMENTS AND LIMITS

There are many expressions used for performance, objectives, and requirements in telecommunications. These expressions are all subject to interpretation according to the nature of the impairment, the characteristic of the channel and the signal, or the way these characteristics interact. Depending on circumstances, digital signal transmission performance or objectives may be expressed in terms of signal–to–noise or notched–noise ratios, noise impairment, various expressions for error ratios, or even percentage of eye closure. Television impairments are usually expressed in terms of signal–to–interference ratios where the signal is measured in peak–to–peak volts; however, the interference may be expressed in rms, peak, or peak–to–peak volts. Random noise and echoes are weighted by frequency and time delay. Telephone objectives and requirements are often expressed in terms of absolute values of interference as measured at specified transmission level points (TLPs).

All of these expressions must be thoroughly understood, since often it is necessary to interpret one in terms of another or to derive one from another. Part of the interpretation process requires a thorough understanding of the TLP concept. When the concept is properly applied, telephone system noise in dBrnc0, for example, can be used in the signal–to–noise ratio for analog data transmission analysis.

Objectives have been defined as desired goals. Requirements are performance parameters that must be met if objectives are to be satisfied. Limits are performance parameters that, when exceeded, indicate a need for some form of corrective action and, in some cases, removal of a circuit from service until the corrective action is completed.

When a new transmission system is being developed, *design objectives* are applied to guide the generation of *design requirements*. The design requirements are maximum or minimum values that must be met in the controlled development environment if the design objectives are to be met.

Upon field installation and connection of a system to provide trunks or other transmission channels, a series of tests is conducted before the circuits are released for service to ensure that the equipment is properly aligned and to verify that the circuit is performing within limits. The minimum and maximum values established for these tests are called circuit–order requirements or limits.

Maintenance requirements are intended to reflect performance that is practical to obtain in the field environment using existing equipment and operating procedures. Maintenance limits are those within which performance is satisfactory; when the limits are exceeded, maintenance action is required. An ML may be established at some value that, if exceeded, requires that a circuit be taken out of service. Such a limit is called a turn–down limit or immediate action limit.

Generally, before divestiture, there were three categories of requirements for testing the voice–frequency transmission parameters of trunks and transmission channels. The requirements covered preservice tests, routine tests, and trouble tests. The corresponding limits were preservice, maintenance, and immediate action. In many cases, the preservice and maintenance limit values were the same.

With divestiture, the above categories are still valid for intraLATA (local access and transport area) connections. For exchange access connections to the intraLATA network, another category was added for acceptance tests and acceptance limits on such connections. These tests are performed at the request of the interexchange carrier (IC). As the network evolves toward digital connectivity, it is expected that suitable requirements will be placed on digital transmission parameters to replace the present voice–frequency requirements.

25–3 ALLOCATION

Objectives are usually established and applied in a format that expresses the overall performance goal for a reference connection. To be useful in development, design, or operation, these objectives must be appropriately allocated to a variety of impairments, to portions of the plant, or to parts of systems. Objectives, performances, allocations, and interfaces for many types of service access connections are being standardized by industry participation in the Exchange Carriers Standards Association (ECSA). This will be described later in this chapter.

Allocation for Analog Transmission

The allocation process requires the exercise of considerable judgment and knowledge of many system performance parameters and cost relationships. The process is seldom arbitrary, but in the absence of data indicating otherwise, it is common to assume, for example, that noise impairments add according to their powers and that similar impairments from different parts of a system also add by power. It does not necessarily follow that several impairments are given equal weight. If economic factors or system parameters are significant, the impairments may be allocated different proportions of an overall objective. For example, because noise is length–dependent in analog systems, long–haul facilities are given a larger share of the noise allocation. Also, echo control techniques are usually applied in long–haul facilities.

If the layout is known and power addition is assumed, the allocation of the objective is a straightforward process of dividing the power of the impairment that just meets the objective among the sources of impairments in the elements that make up the reference connection. When allocation to various parts of the plant or system is being formulated, the relative costs of achieving the allocated objective must be carefully weighed. The relationships between allocations, the characteristics of different parts of the plant, and the relative extent to which the parts of the plant are used must be considered.

Allocation for Digital Transmission

Allocation problems for digital signal transmission are similar in principle to those encountered in analog signal transmission.

However, the problems differ in detail because of the discrete nature of the signal, the regenerative processes used in digital repeaters, and the properties of time–division signal coding. Digital transmission impairments include errors, jitter, misframes, and processing of the signal (such as for quantizing or for a digital pad). Generally, the digital error performance objectives are based on data transfer, which is more stringent than for voice transmission.

In a regenerative repeater system, transmission line noise is noncumulative because signal pulses are reconstructed at each repeater. However, margin must be provided to guard against pulse distortion to the error–occurring point because, at that point, performance deteriorates very rapidly.

Because performance degradation is abrupt and unexpected, digital system components such as repeaters and multiplex equipment normally operate at an error ratio that is near zero. A high system error ratio is usually due to a single repeater operating with excessive pulse distortion or insufficient margin. Therefore, it is common practice to assign the same error–ratio objective to parts of a system as is assigned to the whole system. The probability of exceeding this error ratio is then allocated among the parts.

Certain impairments are cumulative and careful attention must be given to these when objectives are being allocated. Timing jitter is cumulative along a repeatered line. Therefore, the total objective must be allocated so that relatively large margins are maintained.

Misframes occur when the demultiplexing terminal loses synchronism with the incoming bit stream. Communication on all channels is interrupted until synchronism is recovered. The effect is noted at all levels of the digital hierarchy below that at which it occurs. Thus, misframe impairments must be allocated among the various multiplex levels.

Another class of impairment that will accumulate has its sources in the digital processing equipment used to convert analog signals to a digital format, or used for digital pad loss, echo cancelling, or reduction in bit rate for voice. The coding process produces quantizing noise, an impairment that depends on the type of coding used and that increases with the number of times the signal is processed. This process may also cause changes in transmission characteristics in addition to adding noise. Care must be taken to allocate objectives realistically in relation to laws of accumulation that may pertain to given situations.

Transmission Level Point Translations

Objectives and requirements are often expressed in reference to a specific TLP, generally 0 TLP. Sometimes it is necessary to express the same objective in reference to some other TLP. This translation process is usually straightforward, but care must be taken to make the translation properly.

The transmission level of any point other than 0 TLP represents the loss (or gain), in dB, between that point and the reference. Since loss is normally a function of frequency, TLPs are usually defined at 1004 Hz. The original location of 0 TLP was historically the jack of a toll switchboard in the transmitting direction. However, as the toll network evolved, it became more practical to locate the 0 TLP at the trunk appearance on the outgoing, or transmit, side of end offices.

Knowledge of the transmission level at various points in the network facilitates the measurements associated with installation, maintenance, and testing. For example, the difference between two TLPs that have the same reference is the gain or loss between those points. In addition, measurements taken at different TLPs can be referred to a common reference point to compute their relative values. This concept is helpful for the addition of interference signals, the expression of signal–to–noise ratios, and the relation of performance to objectives in system evaluation.

25-4 STANDARDS AND STANDARDS ORGANIZATIONS

There are several organizations that deal in standards [1,2]. The more prominent ones include: the Exchange Carriers Standards Association, the Institute of Electrical and Electronics Engineers (IEEE), the American National Standards Institute

(ANSI), the Electronic Industries Association (EIA), and the International Telecommunications Union (ITU). Only ECSA and the ITU are addressed in the following. The publications of several organizations listing their available standards are noted under the heading "Additional Reading" at the end of this chapter.

The Exchange Carriers Standards Association

Prior to divestiture, AT&T developed national interface information based on internal transmission and signalling objectives and requirements. This information was published as Technical References, principally for designers and manufacturers of business machines, communications systems, and terminal equipment. The references were de facto industry "national standards" for connecting to the telephone plant and for telephone network characteristics.

Upon divestiture from AT&T, seven independent regional companies were established, with most companies owning several operating telephone companies (OTCs). The OTCs were chartered to provide telephone service in exchange areas now called LATAs. In addition, divestiture granted interLATA companies (AT&T, MCI, etc.) equal service access to the intraLATA networks. With this dispersion of ownership, a telephone connection may now traverse the areas of several different and unrelated telephone companies previously owned by a parent company that had established and controlled common standards. During planning for divestiture, it was recognized that to ensure continuing service connection capability and quality, an independent organization would be needed to provide a systematic method to develop interconnection standards to cover the characteristics of new and old services and network elements. This new organization, incorporated as a nonprofit organization in 1983, is the Exchange Carriers Standards Association, Inc. It is an association that was established voluntarily by members of the exchange carrier industry to address exchange access interconnection standards and other technical issues in the postdivestiture era.

ECSA serves three major purposes. First, it represents exchange carrier interests in standards and related technical matters affecting the exchange carrier industry. Second, it sponsors and supports the independent T1 Committee, accredited by the ANSI as a consensus mechanism for developing voluntary interconnection standards. Third, ECSA sponsors the Carrier Liaison (CL) Committee, a group that organizes and coordinates industrywide forums for the discussion and resolution of nationwide problems concerning exchange access services.

The following provides a brief general description of ECSA's membership and committee structure, an overview of the T1 Committee, and a more detailed discussion of the CL Committee and its associated forums.

Open in membership to all exchange carriers, ECSA is governed by a 21–member board of directors. Member carriers are represented on the board of directors according to three classes based on carrier size (i.e., number of subscriber access lines) to ensure diverse representation. All meetings are open to any interested party.

ECSA Committee Structure. Certain ECSA standing committees provide vehicles for addressing interindustry concerns. A standing committee, the Exchange Telephone Group Committee, oversees ECSA's membership and participation in ANSI and other standards organizations.

The committees that sponsor and support forums with industrywide participation are the Standards Advisory Committee and the Liaison Committee.

The Standards Advisory Committee is responsible for ECSA representation and participation in the T1 Committee on interconnection standards. The scope of this committee's activities includes coordinating participation in the T1 Committee and ensuring continuing ANSI accreditation of T1. T1 (not to be confused with the T1 transmission system) is ANSI's coding of standards committees; the "T" stands for telecommunications and the "1" stands for the first entity.

The Liaison Committee provides guidance to ECSA members of the Carrier Liaison Committee, furnishes assistance to the CL Committee's operations, and proposes nominations for CL Committee leadership.

T1 Committee and Subcommittees. This committee was established in February 1984 to develop technical standards and reports supporting the interconnection and interoperability of telecommunications networks at interfaces with end–user systems, carriers, information and enhanced–service providers, and customer–premises equipment (CPE). The T1 Committee has several technical subcommittees that recommend standards and develop technical reports in their areas of expertise. The subcommittees also recommend positions on matters under consideration by other North American and international standards bodies.

The technical subcommittees and their titles are:

T1E1 – Customer–to–carrier installation interfaces. This committee covers the interfaces between CPE and telecommunications networks. There are five working groups:

- T1E1.1 Analog Interfaces
- T1E1.2 Digital Interfaces
- T1E1.3 Special Interfaces
- T1E1.4 Integrated Services Digital Network (ISDN)
- T1E1.5 Editing Group.

T1M1 – Internetwork operations, administration, maintenance, and provisioning. This committee deals with network management including measuring systems and network tones and announcements. There are five working groups:

- T1M1.1 Internetwork Planning and Engineering
- T1M1.2 Internetwork Operations

T1M1.3 Testing and Operations Support Equipment

T1M1.4 Administration Systems

T1M1.5 Architecture, Interfaces, and Protocols.

T1Q1 – Performance. This committee is concerned with end–to–end performance and allocation among portions of the network connections. The six working groups are:

T1Q1.1 4–kHz Voice

T1Q1.2 4–kHz Voiceband Data

T1Q1.3 Digital Services and ISDN

T1Q1.4 Digital Packet

T1Q1.5 Wideband Program

T1Q1.6 Wideband Analog.

T1S1 – Services architecture and signalling. This group deals with emerging network structures and signalling systems. The four working groups are:

T1S1.1 ISDN Architecture and Services

T1S1.2 ISDN Switching and Signalling Protocols

T1S1.3 Common–Channel Signalling (CCS)

T1S1.4 Carrier Interfaces.

T1X1 – Digital hierarchical interfaces and signalling. This committee's concern is digital multiplex hierarchies and synchronization. The four working groups are:

T1X.3 Digital Synchronization Interfaces

T1X.4 Metallic Hierarchical Interfaces

T1X.5 Optical Hierarchical Interfaces

T1X.6 Tributary Analysis.

T1Y1 – Specialized subjects. This committee's concern is the implications of new technology on telecommunication networks. The four working groups are:

T1Y1.1 Specialized Video and Audio Services

T1Y1.2 Specialized Voice and Data Processing

T1Y1.3 Advanced Technologies and Services

T1Y1.4 Environmental Standards for Exchange and Interexchange Carrier Networks.

Membership and full participation in the ECSA–sponsored and ANSI–accredited T1 Committee are open to all parties with a direct and material interest in the T1 process and activities. The ANSI Board of Standards Review verifies that the requirements for due process, consensus, and other criteria for standards approval have been met.

Carrier Liaison Committee and Associated Forums. This committee was proposed by ECSA in 1984 and endorsed by the Federal Communications Commission (FCC) in 1985 to provide interindustry mechanisms for discussions and voluntary resolution of nationwide concerns about the provision of exchange access services. It serves as an umbrella organization for three industry forums: the Industry Carriers Compatibility Forum (ICCF), the Network Operations Forum (NOF), and the Ordering and Billing Forum (OBF). CL Committee members are appointed by the ECSA Board of Directors, each director appointing one member.

In trying to reach consensus, the CL Committee operates on the principle that, although a proposed resolution may not be a participant's first choice, it is one that he or she can accept and support. Significant opposition to a proposal usually stops the resolution process. Furthermore, any resolutions that are adopted are proposals for *voluntary* implementation and are not binding on participating carriers. While reserving judgment as to implementation, however, each ECSA participant is committed to discuss issues and consider proposed resolutions in good faith.

Industry Carriers Compatibility Forum. This committee was created in April 1983 to provide a vehicle for the exchange of information on technical interface and interconnection issues between exchange carriers and ICs.

The original purpose of ICCF was to provide an opportunity for subject–matter experts to speak on such key issues as the interface performance specification for Feature Group D, the philosophy of transmission limits, and numbering and dialing plans. The purpose and structure of the ICCF has been expanded from a lecture format to a combination of tutorial presentations and workshops to ease an exchange of technical information.

The International Telecommunications Union

The ITU is one of the most active organizations in developing consensus recommendations on telecommunications standards for international connections. A greatly simplified overview of this organization is given in the following.

CCITT and CCIR. The ITU and its advisory committees, the International Telegraph and Telephone Consultative Committee (CCITT) and the International Radio Consultative Committee (CCIR), have played a major role in promoting cooperation for developing standards (i.e., recommendations). The ITU originated in 1865. While only national governments may be members, recognized private operating agencies and industrial organizations can participate in its activities.

The CCITT and CCIR hold Plenary Assemblies every four years to review the work of numerous study groups (SGs) that were established on various telecommunication subjects by the Plenary. The SGs divide their work among working parties (WPs) and rapporteurs (for special reports). The SGs submit documents to the Plenary Assembly for adoption, by consensus, as recommendations. The CCITT and CCIR technical organizational structure of the SGs and WPs is shown in Table 25–1, which lists the codes and titles of the entire telecommunications field.

Table 25–1. International Telecommunications Union Advisory Committees

A. CCITT	International Telegraph and Telephone Consultative Committee (Plenary Assembly meets every 4 years: 1984, 1988, 1992)	
SG I	Definition, operation, and quality of service aspects of telegraph, data transmission, and telematic services	
	WP I/1	Telex, telegraph, and mobile services
	WP I/2	Message-handling systems and directory services
	WP I/3	Teletex, ISDN, and teleconferencing
	WP I/4	Facsimile, data, and videotex
SG II	Operation of telephone network and ISDN	
	WP II/1	Operations and services
	WP II/2	Human factors
	WP II/3	Numbering, routing, and interworking
	WP II/4	Traffic engineering, forecasting, and network planning
	WP II/5	Quality of service, network management, and mobile service
	WP II/6	Availability and reliability
	QSDG	Quality of service development group
	NMDG	Network management development group
SG III	General tariff principles including accounting	
	WP III/1	Private leased international circuits
	WP III/2	Public international data networks
	WP III/3	Telegraph telematic services
	WP III/5	World Administrative Telegraph and Telephone Conference
	WP III/6	ISDN
SG IV	Transmission maintenance of international lines, circuits and chains of circuits, maintenance of automatic and semi-automatic origin	
	WP IV/2	Measuring equipment
	WP IV/5	Maintenance of systems
SG VII	Data communication networks	
	WP VII/1	Network services, facilities, and protocol validation
	WP VII/2	Network access interfaces
	WP VII/3	Interworking, switching, and signalling
	WP VII/4	Transmission and message handling
	WP VII/5	Routing, numbering, and the layered model
	SR ISDN	ISDN-related issues
	SR DEFS	Terms and definitions
SG VIII	Terminal equipment for telematic services	
	WP VIII/I	Terminal characteristics
	WP VIII/2	Common protocols and interworking

Table 25-1. International Telecommunications Union Advisory Committees (Continued)

SG IX	Telegraph networks and terminal equipment	
	WP IX/1	Terminals and customer facilities
	WP IX/2	Signalling and interworking
	WP IX/3	Time-division multiplex systems
	WP IX/4	Transmission standards
SG X	Languages and methods for telecommunication applications	
	WP X/1	Man-machine language (MML)
	WP X/2	Environment, software quality assurance, and software reliability
	WP X/3	Specification and description languages, formal descriptive techniques
	WP X/4	CCITT high-level language (CHILL)
SG XI	ISDN and telephone network switching and signalling	
	WP XI/1	Interworking, satellites, and mobile service
	WP XI/2	Signalling System No.7
	WP XI/3	Field trial of digital switching equipment
	WP XI/4	Digital switching
	WP XI/5	Signalling and switching functions
	WP XI/6	Digital subscriber line signalling
	JEG	Joint experts group of WP XI 2 & WP XI 6
SG XII	Transmission performance of telephone networks and terminals	
	WP XII/2	Telephone terminals
	WP XII/3	Transmission quality and opinion models
	WP XII/4	Transmission objectives and planning
SG XV	Transmission systems	
	WP XV/1	Sound program, video, and multiservices transmission
	WP XV/2	Voice processing
	WP XV/3	Digital equipment
	WP XV/4	Optical fiber system planning guide
	WP XV/5	Optical cables and systems
	WP XV/6	Metallic cables and systems
SG XVII	Data transmission over the telephone network	
	WP XVII/1	Modems
	WP XVII/2	ISDN
	WP XVII/3	Transmission, maintenance, and interfaces
SG XVIII	Digital networks including ISDN	
	WP XVIII/1	Service aspects
	WP XVIII/2	Network aspects
	WP XVIII/3	User-network interfaces, layer 1
	WP XVIII/4	Architecture and models
	WP XVIII/5	Maintenance and general aspects
	WP XVIII/6	Performance aspects
	WP XVIII/7	Transmission aspects
	WP XVIII/8	Speech processing

Table 25–1. International Telecommunications Union Advisory Committees (Continued)

GAS9	Transition from analog to digital telecommunication networks	
B. CCIR	International Radio Consultative Committee (Plenary Assembly meets every four years: 1986, 1990, 1994)	
	SG 4	Fixed service using communication satellites
	SG 8	Mobile services
	SG 9	Fixed service using radio relay systems
	SG 10	Broadcasting service (sound)
	SG 11	Broadcasting service (television)
	CMTT	Joint CCIR–CCITT Study Group on Transmission of Sound Broadcasting and Television Systems Over Long Distances
	WP CMTT–AB	Analog transmission of television signals including conversion of standards, multiplexed analog components (MAC) systems, and high–definition television
	WP CMTT–AN	Digital or hybrid analog digital transmission of television signals
	WP CMTT–C	Transmission of sound program signals
CMV	Joint CCIR–CCITT Study Group for Vocabulary	
	WP CMV/A	Terminology common to the CCIR and CCITT
	WP CMV/B	Terminology specific to CCIR
	WP CMV/C	Symbols, units, and abbreviations

In recent years the ITU has published the results of the Plenary meetings in books of different colors, with reference made to them in terms of color. Table 25–2 lists the CCITT books and some of the general technical subjects that are covered in the proceeding period. The work encompasses the world numbering plan, dialing plan, routing plan, transmission systems, etc., and various transmission parameters and recommendations for the international telecommunications network, which involves national and international carriers.

Table 25–2. CCITT Plenary Assembly Books

A. IInd Plenary Assembly, 1960, Red Book – Covers open wire and coaxial frequency–division multiplexing (FDM) carrier systems, two–frequency inband signalling, and other subjects.

B. IIIrd Plenary Assembly, 1964, Blue Book – Covers transistor carrier systems, submarine cable and carrier systems, FDM radio relay, multifrequency (MF) signalling, and other subjects.

C. IVth Plenary Assembly, 1968, White Book – Covers FDM carrier systems, TV transmissions systems, emerging CCS systems, and other subjects.

D. Vth Plenary Assembly, 1972, Orange–Green Book – Covers emerging digital transmission technology, CCS Signalling System No. 6, and other subjects.

E. VIth Plenary Assembly, 1976, Orange Book – Covers pulse code modulation (PCM) systems and multiplexes, digital networks, outline of functional specification and descriptive language (SDL) and man–machine language, and other subjects.

F. VIIth Plenary Assembly, 1980, Yellow Book – Covers high–order PCM multiplex, frame structure for digital switching, digital transmission, higher–speed CCS, standards for SDL, MML, and CCITT CHILL languages, and other subjects.

G. VIIIth Plenary Assembly, 1984, Red Book – Covers emphasis on digital telecommunications, digital data connectivity, Signalling System No. 7, and data and packet protocols.

CCITT—Reference Level and Noise. Of general interest is the CCITT definition of transmission or reference level and message noise, which is described in the following to show differences from North American usage.

Currently in the reports published by ITU, transmission parameters are expressed in decibels. Some parameters are also expressed in decimal units of the international system of units. For example, noise and noise objectives are expressed in picowatts and picowatts per kilometer.

Reference Level. International practices and recommendations include the use of reference level, a term analogous to the 0 TLP used in the North American telecommunications system. The reference level point is sometimes referred to as 0 dBr (dB relative level). For four-wire operation, the transmitting end of the circuit is defined as a −3.5 dBr point at the "virtual" switching point, a theoretical point whose exact location depends on national practice.

Noise. The basic unit of noise measurement used in international practice is the picowatt (pW), i.e., 10^{-12} watt. The picowatt may be expressed in decimal or logarithmic terms: the equivalent values are 1 pW = 10^{-12} W = 10^{-9} mW = −90 dBm = 0 dBrn. It should be noted that for a 1000–Hz signal, this is the same reference as used in North America. In international maintenance practice, the standard test signal may be 800 or 1000 Hz.

Message circuit noise is measured, according to CCITT recommendations, by a noise–measuring set called a *psophometer* (noise meter). The set is equipped with a weighting network that has a characteristic somewhat similar to the C–weighting characteristic used in the North American telecommunications system. The two characteristics are shown for comparison in Figure 25–1. For general conversion purposes, it is usually sufficient to assume that the psophometric weighting of 3–kHz white noise decreases the average power by about 2.5 dB (compared with the 2.0–dB factor for C–message weighting). The term *psophometric voltage* refers to the rms weighted noise voltage, and is usually expressed in millivolts.

The (rounded) conversion factor recommended by the CCITT for practical comparison purposes is that 0 dBm of white noise measured by a psophometer (1951 weighting) is equivalent to 90 dBrn measured on a 3A–type noise–measuring set with C–message weighting. This conversion, which applies to white noise in the 300–to–3400–Hz band, is not valid for other noise shapes because of the differences between psophometric and C–message weighting [3].

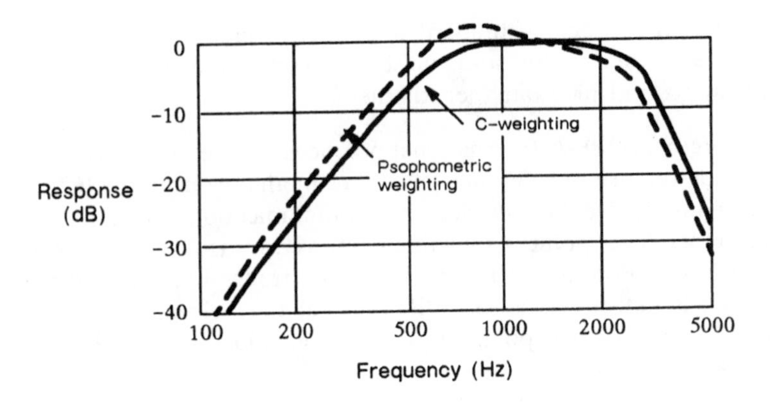

Figure 25-1. Comparison of noise-weighting characteristics.

The relationships between various CCITT and North American noise units are summarized in Table 25-3. The data are particularly useful for conversion from one noise unit to another since an estimate of the frequency spectrum effects can be obtained by comparing the three conditions tabulated. The 1-kHz values are given for comparison of the various conditions used. The 1-kHz psophometric reading appears 1 dB high because the psophometric reference is 1 pW at 800 Hz. The 0-to-3-kHz band of white noise approximates the noise obtained from a message channel. The broadband white noise

Table 25-3. Comparison of Noise Measurements

Noise Unit	Total Power of 0 dBm		White Noise of -4.8 dBm/kHz Not Bandlimited
	1-kHz Tone	0-to-3-kHz White Noise	
dBrnc	90.0 dBrnc	88.0 dBrnc	88.4
dBrn 3 kHz flat	90.0 dBrn	88.8 dBrn	90.3 dBrn
dBrn 15 kHz flat	90.0 dBrn	90.0 dBrn	97.3 dBrn
Psophometric rms voltage (600 ohms)	870 mV	582 mV	604 mV
pWp	1.26×10^9 pWp	5.62×10^8 pWp	6.03×10^8 pWp
dBp	91.0 dBp	87.5 dBp	87.8 dBp

readings are proportional to the total area under the weighting curve and thus give significant information concerning the weighting function above 3 kHz. Similar data for other conditions or weightings can be obtained by integrating the appropriate weighting characteristics over the required frequency band.

References

1. "Telecommunications Standards," *IEEE Communications Magazine*, Vol. 23, No. 1 (Jan. 1985).

2. "An Overview of the Forums for Standards and Regulations for Digital Networks," *Telecommunications* (Oct. 1986).

3. CCITT, Vth Plenary Assembly, *Green Book*, Vol. V, Rec. P.53 (Geneva: International Telecommunications Union, 1973).

Additional Reading

Bell System Technical Reference PUB 10000, *Catalog of Publications,* American Telephone and Telegraph Company (July 1983).

Catalog of Technical Information, Technical Reference TR–TSY–000264, Bellcore (Iss. 3, July 1987).

IEEE Standards Listing (Piscataway, NJ: IEEE Service Center).

1987 Catalog of EIA and JEDEC Standards and Engineering Publications (Washington, DC: Electronic Industries Association, 1987).

Overview: Exchange Carriers Standards Association T1 Committee, Carrier Liaison Committee, and Associated Forums (Parsippany, NJ: Exchange Carriers Standards Association).

Chapter 26

Transmission Objectives and Requirements

Transmission objectives, derived from grade–of–service analysis of the results of subjective tests and performance measurements, are stated in terms of design, performance, or maintenance objectives. Requirements, derived from the objectives, are also given in these same terms and additionally in terms of maintenance limits. These limits define points at which performance is unacceptable, so that circuits must be taken out of service until repairs or adjustments can be made.

The objectives are continuously studied and modified to adapt to the changing environment caused by new equipment and systems, new technology, new services, and changes in customer expectations. The objectives are often established as overall values applicable to terminal–to–terminal connections. They are then allocated in various ways to appropriate portions of the plant, to various impairments, or to a particular type of service.

This chapter discusses some established objectives, with numerical values given wherever possible. Where objectives have not become standard, only general discussion is included to indicate the nature of transmission problems involved. Space does not permit a comprehensive listing of objectives for all types of signals, systems, or services.

The discussion covers voice–frequency (VF) channel objectives for speech volume, loss, noise, talker/listener echo, return loss, and other channel characteristics. It also covers several network loss designs and allocations, as well as operator services transmission objectives.

The transmission of digital signals over analog and digital circuits is discussed, and International Telegraph and Telephone Consultative Committee (CCITT) digital error–ratio limits are included. Video transmission objectives are discussed for analog and digital connections.

26–1 VOICEBAND CHANNEL OBJECTIVES

Transmission objectives for voiceband channels in the message telecommunications service (MTS) network were initially established to satisfy the needs of speech transmission. As new types of signals and services have evolved and as new technology has been applied to all parts of the system, existing objectives have been modified and new objectives have been developed where necessary.

Bandwidth

The bandwidth of telephone loops and trunks has evolved without a specific or consistent set of design objectives. Initially, deficiencies in the bandwidth of such circuits were masked by station–set limitations. The necessity for establishing an acceptable channel bandwidth allocation and carrier separation was recognized when frequency–division multiplexing (FDM) carrier systems were introduced. At that time, the single–sideband (SSB) mode of transmission was established. Also, the 4–kHz spacing of carriers was found to be adequate in view of practical bandwidth limitations and in view of articulation tests conducted for the purpose of establishing bandwidth requirements for intelligibility and naturalness of speech.

As new systems have been introduced and design technology has improved, efforts have been continued to make the effective bandwidth of network channels as wide as is economically feasible within the constraints of (1) the 4–kHz carrier separation and achievable filter designs, (2) the aliasing problems in pulse code modulation (pulse amplitude modulation) carrier with 8–kb/s sampling rate, and (3) the unavoidably poor singing return losses (SRLs) of loaded cable facilities near the high–frequency cutoff region.

More recently, subjective tests have shown that the preferred bandwidth for voice communications is approximately from 200 to 3200 Hz. As a result of these tests and technological advances, efforts are being made to establish standard design objectives for the bandwidth of each major portion of the network so that overall connections can meet the bandwidth objective. The natural increase in attenuation with higher frequencies for nonloaded cable and the sharp high–frequency cutoff of loaded facilities cause both types of facilities to exert considerable influence over the effective bandwidth of loops and VF trunks in an overall connection. In a connection with analog–carrier–derived trunks and with analog switching, the VF bandwidth depends on the number of trunks in tandem because of the conversion to carrier in each trunk. With digital trunking and switching, the bandwidth is not dependent because only one analog–to–digital and one digital–to–analog conversion is needed. Economics must be considered in establishing objectives for each portion of the network. The objectives should be as stringent as necessary when considering other portions. In any event, the 200–3200–Hz goal is becoming more attainable as the digital network evolves.

While this discussion centers about the transmission of speech signals, it should be pointed out that as new voiceband data services are introduced, the continued pressure to transmit at higher data rates creates additional demand for wider bandwidths.

Frequency Response

MTS network design objectives for inband amplitude– and phase–versus–frequency response are not generally available (except for the case of conditioned data loops). However, the same factors that tend to make the effective channel bandwidth as wide as possible also work to make the inband response as uniform as possible. The difficulties have been to express the objective values in a generally acceptable manner and to allocate channel impairment requirements optimally among the many contributors.

Slope–Frequency Response. The only loop design objective for this characteristic applies to dial–up data loops for speeds of 300 bits per second (b/s) or higher. The loop objective is that the loss at 2800 Hz shall be no more than 3 dB greater than the loss at 1000 Hz, which is to be less than 10 dB [1].

Except for loops, there are maintenance requirements on the slope–frequency response of the other links in the MTS network. These are expressed in terms of 404–Hz and 2804–Hz loss deviations from the loss at 1004 Hz. The limits depend on the type of facility.

Phase/Frequency Distortion. MTS network phase/frequency distortion objectives, expressed in terms of envelope delay distortion, are applied to loops and other special–services circuits conditioned for data transmission [1]. For example, if a loop is conditioned for data transmission at a rate of 300 b/s or higher, a performance objective applies of no more than 100 μs of differential delay between any two frequencies over the band from 1000 to 2400 Hz.

Basic Network Loss Designs

Network loss designs are intended to provide a satisfactory compromise among technical needs to provide adequate speech volumes, minimum contrast of received volumes on different calls, and adequate protection against talker echo, listener echo, and near–singing. A discussion of these topics follows. Allocations of loss to various trunks of the network to form loss plans are covered later in this chapter. The loss plans have been based on the network connecting to loops terminated in 500–type telephone station sets or equivalent. This set employs a carbon transmitter, a magnetic receiver, transmit– and receive–level equalization, a four–part hybrid with sidetone path, etc. The 500–type set meets the transmission requirements of Reference 2.

Speech Volume. The basic problem in telephone transmission is to provide satisfactory speech volume at the receiver. The received signal amplitude is a function of many interacting parameters, starting with the transmitted signal amplitude. The latter depends on telephone speaking habits, station–set efficiency of conversion from acoustic to electrical energy, sidetone circuit design of the station set, and losses in the circuits between the transmitter and the receiver.

Received volume differs from many other quality parameters in that its effects are double–ended; volume can either be too low, causing difficulty in understanding the received message, or too high, causing listener discomfort. Subjective tests have been made to determine listener reactions to different volumes [3]. The results of one such series of tests, plotted in Figure 26–1, clearly show the double–ended nature of this parameter. Volumes to the left of the two left–hand curves are judged to be too low to satisfy listeners while volumes to the right of the right–hand curve are too high. Each of the curves, which divide regions of volume rated poor, fair, good, etc., is approximately normal with a standard deviation of about 5 dB. The curves show a fairly wide range (from about −46 vu to about −13 vu at the median values) over which received volumes are rated good.

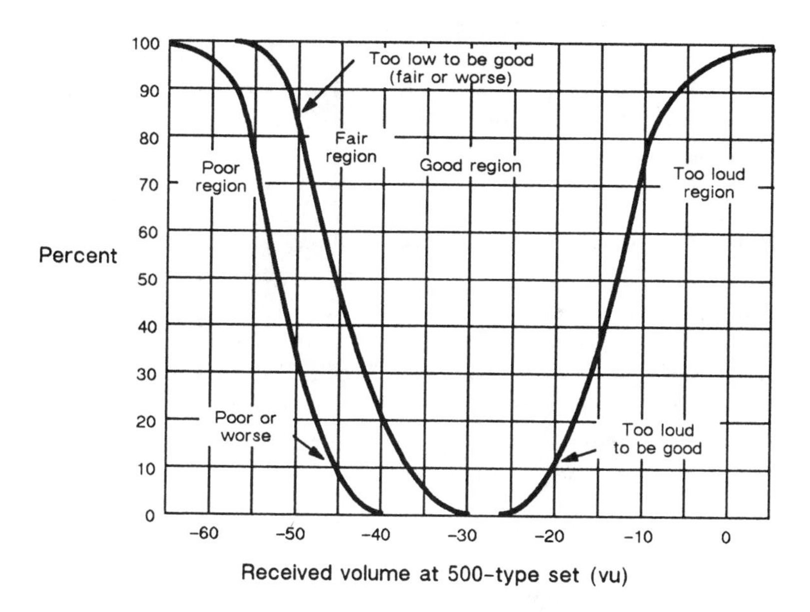

Figure 26–1. Judgment of received volume from subjective tests.

Data of the type shown in Figure 26–1 have been used to help establish allowable circuit losses in end–to–end customer connections. The total loss allowance is allocated to the various parts of the plant in accordance with the results of economic studies and with a satisfactory loss/noise/echo grade of service established by subjective testing.

Loss and Echo. In analog systems, increasing the loss is a practical way of reducing echo and near–singing conditions. Increasing loss, however, introduces an impairment (reduced received volume), and a compromise must be sought to maintain circuit losses at satisfactory values yet reduce echo effects to acceptable values. The parameters of echo amplitude and echo delay contribute significantly to echo performance in the network. Echo amplitude depends on the impedance matches at circuit interfaces and the losses of the circuits involved. The phenomenon is treated in terms of return loss and trunk losses that combine to produce echo path loss.

The most serious source of echo is the low return loss found at end offices where connections are made between loops and trunks. While reasonable control of connecting trunk impedance can be exercised, the impedances of the randomly connected loops vary widely due to varying lengths, bridged taps, and circuit makeup.

The distribution of echo return losses (ERLs) at analog end offices, calculated from loop survey data, has a mean value of 11 dB and a standard deviation of 3 dB, as measured against a standard termination of 900 ohms + 2 μF (without special loop segregation) [4,5]. The distribution of SRL has a mean value of 6 dB and a standard deviation of 2 dB. These return loss distributions were used in the overall process of establishing echo and loss objectives for other parts of the older network.

The transmission objectives for loss and talker echo in the predivestiture analog network were established on the basis of the via net loss (VNL) concept [6,7]. Before divestiture, as the introduction of time–division switching provided for some digital connectivity, a fixed–loss plan (no VNL) was under preparation for all–digital connections.

At divestiture, the overall network structure was changed from a VNL hierarchy to the operating telephone companies' intraLATA (local access and transport area) networks interconnected, by interexchange carriers (ICs), generally via access tandems. As the fixed–loss design was not fully formulated or implemented because of analog plant, a transition loss design was prepared for the intraLATA networks.

A discussion of the VNL, fixed–loss, and transition loss designs follows. Equal access connections are covered in Volume 3.

Via Net Loss Design. The VNL design is based on adding a small loss, depending on length, to a trunk to reduce echo impairment. A significant impairment related to talker echo is echo path delay (the round–trip propagation time of a connection). This can be predicted from propagation times of types of facilities used in making up a connection. The network loss and echo path delay can be predicted statistically from the variations in facility properties and in return loss at the distant end office. Similarly, reaction to talker echo (the third controlling parameter in judging echo performance) can be predicted statistically from the variation in customer tolerance.

The required one–way overall connection loss for satisfactory echo is plotted in Figure 26–2 as a function of the round–trip echo delay and of the number of trunks in the connection. Also shown is a linear approximation for one trunk. This approximation was derived empirically by considering (1) the need for increased loss at low delays to prevent singing or near–singing, (2) the need to control noise, crosstalk, and analog carrier loading, (3) the compromise between sufficient loss to control the effect of echo and the degradation introduced by echo suppressors, and (4) the analytical advantage of having the loss expressed as a linear function of echo path delay.

Linear approximations for more than one trunk may be derived by adding 0.4 dB for each additional trunk to the loss required for a single trunk. This loss is approximately the difference between loss curves at 45 ms delay. The linear approximations may be drawn from the equation

$$\mathrm{OCL} = 0.102D + 0.4N + 4.0 \quad \mathrm{dB} \qquad (26\text{–}1)$$

where OCL is the overall connection loss, D is the echo path delay in milliseconds, and N is the number of trunks in the connection. Note that 0.102 is the slope of the dashed line of Figure 26–2 and that 4.4 ($0.4N$ and 4.0) dB is the zero–delay intercept of this line. Equation 26–1 is used for connections involving round–trip delays up to 45 ms. For delays in excess of 45 ms, one of the trunks is equipped with an echo canceller or suppressor and operated at zero loss.

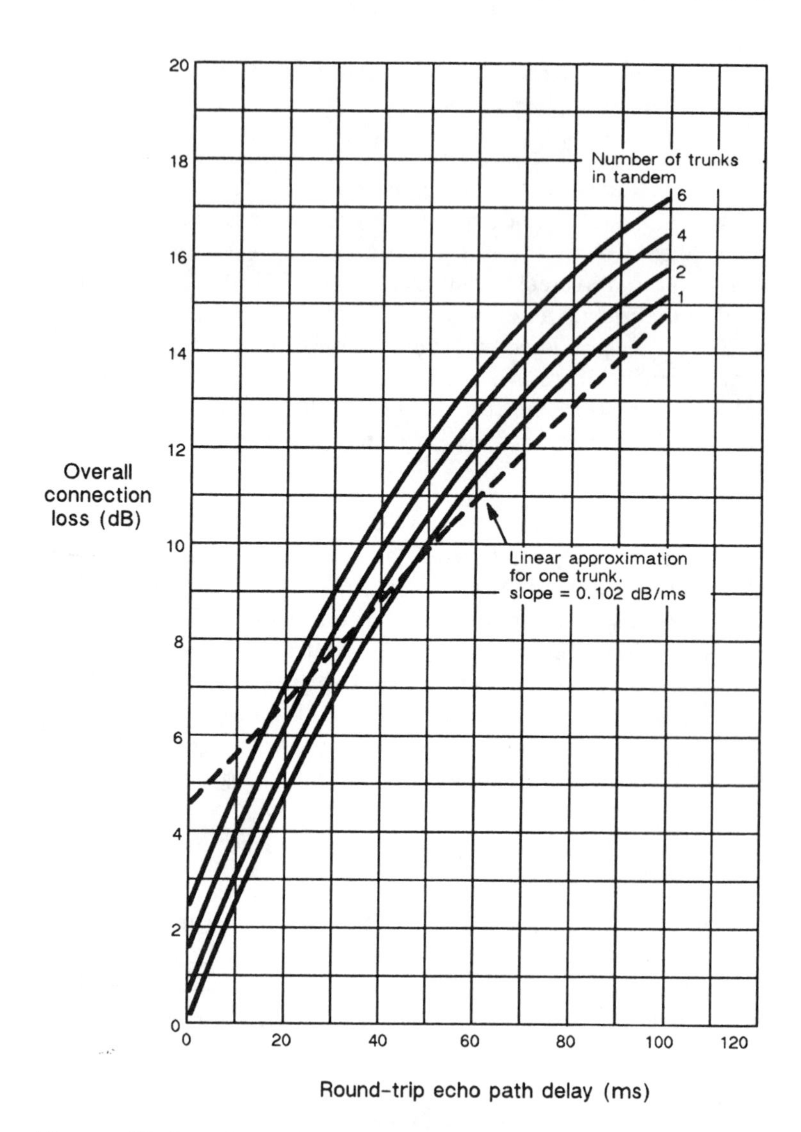

Figure 26–2. One–way loss for satisfactory echo in 99 percent of connections.

The final step in the VNL design process is to assign trunk losses so that each type of trunk in a connection operates at the lowest practicable loss consistent with its length and the type of facility used. In Equation 26–1, 2 dB of the constant is assigned to each tandem–connecting trunk, and the remainder is assigned to each trunk in the connection, including tandem–connecting trunks. The amount added to each trunk is in proportion to the echo path delay of the trunk and is defined as via net loss.* It is

$$\text{VNL} = 0.102D - 0.4 \quad \text{dB} \tag{26–2}$$

where D is the echo path delay in milliseconds.

Since the echo path delay is directly related to the length of the circuit, Equation 26–2 is usually written in terms of length and a via net loss factor (VNLF).

$$\text{VNL} = \text{VNLF}(d) + 0.4 \quad \text{dB} \tag{26–3}$$

where d is the distance in miles. The VNLF is

$$\text{VNLF} = \frac{2 \times 0.102}{v} \tag{26–4}$$

where v is the velocity of propagation in miles per millisecond. The velocity of propagation used in Equation 26–4 must allow for the delay in an average number of terminals as well as for the delay of the medium. Values of VNLF are given in Table 26–1 for facilities, both those that are commonly used and those that are only of historic interest.

Table 26–1. Via Net Loss Factors

Type of Facility	Via Net Loss Factor (dB/Mile)	
	Two-Wire Circuits	Four-Wire Circuits
Toll cable (quadded low-capacity)		
19H88-50	0.03	0.014
19H44-25	0.02	0.010
VF open wire	0.01	---
Carrier (cable, open wire, microwave radio)	---	0.0015
VF local cable (loaded or nonloaded)	0.04	0.017

Although the VNL formula is a continuous function of distance, it has been found desirable administratively to use a step

* The method of defining tandem–connecting trunk loss, which involves the assignment of an additional 0.5 dB of loss (for central-office equipment) to the tandem–connecting trunk, results in a design value of VNL + 2.5 dB. This added loss was previously assigned to the loop; thus, the customer-to-customer loss has not changed.

approximation to this formula for trunks on carrier facilities. It was decided that the first step be 0.5 dB and each step thereafter increase by 0.3 dB, with step values agreeing with the formula at the midpoint of the interval as shown in Table 26–2 [8], which gives values for inserted connection loss (ICL).

Table 26–2. VNL and ICL Values for Trunks Operating on All Carrier Facilities

Trunk Length (Miles)	ICL = VNL dB (0.0015 x Avg. Length + 0.4 dB)
0 - 165	0.5
166 - 365	0.8
366 - 565	1.1
566 - 765	1.4
766 - 965	1.7
966 - 1165	2.0
1166 - 1365	2.3
1366 - 1565	2.6
1566 - 1850	2.9
Any length with echo suppressor/canceller	0.0

Predivestiture Proposed Fixed–Loss Design. With the evolution from analog to digital methods of transmission and switching, it was realized before divestiture that a switched digital network (SDN) could evolve. This type network would need a new loss design [8].

The VNL design described in the preceding paragraphs is not well–suited to the SDN because of the requirement that loss be inserted into each trunk. This would require either (1) decoding of the digital signal followed by insertion of analog loss and recoding, or (2) changing the encoded signal by some digital processing technique to lower the ultimate analog signal level. Both of these techniques add to the cost of facilities and introduce transmission impairments by adding delay, quantizing noise, and distortion to the analog signal. It would also eliminate digital integrity for digital data transmission.

To avoid the penalties of per–trunk insertion of loss in an all–digital environment, the loss for control of talker echo must be inserted at the end offices (EOs). The loss may be implemented either by digital processing before digital-to-analog conversion (which adds delay, quantizing noise, and possible distortion), or by analog padding after conversion. Digital data transmission was to be delivered intact without such processing. Since it is impractical to insert different losses on connections having different end-to-end mileages, a single value of connection loss is desirable. Determining a compromise value of loss for the fixed-loss design in the SDN was based on several considerations. First, there would be no reflection points for echo in the four-wire digital network between end offices. In addition, noise in an all-digital network, with only one encoder and decoder, is negligible when compared to the analog network and has a constant value for all connection lengths. Early studies showed that for a fixed-loss value of 6 dB, assuming the use of echo control on trunks greater than 1850 miles in length, there is a small decrease (compared to the VNL plan) in echo performance with the fixed-loss plan. Additional studies were deferred as the fixed-loss design strictly applies only to an all-digital network and it would be some time before this would occur.

Postdivestiture Transition Loss Design. It was realized that a transition loss design would be necessary to ensure satisfactory performance when analog and digital switching systems are interconnected. The transition design being implemented was designed to make the combined analog-digital network compatible with the characteristics of the present analog network. Following are the principal characteristics and constraints of the combined network.

(1) The expected measured loss and the ICL of each trunk should be the same in both directions of transmission.

(2) The −2 dB TLP (associated with design and testing) at the outgoing side of analog tandem switches and the 0-dB TLP at end offices are retained.

(3) A −3 dB TLP (associated with design and testing) is established for digital tandem offices.

(4) The −16 dB and +7 dB TLPs at input and output of analog carrier systems are retained.

(5) Existing test and lineup procedures and TLPs for digital channel banks are retained.

(6) Combination intertandem trunks, terminating in digital terminals at a digital switching office at one end and in D–type channel banks at an analog switching office at the other end, are designed to have 1–dB ICL.

(7) Analog intertandem trunks are designed according to the VNL plan.

The allocations of loss to the trunks making up the transition design hierarchical network are indicated later in this chapter under the heading "Network Loss Allocations."

A transmission plan for a region digital switched network (RDSN) is being developed to provide digital connectivity and digital integrity. Discussed in Volume 3, this plan considers the influence that various loss plans, loop return losses, and connection delay have upon estimated loss, noise, and echo grade–of–service performance.

Echo, Singing, and Near–Singing

Echo is the signal at any point in a circuit that results from power being reflected in some manner from the primary speech path. This reflection can occur in the transmission path either at a four–to–two–wire junction or at an impedance irregularity on a two–wire circuit. The reflected signal causes three types of phenomenon: talker echo, listener echo (near–singing), and singing.

Figure 26–3 shows examples of where reflections occur on an end–to–end connection [9]. The two–wire path at each end includes the customer loop and may also include two–wire switches and trunks. The four–wire path may include four–wire trunks and switches or may consist of a single four–wire digital end office. The four–wire path is connected to the two–wire path at each end by a hybrid. If the impedance (Z) of the balancing network at the hybrid does not match the impedance of the two–wire path, some of the power arriving on the four–wire path will be reflected. Multiple echo paths can occur on connections that have multiple reflection points, but such connections usually have a single predominant echo.

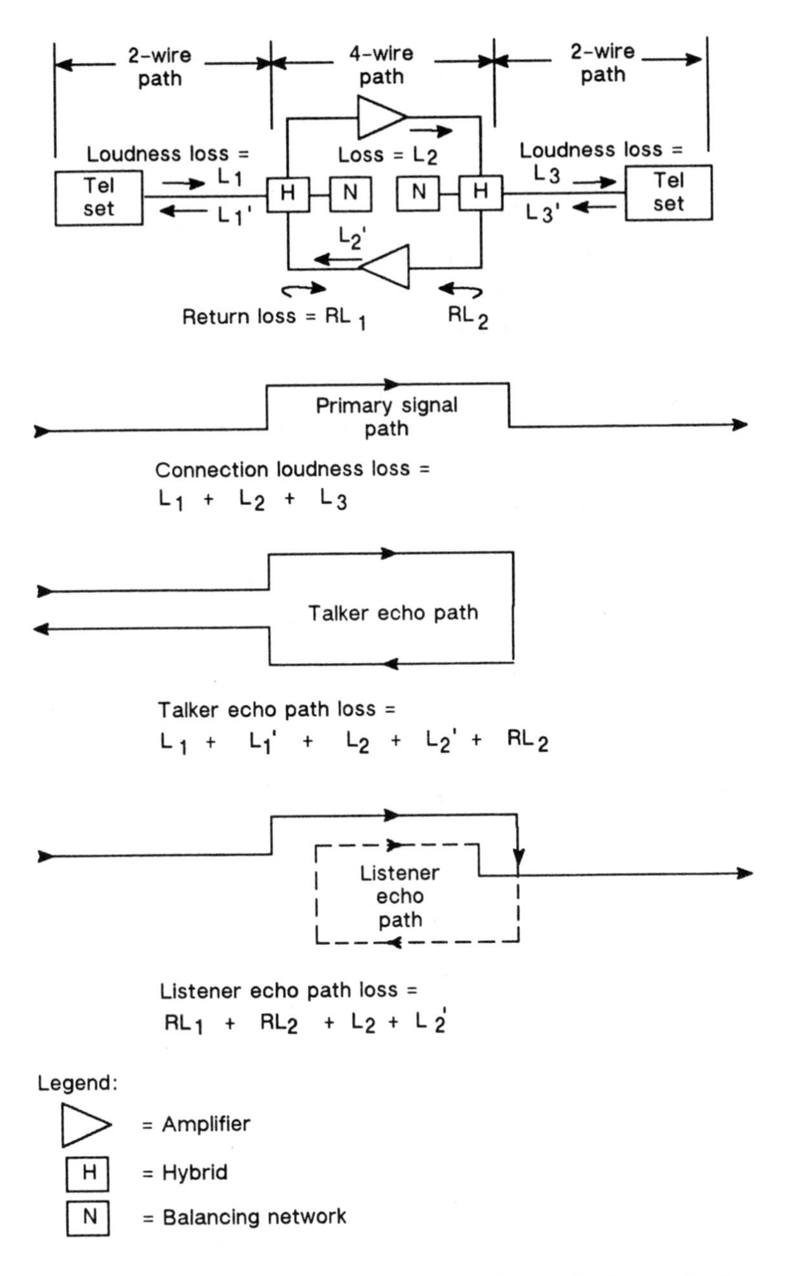

Figure 26–3. Reflection points and echo paths in a telephone connection.

The fraction of power reflected at a two–wire junction depends on the impedance mismatch. It is expressed in terms of return loss in dB as follows:

$$\text{Return loss} = 20 \log_{10} \left[\frac{Z_1 + Z_2}{Z_1 - Z_2} \right] \text{ dB} .$$

This is a formula for the return loss at a mismatch between any two impedances, Z_1 and Z_2. The formula is approximately correct for the echo reflected into the four–wire path at a four–to–two–wire junction, where Z_1 is the impedance of the balancing network and Z_2 is the impedance of the two–wire path as seen from the hybrid. Return loss is a function of frequency.

Figure 26–3 also depicts talker echo and listener echo paths in a four–wire connection. *Talker echo* occurs when primary speech reflected at the far end returns to the talker along the talker echo path as indicated. The talking customer hears his or her own voice delayed by the total delay of the echo path. If the reflected speech has sufficient amplitude and delay, it can be annoying and can interfere with the talker's normal speech process. The amount of annoyance caused by talker echo depends on the amplitude and delay of the echo. Figure 26–4(a) shows a talker echo opinion model [9]. A combined loss/noise/talker echo opinion model is shown in Figure 26–4(b). In either case, as the acoustic echo path loss of a connection decreases or the echo path delay increases, the curves show a decrease in the percent of customers that would rate the connection good or better. The acoustic echo path loss includes the transmit objective loudness rating (TOLR) and receive objective loudness rating (ROLR) (see Table 24–1) of the telephone set plus loop, the round–trip electrical loss from the loop to the reflection point and back, and the echo return loss at the reflection point. For loops with a 500–type telephone set, the sum of the TOLR and ROLR has a mean of about 4 dB and a standard deviation of about 3.5 dB.

If a talker's speech reflected from the far end is again reflected at a near–end impedance mismatch, it is heard twice by the listener. The effect is frequently referred to as near–singing distortion for short delays or, more generally, *listener echo*. For short echo delays, this impairment causes the listener to perceive a hollowness in the talker's speech as if the talker were speaking

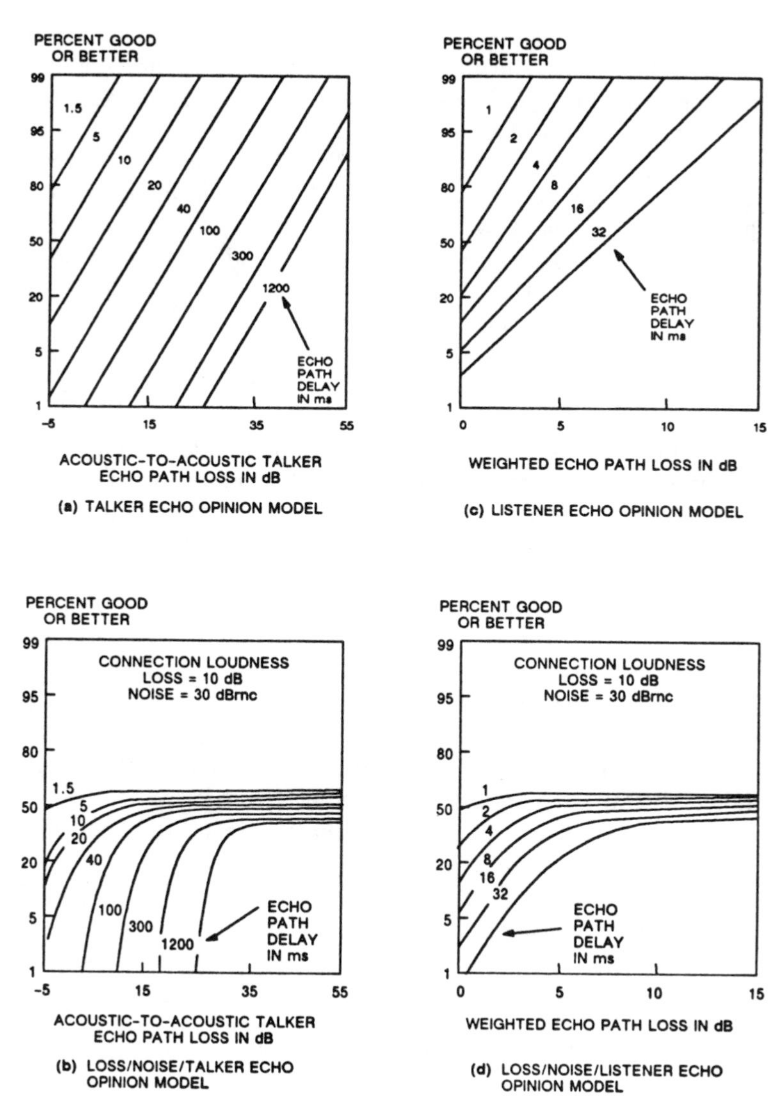

Figure 26-4. Loss/noise/echo grade of service.

into an empty barrel. Listener echo is generally of concern on relatively short-length, two-wire-to-two-wire, intra-digital-office (zero-loss) connections. On longer connections, provisions to control talker echo will also control listener echo.

The listener echo path, shown by the broken line in Figure 26–3, includes the return loss at both ends of the four–wire path and the round–trip electrical loss between the reflection points. Listener echo performance is judged on the basis of a measure called weighted echo path loss (WEPL). WEPL is the reciprocal (expressed in dB) of the average magnitude of the voltage gain of the listener echo path over the band from 200 to 3400 Hz. Figure 26–4(c) shows a subjective opinion model for listener echo as a function of WEPL and echo path delay. A combined loss/noise/listener echo opinion model is shown in Figure 26–4(d).

Loop Segregation. To meet loop hybrid balance objectives in digital end offices, it has been found necessary to segregate the loop population into several subpopulations (loaded, nonloaded, and others), as shown in Table 26–3 [10]. Return loss specifications have been developed for loop hybrids when they are terminated in impedances that represent the loops. Meeting these requirements will ensure that WEPL objectives are met, provided that the delay and loss of listener echo paths are controlled.

Singing is caused by circulating power in the transmission path and occurs in the same way as listener echo. Singing places a sustained loud tone on the connection. Referring to Figure 26–3, singing arises if the gains in repeaters or carrier channels are high enough and the return losses low enough so that the round–trip path gain at some frequency has magnitude greater than one with zero or multiples of 360 degrees phase shift. (This can also occur at frequencies near the upper and lower edges of the voiceband where impedances are not well matched.) Standard return loss measuring sets transmit a shaped spectrum of noise in each of these bands. Singing return loss is used to evaluate singing protection and is defined as the lower of the two return loss measurements (SRL–high at the high end of the band and SRL–low at the low end).

The singing margin of a connection is defined as the amount of gain that, added to the listener echo path, would start it singing. *Near–singing* is the "hollow" condition that occurs just before actual singing takes place. Meeting the objectives for WEPL in Table 26–4 [9] will ensure that singing or near–singing will not occur.

Table 26–3. Digital End–Office Hybrid Balance Requirements

Interface Type	Termination Network	Performance (Hybrid Structural)
Loaded 2W analog loop	1650 Ω in parallel with 0.005 μF + 100 Ω (Note)	Singing ≥ 20 dB (200–3400 Hz) Echo ≥ 25 dB (500–2500 Hz)
Nonloaded 2W analog loop	800 Ω in parallel with 0.005 μF + 100 Ω (Note)	As above
2W analog special service lines	900 Ω + 2.16 μF	As above
Remote terminal 2W analog loop:		
Remote to host loss ≥ 0 dB and < 2 dB	As above	As above for loaded and nonloaded loops
loss ≥ 2dB special service lines	900 Ω	Singing ≥ 15 dB (200–3400 Hz) Echo ≥ 20 dB (500–2500 Hz)
2W analog trunks and 2W analog special service trunks	900 Ω + 2.16 μF	Singing ≥ 25 dB (200–3400 Hz) Echo ≥ 30 dB (500–2500 Hz)

Note: This resistor has little effect at voice frequency. It was included to ensure that at high frequency the impedance of the network will not approach zero ohms, which could lead to stability problems. It may be eliminated if high-frequency instability will not be a problem.

Table 26–4. Objectives For Weighted Echo Path Loss*

Listener Echo Path Delay (ms)	Percent of Connections/ Minimum WEPL (dB)			
	50%	95%	99%	99.9%
2	16	11	9	7
3	17	12	10	8
4	18	13	11	9
5	19	14	12	10
6	20	15	13	11
7	21	16	14	12
8	22	17	14	12

*Function of switching equipment and local distribution loop design.

Talker Echo Objectives. One way to express the network echo design objective is: the trunk loss designs should be such that talker echo (the dominant impairment) is satisfactorily low on more than 99 percent of all telephone connections that encounter the maximum delay likely to be experienced. This way of expressing the overall objective recognizes that (1) echo performance can be controlled by controlling trunk losses, (2) the control of echo on *connections* implies further control of echo on the trunks used to form customer–to–customer connections, and (3) echo impairment is a function of the amount of delay in the connection.

Listener Echo Objectives. Listener echo can be of particular concern on short connections. Digital end offices are potential sources of listener echo when two local subscribers are connected, due to the low loss of the equivalent four–wire path and the potential for poor balance at each loop hybrid. For this reason, WEPL objectives for design of digital switching systems have been formulated. They are shown in Table 26–4 as a function of echo path delay.

Return Losses. The talker echo path includes the return losses at the reflecting points in the network, usually at the two–wire–to–four–wire conversions. Control of echo also means setting objectives for the control of return losses. Return loss objectives in the predivestiture era were specified for connection points in the toll portion of the network; generally the echo and singing return loss objectives for these points were more stringent than for the local portion because toll trunk parameters were more controllable. The following objectives are typical, but not all–inclusive.

Predivestiture:

(1) *Through balance.* For four–wire trunks terminating at two–wire switches in class 1, 2, or 3 switching offices, the ERL objective was 27 dB (minimum 21 dB); the SRL objective was 20 dB (minimum 14 dB).

(2) *Terminal balance.* For the interface between four–wire intertoll trunks and most two–wire toll–connecting trunks at class 1, 2, 3, or 4 switching offices, the ERL objective was 18 dB (minimum 13 dB). For four–wire

toll–connecting trunks, the objective was 22 dB (minimum 16 dB). For both two–wire and four–wire trunks, the SRL objective was 10 dB (minimum 6 dB).

These return losses were measured against standard terminations (600 or 900 ohms + 2.16 μF), the values of which depended on the type of switching office involved. The measurement process and the complexity of impedance adjustment that permitted these return loss objectives to be met led to the expression of objectives in terms of through balance and terminal balance requirements for many types of trunk at various types of toll switching offices.

Postdivestiture:

Through balance and terminal balance limits for the postdivestiture intraLATA networks are given in Table 26–5 [9].

Table 26–5. IntraLATA Office Balance Limits

Measurement Type	ERL (dB)		SRL (dB)	
	Preservice Limit	Immediate Action Limit	Preservice Limit	Immediate Action Limit
TERMINAL BALANCE				
Analog Switch				
2–wire facilities:				
Interbuilding	18	13	10	6
Intrabuilding	22	16	14	10
4–wire facilities	22	16	15	11
Digital Switch				
2–wire facilities:				
Interbuilding	18	16	13	11
Intrabuilding	22	16	15	11
4–wire facilities	22	16	15	11
THROUGH BALANCE	27	21	20	14

Network Loss Allocations

Loop Loss. Transmission objectives for loop loss have been derived on the basis of satisfying an overall loss/noise grade–of–service objective [3,11]. Control of loop loss is accomplished by the application of carefully specified rules in the design and layout that produce a satisfactory distribution of losses. The major rules used prior to 1980 are parts of the *resistance* and *long–route design* plans. Later, these plans were replaced by the revised resistance design (RRD), the concentrated range extension with gain (CREG), and the modified long–route design (MLRD) plans [8]. These three design plans permit straightforward application of the rules to the installation of new cables, inductive loading, and electronic equipment so that overall loss objectives are met, because the objectives are integral parts of the plans. When the plans are properly applied, the resulting distribution of loop losses has a maximum value of about 8.5 dB (including the effects of bridged taps). The trend toward digital loop carrier (DLC) on long routes implies reduced and stable losses. The mean value and the standard deviation of the loss distribution depend on the geographical area served and on the concentration of customers within the area. For all exchange carriers, an average 1004–Hz value of 3.7 dB and a standard deviation of 2.3 dB are typical; these values are used in determining network loss objectives and grade of service.

Because a numerical loss objective (other than the maximum) is not expressed for individual loops, special treatment must be applied (1) when a loop is assigned to a service with data conditioning or similar special–service need, and (2) when transmission complaints still exist after it has been verified that the loops involved have been installed according to appropriate design procedures.

Predivestiture Via Net Loss Allocation. Loss objectives for analog switching machines were generally less than 1 dB. Losses of various types of trunk constituted the remaining major allocation to parts of the plant. For purposes of circuit design, trunk losses were defined to include average switching system loss. Many of the loss values were given in terms of VNL, which varied according to the length and type of facility.

Losses allocated to trunks depended on the position of the trunks in the switching hierarchy and the probability of encountering tandem connections of such trunks in an end–to–end telephone connection. In the toll portion of the network, interregional intertoll trunks were designed on the basis of maximum round–trip echo delay that could occur on connections involving the interregional trunks. If the delay could exceed 45 ms, the interregional trunks were equipped with echo cancellers or suppressors and the trunks were operated at 0 to 0.5 dB loss. (Losses high enough to satisfy echo requirements would generally be too high to satisfy volume and contrast objectives.) If the round–trip echo delays were less than 45 ms, the interregional trunks were operated at VNL, with a maximum of 2.9 dB.

High–usage intertoll trunk groups were operated at VNL where the value of loss was VNL ≤ 2.9 dB, equivalent to a maximum trunk length of about 1850 miles on carrier facilities. If echo requirements called for a loss greater than 2.9 dB, the trunks were operated at 0–dB loss and were equipped with echo cancellers or suppressors unless they were in a final–routing chain. To avoid having more than one echo suppressor in a connection, suppressors were generally permitted only in final groups between regional centers. Secondary intertoll trunks were operated as close to 0 dB as possible, with a maximum of 0.5 dB. Final intertoll trunk groups were operated at VNL, but at a maximum of 1.4 dB loss.

Toll–connecting trunks were usually operated at VNL + 2.5 dB loss with a maximum loss of 4.0 dB. An alternative design allowed a trunk to have 3.0 dB to 4.0 dB loss provided it contained less than 15 miles of VF cable facilities or less than 200 miles of carrier facilities. On long end–office trunks (usually interregional) between class 4 and class 5 offices where echo requirements indicated the need for loss greater than 4 dB, an echo suppressor was permitted, with the loss set at 3 dB.

In the local portion of the network, direct trunks were designed to a nominal loss of 3 dB with a maximum of 5 dB. Toll trunks were operated at a nominal loss of 3 dB and a maximum of 4 dB. Intertoll trunks were operated at VNL. Loss values were assigned similarly to all service and miscellaneous trunks used in the network. Long interregional direct trunks (between class 5 offices) were allowed to operate without echo suppressors at VNL + 6 dB loss (maximum 8.9 dB) over distances of up to 4000 miles.

The loss design objectives for the analog, digital, and combination trunks used in the VNL analog network, including operator service trunks, are listed in Table 26–6 [8].

Table 26–6. Predivestiture Loss Allocation (VNL Plan)

Trunk Type			ICL (dB)	Max ICL (dB)	Echo Supp. (MILES)	Remarks
Analog Trunk	TCT		2.5+VNL			(1) Trunks without gain <200 miles ICL: Min. 2.0 dB Max. 4.0 dB (2) Trunks with gain <200 miles, Max.=4.0 dB >200 miles, Max.=5.4 dB
	ITT	HU	VNL	2.9	1850	With echo supp. ICL = 0 dB
		Final	VNL	1.4		
	EOT: Class 5 - Class 5		6.0+VNL	8.9 (200–4000 route-miles)	No echo suppressors	Trunks without gain < 500 miles ICL: Min. 0 dB Max. 5.0 dB (2) Trunks with gain ICL = 3.0 dB (3) With echo supp. ICL = 3.0 dB
	EOT: Class 5 - higher class		2.5+VNL	5.5 (200–1850 route-miles)	1850	(1) Trunks without gain < 200 miles ICL: Min. 2.0 dB Max. 4.0 dB (2) Trunks with gain ICL = 3.0 dB (3) With echo supp. ICL = 3.0 dB
Digital Trunk	TCT		3.0			
	ITT		0		1850	
	EOT: Class 5 - higher class		3.0	1	850	With echo supp. ICL = 30 dB
	EOT: Class 5 - Class 5		3.0 6.0			0 to 200 miles 201 to 1000 miles
Combin. Trunk	TCT		3.0			
	ITT		1.0		1850	
	EOT: Class 5 - Class 5		3.0 6.0			0 to 200 miles 201 to 1000 miles
	EOT: Class 5 - higher class		3.0		1850	With echo supp. ICL = 3.0 dB

Table 26–6. Predivestiture Loss Allocation (VNL Plan) (Continued)

Trunk Type			Equiv	ICL (dB)	Remarks
Operator Service	TSPS	RTA to TSPS Base unit	ITT	0	
	No. 5 ACD	Local DA: Class 5 - ACD	Tandem trunk	3.0	
		Class 5 - Concentrator	Tandem trunk	3.0	Unless gain transfer is used with 1 trunk concentrator.
		Concentrator - ACD	Intertandem trunk	0.5	
		Intra-NPA DA: Tandem office-ACD	Intertandem trunk	0.5 1.5	From directional tandem From sector tandem
		Toll - DA Class 3 - ACD	ITT	0.8	
	AIS and No. 5 ACD Intercept	Class 5 - ACD	AIS trunk	3.0	
		Class 5 - Concentrator	AIS trunk	3.0	Unless gain transfer is used
		Concentrator - ACD or AIS	ITT	0.8	

Note: VNL = VNLF × (Avg. Length) + 0.4 dB.

Predivestiture Proposed Fixed-Loss Allocation. The switched digital network non–VNL fixed–loss plan proposed in the late 1970s included interconnections to analog switching offices of the existing VNL plan. Interconnections were made by "combination" analog–to–digital office trunks. Allocations of trunk losses for various network connections are shown in Table 26–7 [8]. Proposed digital connection losses were 0 dB for intraoffice, 3 dB for short toll of up to 200 miles, and 6 dB for 200 to 1000 miles. There were problems, however, with intraoffice connections because of listener echo and near–singing impairments, which were being investigated at the time. This plan was not completed and studies were deferred until all–digital connections became more prevalent.

More recently, a transmission plan for an RDSN has been developed to provide digital connectivity and digital integrity for 56 kb/s or 64 kb/s (ISDN) transmission capability. This plan considers the influence that loop plant, loop return losses, transmission delays, and network loss plans have on grade–of–service performance.

Table 26–7. Predivestiture Proposed Fixed–Loss Allocation Plan for SDN

Trunk or Connection	From Switching Office Type	To Switching Office Type	Loss
Intraoffice Connection	Class 5	Class 5	0
Interoffice Trunk	Class 5	Class 5	3
Toll Connection	Class 5	Class 5	3* 6**
Toll–Connecting Trunk	Class 5 Digital Toll Analog Toll	Toll Class 5 Class 5	3 3 3
Intertoll Trunk	Digital Toll Analog Toll Digital Toll	Digital Toll Digital Toll Analog Toll	0 1 1

* 0 to 200 miles.

** 201 to 1000 miles.

Postdivestiture Transition Loss Allocation. In today's environment, access to LATA networks is provided to ICs, as discussed in Volume 3, Chapter 18. Within the LATA, a transitional loss allocation is available. This covers the intraLATA connections and trunks, which are shown in Figure 26–5 [9].The transition transmission loss design was used to achieve an end–office–to–end–office LATA connection loss of approximately 6 dB where a connection could exceed 200 miles. This applies to connections provided by either all–analog, all–digital, or a combination of analog and digital trunks. These trunks are defined as:

(1) *Analog trunks* interface at VF with switching systems at both ends. Also, trunks that use analog facilities, wholly or in part, are considered analog trunks regardless of the switching systems and the manner of interface. Digitally terminated analog (DTA) trunks use both analog and digital facilities, interface at VF with a switching system at one end, and interface digitally with a digital switching system at the other end. DTA trunks are treated as analog trunks.

(2) *Digital trunks* use digital transmission facilities and interface digitally with digital switching systems at both ends.

(3) *Combination trunks* interface at VF with a switching system at one end, use digital transmission facilities, and interface digitally with a digital switching system at the other end.

End–office–to–end–office connection losses through the trunks of the transition network are achieved for analog trunks by the usual adjustment of VF transmission pads in the circuit, for combination trunks by such pads in the analog portion while maintaining an alignment based on a digital reference signal (DRS) in the digital portion, and for digital trunks by maintaining the DRS alignment and providing the necessary loss in the decoding process. The DRS is defined as the assigned pulse code modulation (PCM) digital word code that represents a 0–dBm0 1000–Hz VF signal. In the digital PCM domain, the designations encode level point (ELP) and decode level point (DLP) are used instead of transmission level (TL). The ELP can be defined as that power at VF in dBm0 that when applied to the encoder results in the DRS (i.e., 0 ELP means 0 dBm0 produces the DRS). The DLP can be defined as that power at VF in dBm0 that is produced by the DRS applied to the decoder (i.e., −3 DLP means −3 dBm0 results from decoding the DRS).

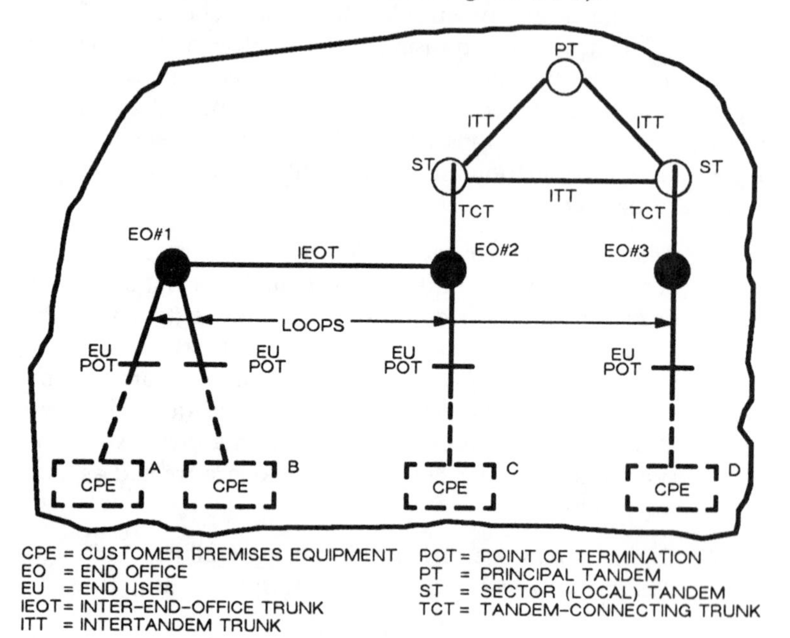

Figure 26–5. IntraLATA connections.

The intraLATA trunk losses [9] for local and metro trunks of less than 200 miles in length (metropolitan) are shown in Table 26–8, and greater than 200 miles (long distance) in Table 26–9. For a metropolitan internal office connection, the design loss will be 3 dB, generally, but can be 5 dB with some analog trunks. For connections involving more than one tandem office, the loss will be 6 dB for all–digital and 6 to 9 dB for all–analog paths.

The test tone levels, TLPs, ELPs, and DLPs used for alignment to provide the losses for the analog and digital trunks are shown in Figures 26–6 through 26–11. With these alignments, an all–digital connection via tandem offices provides for the input analog voiceband signal to be encoded into digital form at the originating end office, passed through the digital connection without change, and finally decoded to the proper voiceband level at the terminating end office. These arrangements vary widely in popularity; four–wire analog tandems (Figures 26–10 and 26–11) are now almost nonexistent, and four–wire digital tandems (Figure 26–8, etc.) are quite common.

The losses for various intra–end–office connections are indicated in Table 7.4–C of Reference 10. For example, the loop–to–loop office loss between distributing frames is expected to be about 0.5 dB average and less than 1.0 dB maximum.

Loss Maintenance Limits. In order to maintain network performance, 1004–Hz measurements of intraLATA trunk losses are made periodically in accordance with maintenance programs described in Volume 3. The percentage of measurements showing deviations from design values in excess of 0.7 or 1.7 dB (the larger deviations carry heavier weighting) determines the index for the group of trunks under study. If the index is 96 or higher, performance is satisfactory and no action is necessary. If the index is below 96, investigation and corrective action are indicated. If the loss of any trunk deviates from its design value by 3.7 dB or more, it should be removed from service.

Table 26–8. IntraLATA Trunk Losses (ICL)—Metropolitan

Sector Trunk	Length (Miles)	Without Gain (dB)	With Gain (dB)		Remarks/ Notes
			Objective	Max(1)	
Inter End Office (IEOT)					(1) Maximums apply only when gain is supplied by E-type repeaters
Analog	-	0.0 to 5.0	3.0	5.0	
Digital	-	-	3.0	-	
Combination	-	-	3.0	-	
Tandem-Connecting (TCT)					
Analog	-	0.0 to 4.0	3.0	4.0	
Digital	-	-	3.0	-	
Combination	-	-	3.0	-	
Intertandem (ITT)					
Analog	-	-	1.5	-	
Analog (Between ST and PT)	-	-	0.5	-	Balance required
Digital	-	-	0.0	-	
Combination	-	-	1.0	-	

Table 26-9. IntraLATA Trunk Losses (ICL)—Long Distance

Long-Distance Trunk	Trunk Length (Miles)	Without Gain (dB)	With Gain (dB)		Remarks/ Notes*
			Objective	Max	
Inter End Office (IEOT)					
Analog	<200	0.0 to 5.0	3.0	-	
	≥200	-	VNL + 6.0	8.9	Balance required
Digital or Combination	<200	-	3.0	-	
	≤200	-	6.0	-	Balance required
Tandem-Connecting (TCT) Analog	<200	2.0 to 4.0	3.0	4.0	Maximum applies only when gain is supplied by E-type repeaters Balance required
	≥200	-	VNL + 2.5	5.4	Balance required
Dig/Combination	-	-	3.0	-	Balance required
Intertandem (ITT)					Trunks must be 4-wire and designed to meet balance requirements
Analog	-	-	VNL	1.4	
Digital	-	-	0.0	-	
Combination	-	-	1.0	-	

* Balance requirements are shown in Table 26-5.

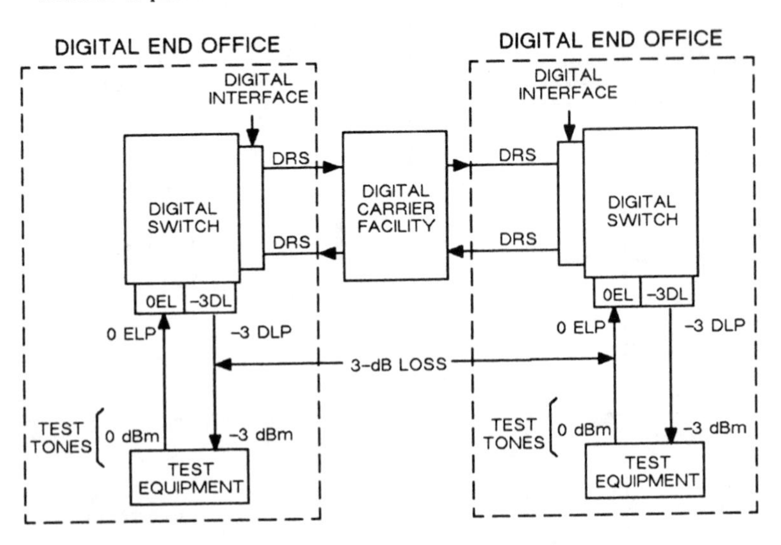

Figure 26-6. Provision of short-haul digital inter-end-office trunks.

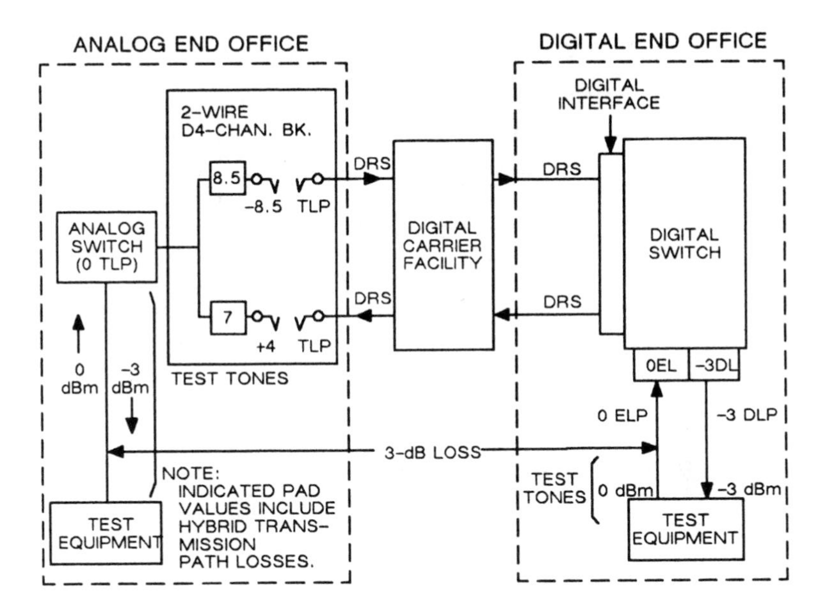

Figure 26-7. Provision of short-haul combination inter-end-office trunks.

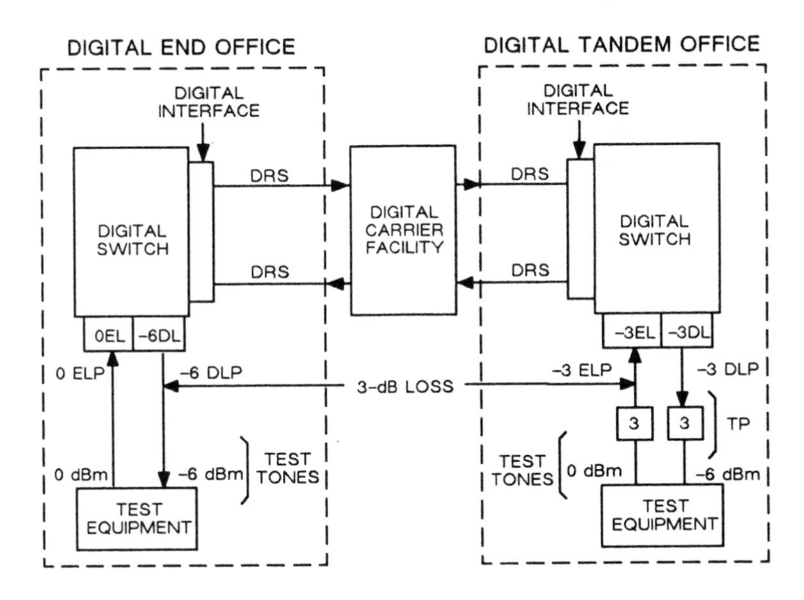

Figure 26-8. Provision of digital tandem-connecting trunks.

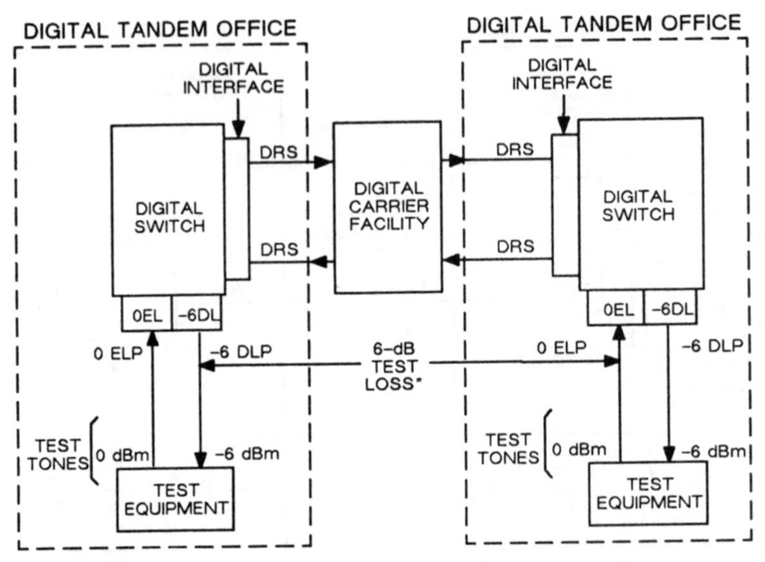

* No loss is introduced in the transmission path.

Figure 26–9. Provision of digital intertandem trunks.

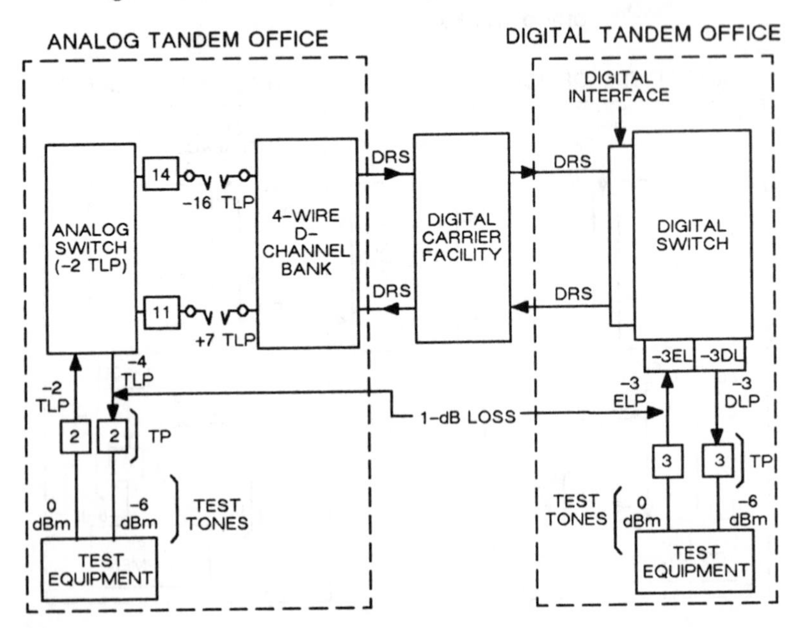

Figure 26–10. Provision of combination intertandem trunks.

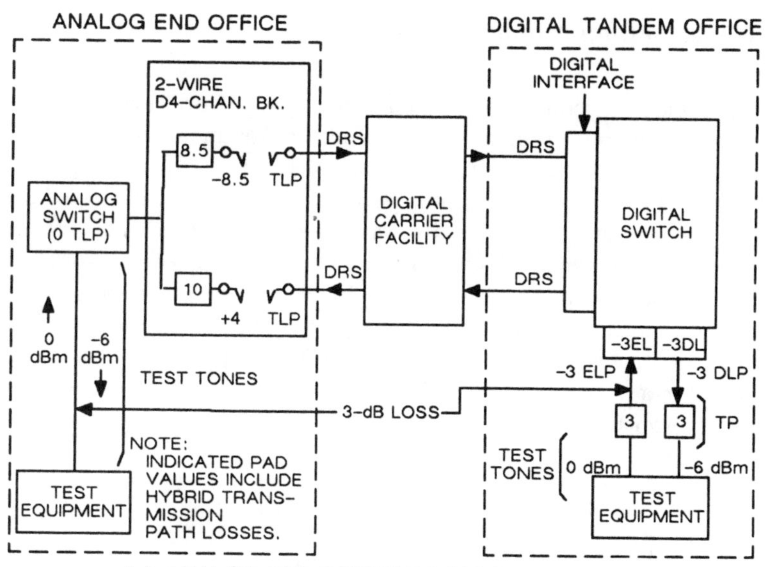

(a) ANALOG END OFFICE TO DIGITAL TANDEM

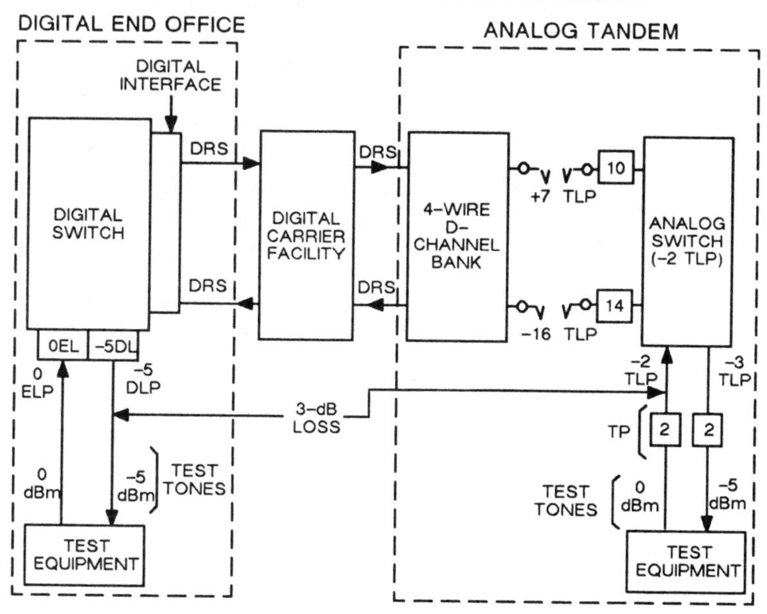

(b) DIGITAL END OFFICE TO ANALOG TANDEM

Figure 26-11. Provision of combination tandem-connecting trunks.

Message Circuit Noise

Message circuit noise is defined as the short–term average noise measured by means of a defined noise–measuring set [11] with a C–message weighting network. Objectives for message circuit noise, allocated to various parts of the network, are based on subjective tests in which noise was evaluated by telephone listeners in the presence of speech signals held at a constant volume. Noise and volume were expressed in dBrnc and vu, respectively, at the line terminals of the station set; observers were asked to rate the performance as excellent, good, fair, poor, or unsatisfactory for a wide range of noise values. The results of these tests are shown in Figure 26–12.

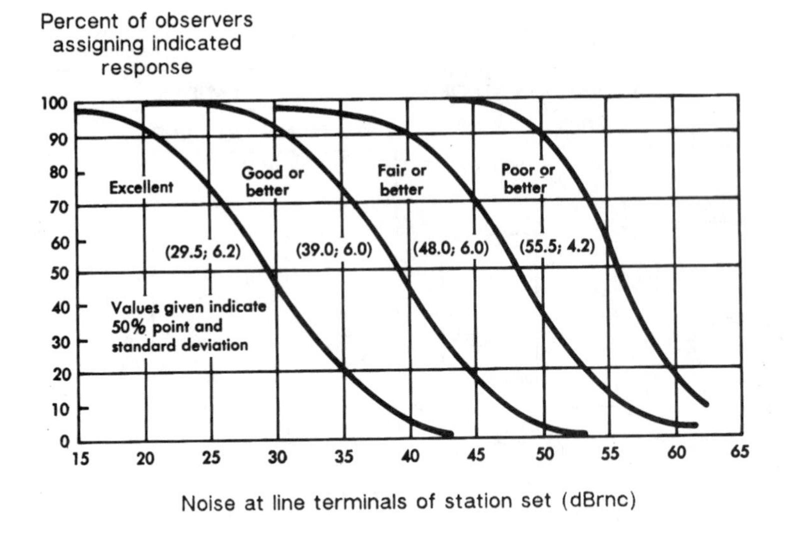

Figure 26–12. Noise opinion curves.

Using this data and the results of a connection noise survey, message circuit noise objectives were established for customer–to–customer toll connections. This assumes little influence of loop noise. The predivestiture noise objectives are shown in Figure 26–13 [9], and are based on ensuring high customer satisfaction as well as economic feasibility.

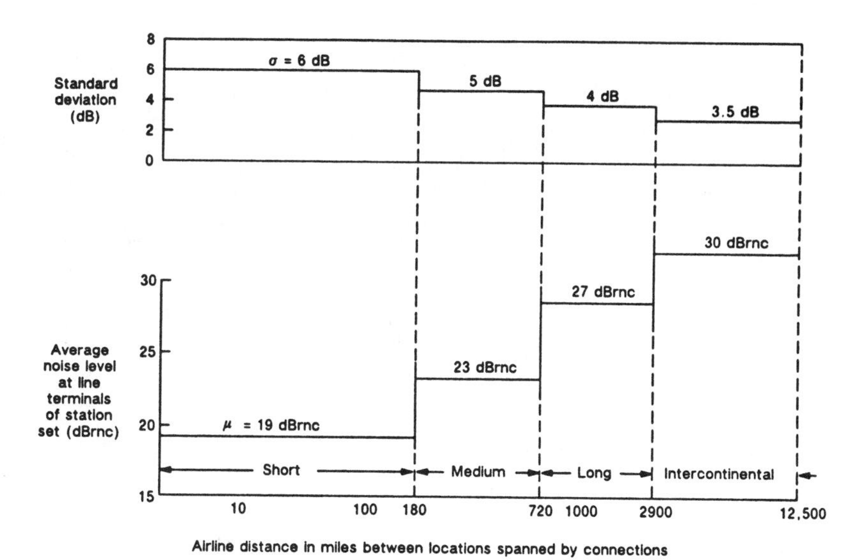

Figure 26–13. Noise objectives for customer-to-customer connections.

Loop Noise Objectives. The message circuit noise objective applied to loops is that noise measured at the line terminals of the station set shall not exceed 20 dBrnc. Most loops have measured noise well below this value. The average on non-carrier loops is about 0 dBrnc [3,4]. Noise at or below this value has little effect on grade of service, but noise in excess of 20 dBrnc deteriorates grade of service appreciably.

In recognition of the special circumstances relating to long routes (those in excess of 1300 ohms controlled by resistance design), the noise objective is made somewhat more lenient. For long routes, the noise objective is administered at 30 dBrnc. For routes on which the limit of 30 dBrnc is exceeded, special treatment (ringer isolators, conversion to DLC, balancing, etc.) must be employed according to circumstances.

Analog Trunk Noise Objectives. The performance objectives for message circuit trunk noise have been allocated to allow for the tendency of noise to accumulate with distance and the smaller number of calls of very long distances compared with those of intermediate and short distances. Trunk noise objectives have been selected to give weighting to these two factors, and allocations have been made for short–haul carrier facilities (for use on trunks less than 250 miles long), and long–haul carrier facilities (for use on trunks over 250 miles long) [3,12]. These allocations, which recognize the inherent variability of performance in the field environment, are expressed in terms of mean values and standard deviations. As shown in Figure 26–14 [12], for short–haul analog carrier, the mean value of the objective is 28 dBrnc0 at 60 route–miles and for long–haul carrier, 34 dBrnc0 at 1000 route–miles. The standard deviation is σ = 4 dB in each case. These allocations allow for a 3–dB increase in noise for each doubling of the distance. This increase is typical of analog carrier system performance but is not experienced in digital carrier systems, which display a fixed noise level of 18 to 21 dBrnc0 for current–vintage terminals. Design objectives for carrier systems are based on these performance objectives, but are normally expressed in terms of worst channel noise in a nominal environment. The design objective for 4000–mile coaxial cable systems, including multiplex equipment, was 40 dBrnc0; it was 41 dBrnc0 for long–haul radio systems of the same length. Short–haul analog radio design objectives are the same as shown in Figure 26–14. These objectives do not apply either during deep radio fading of short duration or to protection switching when noise may briefly reach 55 to 60 dBrnc0. Fading and outages are discussed in Volume 2.

Generally, where message circuit noise objectives are being met, noise objectives for voiceband data transmission are also met. Some analog carrier trunks, however, include syllabic compandors, which compress the range of amplitude of an input channel signal for transmission on the carrier and expand the channel output to a signal that matches the original input. This provides better signal–to–noise margins for the carrier transmission path. For data service, the noise is measured as C–notched noise by applying a tone signal (–13 dBm0) at one end of the circuit to activate the compandor and a narrowband notch filter at the other end of the circuit to eliminate the tone ahead of the measuring set. A tone signal to a C–notched noise ratio of at least 24 dB should be maintained.

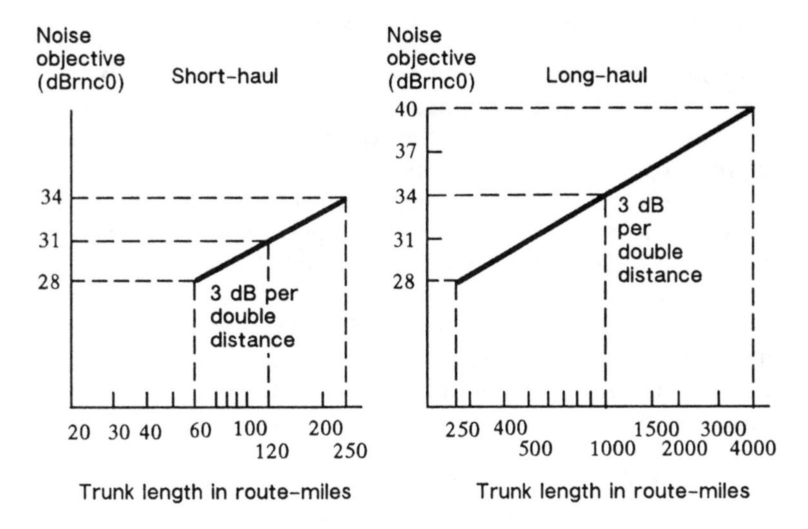

Figure 26–14. Trunk message circuit noise performance objectives.

Digital Trunk Noise Objective. Digital trunk noise depends primarily on the voice coding method employed and the associated processing. It is virtually independent of trunk length, depending instead on how often the coding–decoding function is required. For a single coding function, the noise should not exceed 23 dBrnc0 for current–vintage back–to–back terminals. The insignificant amount of noise of a digital carrier dependent on distance is due to regenerator errors and jitter, a matter of span design.

Impulse Noise. Impulse noise is any burst of noise that produces a voltage in excess of about 12 dB above the rms noise measured by a 3–type noise–measuring set with C–message weighting; in a speech channel, these bursts are usually less than 5 ms but may very rarely be as long as 45 or 50 ms in duration. The ratio of the voltage excess to the rms noise voltage is nominally at least 12 dB for a 3–kHz bandwidth; it may be as great as 40 dB in some systems, particularly analog microwave radio. Impulse noise is superimposed on background message circuit noise [13]. Objectives are dominated by requirements for digital data signal transmission. Circuits that are satisfactory for data are satisfactory for speech transmission.

Impulse noise objectives are usually established on the basis of the number of counts obtained on an impulse noise counter or equivalent during a prescribed measurement interval and may be expressed for loops, trunks, or customer–to–customer connections. The objective for any loop or single voice channel is that there should be no more than 15 impulse noise counts in 15 minutes at a given threshold. For a sampled trunk group, there should be a maximum of 5 counts in 5 minutes at a given threshold. Sampling plans are specified and the noise thresholds are set at different values for loops, for VF trunks, and for compandored and noncompandored carrier trunks. The trunk impulse noise thresholds are shown in Table 26–10.

Table 26–10. Impulse Noise Thresholds for Trunks

	Facility Type		
Trunk Length (Miles)	VF Trunks (dBrnc0)	Compandored* Carrier and Mixed Compandored–Noncompandored (dBrnc0)	Noncompandored Carrier (dBrnc0)
0–60	54	68	58
60–125	54	68	58
125–250	54	68	59
250–500		68	59
500–1000		68	59
1000–2000		68	61
Over 2000		68	64

* Compandored trunks, including those with digital carrier, are measured with a -10 dBm0 tone transmitted from the far end and filtered out ahead of the measuring set by a C–notched filter or equivalent. The C–notched filter is a C–message weighting network with a narrowband suppression section to provide at least 30 dB of attenuation at the tone frequency.

Other Impairments

Intelligible Crosstalk. Intelligible crosstalk objectives are generally expressed in terms of the crosstalk index, a measure of the probability of receiving intelligible crosstalk. Objectives have been established for most types of trunk. A crosstalk index of 0.5 is applied to tandem–connecting, intertandem, and direct trunks. No index objective has yet been established for loops.

Crosstalk objectives for central–office equipment are usually expressed in terms of equal level coupling loss. In four–wire analog offices, the objective for minimum coupling loss between the two sides of one circuit is 65 dB. The coupling objective for different circuits is 80 dB in two–wire and four–wire offices.

Single–Frequency Interference. Generally accepted transmission objectives for single–frequency interferences are not available. When new systems have been designed, design objectives have been applied in a generally conservative manner. The factors that have made it difficult to derive acceptable objectives include the frequency and amplitude of the interference, the stability or variability of frequency and amplitude, the harmonic content of the interference, the presence or absence of masking message circuit noise or other interferences, the possible presence of other single frequencies, and the constancy or intermittency of the interference. As a rule of thumb, single–frequency interference design objectives are 10 to 12 dB below message circuit noise. These objectives apply to speech signal transmission and, when met, result in satisfactory transmission of other voiceband signals. Fortunately, such interference is rare with digital carrier systems.

Frequency Offset. Frequency offset objectives are set primarily to satisfy the needs of program signal transmission. While the determination of the threshold for frequency offset is almost as critical to speech transmission as it is to music transmission, subjective tests have shown that listeners are more tolerant of offset in speech signals than in music signals. The overall performance objective for offset is a maximum value of ±2 Hz; the maintenance objective is ±5 Hz. Digital carrier does not cause frequency offset.

Overload. Overload of broadband or single–channel electronic systems produces signal impairments in the form of noise and distortion. The objective for overload is expressed as a degradation of the grade of service in an individual channel. While objectives have not been firmly established, a reduction of about one percent in good–or–better and an increase of about 0.1 percent in poor–or–worse grades of service appear to be reasonable performance objectives for the overload phenomenon. These criteria, when applied to digital systems, have resulted in the objective that these systems transmit a 3–dBm0 sine–wave signal without overload impairment.

A signal transmitted at higher amplitude than the design value on analog systems may cause intelligible crosstalk or single–frequency tone interference as a result of intermodulation or other crosstalk paths. This impairment is not considered as overload unless it is so extreme that the entire system is affected.

Miscellaneous Impairments. A number of miscellaneous impairments are recognized as having degrading effects on VF data transmission; they include phase and gain hits, phase and gain jitter, and dropouts.

Telephone Station Sets. The transmission performance of station sets is controlled primarily by design, and there are no specific transmission performance or maintenance objectives. The majority of sets in service are the 500–type, which were developed to meet a set of stringent design objectives [2,14,15]. There are no transmission options or adjustments on these sets. It should be noted that network performance objectives had been developed based on station sets meeting these objectives.

26–2 OPERATOR SERVICES TRANSMISSION OBJECTIVES

Operator services are treated separately from message services because of their special functions and their point of access to the network. This section indicates the transmission performance objectives established for operator services [16] and gives detailed objectives for the operator circuit, which provides interconnection between the operator headset and the network.

Operator services are provided by personnel working in centrally located position groups. These services include directory assistance, intercept (both are called "number services"), and toll and assistance functions. To accomplish these services, several operator services networks have been established because of the different service functions and call routing required for operator services. There are several transmission plans to provide satisfactory communication for customer–operator, operator–operator and automated functions. These are described in Volume 3. In the future, all operator functions will be provided by a single operator services network using a universal operator position.

The general transmission performance objectives for the customer–to–customer intra– or interLATA connections of an operator–assisted call should be a loss/noise/echo performance not significantly different from that of an unassisted call spanning the same distance [16]. The objective for intraLATA operator–to–customer connections should be a loss/noise/echo performance equivalent to that provided customers on intraLATA connections.

The acoustic signal–to–noise objective is that all connections are to achieve a ratio of 29 dB at the operator's ear when spoken to by an average customer or operator. This acoustic performance is based on operator telephone sets meeting the specifications of References 17 (older) or 18 (newer). Some other important objectives follow. These cover the operator's transmitted speech through the headset circuits and operator console circuits into the network and the customer's speech from the network through the console and headset to the operator's ear.

Average Voice Level. Many requirements stated in this section are derived from the average customer voice power available on an operator–handled call. The reference point for determining this average speech power is the end–office outgoing switch appearance or 0 TLP. The average speech power at this point has been measured to be equivalent to a –21 dBm0, 1004–Hz tone.

Operator Transmit Levels. The objective of the transmit level requirements is to have the operator speech appear in the network at an average level equivalent to the speech arriving at that point from customers. This objective is intended to provide the means to create a conference call with equal voice level participants. To achieve this condition, operator speech should arrive at any 0 ELP at a −21 dBm average power level and at any analog point equal to −21 dBm0.

It has been determined that operators, under actual use conditions, produce an average speech power of approximately −17 dBm at the transmit headset jack when using a headset with a TOLR of −55.5 dB and conforming to requirements given in Reference 18. As a result, the transmit jack TLP is set at +4.0 to achieve the −21 dBm0 objective. An adjustment of ±6 dB is to be provided in the transmit circuitry, independent of sidetone adjustment or level, to accommodate headsets having different transmit efficiencies (i.e., TOLR).

Operator Receive Levels. The objective of the receive level requirements is to deliver to the ear as loud a voice signal as can be comfortably accommodated by the operator with an acoustic signal–to–noise ratio of 29 dB or better. An operator experiences circuit noise and room noise.

The receive headset jack TLP is set to −8 dB when headsets are used that have an ROLR of 36.5 dB and conform to requirements given in Reference 18. The TLP is to have an adjustment range of at least ±6 dB, independent of sidetone, to accommodate other headsets of differing receive efficiencies (i.e., ROLR). The TLP is set so that a −29 dBm electrical signal at the receiver jack converts to an 88–dBSPL acoustic signal, representing average customer speech, at the output of the receiver. Studies of customer speech power taken at the operator receive headset jack have indicated that the eight–hour time–weighted average noise dosage is well below the Occupational Safety and Health Administration (OSHA) requirement of 85 dBA when using these receive requirements.

Sidetone. Sidetone is provided by an electrical circuit outside the operator's headset. The circuit feeds part of the operator's speech from the transmit path into the receive path. For sound pressure levels, the received sidetone objective is 12 dB below the operator's speech. Sidetone should not be affected by receive volume control adjustments. If other alignment adjustments change the sidetone level, the sidetone level should be readjusted. Sidetone operation and level should not be affected by automatic gain control or echo control operation.

Echo Control. Operator talker echo control should be provided in the position circuitry. The echo control should provide additional loss in the operator receive path whenever speech is present in the transmit path. The additional loss should be provided by a voice–switched attenuator of the specified range (R) activated by speech of 79 ±2 dBSPL or at a rate of $0.1R$ dB of added receive loss for every 1 dB of increased operator speech level. The range of loss control should be variable from 10 to 25 dB, yielding increased loss rates of 1 to 2.5 dB per 1–dB increase in operator speech level. Maximum loss should be inserted for speech levels of 89 dBSPL or higher. The attack time should be 100 ±20 ms and the release time should be 150 ±30 ms.

Amplitude Limiting. To prevent annoying high–level acoustic signals from reaching the operator, limiting is required to control the levels of high–amplitude signals that may occur on position circuits. The maximum amplitude of any steady–state received signal at the earpiece of an operator headset should not be greater than 94 dBSPL. To allow for tolerances, it is desirable that the limiting level be adjusted so that all positions limit at the electrical equivalent of 91 dBSPL for a nominal headset (–26 dBm at the receive jack). The attack time should be 10 ± 5 ms and the release time should be 150 ± 30 ms.

Headset Impedances. The balanced–to–ground impedances of the transmit and receive ports are 50 ohms and 300 ohms respectively.

Room Noise. Room noise has been found to be the largest contributor to the acoustic signal–to–noise deterioration in past operator services. Headset design and usage play a large role in determining the effect of room noise. The room noise requirements call for a maximum average room noise usually assumed to be the average noise during the busy hour and a maximum peak room noise usually assumed to be a maximum five–minute average.

	Over–Ear Earpiece	In–Ear Earpiece
Maximum average room noise	55 dBA	52 dBA
Peak room noise	62 dBA	59 dBA

The noise is to be measured near the operator's head and may vary position by position depending on the layout of the operator office. Operator density, room acoustics, position display noise, and keyboard noise should all be considered when operator rooms are designed. The acoustic levels specified are all referenced to 20 μPa with A–weighting with the noise meter adjusted for slow response.

Reference 18 covers other requirements such as compression, harmonic distortion, return loss, radio interference, etc., and gives more detail on attack and release times for the various circuits.

26-3 DIGITAL SIGNALS ON ANALOG FACILITIES

As in the case of transmission objectives for voice–frequency channels, the expanding use of existing channels for new signals and services has made it necessary to refine and redefine channel transmission objectives. Similarly, the adaptation of analog systems and portions of analog systems for wideband digital transmission has led to new objectives for wideband channel applications. Frequency bands that were originally provided only as parts of the voice transmission network have been adapted for wideband use. As a result, transmission objectives for the wider bands and new signals were developed.

The transmission of digital signals over analog facilities requires the use of data modems or terminals to convert the signals into a form that can be transmitted over the facility.

The transmission objectives to be established and the manner of adapting systems and signals for compatibility depend on the signal format, the sensitivity of the signal to various impairments, and the characteristics of the system or channel involved. The parameters involved include load capacity, bandwidth, signal–to–noise performance, jitter, error ratios, and the rate of digital transmission.

The wide range of bandwidths, signal formats, impairments, services, and digital systems makes it difficult to present a complete set of wideband digital transmission objectives. Therefore, this discussion is limited to a number of examples of objectives that have been established for specific signal formats and to the approach used in several digital system designs. In most cases, the determination of the objective ultimately rests on the required grade of service.

There are two types of wideband digital signal commonly transmitted on analog systems: (1) the 1A Radio Digital System (1A–RDS) signal, a 1.544–Mb/s signal transmitted at baseband (0 to 500 kHz) over microwave radio systems in a multilevel signal format containing seven discrete levels, and (2) a family of binary digital data signals that may be transmitted at 19.2 kb/s, 50.0 kb/s, or 230.4 kb/s in the half–group, group, or supergroup bands, respectively, of FDM equipment [19,20]. Digital radio for digital signal transmission is covered in Volume 2.

Performance Evaluation

Transmission objectives for wideband digital signals are expressed variously in terms of error ratio, noise impairment, and eye diagram parameters. In addition, objectives must be expressed for signal power when digital signals are to be transmitted on analog systems.

Error Ratio. A specialized design objective for wideband digital signal transmission is an error ratio of 10^{-6} (i.e., the terminal–to–terminal error ratio shall not exceed one error in 10^{6} bits). Error–ratio counters are routinely used with many systems to determine error performance for the complete end–to–end connection or for a link in the connection. Violations of a predetermined code format are counted and compared with the objective, which must be expressed in the same terms. The objective is the value allocated to the particular link under test.

Noise Impairment. The expression of an objective in terms of noise impairment is used to equate the degradation of channel performance by various impairments to an equivalent degradation due to Gaussian noise. This equivalence can be explained in another way. A certain error ratio can be expected from a given channel whose characteristics are ideal in all respects except for the presence of Gaussian noise. The noise impairment due to the introduction of some other degradation, such as delay distortion, is measured by the improvement in Gaussian noise (improved signal–to–noise ratio) that would be required for the same channel performance as in the channel impaired only by the original value of Gaussian noise.

Two goals are met by expressing objectives in terms of noise impairment. First, objectives can be allocated to a variety of impairments in an orderly manner that lends itself readily to changes necessary to meet specific conditions. Second, a straightforward method is provided for determining how good the channel signal–to–noise ratio must be to meet a specified error–ratio objective. Both advantages are especially desirable for studies of digital signal transmission on analog channels.

Eye Diagram Closure. When a random stream of digital pulses is properly impressed on an oscilloscope, the successive pulses can be made to form a pattern, called an eye diagram. As the pulse stream is impaired by channel imperfections (such as noise, gain and delay distortion, and crosstalk), the opening in the eye (or eyes for multilevel signals) is reduced by predictable amounts. Thus, the eye pattern may be used as a measure of performance, and transmission objectives can be expressed in terms of the percentage of eye closure.

This manner of stating objectives is not particularly useful in operating or maintaining systems, but has found considerable use in system design where measurements are made under laboratory conditions [20]. The approach has been used to compare performance and objectives; it has also been used as a means of allocating objectives among different impairments, each being allowed a certain percentage of eye closure in the horizontal (timing) or vertical (amplitude) dimension or both.

Signal Power. When a signal is impressed upon a transmission channel, the channel must be capable of transmitting the signal satisfactorily; in addition, the signal cannot be allowed to degrade other signals that share the same transmission system. Overload performance is one criterion that must be satisfied in both respects.

The impressed signal amplitude must be limited so that the signal itself is not degraded by overloading the channel. The degradation would fall between two extremes, one in the form of peak clipping that might be relatively innocuous and the other in the form of excessive distortion that would render the signal useless. The limiting value depends in each case on the characteristics of the channel or system to be used.

Simultaneous transmission of digital and other kinds of signal on analog facilities further requires that the load imposed by the digital signals not seriously impair the other signals. The usual criteria for the loading objective are (1) that the average power in the digital signal shall not exceed the average power allotted to the displaced speech channels (−16 dBm0 per 4–kHz band), and (2) that any single–frequency component of the digital signal be below −14 dBm0. The latter criterion is sometimes relaxed if the component is not a multiple of 4 kHz or if the amplitude variability results in a low probability of its exceeding −14 dBm0.

Design Applications

It will be helpful to illustrate for specific cases the ways objectives evolve, are derived, and are applied.

Bit Rate and Bandwidth. In the design of a new digital transmission system or the adaptation of analog facilities to the transmission of digital signals, the first consideration is the overall system design problem of relating available bandwidth to the desired transmission rate. First–order effects on the design include: (1) the achievable signal–to–noise ratio of the proposed facility, (2) the desirability of designing a synchronous system that permits regeneration, (3) the cost involved in terminal and regenerator equipment, (4) the feasibility and cost of equalizing the medium, and (5) the transmission objectives that must be satisfied if the service needs are to be met. While the concern here is primarily with the objectives, all of these effects interact in ways that make discussion of objectives meaningless unless the interactions are explored as well.

The need for digital transmission over the analog microwave radio network evolved partly from the Digital Data System (DDS) program. The feasibility of transmitting a DS1 signal directly on a TD–type radio system was established, but was deemed undesirable because the DS1 signal carries significant energy at frequencies up to 1.544 MHz. A substantial number of telephone channels would thus have to be dropped to accommodate the digital signal. It was also shown that the upper half of the DS1 spectrum might be filtered or the signal might be coded as a three–level, class IV, partial response signal with spectral nulls at 0 and 772 kHz. The former approach was more theoretical than practical; the latter still appeared too costly because about 120 message channels would have to be dropped to provide a rolloff band.

A seven–level, class IV, partial response signal with a 15–percent rolloff band was chosen and used in 1A–RDS. The signal has spectral nulls at 0 and 386 kHz and extends only to 444 kHz, well below the 564–kHz multiplex low–end frequency. Thus, no message channels are displaced.

Performance Objectives. Objectives for 1A–RDS were derived from those established for DDS. They were based on a level of performance that was judged to provide a high–quality service at the customer rates of 56 kb/s and below. The basic criterion was stated in terms of percentage of error–free seconds. Allowances were included for known sources of hits, such as protection switching. A subset of objectives covers the number of errored–seconds that occur in shorter periods of time and the number and length of error bursts.

Designs Based on Noise Impairments. In setting objectives for transmitting wideband digital data signals in the half–group, group, and supergroup bands of the broadband systems, a major concern was the equalization of the gain and delay distortion in those bands. The objectives for these services were derived initially from the basic goal of achieving an error ratio of 10^{-6} or better 95 percent of the time. Portions of this objective were then allocated to impairments (random and impulse noise, for example); the remainder was allocated to misequalization, data set limitations, net loss variations, and jitter.

These allocations first involved the derivation of a required signal–to–noise ratio of 12.7 dB. After noise impairments had been assigned to each of the principal sources of degradation anticipated, it was concluded that an overall signal–to–noise ratio (Gaussian noise) of 22 dB would be required to meet the service objective; this signal–to–noise ratio was used as a design objective.

26–4 DIGITAL SIGNALS ON DIGITAL CIRCUITS—ERROR–RATIO OBJECTIVES

As the digital network evolves and supplants or replaces the present mixed analog/digital network, end–to–end switched digital connectivity will be provided between end offices or between customers for speech or data transfer at 56 or 64 kb/s. A measure of the quality of the digital connection can be made in the digital realm by determining objectives for various bit error ratios. For speech (56 kb/s and 64 kb/s), tests indicate that a random long–term bit error ratio of 10^{-6} or less is not detectable, 10^{-5} or less is just discernible, 10^{-4} or less is annoying, and 10^{-3} or less is a serious impairment [21].

Because errors are usually bursty, however, additional qualifications are necessary and were included in CCITT deliberations [22]. In the study recommendation *Rec. G.821*, the overall error performance objective of an integrated services digital network (ISDN) connection was allocated to local–, medium–, and high–grade circuit classifications. The overall objectives are shown in Table 26–11 for a 64–kb/s circuit switched connection used for voice or as a "bearer channel" for data services. The overall connected objectives were subdivided into their constitutional parts. It was assumed that the reference connection would consist of two local–grade end links, two intermediate medium–grade links, and a high–grade international link. Allocations of the objectives were made for these links and are shown in Table 26–12. These objectives are still subject to change and, along with block error ratios, are under study by the CCITT and the Exchange Carriers Standards Association (ECSA).

Table 26–11. Overall Error Performance Objective

Performance	Objective
(a) Degraded minutes	Fewer than 10% of one–minute intervals to have a bit error ratio worse than 1×10^{-6}.
(b) Severely errored seconds	Fewer than 0.2% of one–second intervals to have a bit error ratio worse than 1×10^{-3}.
(c) Errored seconds	Fewer than 8% of one–second intervals to have any errors.

Table 26–12. Allocation of Error Performance to Links

Link	Allocated Performance Objectives (64 kb/s)		
	% Degraded Minutes of One-Minute Intervals	% Errored Seconds of One-Second Intervals	Severely Errored Seconds of One-Second Intervals
Local	Less than 1.5% having a bit error ratio worse than 1 in 10^6	Less than 1.2% having any errors	Less than 0.015% having a bit error ratio worse than 1 in 10^3
Medium	Same	Same	Same*
High	4.0%	3.2%	0.04%*

* Can be increased by 0.05% for a reference radio relay system.

Reference circuit lengths:

Each end—local link plus medium link = 1,250 km.
Each end—local link plus high link = 25,000 km.
Overall = 27,500 km.

26–5 VIDEO TRANSMISSION OBJECTIVES

This section notes two older published video transmission objectives that include methods of measurement and technical interface parameters. It also gives, in table form, the latest technical specifications and interfaces for transmission facilities that may be offered by the exchange carriers for access to an IC or for point–to–point service within a LATA. This is followed by a discussion of some impairments caused by transmission characteristics such as noise. The objectives and specifications are based on one–way transmission of a standard 525–line/60–field monochrome or National Television System Committee (NTSC) color video signal with or without an associated audio signal for a 4000–mile end–to–end connection consisting of a long–haul facility and two local facilities in tandem.

Video Standards

End–to–end performance objectives were formulated and published by the Network Transmission Committee (NTC) under sponsorship of the parent Joint Committee of Television Network Broadcasters and the predivestiture Bell System [23]. The objectives and methods of testing for video transmission analog facilities presented were to cover Bell System services.

The Electronic Industries Association (EIA) published video standards [24] that deal primarily with analog microwave radio systems using angle modulation and performance standards for the picture and audio channels, methods of measurement, and baseband interface parameters. It expanded and refined the information from Reference 23; for example, in addition to end–to–end standards, allocations of impairments were made for short–haul (single hop), medium–haul (up to 10 hops), long–haul (10 to 150 hops), and satellite (one repeater) relay systems.

The allocations were made with consideration of the way impairments were expected to accumulate for tandem connections. Some impairments accumulate on a linear basis where the total = $X_1 + X_2 + \ldots$, some on a 2/3 power basis where the total = $(X_1^{3/2} + X_2^{3/2} + \ldots)^{2/3}$, and some on a root sum square (rss) basis where the total = $(X_1^{2} + X_2^{2} + \ldots)^{1/2}$.

At divestiture, it was clear that specifications were needed for video services in connections accessing ICs and in local LATA connections. Reference 25 covers technical specifications for video and audio service channels and includes interface illustrations. As an example, Table 26–13 lists the technical characteristics of service for a full– or part–time video channel versus the length of the facility in route–miles in three bands: (a) equal to or less than 20 route–miles, (b) greater than 20 but equal to or less than 150 route–miles, and (c) greater than 150 route–miles. The characteristics are in terms of IRE units, which are defined in Figure 26–15; this figure shows that a 1V peak–to–peak composite signal has been assigned 140 IRE units, 100 above and 40 below zero.

The technical characteristics for the associated audio channels are listed in Table 26–14 for a 15–kHz frequency response and in Table 26–15 for a 5–kHz frequency response.

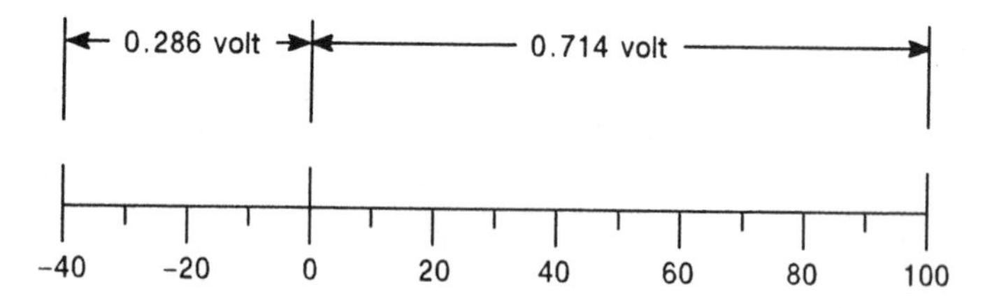

Figure 26–15. The IRE scale units for a 1V peak–to–peak composite signal.

Some of the transmission impairments caused by noise, single–frequency interference, echo, crosstalk, phase and delay, and audio/video delay are discussed in the following. Objectives are included for the overall path from camera to receiver.

Random Noise

The degree of noise impairment to television signals is a complex function of the distribution of noise power versus frequency and the characteristics of the impaired signal (for example, monochrome versus color signal). When the noise is at a high enough amplitude, it may appear as fine, closely packed dots in rapid, random motion. When observed in monochrome signal transmission, the dots appear to have the characteristics of a swirling snowstorm; as a result, the impairment has commonly been referred to as "snow."

Table 26–13. Video Channel Acceptance Limits

Parameter	Acceptance Limits		
	Route-Miles ≤20	**Route-Miles >20 ≤ 150**	**Route-Miles >150**
Insertion Gain	± 3.0 IRE	±3.0 IRE	±3.0 IRE
• One-Second Variation	±1.1 IRE	±1.7 IRE	+2.9/-2.8 IRE
• Hourly Variation	±1.7 IRE	+3.5/-3.3 IRE	+5.3/-5.0 IRE
Field-Time Distortion	3.0 IRE p-p [1]	3.0 IRE p-p	3.0 IRE p-p
Line-Time Distortion	0.5 IRE p-p	1.0 IRE p-p	1.5 IRE p-p
Short-Time Distortion (Bar) [2]	2.0%	2.0%	3.0%
Long-Time Waveform Distortion (Bounce)	8.0 IRE peak 3.0 seconds settling time	8.0 IRE peak	8.0 IRE peak
Chrominance-Luminance			
Gain Inequality	± 2.0 IRE	±4.0 IRE	± 7.0 IRE
Delay Inequality	20 ns	33 ns	54 ns
Intermodulation	1.0 IRE	2.0 IRE	4.0 IRE
Amplitude/Freq. Characteristic	Fig. 4-1 [25]	Fig. 4-2 [25]	Fig. 4-3 [25]
Luminance Nonlinear Distortion	2.0 IRE	4.0 IRE	8.0 IRE
Chrominance Nonlinear			
Gain Distortion	±1.0 IRE	±2.0 IRE	± 4.0 IRE
Phase Distortion	1°	2°	4°
Sync Signal Nonlinearity	1.0 IRE	2.0 IRE	4.0 IRE
Dynamic Gain Distortion 10% and 90% APL			
• Picture Signal	2.0 IRE p-p	3.0 IRE p-p	5.0 IRE p-p
• Sync Signal	1.2 IRE p-p	1.6 IRE p-p	2.4 IRE p-p
Differential Gain	2.0 IRE	5.0 IRE	8.0 IRE
Differential Phase	0.7°	1.3°	2.5°
Signal-to-Noise, Weighted [3]	67 dB [4]	60 dB	55 dB
0 to 10 kHz	53 dB	48 dB	44 dB
300 Hz to 4.2 MHz	67 dB	62 dB	58 dB

Notes:

1. For wire circuits of > 10 miles and multihop microwave circuits, the objective is 1 IRE until Jan. 1, 1992.

2. *IEEE Std. 511–1979* graticule.

3. Noise weighting per Reference 22.

4. For wire circuits > 10 miles and multihop microwave circuits, the weighted signal-to-noise objective is 64 dB until Jan. 1, 1992.

Table 26–14. Audio Channel Acceptance Limits—15 kHz

Parameter	Acceptance Limits
Insertion Gain, 404 Hz	0dB ± 0.5 dB
Amplitude/Freq. Characteristic relative to 404Hz	±0.5 dB, 100–7500 Hz; + 0.5, -1.0 dB, 50–15,000 Hz
Total Harmonic Distortion + Noise @ +18 dBm0[1]; 404 Hz	0.5% ≤ 20 Route-Miles[2] 1% >20 Route-Miles
Maximum Steady-State Test Levels	
50 Hz ≤ f ≤ 404 Hz	+8 dBm (Averaged over 1 second)
404 Hz < f ≤ 15 kHz	0 dBm (Averaged over 1 second)
Gain Difference Between Channels	
50 Hz ≤ f ≤ 15 kHz	≤1.0 dB
Phase Difference Between Channels	
50 Hz < f ≤ 100 Hz	≤10°
100 Hz < f ≤ 7.5 kHz	≤3°
7.5 kHz < f ≤ 15 kHz	≤10°
Crosstalk	
50 Hz to 15 kHz	Shall not degrade measured S/N ratio by more than 0.5 dB
Signal-to-Noise, 15-kHz Flat Weighting (Referenced to +18 dBm peak level)	
≤20 Route-Miles	66 dB
>20 ≤ 150 Route-Miles	65 dB
>150 Route-Miles	61 dB
Audio to Video Time Differential	25 ms lead, 40 ms lag

Notes:

1. This level may not be continuously applied.

2. Until Jan. 1992.

If the noise is concentrated at the lower video frequencies, the dots are relatively large or appear as streaks in the picture. If the noise is concentrated at high frequencies, the dots are much finer and harder to see. Hence, equal powers of noise are judged to be more annoying at low than at high frequencies. When the noise is concentrated in relatively narrow bands, it produces fleeting herringbone patterns in the receive pictures. If the band is made narrower, the pattern approaches that of a single-frequency interference. Thus, equal powers of noise tend to be more objectionable as the bandwidth of the noise is decreased.

Table 26–15. Audio Channel Acceptance Limits—5 kHz

Parameter	Acceptance Limits
Insertion Gain at 404 Hz	0 ± 0.5 dB
Amplitude/Freq. Characteristic Relative to Gain at 404 Hz 100–5000–Hz Frequency Band	± 1.0 dB
Signal–to–Noise, 15–kHz Flat Weighting (Referenced to +18 dBm[1] Peak Level)	62 dB
Total Harmonic Distortion + Noise @ +18 dBm0,[1] 404 Hz	2.5%
Maximum Steady–State Test Levels (Averaged over 1 second)	
100 Hz $\leq f \leq$ 404 Hz	+ 8 dBm
404 Hz $< f \leq$ 5 kHz	0 dBm
Crosstalk	Note 2
Audio to Video Time Differential	25 ms lead, 40 ms lag

Notes:

1. This level may not be continuously applied.

2. The acceptance limit for crosstalk is that no audible conversation, bits of conversation, or tones are detectable in the noise–measuring set monitor receiver. The noise–measuring set gain is adjusted to the noise requirement at the point of measurement. The monitoring tests should be made during heavy traffic periods by listening for at least ten minutes.

These observations have led to the expression of random noise objectives in terms of a single weighted value applicable to monochrome or color signals. The weighting, which takes into account the more objectionable nature of low–frequency noise, makes possible the use of a single number as an objective (i.e., equal measured values mean equal subjective effects, regardless of the type of noise). The effect of narrowband noise is accounted for simply by weighting its effect with that of broadband noise on the basis of total power. Thus, if single–frequency interference is present in a channel, the random noise objective must be made more stringent by an amount that makes the power sum of random and single–frequency noises meet the random noise objective. In addition, the single–frequency objective must also be met.

The random noise weighting characteristic is shown in Figure 26–16. In spite of some differences in annoying effects in monochrome and color signal transmission, it is found that satisfactory results are obtained when this single weighting curve is used to evaluate noise on facilities used for both types of signal [26]. The objective is in terms of a signal–to–noise ratio of the peak–to–peak composite signal voltage (including synchronizing pulses) to the weighted root mean square (rms) noise voltage in the frequency range of 10 kHz to 4.2 MHz. The noise from 0 to 10 kHz is treated separately.

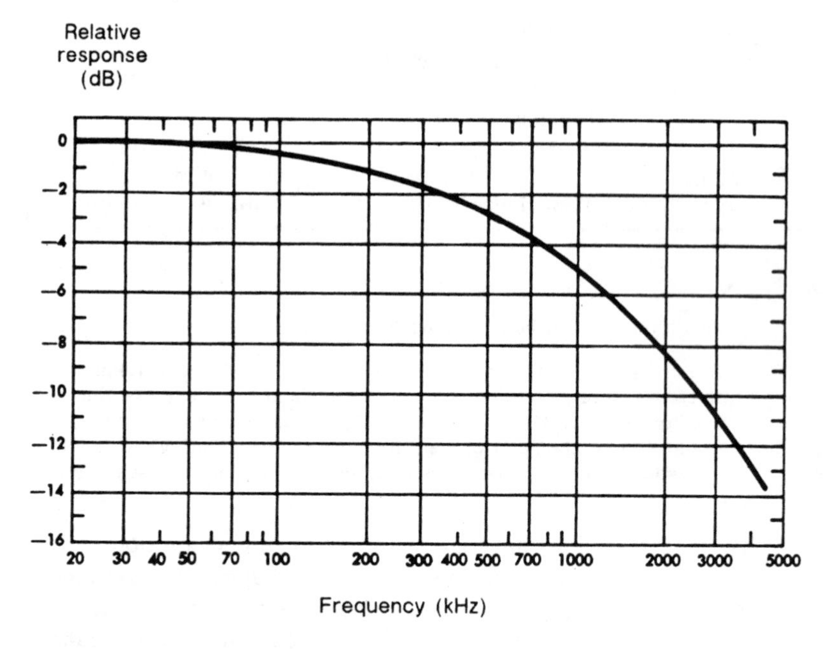

Figure 26–16. Random noise weighting for broadcast television signals.

Low–Frequency Noise

Noise in the band from 0 to 10 kHz is measured with a prescribed low–pass filter in a manner similar to that for random noise. It is treated separately because of the possibility of the presence of power–frequency interference, which causes a bar pattern in the received picture.

The objective for low–frequency noise is expressed in terms of the ratio of the peak–to–peak signal voltage to the rms interference voltage in the 0–to–10–kHz band.

Single–Frequency Interference

A single–frequency interference usually appears on a television receiver as a discernible bar pattern that may be stationary or in motion. If the interference is an integral multiple of the nominal 60–Hz field frequency, it appears as a broad, stationary, horizontal pattern. If the interference differs slightly from a 60–Hz multiple, the bars travel up or down the picture. If the interference is weak, the impairment may more nearly resemble a flickering than a bar pattern, an impairment much more annoying than a stationary pattern. The effect depends on the flicker rate.

For frequencies at or near multiples of the line–scanning frequency, the patterns are stationary or moving, vertical or diagonal bars. The bar structures become finer as the interfering frequency increases; the most critical frequencies are in the range of 100 to 300 kHz.

Similar phenomena are produced by single frequencies near the color carrier frequency. The high– and low–frequency characteristics must be determined as high or low frequencies relative to (i.e., displaced from) the color carrier frequency of 3.579545 MHz.

While there is a wide variation of subjective reaction to single–frequency interferences according to their frequency, stability, multiplicity, etc., the objective is usually stated as two simple numbers. First, the objective for a single interferer is taken as a signal–to–noise ratio of 69 dB or more where the signal amplitude is expressed in peak–to–peak volts (including the synchronizing pulse) and the interference is expressed as an rms voltage. The second expression for the interference is that the total weighted interference (including random noise) is to be 55 dB or more below the signal, which is the same value as that for weighted random noise.

Flat and Differentiated Echoes

Echoes are complex phenomena whose interfering effects depend on echo amplitude, time separation from the main signal, the nature of the original signal, and the frequency characteristic of the echo source. If the echo essentially covers the entire transmitted band, it is referred to as a flat echo. If it has a sharp frequency characteristic with stronger reflections at high frequencies, it is known as a differentiated echo. Differentiated echoes are generally less interfering than flat echoes. If the echo path accentuates the high–frequency echo components at a rate of 6 dB per octave, the echo is less interfering than flat echo by about 15 dB.

Echo Objective

The echo objective for video transmission is a 40–dB signal–to–echo ratio. It is expressed in terms of a single, well–defined, long–delayed (10 μs or more) echo. In practice, many echoes are usually present, and each component echo must be weighted in accordance with a weighting function that represents the change in subjective effect with the time displacement of the echo. The weighted components are then combined on a power basis for comparison with the objective. A typical time–weighting function is shown in Figure 26–17. Further analysis of subjective test data has shown that the function also varies according to picture content and the polarity of the echo [27].

Any departure from flat amplitude response or linear phase response of a transmission channel can be expressed in terms of the Fourier components of the response functions. These components are expressed as cosine functions of the amplitude response and as sine functions of the phase response. The Fourier components can then be regarded as generating echoes that may be summed by power after the weighting function has been applied.

Crosstalk

Video crosstalk occurs when an undesired signal interferes with a desired signal. The objectives for crosstalk are expressed

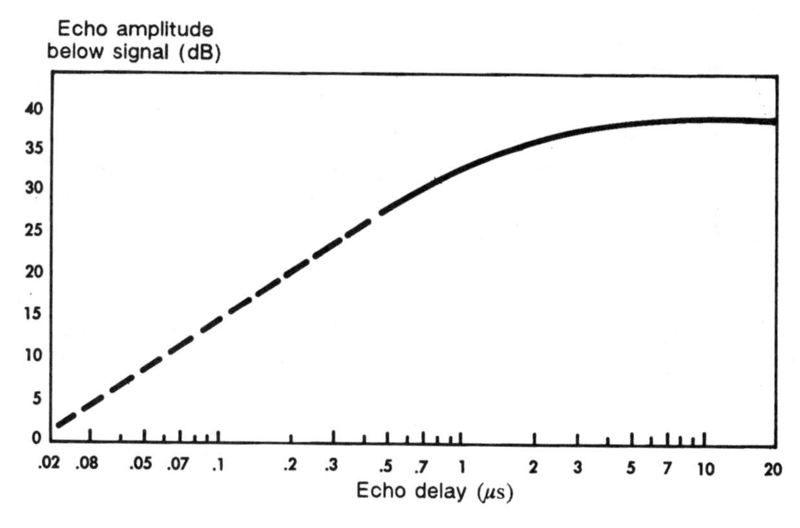

Figure 26–17. Single flat echo objective (echo time-weighting curve).

in terms of dB of loss in the coupling path between the two signals at 4.2 MHz at equal TLPs. When the coupling path is flat with frequency, the crosstalk is called flat crosstalk. When the crosstalk path loss decreases with frequency at a specified rate in dB per octave, the coupling is called x dB differentiated crosstalk where x is the rate of loss decrease, typically 6 dB/octave.

Where crosstalk can be seen, the undesired picture may be stationary or may move erratically across the wanted image, depending on whether the two are synchronized. If the crosstalk image moves across the picture, it appears to be framed. The apparent framing is formed by the synchronizing pulses of the interfering signal. The framing tends to be more noticeable than any feature in the image. The side frames, which extend from the top to the bottom of the wanted picture, interfere with the total wanted picture. The effect is similar to a windshield wiper moving across the picture; the term "windshield wiper effect" is sometimes applied.

If the crosstalk is weak (high coupling loss), neither the frame nor the image is discernible. At such a near–threshold point, only a slight flicker can be seen as the frame moves across certain portions of the desired picture. The subjective effect is more dependent on flicker rate than on crosstalk magnitude.

If the coupling loss varies with frequency, resulting in differentiated crosstalk, the interfering image may appear to be in bas-relief. However, the synchronizing pulses are still the most prominent feature in the crosstalk image because they have the largest rate of change.

The overall objective for crosstalk coupling loss between equal level points is dependent on the nature of the coupling path. Some typical path characteristics that may be encountered in practice are illustrated in Figure 26–18. The applicable objectives, expressed in dB of loss at 4.2 MHz, are as follows:

Crosstalk Path	Objective (dB)
Flat	58
6 dB/octave	37
12 dB/octave	21
24 dB/octave	17.5

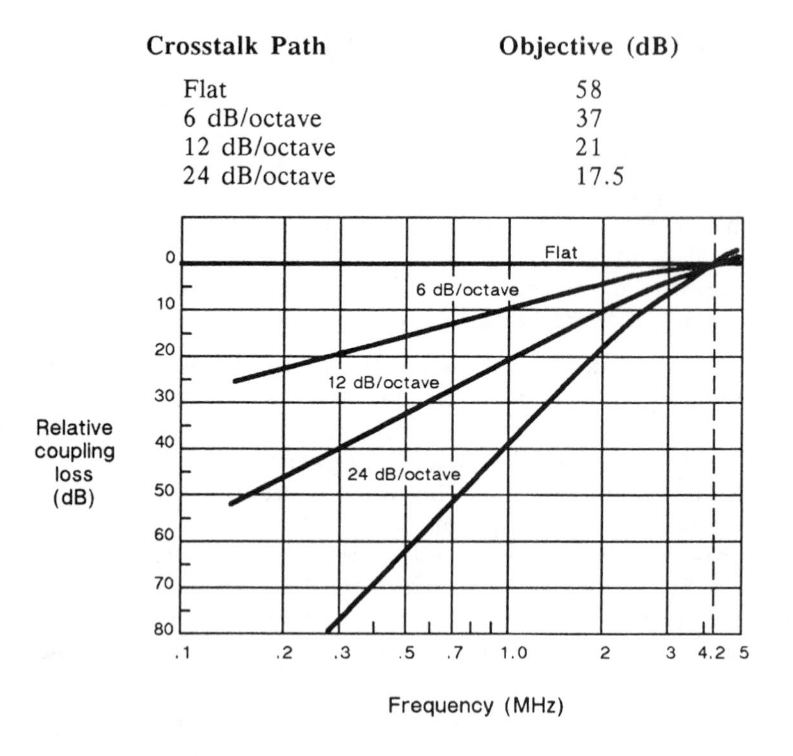

Figure 26–18. Coupling path loss characteristics.

Differential Gain and Phase

These impairments, which have serious effects on color transmission, are described in Chapter 18, Part 4. Differential gain produces undesired changes in color saturation. Differential phase produces changes in color hue.

Luminance/Chrominance Delay

While the luminance and chrominance information in a video signal is transmitted over the same channel, the dominant components of one part of the signal are so far removed in frequency (over 3 MHz) from the other part that there can be a significant delay difference between the two. When this delay difference is excessive, the color portions of the signal are shifted relative to the luminance portions; i.e., there is a misregistration of color. This type of effect is most noticeable at sharp vertical edges of highly saturated color areas that are bounded by low–saturated color areas relatively free of detail [28].

Audio/Video Delay

It was customary at one time to transmit video and associated sound signals over separate transmission paths. If the difference in absolute delay between the two paths is excessive, an impairment results because the picture and sound are out of synchronism; the sound is heard before or after the producing action in the picture.

Digital Video

To convert a commercial color video signal into digital form requires coding into a digital stream of 45 Mb/s or more, depending upon the quality desired. Digital form at rates of 100 Mb/s or more (full rate) is desirable in the studio in order to provide for the application of digital techniques for special effects such as freezing, split screen, etc. For transmission, however, the full–rate digital signal does not fit into the digital rate hierarchy. In order to reduce the bit rate required for comparable performance, various schemes have been studied, for example, the effort to reduce the redundant information in each frame and thereby reduce the bit rate. With the reduction, however, new impairments are introduced when faced with transmission impairments and with multiple coding–decoding processes. At present, there are no accepted standard objectives for less than full–rate digital video transmission that are comparable to analog transmission.

Proposed performance requirements for video digital terminals for the transmission over a 44.736–Mb/s, DS3 digital channel are presented in Reference 29. These requirements apply to digitizing the video analog signal by using a nine–bit linear analog–to–digital conversion with sampling at 10.74 MHz synchronized to three times the color subcarrier and with adaptive differential coding having three predictors (same line, previous line, and previous field). The reduced bit rate can be transmitted over digital radio or fiber optic facilities. Forward error correction and a means of dealing with uncorrected error bursts are to be included.

These requirements are for the access link to IC channels and agree with the analog requirements of Table 26–13 for a 20–to–150–route–mile connection. They are not applicable to a 0–to–20–route–mile connection (usually the link to the studio) because the requirements for this link are more stringent.

References

1. Bell System Technical Reference PUB 41005, *Data Communications Using the Switched Telecommunications Network,* American Telephone and Telegraph Company (May 1971).

2. *EIA Standard EIA–470,* "Telephone Instruments with Loop Signalling for Voiceband Applications" (Washington, DC: Electronic Industries Association, Iss. 1, Jan. 1981).

3. Lewinski, D. A. "A New Objective for Message Circuit Noise," *Bell System Tech. J.*, Vol. 43 (Mar. 1964), pp. 719–740.

4. Gresh, P. A. "Physical and Transmission Characteristics of Customer Loop Plant," *Bell System Tech. J.*, Vol. 48 (Dec. 1969), pp. 3337–3385.

5. Manhire, L. M. "Physical and Transmission Characteristics of Customer Loop Plant," *Bell System Tech. J.*, Vol. 57, No. 1 (Jan. 1978).

6. Andrews, F. T., Jr. and R. W. Hatch. "National Telephone Network Transmission Planning in the American Telephone and Telegraph Company," *IEEE Transactions on Communications Technology*, Vol. COM–19 (June 1971), pp. 302–314.

7. Huntley, H. R. "Transmission Design of Intertoll Trunks," *Bell System Tech. J.*, Vol. 32 (Sept. 1953), pp. 1019–1036.

8. American Telephone and Telegraph Company. *Notes on the Network* (Greensboro, NC: Western Electric Company, Inc., 1980).

9. *Notes on the BOC IntraLATA Networks—1986*, Technical Reference TR–NPL–000275, Bellcore (Iss. 1, Apr. 1986), Section 7.

10. *LSSGR, LATA Switching Systems Generic Requirements*, Technical Reference TR–TSY–000064, Bellcore (Iss. 2, July 1987), Vol. 1, Section 7 and Vol. 2, Section 15.

11. *IEEE Std. 743–1984*, "IEEE Standard Methods and Equipment for Measuring the Transmission Characteristics of Analog Voice–Frequency Circuits" (New York: Institute of Electrical and Electronics Engineers, Inc., 1984), pp. 12 and 18–24.

12. Members of Technical Staff. *Transmission Systems for Communications*, Fifth Edition (Murray Hill, NJ: AT&T Bell Laboratories, Inc., 1982), Chapter 8.

13. Fennick, J. H. "Amplitude Distributions of Telephone Channel Noise and a Model for Impulse Noise," *Bell System Tech. J.*, Vol. 48 (Dec. 1969), pp. 3243–3263.

14. Inglis, A. H. and W. L. Tufnell. "An Improved Telephone Set," *Bell System Tech. J.*, Vol. 30 (Apr. 1951), 239–270.

15. Bell System Technical Reference PUB 48005, *Functional Product Class Criteria—Telephones,* American Telephone and Telegraph Company (Jan. 1980).

16. *OSSGR, Operator Services Systems Generic Requirements,* Technical Reference TR–TSY–000271, Bellcore (Iss. 2, Dec. 1986 with Rev. 3, Mar. 1988), Vol. 1, Section 7 and Vol. 2, Section 21.

17. Bell System Technical Reference PUB 48006, *Functional Product Class Criteria—Head Telephone Sets (Headsets),* American Telephone and Telegraph Company (Jan. 1980).

18. *Generic Requirements for Telephone Headsets Used at Operator Consoles,* Technical Reference TR–NPL–000314, Bellcore (Iss. 1, Dec. 1987).

19. Seastrand, K. L. and L. L. Sheets. "Digital Transmission Over Analog Microwave Radio Systems," *Conference Record,* IEEE International Conference on Communications (Philadelphia, PA: June 1972).

20. Mahoney, J. J., Jr. "Transmission Plan for General Purpose Wideband Services," *IEEE Transactions on Communications Technology,* Vol. COM–14, No. 5 (Oct. 1966), pp. 641–648.

21. Gruber, J. G. and H. LeNguyen. "Performance Requirements for Integrated Voice/Data Networks," *IEEE Journal on Selected Areas in Communications,* Vol. SAC–1, No. 6 (Dec. 1983).

22. CCITT, VIIIth Plenary Assembly, *Red Book,* Vol. III, Fascicle III.3, Rec. G.829 (Geneva: International Telecommunications Union, 1985).

23. Network Transmission Committee. "Video Facility Testing, Technical Performance Objectives," *NTC Report No. 7,* revised (Washington, DC: The Public Broadcasting System, Jan. 1976).

24. *EIA Standard EIA–250–B,* "Electrical Performance Standards for Television Relay Facilities" (Washington, DC: Electronic Industries Association, Sept. 1976).

25. *Television Special Access and Local Channel Services—Transmission Parameter Limits and Interface Combinations,* Technical Reference TR–NPL–000338, Bellcore (Iss. 1, Dec. 1986).

26. Cavanaugh, J. R. "A Single Weighting Characteristic for Random Noise in Monochrome and NTSC Color Television," *Journal of the SMPTE,* Vol. 79 (Feb. 1970), pp. 105–109.

27. Lessman, A. M. "The Subjective Effects of Echoes in 525–Line Monochrome and NTSC Color Television and the Resulting Echo–Time Weighting," *Journal of the SMPTE,* Vol. 81 (Dec. 1972), pp. 907–916.

28. Lessman, A. M. "Subjective Effects of Delay Differences Between Luminance and Chrominance Information of the NTSC Color Television Signal," *Journal of the SMPTE,* Vol. 80 (Aug. 1971), pp. 620–624.

29. *Broadcast Quality Digital Television Terminals,* Technical Advisory TA–TSY–000195, Bellcore (Iss. 2, Aug. 1987).

Chapter 27

Economic Trends

The quality of service provided by a telecommunications network must be based on an appropriate balance between customer satisfaction and the cost of service. To make service objectives meet the criterion of reasonable cost, compromises must often be made among the objectives or between objectives and system development or application parameters. This chapter discusses the broad influence of economic trends upon network development.

A number of compromises may be used to illustrate the process of adjusting designs, applications, and objectives for economic reasons. There are, of course, no unchanging and absolute relationships among these factors. Guidelines tend to change with time because new systems, new services, and changing customer perceptions bring about changes in the objectives. Furthermore, economic relationships are significantly affected by local and national economic factors such as inflation.

Examples are given of general quality improvement versus cost increase, allocation of the overall quality objective to various links, and the estimated costs (at the time of the studies) of several alternatives of serving the loop–feeder plant, including digital carrier and fiber optics. Information is included also on a method to minimize estimated costs for the application of fiber optics, and on a reference to a proposed concept to better estimate economic lives and depreciation rates based on experience in applying a similar concept in computer technology.

27–1 OBJECTIVES

The derivation and application of transmission objectives often involve judgment as to what can be accomplished within reasonable cost constraints, the compromises that result from such judgments, the reconciliation of one set of objectives with another, and the existing economic, environmental, and human resource factors. Consider first the determination of transmission objectives and the economic factors involved.

Determination of Objectives

The determination of transmission objectives requires subjective testing to establish the relationship between an impairment and observer opinions of its effect. The test results are then related to measured or estimated performance parameters to calculate the grade of service that can be expected for the combination of parameters involved in modeling the various assumed connections. It is often possible at this point to estimate the cost of achieving this grade of service and the effects of changing the objectives or the performance parameters. These changes can then be evaluated economically by comparing the results with the estimated initial cost of the assumed models.

Qualitatively, the results are usually predictable. In nearly all cases, costs increase when objectives are made more stringent or when performance is improved. The characteristics of a cost/grade-of-service curve are obviously important, and the judgment that must be exercised in establishing the objectives is influenced by the nature of this curve. Sometimes, however, advances in technology bring both better performance *and* reduced cost. Digital networking has captured the industry because of superior economics, yet has also brought low and fixed noise level, reduced delay distortion, and other transmission improvements.

In Figure 27–1, curve A shows a gradual increase in cost with improving grade of service and demonstrates many situations in which the simple prediction of increasing cost with improving grade of service is verified. Since the simple prediction does little to support engineering judgement, the establishment of the objective must be based on other criteria. On the other hand, curves *B–B′* and *B–B′′* represent very different sets of circumstances.

Curve *B–B′* shows that a relatively small increase in cost yields a substantial improvement in grade of service up to a good–or–better (GoB) rating of 98 percent and that, regardless of cost, the grade of service cannot be increased beyond 98 percent. Thus, from the point of view of economic effects, any attempt to achieve higher than a 98–percent GoB grade of service would be wasteful.

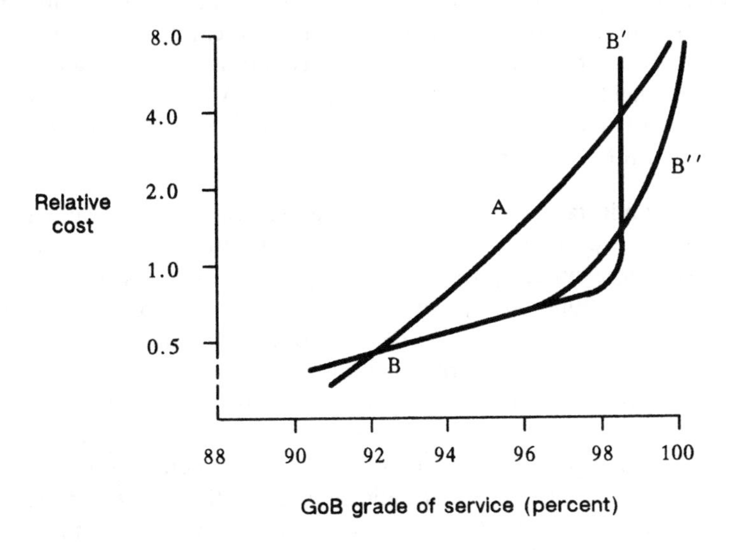

Figure 27–1. Typical cost curves.

Curve *B–B''* shows another type of relationship in which costs increase somewhat faster above 96–percent GoB grade of service than below. The change is not nearly as abrupt as for curve *B–B'*, and the achievement of a 96– or 97–percent grade of service (considered satisfactory) would be justified. The important point to note is that the derivation of cost curves such as those illustrated often provides strong support for determining objectives. Such curves may also be used to judge the cost of improving performance.

Allocation of Objectives

Objectives may be allocated to different sources of an impairment. For example, the total objective for one type of impairment may be allocated to different parts of the plant (e.g., local or tandem), or it may be allocated to various parts of a transmission system. Each method of allocation is either directly dependent on or indirectly tempered by economic factors.

As an illustration of how economic factors influence allocation of the objective for one impairment to various parts of the plant, the noise allocated to loops, 20 dBrnc, is such that no amount of expenditure in the loop plant could significantly improve the noise grade of service of a trunk connection unless both loop and trunk objectives were made more stringent. At present, the noise resulting from trunks (carried predominantly on carrier systems) controls the noise grade of service.

Economic factors may affect the allocation of an impairment to different parts of a system. For example, in long analog cable transmission systems, the design objective for message circuit noise for a 4000–mile system is 40 dBrnc0. A possible allocation of this objective might be 37 dBrnc0 to the line repeaters and 37 dBrnc0 to terminal multiplex equipment. However, the difficulty and cost of achieving higher performance in line repeaters (used in large numbers compared to the number of terminals) is recognized by a higher allocation to the line equipment. In a typical system, the line repeaters are allocated 39.4 dBrnc0 and the terminal equipment 31.2 dBrnc0. This allocation, when further translated to individual units (repeaters or terminals), still results in a per–unit allocation that is more stringent for a repeater than for the terminal. However, the economic balance is such that a further allocation of the objective to the repeater (which already has been allocated about 87 percent of the total) would not result in significantly lower overall costs.

An example of the allocation of quality objectives for an all–digital system is shown in Figure 27–2 for the Digital Data System (DDS) [1]. One measure for an overall quality objective was 99.5–percent error–free seconds at 56 kb/s for an end–to–end connection consisting of what are now intra– and interLATA (local access and transport area) links. The intraLATA links include the loops and tandem connections to the long–haul interLATA link. An allocation of 99.75–percent error–free seconds, a major part of the overall objective, was made for the long–haul link because, in part, of the performance of multihop radio links and the technical and economic feasibility, at the time, of improving the radio performance for this service. The allocation of

99.975–percent error–free seconds for each of the other links is very stringent in order to meet the overall objective.

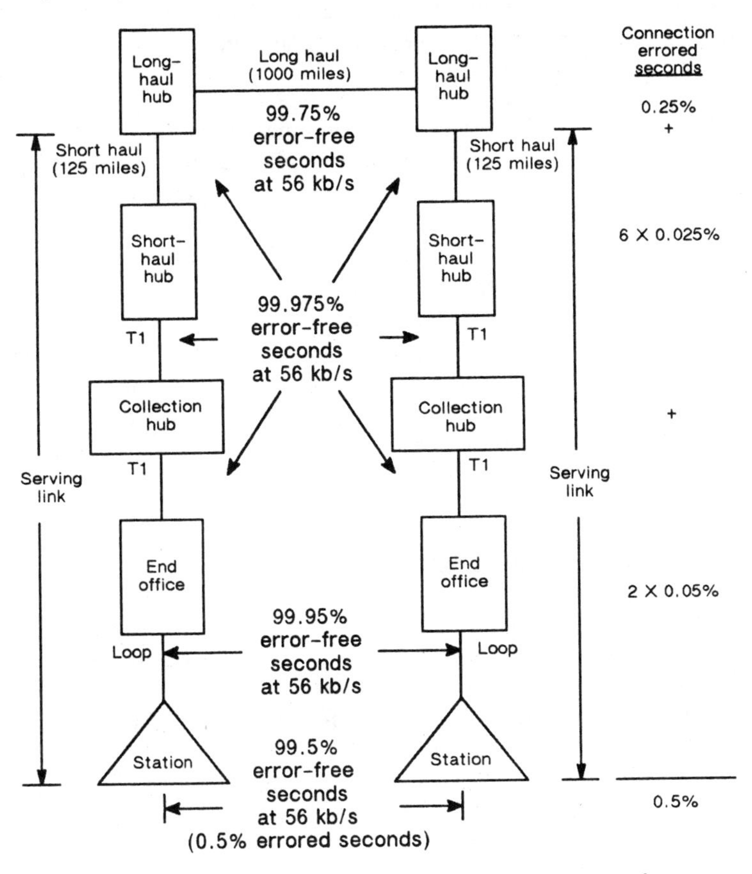

Figure 27–2. Digital Data System error performance allocation model.

These allocations were based technically on information [2] that was available prior to 1982 on the performance of individual transmission systems and then projected to end–to–end performance.

Economic Objectives

At certain times and under certain circumstances, economic objectives may supersede all others. In times of economic stress, the desirability of improving performance or increasing route capacity may have to be subordinated to the necessity of reducing capital and operating expenditures. Such circumstances, undesirable as they seem, must be recognized and improvements or expansion must be deferred.

In addition to the effects of economic stress, other less dramatic effects must be considered. Among the most significant of these is the availability of capital funds versus anticipated revenue. Sometimes it is necessary to keep outmoded equipment in service by paying for its maintenance from operating funds, even though the results of engineering economy studies have demonstrated the desirability of replacing the old equipment with new. When capital funds are in short supply, it is impossible to update equipment in the desired manner. This type of situation may be disclosed by engineering economy studies that compare initial capital outlays and estimated operating costs to available capital funds and anticipated revenue.

27-2 DESIGN COMPROMISES

Most of the compromises that must be made between objectives and cost are made during the development and design of new systems and new equipment. These compromises are made at every stage of development and design; the type of system to be developed, the features to be provided, the choice of circuits and physical designs, and the selection of components all relate to the balance between objectives (grade of service) and cost.

Circuit Devices

Devices used in electronic circuits include large-scale integrated circuits supported by such elements as resistors, capacitors, inductors, transformers, transistors, and diodes. Each device or element selected for the circuit under design must obviously meet the requirements imposed by its function in the circuit. It must be of the correct value, capable of dissipating a certain amount of power, characterized by input and output relationships that are adequately linear, sufficiently reliable, etc. Even with these constraints, there is often a wide choice within which circuit needs can be met. Making that choice with good judgment involves consideration of costs and their relationship to the circuit requirements. Two significant factors are the cost of the devices used and the ingenuity of the designer in using a readily available device to serve more than one function.

The benefits of mass production are evident in the reduced cost of devices. Also, economic benefits are usually effected when a device can be made to serve multiple functions. In applied cost reduction studies, careful attention is given to every aspect of the design, including the environmental conditions that are found in the operating plant (heat, humidity, voltage, handling, etc.) as well as the circuit requirements. The cost of circuit protection against induced voltage transients and radio frequency radiation/susceptibility should be considered also.

As discussed here, circuits are packaged entities of interconnected electronic devices that provide a specific function such as modulation, multiplexing, or amplification. A digital or analog circuit may include electronic networks, filters, equalizers, microprocessors, memory, logic, etc.

The design of circuits has progressed rapidly in recent years from point–to–point connection of devices through printed wiring techniques to a gamut of thin–film, thick–film, surface–mount, and large–scale integrated circuit arrangements that have evolved with the development of solid–state technology. With the wide choice of circuit arrangements available, careful attention must again be paid to economic factors. If large numbers of identical circuits are to be built and close control of circuit performance is required, integrated circuits are likely to be a good choice. Sometimes, the added expense of integrated circuits in small quantities is justified because the reproducibility of interexchange carrier performance is high.

Physical Design

The physical design of equipment and facilities is greatly influenced by the costs of maintenance and operation as well as by the costs of manufacture and installation. Recent trends in physical design have been influenced by new standards in building design and by the recognition that both transmission and operation could be improved by integrated designs of equipment bays [3,4]. These integrated designs, sometimes called unitized bays, include many more combinations of transmission, signalling, and switch interfacing than were formerly provided in a single bay. Some of these combinations have been made possible by the development of miniature devices and some by improved techniques of bay wiring and functional circuit interconnection. The new designs result in a significant reduction in office wiring, the elimination of a number of cross–connect frames, reduced congestion of cable racks and cross–connect frames, and the general elimination of jack fields.

The most significant feature of the current building standards is the reduction of ceiling height and the concomitant standardization of seven–foot equipment bay heights. The packaging of electronic circuits must now be consistent with the seven–foot standard, but in order to serve existing buildings with reasonable efficiency, bays are also designed to old standards. The necessity for designing equipment for several bay heights has led to a number of design compromises that may eventually be unnecessary. As buildings of new design become predominant, bay designs for the older buildings will no longer be economically justifiable.

The advent of solid–state technology has also led to situations in which the solutions to design problems have resulted in various compromises. For example, many more channel terminations can be accommodated in the same volume as before. The result is higher heat dissipation per unit of volume, so temperature control has become a major issue in packaging solid–state devices. Since the higher density of components may lead to higher weight per unit of volume in many designs, floor loading may have to be reconsidered. Thus, physical designs have interacted with circuit and system designs to bring adjustments in objectives and design features. The process of adjustment and compromise is continuous and parallels the development of all aspects of new technology.

Systems

The application design of systems follows the same pattern of compromise as has been outlined for circuits and physical design.

System features and design criteria must be considered with respect to feasibility and cost. Reliability, maintainability, restoration of service, automatic–versus–manual testing, remote control and alarms, and other operational features must all be weighted carefully in terms of service and cost.

Carrier on Cable vs. Voice Cable. The balance among system alternatives and cost factors plays an important role in determining whether to apply a new system. An old example, illustrated in Figure 27–3, involves the cost of a carrier system relative to the cost of copper pairs for voice–frequency (VF) transmission. Estimated 1975 costs for carrier and VF transmission are normalized to a value of unity at the point where the two costs are equal. As illustrated, the cost of carrier transmission has a base, *A*, representing the fixed cost of the terminal equipment. To this base cost is added the line cost (cable and electronics), which increases approximately linearly with distance. The cost of VF transmission increases linearly from a base of zero except for discontinuities, designated *B*, introduced by possible gauge changes and the periodic need for VF repeaters. The slope of the VF facility cost curve is directly affected by the total cable cost and the number of pairs per circuit.

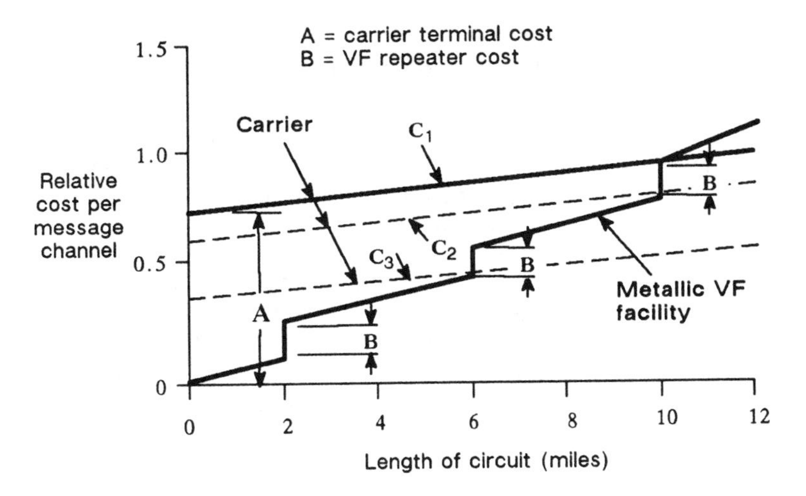

Figure 27–3. Comparison of estimated 1975 costs for very short circuits.

It can be seen that the cost/distance curve for the carrier system is less steep than for VF transmission. Further, it is evident that even if this slope is significantly reduced, the cost of a circuit is not materially affected because of terminal costs. The conclusion is that only a significant reduction of the terminal cost, A, can be expected to improve the position of carrier transmission relative to that of VF transmission. Curve C_2 shows the effect of a terminal cost reduction of about ten percent, a reduction that has no effect on the relative markets for the two transmission modes because the crossover point of the cost curve is still at ten miles. However, with a different set of curves and crossover points, a ten–percent reduction might be very significant and lead to a different conclusion.

Curve C_3 shows the effect of a terminal cost reduction of about 50 percent. This would encourage the development of new carrier terminals if the cost reduction appears to be possible, because the crossover point of the carrier and VF transmission cost curve is now at six miles. In addition, it would be necessary to show that there are large numbers of circuits in the range of six to ten miles and that there could be a high expectancy of achieving the 50–percent cost reduction by terminal redesign. Facilities integrated with digital switching would further reduce the break–even distance.

Many studies of the type described were made to guide the development of digital carrier systems. The curves of Figure 27–3 are representative, but are not based on any specific study results. Many other details must be included in a transmission system development study, such as the gauge of wire and the loss to which the circuits are designed.

Economic studies carried out as early as 1980 on the application of electronic digital carrier systems versus copper pairs in the loop feeder area (carrier serving area) indicated that the economics favored digital transmission. When the digital facility is directly interfaced with a digital central office, studies [5] have indicated that such an arrangement would be economical for about 35 percent of the growth lines of the office. With decreasing digital costs since these studies, it is expected that a greater percentage of the growth lines would be provided via digital distribution.

Fiber Optic Loop Feeder vs. Cable Systems. The rapid development and decreasing costs of fiber optics have opened up the possibility of serving the loop feeder area with fiber optics, assuming digital central offices. A study performed in the Southern Bell Telephone Company [6] compared the present worth of expenditures for several alternative transmission facilities that could be used in an assumed model. The alternatives were employment of single–mode optical fiber operating at the DS3 (45 Mb/s) rate, multimode at DS2 (6.3 Mb/s) rate, screened copper cable with digital carrier at DS1 (1.5 Mb/s) rate, reconditioned copper pairs with digital carrier, and VF pairs. The model assumes that 64–kb/s channels are required, and that the complete cost of the fiber optic facilities and the conditioning costs (but not existing copper pairs) for electronic digital carrier were included. Inflation factors, broad gauge factors, cost of money, historical costs, etc., were considered for a 20–year study plan, assuming 340 new access lines per year.

One result of the study for a digital central office is indicated in Figure 27–4 [6], which shows the present worth of expenditures for several alternatives. Conditioning existing copper pairs with digital carrier appears to be the most cost effective for this assumed model. The author, however, cautions that as cost factors are rapidly decreasing from the time of the study, and with newly emerging technology, that the analysis is obsolete for future use. It is included here as an example showing the alternatives that were considered and their cost effectiveness for a particular model at a particular point in time. It should be noted that the study was based on providing a basic service without consideration of future desired capability, which may offset the results of the cost study.

Method—Fiber Optic Loop Distribution

Another study on the application of fiber optics to the loop distribution plant has been reported [7]. The objective of this study was to lay the groundwork for developing a cost–effective design method for fiber optic systems tailored to the application. A mathematical model involving seven decision variables and including cost and performance evolved. The key decision variables were the wavelength, fiber loss, splice loss, transmitted power, detector sensitivity, number of splices, and the splicing length of cable. The model used optimization techniques to indicate how the key variables could be cost–perfected. The cost functions were presented in several figures showing how the assumed cost varied with the decision variables.

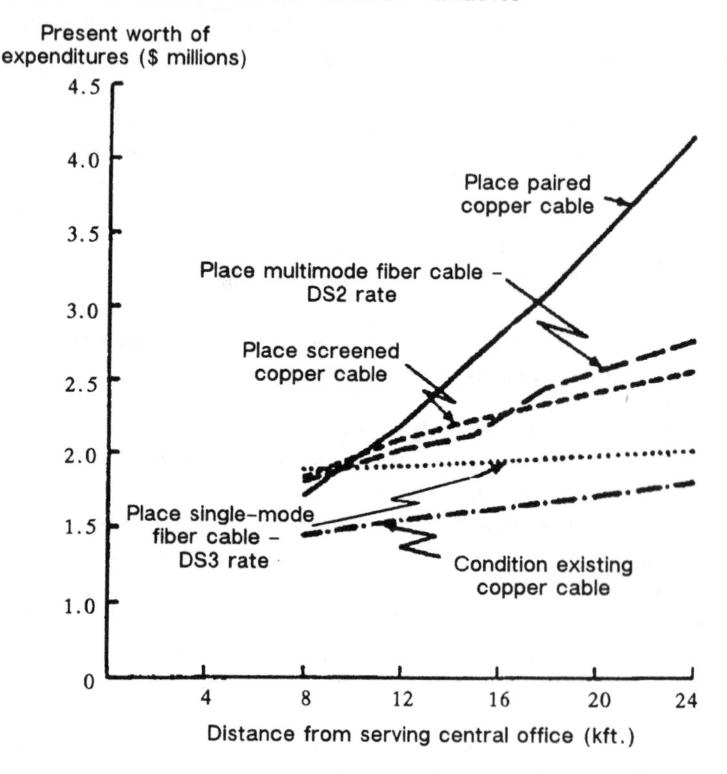

Figure 27–4. 340 access lines per year—digital central office.

The method was applied to a sample circuit for transmitting 135 Mb/s over a distance of 20 km with a bit error ratio (BER) of 10^{-9} or better. The results of the study show that, for typical applications, expensive fiber with very low attenuation and expensive low–loss splices may not be required, depending upon the other variables, but can be tailored for a solution. The results are conditioned by the cost functions assumed; caution should be taken in trying to generalize the results of the study. The cost functions should be developed for each case until costs become stabilized.

27-3 EXPLODING DIGITAL TECHNOLOGY

The digital technology explosion coupled with rapidly decreasing costs are making it difficult to use economic modeling to help decide whether to repair or update older technology plant or to replace it with new technology plant. The new plant will provide for the old services and for greater future capability.

The expanding evolution is fueled by rapid advances in digital technologies, such as increasingly higher processing speeds, very–large–scale integration, high–capacity optical fiber, greater repeater spacing, etc. In addition, greater reliability, better performance, reduced cost, improved maintenance, reduced space, and lower power are provided.

Beginning this evolution was the current integration of digital carrier with digital electronic switching to form a 56/64 kb/s communication path with customer control. Now, with fiber optics and the possibility of future photonic switching [8], digitized services with even greater capacity and lower cost will be possible.

With this burgeoning expansion and rapid cost changes, determining appropriate depreciation rates by traditional economic modeling based on historic data on each type of plant is no longer suitable, as the future will not closely resemble the past. Assumptions based on repetition and continuity of the stable plant of the past are no longer valid.

New Economic Model

A new economic model has been proposed [9] for forecasting equipment lines and for calculating depreciation rates based on economic trends in similar technologies such as that used in the microcomputer field. Evidence of the changes in decreasing costs and other trends are shown in the following, which supports the new modeling technique.

Rapid advances in technology and economics are taking place in fiber optics and other fields. The cost trends are shown by the optical fiber experience curve in Figure 27–5 in terms of dollars per meter per 100 MHz–km bandwidth, versus sales (production). The figure indicates costs have decreased greatly with increased production (learning how to produce better at lower cost). In addition, the performance of fiber optics has increased rapidly as indicated in Figure 27–6, which shows the increasing bit rate times length versus year.

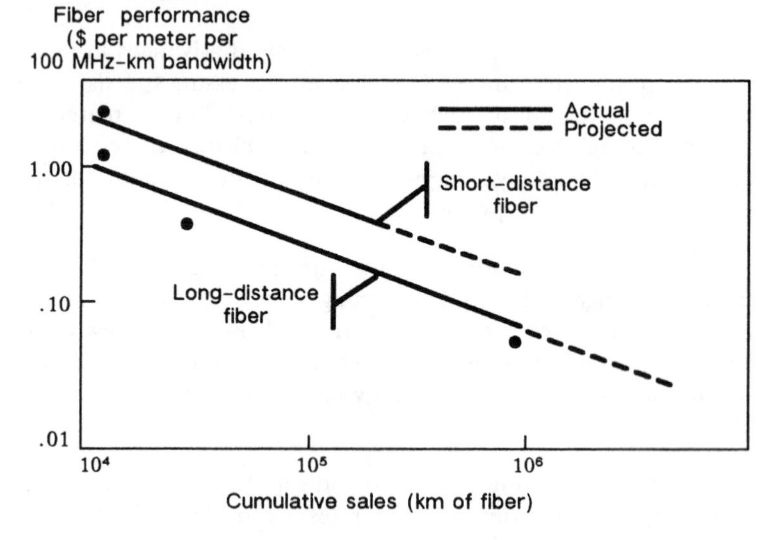

Figure 27–5. Optical fiber experience curve.

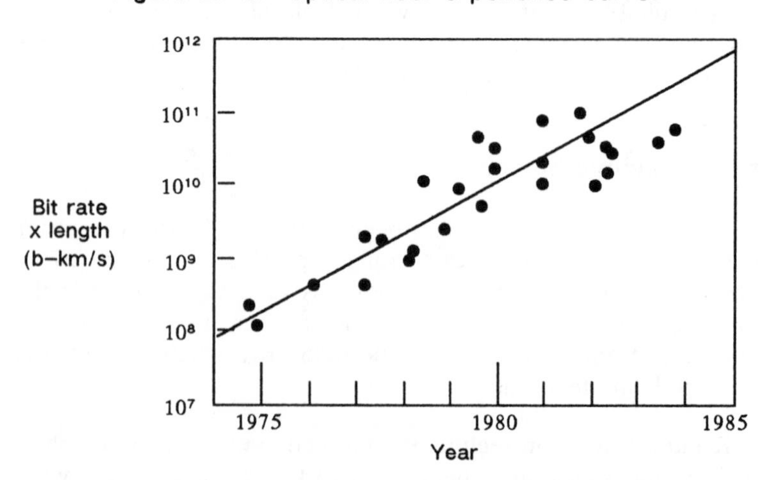

Figure 27–6. Progress in fiber systems performance.

An example of projected optical fiber costs for transmission on interoffice trunking costs over a ten–year interval was given in Reference 9. Many assumptions were made for the example, such as a span of 10 km, 100–percent fill, costs of fiber to decrease by 5 to 1, electric–optic interface costs to decrease by 10 to 1, multiplexing costs to decrease by 15 percent per year, and operations costs to decrease by about 7 percent per year. Also, it was assumed that the 1–Gb/s rate would increase to 5 Gb/s over the time period. The projected price trend as a percentage of the start year is indicated by Figure 27–7, which shows a rapid decrease in costs for 1 Gb/s worth of transmission.

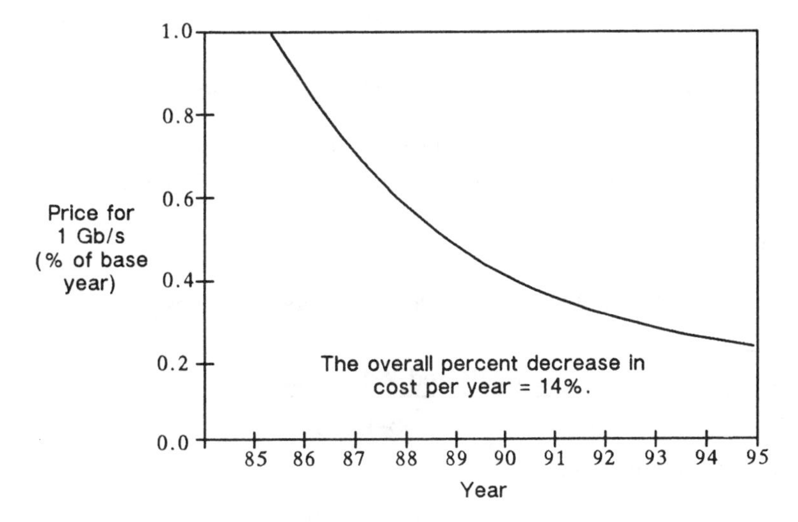

Figure 27–7. Lightwave system price trends (projected).

Reliability is an important quality that is also changing rapidly and increases year by year as manufacturing techniques improve. This leads to higher levels of processing integration. In microcomputer technology, the defects per million integrated circuit chips have decreased enormously in a few years as shown in Figure 27–8, reducing costs accordingly.

All of these trends must be considered, including enhancements for new services and capabilities, as part of an economic modeling process. Reference 9 introduces the concept and proposes a model for estimating economic lives and economic depreciation rates, which include a measure of these trends.

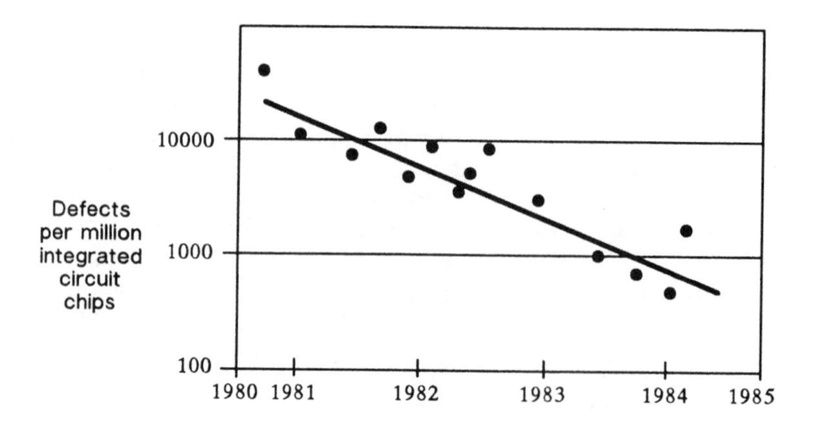

Figure 27–8. Microcomputer product quality levels.

Some of the projected advances in technology and network design, and the implications are listed in Table 27–1. The application of these advances will require extensive economic modeling as designs proceed.

Table 27–1. Technology and Network Design

Projected Advances
• Fiber transmission > 100 Gb–km/s • Switching speeds >100 Mb/s • Self–routing nonblocking switch fabrics • High–capacity remote electronics • Integration of transmission and switching
Implications
• High–capacity transport at low per–circuit cost • Few outside plant repeaters • Merging of facility/circuit switching for high data rates • High–speed packet switching for dynamic allocation of network resources • Fiber distribution added to fiber feeder deployment

References

1. Malek–Zavarei, M. and A. Tabatabai. "Generic Design of Digital Services Circuits," *Globecom 1984 Conference Record* (Atlanta, GA: Nov. 1984), Vol. 1, pp. 13.6.1–13.6.4.

2. Ritchie, G. R. and P. E. Scheffer. "Projecting the Error Performance of the Bell System Digital Network," *Conference Record*, IEEE International Conference on Communications (Philadelphia, PA: 1982), pp. 2D.2.1–2D.2.6.

3. *Network Equipment–Building System (NEBS) Generic Equipment Requirements*, Technical Reference TR–EOP–000063, Bellcore (Iss. 3, May 1988).

4. Giguere, W. J. and F. G. Merrill. "Getting It All Together with Unitized Terminals," *Bell Laboratories Record*, Vol. 51 (Jan. 1973), pp. 13–18.

5. "Inside the New Digital Subscriber Loop System," *Bell Laboratories Record*, Vol. 58 (Apr. 1980), pp. 111–115.

6. Bergen, R. S. "Economic Analysis of Fiber versus Alternative Media," *IEEE Journal on Selected Areas in Communications*, Vol. SAC–4, No. 9 (Dec. 1986).

7. Misra, R. B. "Loop Optic Design Optimization Study," *IEEE Journal on Selected Areas in Communications*, Vol. SAC–4, No. 5 (Aug. 1986).

8. Williamson, S., Technical Editor. "Photonic Switching," *Telephony* (Jan. 19, 1987), p. 94.

9. *Technological and Market Obsolescence of Telephone Network Equipment*, Science and Technology Series ST–BEL–000029, Bellcore (Iss. 1, June 1986).

Acronyms

The acronyms listed here reflect usage in this book. They may be used differently in other contexts.

ABBH Average Bouncing Busy Hour
ABS Average Busy Season
ABSBH Average Busy–Season Busy Hour
ACD Automatic Call Distributor
ACRS Accelerated Cost Recovery System
ACXT Apparatus Case Crosstalk
ADM Adaptive Delta Modulation
ADPCM Adaptive Differential Pulse Code Modulation
ADR Asset Depreciation Range
ADTS Automated Digital Termination System
AGC Automatic Gain Control
AIOD Automatically Identified Outward Dialing
AIS Alarm Indication Signal (or)
AIS Automatic Intercept System
AL Acceptance Limit
ALBO Automatic Line Build–Out
AM Amplitude Modulation
AMI Alternate Mark Inversion
AML Actual Measured Loss
ANSI American National Standards Institute
AOS Alternate Operator Services
APC Automatic Power Control
APD Avalanche Photodetector Diode
ARSB Automated Repair Service Bureau
ASK Amplitude Shift Keying
AT Access Tandem
AT&T American Telephone and Telegraph Company

AUTOVON Automatic Voice Network
AWG American Wire Gauge
AXPIC Adaptive Cross–Polar Interference Canceller

B3ZS Bipolar Format with Three–Zero Substitution
B6ZS Bipolar Format with Six–Zero Substitution
B8ZS Bipolar Format with Eight–Zero Substitution
BDDS Basic Dedicated Digital Service
BER Bit Error Ratio
BETRS Basic Exchange Telecommunications Radio Service
BITS Building Integrated Timing Supply
BLER Block Error Ratio
BOC Bell Operating Company
(or)
BOC Build–Out Capacitor
BOL Build–Out Lattice
BOR Build–Out Resistor
BPF Bandpass Filter
BRI Basic Rate Interface
BSRF Basic Synchronization Reference Frequency

CAC Compandored Analog Carrier
CAMA Centralized Automatic Message Accounting
CARL Computerized Administrative Route Layouts
CAROT Centralized Automatic Reporting on Trunks
CATV Cable Television
CCC Clear–Channel Capability
CCG Composite Clock Generator
CCIR International Radio Consultative Committee
CCIS Common–Channel Interoffice Signalling
CCITT International Telegraph and Telephone Consultative Committee
CCS Common–Channel Signalling
(or)
CCS Hundred Call Seconds (Per Hour)
CCSA Common–Control Switching Arrangement
CDCF Cumulative Discounted Cash Flow

CDMA Code–Division Multiple Access
CEPT European Conference of Posts and Telecommunications
CFA Carrier Failure Alarm
CGA Carrier Group Alarm
CHILL CCITT High–Level Language
CL Carrier Liaison
CLRC Circuit Layout Record Card
CMC Cellular Mobile Carrier
CMD Circuit–Mode Data
CMOS Complementary Metal–Oxide Semiconductor
CMTT Joint CCIR–CCITT Study Group on Transmission of Sound Broadcasting and Television Systems Over Long Distances
CMV Joint CCIR–CCITT Study Group for Vocabulary
CO Central Office
COD Central–Office District
COT Central–Office Terminal
CPE Customer–Premises Equipment
CRC Cyclic Redundancy Check
CREG Concentrated Range Extension with Gain
CS Channel Switching
CSA Carrier Serving Area
CSDC Circuit Switched Digital Capability
CSP Control Switching Point
CSU Channel Service Unit
CTX Centrex
CUCRIT Capital Utilization Criteria
CX Composite (Circuit)

DA Directory Assistance
DART Distribution Area Rehabilitation Tool
DAVID Data Above Video
DBOC Drop Build–Out Capacitor
DCE Data Circuit–Terminating Equipment
DCF Discounted Cash Flow
DCMS Digital Circuit Multiplication System
DCS Digital Cross–Connect System
DCT Digital Carrier Terminal (or)
DCT Digital Carrier Trunk
DDB Digital Data Bank

DDD Direct Distance Dialing
DDS Digital Data System
DEMS Digital Electronic Message Service
DFB Distributed Feedback
DFSG Direct Formed Supergroup
DIC Direct InterLATA Connecting
DILEP Digital Line Engineering Program
DLC Digital Loop Carrier
DLP Decode Level Point
DM Digital Multiplex
DNHR Dynamic Nonhierarchical Routing
DP Dial Pulse
DPP Discounted Payback Period
DRS Digital Reference Signal
DS Digital Signal
DS0 Digital Signal Level 0 (64 kb/s)
DS1 Digital Signal Level 1 (1.544 Mb/s)
DS2 Digital Signal Level 2 (6.312 Mb/s)
DS3 Digital Signal Level 3 (44.736 Mb/s)
DS4 Digital Signal Level 4 (274.176 Mb/s)
DSB Double Sideband
DSBSC Double-Sideband Suppressed Carrier
DSBTC Double Sideband with Transmitted Carrier
DSI Digital Speech Interpolation
DSL Digital Subscriber Line
DSN Defense Switched Network
DSS Data Station Selector
DSU Data Service Unit
DSX Digital Signal Cross-Connect
DTA Digitally Terminated Analog
DTE Data Terminal Equipment
DTMF Dual-Tone Multifrequency
DTS Digital Termination System
DUV Data Under Voice
DX Duplex (Signalling)

EA Equal Access
EAEO Equal-Access End Office
EAS Extended Area Service
ECSA Exchange Carriers Standards Association
EDD Envelope Delay Distortion
EFS Error-Free Seconds
EHD Expected High Day
EIA Electronic Industries Association

EIRP Effective Isotropic Radiated Power
ELCL Equal Level Coupling Loss
ELEPL Equal Level Echo Path Loss
ELERL Equal Level Echo Return Loss
ELP Encode Level Point
ELSRL Equal Level Singing Return Loss
EML Expected Measured Loss
EO End Office
EPL Echo Path Loss (or)
EPL Equivalent Peak Level
ERL Echo Return Loss
ES Errored Seconds
ESD Electrostatic Discharge
ESF Extended Superframe Format
ESM Economic Study Module
ET Exchange Termination
ETL Equipment Test List
ETN Electronic Tandem Network
ETV Educational Television
EU End User
EVE Extreme Value Engineering

FAX Facsimile Communication
FCC Federal Communications Commission
FCOD Foreign Central–Office District
FDM Frequency–Division Multiplexing
FDMA Frequency–Division Multiple Access
FET Field–Effect Transistor
FEXT Far–End Crosstalk
FFM First Failure to Match
FFT Fast Fourier Transform
FG Feature Group
FM Frequency Modulation
FMAC Facility Maintenance and Administration Center
FSK Frequency Shift Keying
FSL Free Space Loss
FX Foreign Exchange
FXS Foreign Exchange Station

GDF Group Distributing Frame
GoB Good or Better
GOSCAL Grade–of–Service Calculation
GS Ground Start

HAIS Host Automatic Intercept System
HC High Capacity

HCDS High–Capacity Digital Service
HDBH High–Day Busy Hour
HDTV High–Definition Television
HF High Frequency
HU High Usage

IAL Immediate Action Limit
IC Interexchange Carrier
ICCF Industry Carriers Compatibility Forum
ICL Inserted Connection Loss
IDF Intermediate Distributing Frame
IDLC Integrated Digital Loop Carrier
IEEE Institute of Electrical and Electronics Engineers
IEOT Inter–End–Office Trunk
IF Intermediate Frequency
IMTS Improved Mobile Telephone Service
INA Integrated Network Access
INMD In–Service Nonintrusive Measurement Device
IROR Internal Rate of Return
ISC Intercompany Service Coordination
ISDN Integrated Services Digital Network
ISMX Integrated Subrate Multiplexer
ITC International Teletraffic Congress
ITT Intertandem Trunk
ITU International Telecommunications Union
ITV Industrial Television
IXT Interaction Crosstalk

JEG Joint Expert Group
JFS Jumbogroup Frequency Supply
JIS Jurisdictionally Interstate Service
JMX Jumbogroup Multiplex

KEMAR Knowles Electronic Manikin for Acoustic Research
KSI Key Service Indicator
KTS Key Telephone System
KTU Key Telephone Unit

LAD Loop Activity Data
LAMA Local Automatic Message Accounting
LAN Local Area Network
LATA Local Access and Transport Area

LATIS Loop Activity Tracking Information System
LBO Line Build-Out
LBR Local Business Radio
LBRV Low-Bit-Rate Voice
LCD Liquid Crystal Display
LD Long Distance
LDR Local Distribution Radio
LEAD Loop Engineering Assignment Data
LED Light-Emitting Diode
LEIM Loop Electronics Inventory Module
LFACS Loop Facility Assignment and Control System
LIU Line Interface Unit
LMX L-Type Multiplex
LOCAP Low Capacitance
LPC Linear Predictive Coding
LPF Low-Pass Filter
LRD Long-Route Design
LRE Loop Range Extender
LSB Lower Sideband
LSSGR LATA Switching Systems Generic Requirements
LT LATA Tandem
LTEE Long-Term Economic Evaluator

MAC Multiplexed Analog Component
MAN Metropolitan Area Network
MARR Minimum Attractive Rate of Return
MAT Metropolitan Area Trunk
MCVD Modified Chemical Vapor Deposition
MDF Main Distributing Frame
MF Multifrequency
MFD Mode Field Diameter
MFT Metallic Facility Terminal
MG Mastergroup
MGDF Mastergroup Distributing Frame
MGT Mastergroup Translator
M-JCP Multiplexer Jack and Connector Panel
MJU Multipoint Junction Unit
ML Maintenance Limit
MLDS Microwave Local Distribution System
MLRD Modified Long-Route Design
MMGT Multimastergroup Translator
MMIC Monolithic Microwave Integrated Circuit
MML Man-Machine Language
MMX Mastergroup Multiplex
MNRU Modulated Noise Reference Unit

MRSELS Microwave Radio and Satellite Engineering and Licensing System
MSS Mobile Satellite Service
MTBF Mean Time Between Failures
MTS Message Telecommunications Service
MTSO Mobile Telephone Switching Office
MTU Maintenance Terminating Unit

NA Numerical Aperture
NANP North American Numbering Plan
NAS North American Standard
NBOC Network Building–Out Capacitor
NCAC Noncompandored Analog Carrier
NCF Net Cash Flow
NCTE Network Channel Terminating Equipment
NE Network Element
NEBS Network Equipment Building System
NEXT Near–End Crosstalk
NI Network Interface
NIC Nearly Instantaneously Compandored
NMDG Network Management Development Group
NOF Network Operations Forum
NOTE Network Office Terminal Equipment
NPA Numbering Plan Area
NPV Net Present Value
NPWE Net Present Worth of Expenditures
NRZ Non–Return to Zero
NT Network Termination
NT1 Network Termination Type 1
NTC Network Transmission Committee
NTIA National Telecommunications & Information Administration
NTS Network Technical Support
NTSC National Television System Committee
NTT Nippon Telegraph and Telephone
NTTMP Network Trunk Transmission Measurements Plan
NXX End–Office Code

OAM Once A Month
OBF Ordering and Billing Forum

OCL Overall Connection Loss
OCU Office Channel Unit
OCUDP Office Channel Unit Dataport
OCVD Outside Chemical Vapor Deposition
OGT Outgoing Trunk
OMFS Office Master Frequency Supply
ONAL Off–Network Access Line
ONI Operator Number Identification
ONS On–Premises Station
OPS Off–Premises Station
OPX Off–Premises Extension
ORB Office Repeater Bay
OS Operations System
OSHA Occupational Safety and Health Administration
OSI Open Switch Interval
OSS Operations Support System (or)
OSS Operator Services System
OSSGR Operator Services Systems Generic Requirements
OTDR Optical Time Domain Reflector

PAM Pulse Amplitude Modulation
PAR Peak–to–Average Ratio
PBX Private Branch Exchange
PC Personal Computer (or)
PC Primary Center
PCM Pulse Code Modulation
PDM Pulse Duration Modulation
PEVL Polyethylene Video Line (16–gauge cable)
PFM Pulse Frequency Modulation
PFS Primary Frequency Supply
PIC Polyethylene–Insulated Conductor
PIN Positive Intrinsic Negative
PLAR Private–Line Automatic Ringdown
PLL Phase–Locked Loop
PM Phase Modulation
PMD Packet–Mode Data
POT Point of Termination
POTS Plain Old Telephone Service
PoW Poor or Worse
PPM Pulse Position Modulation
PPSN Public Packet–Switched Network
PPSNGR Public Packet–Switched Network Generic Requirements
PR Protective Relaying

PRR Project Rate of Return
PSAP Public Safety Answering Point
PSDS Public Switched Digital Service
PSK Phase Shift Keying
PSN Public Switched Network
PSS Packet Switching System
PT Principal Tandem
PVN Private Virtual Network
PWAC Present Worth of Annual Charges
PWE Present Worth of Expenditures

QAM Quadrature Amplitude Modulation
QPRS Quadrature Partial Response Signalling
QSDG Quality of Service Development Group

RAIS Remote Automatic Intercept System
RC Regional Center (or)
RC Resistor–Capacitor
RD Resistance Design
RDS Radio Digital System
RDSN Region Digital Switched Network
RDT Remote Digital Terminal
REA Rural Electrification Administration
REG Range Extender with Gain
RF Radio Frequency
RG Ringing Generator
RHC Regional Holding Company
RL Return Loss
ROH Receiver Off Hook
ROLR Receive Objective Loudness Rating
RRD Revised Resistance Design
RSB Repair Service Bureau
RSM Remote Switching Module
RSU Remote Switch Unit
RT Remote Terminal
RTA Remote Trunk Arrangement
RTU Remote Test Unit
RZ Return to Zero

SAGM Separate Absorption Grating and Multiplication
SAI Serving Area Interface
SAS Switched Access Service
SB Sideband
SC Sectional Center
SCA Subsidiary Communications Authorization
SCPC Single Channel Per Carrier
SDE Synchronization Distribution Expander

SDL	Specification and Descriptive Language
SDM	Standard Error of Mean
SDN	Switched Digital Network
SES	Severely Errored Seconds
SF	Single Frequency
SG	Study Group
SGDF	Supergroup Distributing Frame
SL	System Level
SMDR	Station Message Detail Recording
SMDS	Switched Multi–Megabit Data Service
SMRS	Special Mobile Radio Service
SMSA	Standard Metropolitan Statistical Area
SOLR	Sidetone Objective Loudness Rating
SONAD	Speech–Operated Noise–Adjusting Device
SONET	Synchronous Optical Network
SP	Signal Processor
SPC	Stored Program Control
SPCS	Stored Program Controlled System
SR	Sidetone Response
SRDM	Subrate Data Multiplexer
SRE	Signalling Range Extender
SRL	Singing Return Loss
SSB	Single Sideband
SSBAM	Single–Sideband Amplitude Modulation
SSC	Special–Services Center
SSMA	Spread Spectrum Multiple Access
SSN	Switched Services Network
ST	Sector Tandem
STC	Serving Test Center
STDM	Statistical Time–Division Multiplexer
STD–RL	Standard Return Loss
STE	Signalling Terminal Equipment
STL	Studio–to–Transmitter Link
SW	Sync Word
SX	Simplex
SYNTRAN	Synchronous Transmission
T1	Carrier System or Standards Committee
T1DM	T1 Data Multiplexer
T1OS	T1 Outstate
TAS	Telephone Answering Service
TASI	Time–Assignment Speech Interpolation
TC	Time Consistent (or)
TC	Toll Center
TCM	Time–Compression Multiplexing

TCT Tandem–Connecting Trunk
TD Terminal Digit (or)
TD Test Distributor
TDM Time–Division Multiplexing
TDMA Time–Division Multiple Access
TE Transverse Electric
TEM Transverse Electromagnetic
THDBH Ten High–Day Busy Hour
TIC Tandem InterLATA Connecting
TL Transducer Loss (or)
TL Transmission Level
TLP Transmission Level Point
TM Transverse Magnetic
TMRS Telephone Maintenance Radio Service
TOC Television Operating Center
TOLR Transmit Objective Loudness Rating
TRA Tax Reform Act
TSA Transmission Surveillance Auxiliary
TSC Transmission Service Center
TSGR Transport Systems Generic Requirements
TSI Time–Slot Interchange
TSPS Traffic Service Position System
TT&C Tracking, Telemetry, and Control
TWT Traveling Wave Tube

UDPC Universal Digital Portable Communications
UHF Ultra High Frequency
UI Unit Interval
UNICCAP Universal Cable Circuit Analysis Program
USB Upper Sideband
USOA Uniform System of Accounts

VBD Voiceband Data
VF Voice Frequency
VG Voice Grade
VHF Very High Frequency
VITS Vertical Interval Test Signal
VLSI Very–Large–Scale Integration
VNL Via Net Loss
VNLF Via Net Loss Factor
VOGAD Voice–Operated Gain–Adjusting Device
VSA Voice–Switched Attenuator
VSAT Very–Small–Aperture Terminal
VSB Vestigial Sideband

VSWR	Voltage Standing Wave Ratio
VT	Virtual Tributary
VU	Volume Unit
WAL	WATS Access Line
WAN	Wide Area Network
WATS	Wide Area Telecommunications Service
WDM	Wavelength-Division Multiplexing
WEPL	Weighted Echo Path Loss
WLEL	Wire-Line Entrance Link
WORD	Work Order Record and Details
WP	Working Party
ZBTSI	Zero-Byte Time-Slot Interchange

Index

A

B

C

D

E

G

H

I

J

K

L

M

N

O

P

Q

R

U

V

W